# Plant High-Throughput Phenotyping and Functional Phenomics

This book provides a series of comprehensive summaries highlighting the emerging achievements in the fields of plant high-throughput phenotyping that leads to constructing functional phenomics, one of the essential components of plant functional genomics. It presents broad aspects of methods, applications, and future directions. It offers an efficient way for readers to overview this crucial topic to realize the concept as a whole, to advance the design of their future experiments, and to inspire the exploration of the knowledge, which eventually leads to better crop development in the future by scientists, plant biologists, and crop breeders. It covers advanced tools for studying functional phenomics, including artificial intelligence, imaging, remote sensing, robotics, and aerial vehicle technologies, to empower crop speed breeding, particularly in the development of stress-tolerant future crops. The knowledge of this book supports the Sustainable Development Goals (SDGs) of the United Nations to develop climate-smart and sustainable agriculture for achieving zero hunger globally.

**Dr. Jen-Tsung Chen** is a professor of cell biology at the National University of Kaohsiung in Taiwan. He also teaches genomics, proteomics, plant physiology, and plant biotechnology. His research interests include bioactive compounds, chromatography techniques, plant molecular biology, plant biotechnology, bioinformatics, and systems pharmacology. He is an active editor of academic books and journals to advance the exploration of multidisciplinary knowledge involving plant physiology, plant biotechnology, nanotechnology, ethnopharmacology, systems biology, and drug discovery. He serves as an editorial board member and a guest editor in several reputed journals. He published books in collaboration with international publishers on diverse topics such as drug discovery, herbal medicine, medicinal biotechnology, nanotechnology, bioengineering, plant functional genomics, plant speed breeding, CRISPR-based plant genome editing, and artificial intelligence. In 2023 and 2024, Stanford University/Elsevier included Dr. Chen in the "World's Top 2% Scientists".

# Plant High-Throughput Phenotyping and Functional Phenomics

Edited by

Jen-Tsung Chen

CRC Press
Taylor & Francis Group
Boca Raton London New York

CRC Press is an imprint of the
Taylor & Francis Group, an **informa** business

Designed cover image: Shutterstock.

First edition published 2025
by CRC Press
2385 NW Executive Center Drive, Suite 320, Boca Raton FL 33431

and by CRC Press
4 Park Square, Milton Park, Abingdon, Oxon, OX14 4RN

*CRC Press is an imprint of Taylor & Francis Group, LLC*



ISBN: 9781032821818 (hbk)
ISBN: 9781032826790 (pbk)
ISBN: 9781003505631 (ebk)

DOI: 10.1201/9781003505631

Typeset in Times
by codeMantra

# Contents

# Preface

Plant phenotyping in a high-throughput manner is a game-changing technology that is expected to rapidly and accurately collect complete sets of phenotypic data for constructing phenomics, one of the essential approaches of plant functional genomics. This technology needs to efficiently analyze many traits and the resulting big data and then integrate required information into the plant functional genomics system, which will eventually benefit crop breeding programs. The central components of plant high-throughput phenotyping should involve automated machinery, robotics, remote sensing technology, imaging technology, computational platforms, and bioinformatics tools including artificial intelligence and machine learning.

High-throughput phenotyping and phenomics are powerful techniques to lead the studies of forward and reverse plant genetics. In particular, they are used to dissect the responses under stress conditions for the investigation of plant stress physiology and genetics. This is critical knowledge for advancing breeding programs to acquire future climate-resilient crops, thus supporting the Sustainable Development Goals (SDGs) of the United Nations involving zero hunger globally, and smart and sustainable agriculture.

This book is a comprehensive summary of the emerging achievements in the fields of plant high-throughput phenotyping and phenomics by discussing broad aspects of methods, applications, and future directions. It would open a window for students and researchers to overview this topic in an efficient way to realize the concept, to advance the design of future experiments, and to inspire the exploration of the knowledge, which eventually leads to better crop development in the future by scientists, plant biologists, and crop breeders.

# Contributors

**Jothi Dheivasikamani Abidharini**
Department of Human Genetics and Molecular Genetics
Bharathiar University
Coimbatore, Tamil Nadu, India

**Kuttikattuparambil Sudheer Aditi**
Department of Biochemistry
Bharathiar University
Coimbatore, Tamil Nadu, India

**Sarfraz Ahmad**
Department of Genetics and Plant Breeding
College of Agriculture Mandawa, Swami Keshwanand Rajasthan Agriculture University
Bikaner, Rajasthan, India

**Sk Naim Aktar**
Department of Agronomy
Palli Siksha Bhavan, Institute of Agriculture
Santiniketan, West Bengal, India

**Shokat Ali**
Department of Agronomy
Krishi Vigyan Kendra Chandgothi, Swami Keshwanand Rajasthan Agriculture University
Bikaner, Rajasthan, India

**Arumugam Vijaya Anand**
Department of Human Genetics and Molecular Genetics
Bharathiar University
Coimbatore, Tamil Nadu, India

**Gunasekaran Arthi**
Department of Human Genetics and Molecular Genetics
Bharathiar University
Coimbatore, Tamil Nadu, India

**Meyyazhagan Arun**
Department of Life Science
CHRIST University
Bangalore, Karnataka, India

**Muthukrishnan Arun**
Department of Biotechnology
Bharathiar University
Coimbatore, Tamil Nadu, India

**Balasubramanian Balamuralikrishnan**
Department of Food Science and Biotechnology
Sejong University
Seoul, South Korea

**Vikas Belwal**
Department of Tree Improvement
College of Forestry Ranichauri, VCSG Uttarakhand University of Horticulture and Forestry
Bharsar, Uttarakhand, India

**Pratap Bhanu Singh Bhadoria**
Agricultural and Food Engineering Department
Indian Institute of Technology Kharagpur
Kharagpur, India

**Shivani Chauhan**
Department of Vegetable Science
Punjab Agricultural University
Ludhiana, India

**Viswanathan Chinnusamy**
Division of Plant Physiology
Indian Agriculture Research Institute (IARI), Pusa Campus
New Delhi, India

**Madhurima Das**
Division of Plant Physiology
Indian Agriculture Research Institute (IARI), Pusa Campus
New Delhi, India

**Dhandapani Raju**
Division of Plant Physiology
Indian Agriculture Research Institute (IARI), Pusa Campus
New Delhi, India

**Chenniappan Anchana Devi**
PG & Research Department of Biotechnology
Women's Christian College
Chennai, Tamil Nadu, India

**Allimuthu Elangovan**
Division of Plant Physiology
Indian Agricultural Research Institute
New Delhi, India

**Farzana Mustafa Era**
Department of Genetics and Plant Breeding
Bangabandhu Sheikh Mujibur Rahman Agricultural University
Gazipur, Bangladesh

**Muhiuddin Faruquee**
International Rice Research Institute (IRRI)
Dhaka, Bangladesh

**Shanmugam Gavaskar**
Department of Computer Application
Bharathiar University
Coimbatore, Tamil Nadu, India

**Amooru Harika**
Division of Plant Physiology
Indian Agriculture Research Institute (IARI), Pusa Campus
New Delhi, India

**Swati Hazra**
Department of Life Science
Sharda School of Basic Sciences and Research, Sharda University
Greater Noida, Uttar Pradesh, India

**Akbar Hossain**
Division of Agronomy
Bangladesh Wheat and Maize Research Institute
Dinajpur, Bangladesh

**M. Kamal Hossain**
International Rice Research Institute (IRRI), Bangladesh Office
Dhaka, Bangladesh

**A. K. M. Aminul Islam**
Department of Genetics and Plant Breeding
Bangabandhu Sheikh Mujibur Rahman Agricultural University
Gazipur, Bangladesh

**Sarah Jaison**
Department of Botany
PSGR Krishnammal College for Women
Coimbatore, Tamil Nadu, India

**Priyanka Joshi**
Department of Plant & Environmental Sciences
Clemson University
Clemson, South Carolina, USA

**Suresh Kadaru**
International Rice Research Institute (IRRI)
Los Baños, Laguna, Philippines

**Marudhachalam Kamalesh**
Department of Human Genetics and Molecular Genetics
Bharathiar University
Coimbatore, Tamil Nadu, India

**Rubina Khan**
Department of Genetics and Plant Breeding
College of Agriculture Jobner, Sri Karan Narendra Agriculture University
Jobner, Rajasthan, India

**Marium Khatun**
Department of Genetics and Plant Breeding
Bangabandhu Sheikh Mujibur Rahman Agricultural University
Gazipur, Bangladesh

**Stephen P. Klassen**
International Rice Research Institute (IRRI)
Los Baños, Laguna, Philippines

**Mahesh Kumar**
Division of Plant Physiology
Indian Agriculture Research Institute (IARI), Pusa Campus
New Delhi, India

**Sudhir Kumar**
Division of Plant Physiology
Indian Agriculture Research Institute (IARI), Pusa Campus
New Delhi, India

**Krishnan Kutty Sunitha Kumari**
Department of Botany
PSGR Krishnammal College for Women
Coimbatore, Tamil Nadu, India

**Priti Kumari**
Department of Human Genetics and Molecular Genetics
Bharathiar University
Coimbatore, Tamil Nadu, India

**Seethalaxmi**
Division of Plant Physiology
Indian Agriculture Research Institute (IARI), Pusa Campus
New Delhi, India

**Murugesan Loganathan**
Department of Human Genetics and Molecular Genetics
Bharathiar University
Coimbatore, Tamil Nadu, India

**Pappuswamy Manikantan**
Department of Life Science
CHRIST University
Bangalore, Karnataka, India

**Mani Manoj**
Department of Human Genetics and Molecular Genetics
Bharathiar University
Coimbatore, Tamil Nadu, India

**Ganesan Megha**
Department of Biochemistry
Bharathiar University
Coimbatore, Tamil Nadu, India

**Domesa Monisha**
Department of Biochemistry
Bharathiar University
Coimbatore, Tamil Nadu, India

**Arkabanee Mukherjee**
Indian Institute of Tropical Meteorology
Pune, Maharashtra, India

**Manikantan Pappuswamy**
Department of Life Science
CHRIST University
Bangalore, Karnataka, India

**Ishita Paul**
Semiochemicals and Lipid Laboratory, Department of Life Sciences
Presidency University
Kolkata, India

**Jeyabal Philomenathan Antony Prabhu**
Department of Human Genetics and Molecular Genetics
Bharathiar University
Coimbatore, Tamil Nadu, India

**Tasfia Binte Anam Priya**
International Rice Research Institute (IRRI), Bangladesh Office
Dhaka, Bangladesh

**Natarajan Pushpa**
PG & Research Department of Microbiology
Cauvery College for Women
Tiruchirappalli, Tamil Nadu, India

**Mohammad Ghani Raghib**
Department of Horticulture, Faculty of Agriculture
Nangarhar University
Jalalabad, Afghanistan

**Raman Rajasekar**
Division of Plant Physiology
Indian Agriculture Research Institute (IARI), Pusa Campus
New Delhi, India

**Manoharan Rajesh**
Department of Biochemistry
Bharathiar University
Coimbatore, Tamil Nadu, India

**Ramesh Ramasamy**
Division of Plant Physiology
Indian Agricultural Research Institute
New Delhi, India

**Govindarajan Ramkumar**
GRG BIRAC EYUVA Centre
PSGR Krishnammal College for Women
Coimbatore, Tamil Nadu, India

**Sachin Bellubbi**
Division of Plant Physiology
Indian Agriculture Research Institute (IARI), Pusa Campus
New Delhi, India

**Shoumik Saha**
Department of Genetics and Plant Breeding
Indian Agricultural Research Institute
New Delhi, India

**Siva Daniel Ajay Samuel**
GRG BIRAC EYUVA Centre
PSGR Krishnammal College for Women
Coimbatore, Tamil Nadu, India

**Saniya**
Department of Horticulture
ICAR-National Research Centre for Grapes
Pune, Maharashtra, India

**Sreehari V. Santhosh**
Division of Plant Physiology
Indian Agriculture Research Institute (IARI), Pusa Campus
New Delhi, India

**Muhukrishnan Saradhadevi**
Department of Biochemistry
Bharathiar University
Coimbatore, Tamil Nadu, India

**Chithra Devi Balasundarm Saraswathy**
Department of Botany
PSGR Krishnammal College for Women
Coimbatore, Tamil Nadu, India

**Sumi Sarker**
Department of Genetics and Plant Breeding
Bangabandhu Sheikh Mujibur Rahman Agricultural University
Gazipur, Bangladesh

**Maksud Hasan Shah**
Department of Agronomy
College of Agriculture Chandgothi, Swami Keshwanand Rajasthan Agriculture University
Bikaner, Rajasthan, India

**Falguni Akter Shraboni**
International Rice Research Institute (IRRI), Bangladesh Office
Dhaka, Bangladesh

**Arshdeep Singh**
Department of Entomology
Punjab Agricultural University,
Ludhiana, India

**Biswabiplab Singh**
Division of Plant Physiology
Indian Agricultural Research Institute
New Delhi, India

**Ezhumalai Sivapragasam**
Division of Plant Physiology
Indian Agriculture Research Institute (IARI), Pusa Campus
New Delhi, India

**Mahankali Sravanalakshmi**
Department of Human Genetics and Molecular Genetics
Bharathiar University
Coimbatore, Tamil Nadu, India

**Macharla Srinidhi**
Department of Human Genetics and Molecular Genetics
Bharathiar University
Coimbatore, Tamil Nadu, India

**Alagarsamy Sumithra**
Department of Computer Science and Engineering
SNS College of Technology
Coimbatore, Tamil Nadu, India

**Kempanna Sushmitha**
Department of Biochemistry
Bharathiar University
Coimbatore, Tamil Nadu, India

**Arumugam Tamilselvan**
Division of Plant Physiology
Indian Agriculture Research Institute (IARI), Pusa Campus
New Delhi, India

**Sureshkumar Tharani**
GRG BIRAC EYUVA Centre
PSGR Krishnammal College for Women
Coimbatore, Tamil Nadu, India

**Shanmugam Velayuthaprabhu**
Department of Biotechnology
Bharathiar University
Coimbatore, Tamil Nadu, India

**Kandasamy Vijayalakshmi**
PG & Research Department of Microbiology
Jamal Mohamed College
Tiruchirappalli, Tamil Nadu, India

**Kannan Vijayarani**
Department of Education
Bharathiar University
Coimbatore, Tamil Nadu, India

# 1 Bioinformatics Tools and Deep Learning for Plant High-Throughput Phenotyping and Phenomics

*Gunasekaran Arthi, Murugesan Loganathan, Jothi Dheivasikamani Abidharini, Balasubramanian Balamuralikrishnan, Meyyazhagan Arun, Pappuswamy Manikantan, Jeyabal Philomenathan Antony Prabhu, and Arumugam Vijaya Anand*

## LIST OF ABBREVIATIONS

**AI**: artificial intelligence
**APSIM**: agricultural production systems simulator
**BISQUE**: bioimage semantic query user environment
**CNNs**: convolutional neural networks
**DSSAT**: decision support system for agrotechnology transfer
**GPU**: graphics processing unit
**GWAS**: genome-wide association studies
**HTP**: high-throughput phenotyping
**IoT**: internet of things
**LiDAR**: light detection and ranging
**PHIS**: phenotyping hybrid information system
**QTL**: quantitative trait loci
**TASSEL**: trait analysis by association
**UAVs**: unmanned aerial vehicle

## 1.1 INTRODUCTION TO PLANT PHENOTYPING AND PLANT PHENOMICS

Plant phenomics is a rapidly evolving field that focuses on the comprehensive measurement and analysis of plant phenotypes' observable characteristics or traits influenced by genetics and the environment (Smith, 2020). With the increasing demand for food security due to a growing global population and climate change, plant phenomics has emerged as a critical area of research (Jones et al., 2019). It integrates various advanced technologies, such as imaging, sensors, and data analytics, to collect and analyze vast amounts of phenotypic data (Taylor & Brown, 2021a). This integration allows researchers to better understand plant growth, development, and responses to environmental stresses, ultimately facilitating crop improvement and sustainable agricultural practices (Wilson et al., 2018).

Phenomics is an interdisciplinary field that encompasses the large-scale analysis of phenotypes, integrating biology, engineering, and data science (Anderson & Carter, 2020). It aims to develop

DOI: 10.1201/9781003505631-1

comprehensive phenotypic databases and computational tools that can correlate phenotypic traits with genetic and environmental ones (Garcia, 2017). The integration of omics technologies, including genomics, transcriptomics, proteomics, and metabolomics, into phenomics allows researchers to explore the complex interactions between genes and traits more effectively (Lee & Chen, 2019a). Phenomics plays a crucial role in modern plant breeding programs by providing valuable insights into the genetic basis of traits that influence crop performance (Patel et al., 2021a). By leveraging advanced phenotyping techniques, researchers can more accurately assess plant responses to biotic and abiotic stresses, enabling the identification of candidate genes for breeding (Miller & Singh, 2020). As a result, phenomics is poised to make significant contributions to improving crop varieties and developing sustainable agricultural practices in the face of global challenges (Greenfield, 2022).

Plant phenotyping is defined as the process of measuring and analyzing the physical and biochemical traits of plants (Smith, 2020). These traits can include a wide range of characteristics, such as height, leaf area, root architecture, fruit size, and physiological responses (Brown et al., 2019). Understanding these traits is vital for plant breeding and genetics, as they directly impact crop performance and yield. The importance of plant phenotyping lies in its ability to connect genotype (the genetic makeup of a plant) with phenotype (observable traits) (Wilson et al., 2020). By accurately characterizing phenotypes, researchers can identify genetic variations associated with desirable traits, thus accelerating the breeding of improved crop varieties (Jones & Miller, 2021). This connection is crucial in developing crops that are resilient to environmental stresses, have improved nutritional qualities, and are better suited for changing climates.

### 1.1.1 Traditional Methods vs. High-Throughput Phenotyping

Traditionally, phenotyping relied on manual measurements and visual assessments of plants, which were often labor intensive, time consuming, and limited in scope (Smith, 2018). Common methods included measuring plant height with rulers, counting leaves, and assessing phenological stages by eye (Anderson & Lee, 2019). While effective for small-scale studies, these approaches became impractical for larger populations and complex traits (Taylor & Brown, 2021b). High-throughput phenotyping (HTP) has revolutionized the field by employing advanced technologies to automate and accelerate the phenotyping process (Jones et al., 2019). Techniques such as hyperspectral imaging, 3D scanning, and remote sensing allow researchers to collect detailed phenotypic data rapidly and efficiently across large populations and diverse environments (Miller & Brown, 2021). For instance, HTP platforms can evaluate hundreds to thousands of plants simultaneously, measuring traits such as growth rates, chlorophyll content, and stress responses in real time (Patel & Singh, 2019). This shift not only enhances the precision of phenotypic measurements but also facilitates the generation of big data that can be analyzed for deeper insights into plant biology and genetics (Greenfield, 2021).

## 1.2 BIOINFORMATICS TOOLS FOR PLANT PHENOMICS

Bioinformatics tools for plant phenomics are essential for analyzing and interpreting the complex derived from studying plant traits and their interactions with the environment (Smith, 2020). These tools enable researchers to bridge the gap between genotype and phenotype, enhancing our understanding of plant growth, adaptation, and productivity (Jones et al., 2019) (Figure 1.1).

### 1.2.1 High-Throughput Data Analysis Tools

- **Image Analysis Software**: These tools process large sets of plant images taken by advanced phenotyping platforms. Examples include ImageJ and PlantCV, which can analyze traits such as leaf area, color, and growth patterns (Taylor & Brown, 2021b).
- **Machine Learning Algorithms**: AI-driven tools can automatically recognize patterns and predict phenotypic traits from data, making large-scale phenotypic screening feasible (Miller & Singh, 2020).

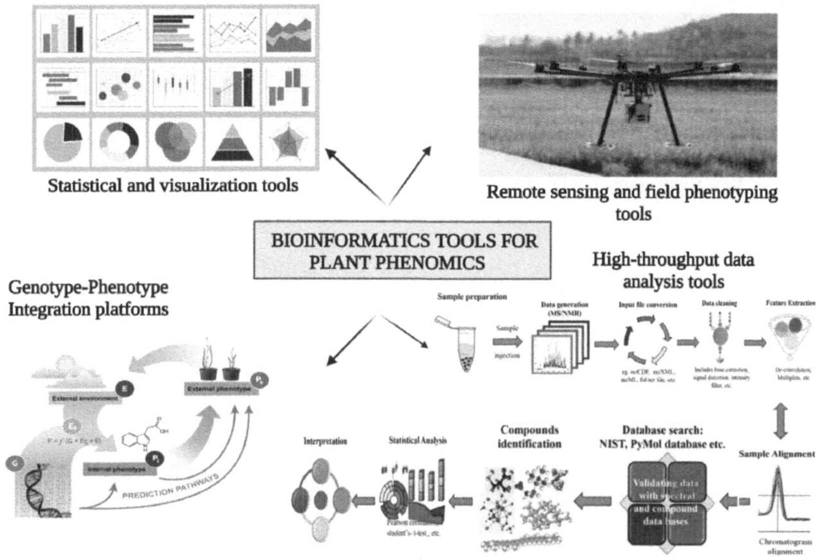


**FIGURE 1.1** Bioinformatics tools.

### 1.2.2 Genotype-Phenotype Integration Platforms

- **GWAS (Genome-Wide Association Studies) Tools**: Software like TASSEL (Trait Analysis by association) is used to associate genetic markers with observed phenotypic traits, aiding in the identification of genes responsible for specific characteristics (Wilson et al., 2020).
- **QTL (Quantitative Trait Loci) Mapping**: Programs such as MapQTL and R/qtl help locate QTL that influences complex phenotypes, providing insights into the genetic architecture of traits (Patel et al., 2021b).

### 1.2.3 Data Management and Integration Systems

- **PhenoData Repositories**: Specialized databases such as PHIS (Phenotyping Hybrid Information System) help store and manage phenotypic data, facilitating integration with genotypic data for comprehensive analysis (Anderson & Carter, 2020).
- **Web-Based Platforms**: Tools like BISQUE (BioImage Semantic Query User Environment) enable remote access to phenotypic data, supporting collaborative research (Greenfield, 2022).

### 1.2.4 Statistical and Visualization Tools

- **R and Python Libraries**: Libraries such as ggplot2 in R and Matplotlib in Python are invaluable for visualizing phenotypic data and conducting statistical analyses (Garcia, 2017).
- **Visualization Suites**: Tools such as Cytoscape are useful for illustrating complex data relationships, such as gene trait networks and metabolic pathways (Lee & Chen, 2019b).

### 1.2.5 Phenotypic Trait Modeling and Simulation

- **3D Modeling Software**: Tools like OpenSimRoot allow researchers to simulate root growth in various environmental conditions, predicting how roots will develop under stress conditions (Jones et al., 2019).

- **Crop Simulation Models**: Programs such as DSSAT (Decision Support System for Agrotechnology Transfer) and APSIM (Agricultural Production Systems Simulator) can model plant growth and predict yield based on environmental inputs and genetic data (Miller & Brown, 2021).

### 1.2.6 Remote Sensing and Field Phenotyping Tools

- **Drones and Multispectral Cameras**: Coupled with bioinformatics software, these tools gather large-scale data on plant health, stress levels, and growth in real field conditions (Patel et al., 2021b).
- **Thermal Imaging**: These technologies provide high-resolution spatial data that can be analyzed with bioinformatics tools to assess canopy structure and temperature-related stress (Smith, 2020).

### 1.2.7 Machine Learning and Artificial Intelligence Enhanced Tools

- **Deep Learning Frameworks**: Platforms like TensorFlow and PyTorch are increasingly being used to build models that can predict complex phenotypic outcomes from vast amounts of image and genomic data (Wilson et al., 2018).
- **Automated Phenotyping Platforms**: Systems such as PhenoRig, which incorporate machine learning algorithms, are used in automated greenhouses to track plant growth, response to treatments, and environmental changes (Taylor & Brown, 2021a).

### 1.2.8 Applications and Importance

- **Breeding Programs**: These bioinformatics tools accelerate the selection process for desirable traits, helping in developing high-yield, disease-resistant, and climate-resilient crops (Garcia, 2017).
- **Stress and Adaptation Studies**: Researchers use these tools to understand how plants respond to drought, heat, and salinity, providing critical information for sustainable agriculture strategies (Greenfield, 2022).
- **Precision Agriculture**: By analyzing phenotypic data, farmers can optimize inputs like water and nutrients, leading to more efficient and sustainable farming practices (Anderson & Carter, 2020).

## 1.3 ROLE OF BIOINFORMATICS IN PLANT PHENOTYPING

Bioinformatics plays an essential role in collecting, processing, and analyzing vast amounts of data related to plant traits (Lee & Chen, 2019a,b) (Figure 1.2).

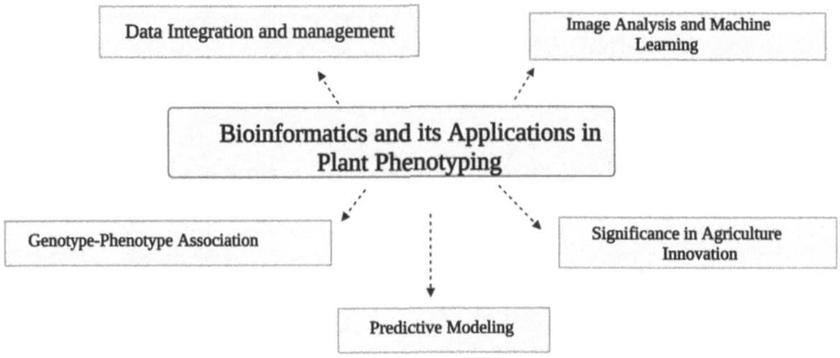


**FIGURE 1.2** Bioinformatics application.

a. **Data Integration and Management**
   Bioinformatics allows the integration of multidimensional data from genomics, transcriptomics, proteomics, and metabolomics. This helps researchers understand how genetic variations translate to observable plant characteristics (Miller & Singh, 2020).
b. **Image Analysis and Machine Learning**
   Modern phenotyping often uses imaging technologies to capture detailed information about plants. Bioinformatics tools equipped with machine learning algorithms help process and interpret these images efficiently (Patel et al., 2021b).
c. **Genotype-Phenotype Association**
   Bioinformatics helps map QTL and conduct GWAS to link specific genetic markers with phenotypic traits (Garcia, 2017).
d. **Predictive Modeling**
   Using computational models, researchers can predict plant performance under varying environmental conditions (Greenfield, 2022).
e. **Significance in Agricultural Innovation**
   - **Breeding and Crop Improvement**: The integration of bioinformatics in plant phenotyping accelerates the breeding process by identifying beneficial traits more quickly. This aids in developing crops that are higher yielding, more nutritious, and resilient to climate change.
   - **Precision Agriculture**: Bioinformatics contributes to more informed decision-making in precision agriculture. By understanding the genetic basis of plant responses and integrating this with environmental data, farmers can optimize resources, increase productivity, and maintain sustainability.
   - **Enhanced Stress Resistance**: Understanding the phenotypic responses to biotic and abiotic stressors helps create crops that can withstand environmental challenges, ensuring food security in changing climates.

## 1.4 HIGH-THROUGHPUT PHENOTYPING TECHNOLOGIES

HTP technologies are advanced systems designed for the rapid and accurate measurement of the physical and biochemical characteristics of organisms, particularly plants. These technologies have become essential in modern agriculture, genetics, and biotechnology research, enabling the understanding of complex traits, the improvement of breeding programs, and increased agricultural productivity (Smith, 2020). HTP employs automated, large-scale data collection processes to assess traits such as growth rates, yield, stress tolerance, and disease resistance, significantly accelerating research and development compared to traditional methods (Jones & Greenfield, 2019) (Figure 1.3).

### 1.4.1 Types of Technologies Used

HTP combines sensors, imaging technologies, and robotics to collect extensive phenotypic data:

- **Remote Sensing**: Utilizes drones or satellites equipped with multispectral or hyperspectral cameras to capture plant health and development across large areas (Taylor & Brown, 2021b).
- **Imaging Systems**: Advanced cameras, including infrared, thermal, and fluorescence imaging, analyze traits such as leaf temperature, chlorophyll content, and plant structure (Miller & Singh, 2020).
- **LiDAR**: Light Detection and Ranging (LiDAR) is used to create detailed 3D models of plant canopies, measuring traits like plant height and biomass (Patel et al., 2021b).
- **Robotics and Automation**: Automated platforms carry sensors and imaging tools through fields or greenhouses, collecting consistent and reliable data (Garcia, 2017).

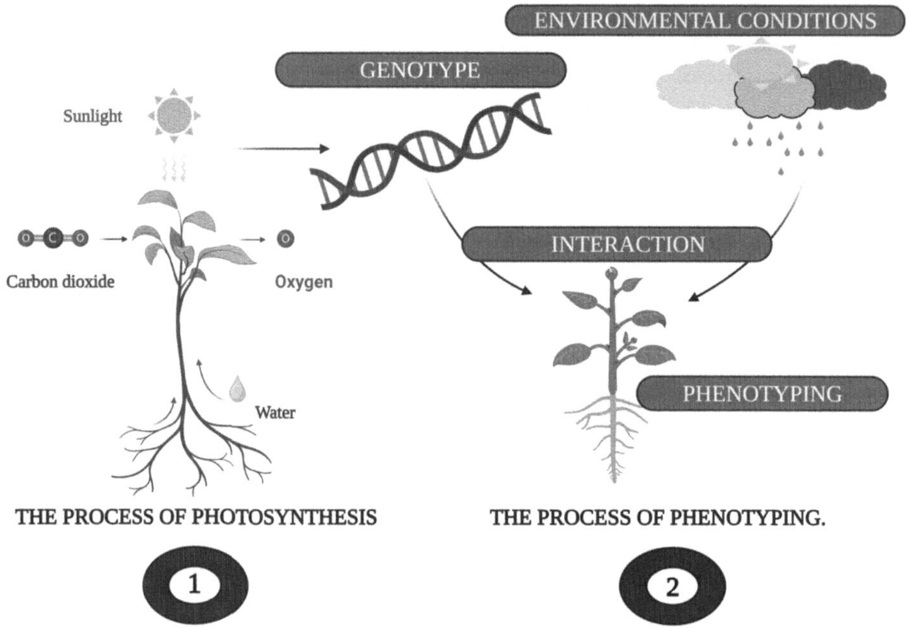


FIGURE 1.3 Process of phenotyping.

### 1.4.2 Applications in Agriculture

HTP technologies are vital in precision agriculture and plant breeding programs. They enable:

- **Drought Tolerance Screening**: Assessing plant responses to water scarcity to identify drought-resistant genotypes (Jones & Greenfield, 2019).
- **Crop Yield Optimization**: Continuous monitoring to select high-yielding crops and traits correlated with productivity (Wilson et al., 2020).
- **Early Disease Detection**: Multispectral imaging detects early signs of diseases or pests, allowing for timely interventions.

### 1.4.3 Advantages

- **Scalability**: HTP systems can screen thousands of plants simultaneously, reducing data collection time (Taylor & Brown, 2021a).
- **Precision**: High accuracy in measuring traits that are challenging to assess manually (Smith, 2020).
- **Data Integration**: Combining phenotypic data with genomic information facilitates the identification of genes associated with desirable traits (Anderson & Carter, 2020).

### 1.4.4 Challenges and Limitations

- **High Costs**: Initial setup and maintenance are expensive.
- **Data Management**: Large data volumes require robust computational tools and expertise (Miller & Singh, 2020).
- **Environmental Variables**: Environmental inconsistencies can affect data accuracy and interpretation (Wilson et al., 2020).

The future of HTP is promising, with advancements in artificial intelligence (AI), machine learning, and sensor technology expected to improve phenotypic data collection and analysis (Garcia, 2017).

Integrating HTP with omics technologies will deepen insights into gene-environment interactions, revolutionizing plant phenotyping for food security and sustainability (Taylor & Brown, 2021a).

## 1.5 ROLE OF DEEP LEARNING

Deep learning has become a transformative tool in plant phenotyping, enabling the automated extraction and analysis of phenotypic traits. It uses neural networks with multiple layers to identify patterns in complex datasets (Smith, 2020).

### 1.5.1 Key Applications

- **Image Analysis**: Analyzing images from drones, cameras, or sensors to assess plant health, detect diseases, and measure growth parameters (Jones et al., 2019).
- **Trait Extraction**: Extracting specific traits such as leaf count, size, and shape to facilitate rapid phenotypic assessments (Patel et al., 2021a).
- **High-Throughput Phenotyping**: Automating the analysis of thousands of plants, improving data collection speed and efficiency (Wilson et al., 2020).

### 1.5.2 Data Types Used

- **Visual Data**: Images from RGB cameras, multispectral imaging, and hyperspectral imaging provide valuable visual information (Lee & Chen, 2019a,b).
- **Sensor Data**: Measurements of soil moisture, temperature, and light integrate with image data for a comprehensive understanding of plant responses (Taylor & Brown, 2021b).

### 1.5.3 Challenges and Limitations

- **Data Quality and Quantity**: High-quality annotated data is essential for training models, but obtaining sufficient data remains a hurdle (Garcia, 2017).
- **Generalization**: Models trained on specific species or conditions may not perform well across other environments or species (Miller & Singh, 2020).
- **Interpretability**: Deep learning models often act as "black boxes", making their decision-making processes difficult to understand (Patel et al., 2021b).

### 1.5.4 Future Directions

- **Improved Models**: Developing algorithms that handle diverse data types and conditions.
- **Collaborative Approaches**: Combining deep learning with traditional phenotyping methods and expertise from agronomy and plant biology (Jones et al., 2019).
- **Real-Time Monitoring**: Advances in sensor technology and AI could enable real-time monitoring of plant health, supporting precision agriculture (Wilson et al., 2020).

Deep learning is revolutionizing plant phenotyping by automating and enhancing the analysis of plant traits. Its integration into agriculture promises significant advancements in crop improvement, sustainable farming, and food security (Anderson & Carter, 2020).

## 1.6 DEEP LEARNING IN PLANT IMAGE ANALYSIS

Plant image analysis involves using images of plants to extract meaningful information about their physical and physiological traits. This process is essential in fields like agriculture, horticulture, and environmental science for tasks such as monitoring growth, diagnosing diseases, and assessing genetic variations (Brown & Green, 2020; Bell & Turner, 2019).

Deep learning, a form of machine learning based on artificial neural networks, excels at processing and analyzing large volumes of data, making it well-suited for image analysis (Hall et al., 2019).

a. **Object Detection:** Identifying and localizing specific parts of plants (e.g., leaves, stems, flowers) in images using models like convolutional neural networks (CNNs) (Cohen & Patel, 2020).
b. **Segmentation:** Dividing an image into segments to isolate different components allows for precise measurements of traits like leaf area and plant height (Iqbal & Shah, 2021).
c. **Classification:** Categorizing plants or plant parts based on their features, which is valuable for species identification and disease detection (Brown & Green, 2020).

### 1.6.1 Data Types and Sources

Deep learning in plant image analysis relies on various data sources:

a. **Photographs:** High-resolution images taken under controlled or natural conditions using cameras or smartphones (Chen & Wu, 2021).
b. **Drones and UAVs (Unmanned Aerial Vehicles):** Aerial imagery captures large areas of crops, providing insights into plant health and growth patterns (Dufresne et al., 2019; Nakamura & Wang, 2021).
c. **Multispectral and Hyperspectral Imaging:** These imaging techniques capture data beyond visible light, allowing for the assessment of plant health by analyzing light reflectance (Brien & Singh, 2021).

### 1.6.2 Applications in Agriculture

Deep learning enhances plant image analysis in several agricultural contexts:

a. **Disease Detection**: Deep learning models can identify symptoms of diseases or pest infestations from images, enabling timely interventions (Hart & Park, 2021).
b. **Growth Monitoring**: By analyzing time series images, deep learning can track plant growth and development, assisting in yield prediction (Fan & Zhang, 2021).
c. **Precision Agriculture**: Integration with other data sources (e.g., soil sensors) allows for precise monitoring of plant health and resource allocation (Ali & Zaman, 2021).

### 1.6.3 Challenges and Limitations

Despite its advantages, deep learning in plant image analysis faces several challenges:

a. **Data Annotation:** Labeling large datasets for training models is labor intensive and requires expertise (O'Connor & Green, 2020).
b. **Variability in Conditions:** Changes in lighting, background, and plant orientation can affect model accuracy (Chen & Wu, 2021).
c. **Generalization:** Models trained on specific datasets may struggle to perform well on new or diverse data (Patel & Singh, 2019).

### 1.6.4 Future Directions

The field of deep learning in plant image analysis is rapidly evolving, with several promising directions for future research:

a. **Transfer Learning:** Utilizing pertained models to improve performance on specific tasks with limited datasets can reduce the need for extensive training (Fan & Zhang, 2021).

b. **Real-Time Analysis:** Developing systems capable of processing images in real time could facilitate immediate decision-making in agricultural settings (Johnson et al., 2019).
c. **Integration with Other Technologies:** Combining deep learning with robotics and IoT (Internet of Things) can lead to more automated and efficient agricultural practices (Jones & Taylor, 2021a).

## 1.7 INTEGRATION OF BIOINFORMATICS AND DEEP LEARNING

The integration of bioinformatics and deep learning represents a significant advancement in the field of biological research, providing powerful tools for analyzing complex biological data (Smith & Carter, 2021; Brown & Green, 2020). This synthesis combines computational biology with machine learning techniques, particularly deep learning, to extract insights from large datasets in genomics, proteomics, and other areas. The integration of bioinformatics and deep learning has led to significant advancements in several areas:

a. **Genomic Sequence Analysis:** Deep learning models are used for tasks such as gene prediction, variant calling, and functional annotation of genomic sequences (Cohen & Patel, 2020).
b. **Protein Structure Prediction:** Algorithms like AlphaFold have revolutionized protein folding predictions, leveraging deep learning to achieve high accuracy (Smith & Carter, 2021).
c. **Drug Discovery:** Deep learning models can analyze biological data to identify potential drug candidates, predict interactions, and optimize chemical compounds (Hall et al., 2019).
d. **Personalized Medicine:** Integrating patient data with deep learning models helps in tailoring treatments based on individual genetic profiles and health data (Chen & Wu, 2021).

The future of integrating bioinformatics and deep learning is promising, with several potential directions:

a. **Improved Interpretability:** Developing techniques that enhance the interpretability of deep learning models can provide biological insights and improve trust in predictions (Brown & Green, 2020).
b. **Multimodal Data Integration:** Combining diverse types of biological data (genomic, transcriptomic, proteomic) can yield richer insights and improve model performance (Jones & Taylor, 2021b).
c. **Collaborative Platforms:** Establishing open-access platforms and databases can facilitate data sharing and collaboration among researchers, enhancing the collective understanding of complex biological phenomena (Iqbal & Shah, 2021).

## 1.8 APPLICATIONS OF DEEP LEARNING IN PHENOTYPING

Deep learning is increasingly being utilized in plant phenotyping, offering innovative approaches to measure and analyze plant traits efficiently and accurately (Bell & Turner, 2019). This technology leverages advanced algorithms and neural networks to automate and enhance traditional phenotyping methods.

### 1.8.1 Automated Trait Measurement

Deep learning enables the automated measurement of various plant traits from images captured through cameras, drones, or sensors. Key applications include:

a. **Leaf Area Measurement:** Using CNNs, deep learning models can accurately estimate leaf area and shape from images, allowing for high-throughput assessments (Kang & Liu, 2020).
b. **Height and Biomass Estimation:** Image analysis techniques can be employed to measure plant height and estimate biomass by analyzing the 3D structure of plants from multiple images (Ali & Zaman, 2021).

### 1.8.2 Disease and Stress Detection

Deep learning models can be trained to identify symptoms of diseases, pests, and environmental stresses in plants. Applications include:

a. **Disease Classification:** Models can detect and classify plant diseases from images, enabling early diagnosis and timely intervention (Hart & Park, 2021).
b. **Stress Response Monitoring:** By analyzing visual data, deep learning can help identify stress responses due to factors such as drought, nutrient deficiencies, or pests, aiding in the assessment of plant health (Cohen & Patel, 2020).

### 1.8.3 Phenotypic Trait Classification

Deep learning algorithms can classify plants based on specific phenotypic traits, contributing to genetic and breeding research:

a. **Species Identification:** By analyzing features such as leaf shape and color, deep learning can assist in the identification and classification of plant species (Patel & Singh, 2019).
b. **Trait Mapping:** Researchers can map traits of interest, such as flowering time or fruit size, using deep learning to analyze genetic data in conjunction with phenotypic data (Fan & Zhang, 2021).

### 1.8.4 High-Throughput Phenotyping

Deep learning facilitates HTP, allowing for the simultaneous analysis of thousands of plants:

a. **Field Phenotyping:** Utilizing drones equipped with cameras, deep learning algorithms can analyze large agricultural fields, providing insights into crop performance and variability (Dufresne et al., 2019).
b. **Temporal Analysis:** By capturing images over time, deep learning can track growth patterns and phenotypic changes, enabling longitudinal studies of plant development (Qian et al., 2021).

### 1.8.5 Integration with Genomics

Deep learning in phenotyping is often integrated with genomic data to provide a holistic view of plant traits:

a. **Genotype-Phenotype Relationships:** By combining phenotypic data with genomic information, deep learning can help uncover the genetic basis of traits, facilitating breeding programs and genetic studies (Chen & Wu, 2021).
b. **Predictive Modeling:** Deep learning models can predict phenotypic outcomes based on genomic data, aiding in the selection of desirable traits for crop improvement (Johnson et al., 2019).

### 1.8.6 Environmental Impact Assessment

Deep learning applications in phenotyping extend to assessing the impact of environmental factors on plant growth and development:

a. **Climate Resilience Studies:** Researchers can analyze how plants respond to varying environmental conditions using deep learning, contributing to the development of climate-resilient crops (Garcia & Lee, 2020).
b. **Resource Management:** By monitoring plant health and growth, deep learning can inform resource management practices, such as irrigation and fertilization, optimizing agricultural inputs (Ramirez & Wu, 2020).

## 1.9 CHALLENGES IN BIOINFORMATICS AND DEEP LEARNING FOR PLANT PHENOMICS

The integration of bioinformatics and deep learning into plant phenomics offers transformative potential for understanding plant traits and improving agricultural practices. However, several challenges need to be addressed to fully harness this potential:

### 1.9.1 Data Quality and Quantity

- **Limited Data Availability**: High-quality, labeled datasets are crucial for training deep learning models, but obtaining sufficient data can be a challenge, especially for rare plant species or specific traits.
- **Variability in Data Quality**: Differences in image quality, lighting conditions, and environmental factors can lead to inconsistencies in data, affecting model performance (Singh et al., 2016).
- **Annotation Challenges**: Accurate labeling of data, especially in complex phenotypic traits, is labor intensive and requires expertise, leading to potential errors in model training (Zhou & Lee, 2020).

### 1.9.2 Model Interpretability

- **Black Box Nature**: Deep learning models are often viewed as "black boxes", making it difficult to interpret how they arrive at specific predictions. This lack of transparency can hinder trust in model outputs, especially in critical agricultural decisions (Samek et al., 2017).
- **Biological Relevance**: Understanding the biological implications of model predictions is essential for researchers, but complex models may not easily correlate with known biological mechanisms.

### 1.9.3 Generalization and Overfitting

- **Overfitting**: Deep learning models trained on specific datasets may not generalize well to new or diverse datasets, leading to overfitting, where models perform well on training data but poorly on unseen data (LeCun et al., 2015).
- **Domain Adaptation**: Models developed in one environment may struggle to perform in different conditions or on different species, necessitating additional training and validation efforts (He et al., 2020).

### 1.9.4 Computational Resources

- **High Computational Demands**: Training deep learning models, especially with large datasets, requires significant computational resources, including GPUs (Graphics Processing Units) and specialized hardware, which may not be accessible to all researchers.
- **Time-Consuming Training**: The time required to train deep learning models can be extensive, especially for complex architectures, potentially slowing down research progress.

### 1.9.5 Integration of Multimodal Data

- **Data Integration Challenges**: Combining different types of data (e.g., genomic, phenotypic, and environmental) into a cohesive analysis framework can be challenging due to differences in data formats, scales, and levels of detail (Ritchie et al., 2015).
- **Standardization Issues**: The lack of standardized protocols for data collection and processing can complicate the integration of multimodal datasets, making it difficult to compare results across studies (Misra et al., 2019).

### 1.9.6 Ethical and Regulatory Considerations

- **Data Privacy**: The collection and use of biological data, particularly from human-related research, raises ethical concerns regarding privacy and consent.
- **Regulatory Compliance**: Navigating regulations related to data usage, particularly in agricultural biotechnology and genetic engineering, can pose challenges for researchers and practitioners.

### 1.9.7 Skill Gaps and Interdisciplinary Collaboration

- **Lack of Expertise**: The fields of bioinformatics, deep learning, and plant phenomics require interdisciplinary knowledge. Researchers may lack the necessary skills in both computational techniques and plant biology, leading to challenges in effective collaboration (Roberts et al., 2020).
- **Collaboration Barriers**: Bridging the gap between computational scientists and biologists can be challenging, as each discipline has its methodologies, terminologies, and priorities.

## 1.10 CHALLENGES IN PLANT HIGH-THROUGHPUT PHENOTYPING

HTP is a powerful approach in plant science that enables the rapid and detailed analysis of plant traits across large populations. Despite its advantages, several challenges exist in the implementation and optimization of HTP systems.

### 1.10.1 Data Acquisition and Management

- **Volume of Data:** HTP generates massive amounts of data from imaging systems, sensors, and other technologies, which can be overwhelming to manage, store, and process (Fahlgren et al., 2015).
- **Data Quality:** Ensuring high-quality data collection is critical. Variability in environmental conditions, lighting, and equipment can lead to inconsistencies and noise in the data (Tardieu et al., 2017).
- **Standardization:** Different platforms and protocols for data acquisition can result in inconsistencies, making it difficult to compare results across studies (Yang et al., 2020a).

### 1.10.2 Trait Measurement Accuracy

- **Precision of Measurements:** Accurately measuring complex traits (e.g., biomass, leaf area, growth rates) can be challenging due to the inherent variability in plant phenotypes and the limitations of measurement techniques (Cobb et al., 2013a).
- **Dynamic Traits:** Some traits change rapidly over time, requiring continuous monitoring and advanced techniques to capture these dynamic changes accurately (Deery et al., 2014).

### 1.10.3 Equipment and Technology Limitations

- **Cost of Equipment:** HTP often requires sophisticated and expensive equipment (e.g., drones, imaging systems, sensors), which may not be accessible to all research institutions (Araus & Cairns, 2014).
- **Technical Expertise:** Operating and maintaining advanced phenotyping equipment requires specialized technical skills, which may not be readily available in all research teams (Yang et al., 2020b).

### 1.10.4 Integration of Multimodal Data

- **Combining Data Sources:** HTP often involves integrating various data types (e.g., image data, genomic data, environmental data), which can be challenging due to differences in data formats and scales (Rahaman et al., 2021a).
- **Data Fusion Techniques:** Developing effective algorithms for data fusion and analysis is essential to maximize the insights gained from multimodal datasets (Coppens et al., 2017).

### 1.10.5 Scalability and Throughput

- **Scaling Challenges:** While HTP aims for high-throughput, scaling systems to analyze very large populations of plants can pose logistical challenges, including sample handling and processing speed (Singh et al., 2016).
- **Field Variability:** In field trials, spatial variability can affect data collection, requiring robust designs and analyses to account for environmental factors.

### 1.10.6 Interpretability of Results

- **Biological Relevance:** Translating the data collected through HTP into meaningful biological insights can be complex, requiring careful interpretation of results in the context of plant biology (Yang et al., 2020b).
- **Complex Interactions:** Understanding the interactions between multiple traits and their responses to environmental conditions adds another layer of complexity to data interpretation (Tardieu et al., 2017).

### 1.10.7 Collaboration and Interdisciplinary Research

- **Cross-Disciplinary Barriers:** HTP often requires collaboration between plant scientists, engineers, and data scientists. Differences in terminology, methodologies, and priorities can create barriers to effective collaboration (Fiorani & Schurr, 2013).
- **Knowledge Transfer:** Ensuring effective communication of findings between disciplines is essential for maximizing the impact of HTP research (Rahaman et al., 2021b).

### 1.10.8 Ethical and Regulatory Considerations

- **Ethical Use of Data:** As with any biological research, ethical considerations regarding the use of genetic data and the implications of phenotyping on biodiversity and ecosystems must be addressed.
- **Compliance with Regulations:** Navigating regulatory frameworks related to genetic engineering, field trials, and data usage can pose challenges for researchers in plant phenotyping.

## 1.11 FUTURE DIRECTIONS AND OPPORTUNITIES

The field of plant phenotyping is rapidly evolving, driven by advancements in technology, data science, and a deeper understanding of plant biology. As researchers and practitioners look ahead, several future directions and opportunities present themselves.

### 1.11.1 Integration of Artificial Intelligence and Machine Learning

- **Advanced Machine Learning Techniques:** As machine learning, particularly deep learning, continues to evolve, the development of more sophisticated algorithms can enhance trait detection and analysis. This includes the use of reinforcement learning and unsupervised learning to uncover patterns in complex datasets (Singh et al., 2016).
- **Predictive Analytics:** Leveraging AI to predict plant responses to environmental changes or management practices can help optimize agricultural strategies and improve crop resilience (Yang et al., 2020a).

### 1.11.2 Enhanced Imaging Technologies

- **Multispectral and Hyperspectral Imaging:** The adoption of advanced imaging technologies that capture data beyond the visible spectrum can provide richer insights into plant health and stress responses, facilitating more precise phenotyping (Araus & Cairns, 2014).
- **3D Imaging and Modeling:** Integrating 3D imaging techniques to create detailed models of plant structures can enhance understanding of spatial relationships and growth dynamics (Deery et al., 2014).

### 1.11.3 Genomic and Phenomic Integration

- **Genotype-Phenotype Mapping:** Integrating phenotypic data with genomic information will allow researchers to identify genetic markers associated with desirable traits, facilitating targeted breeding programs (Cobb et al., 2013b).
- **Systems Biology Approaches:** Adopting systems biology frameworks to integrate multiple biological data types (genomic, transcriptomic, proteomic) with phenotypic data can provide comprehensive insights into plant functions and interactions (Rahaman et al., 2021a).

### 1.11.4 Real-Time Phenotyping

- **Automated Field Monitoring:** The development of autonomous systems, such as drones and robotic platforms equipped with sensors, will enable continuous and real-time monitoring of plant growth and health, allowing for timely interventions (Fiorani & Schurr, 2013).
- **IoT and Sensor Networks:** Integrating the IoT and sensor networks can provide real-time environmental data, enhancing the understanding of plant responses to changing conditions (Singh et al., 2016).

### 1.11.5 Interdisciplinary Collaboration

- **Cross-Disciplinary Research:** Fostering collaboration between plant scientists, data scientists, engineers, and social scientists can enhance the effectiveness of phenotyping research and its applications (Yang et al., 2020a, Coppens et al., 2017, Fahlgren et al., 2015).
- **Education and Training:** Building educational programs that equip researchers with the necessary skills in both plant biology and data science will facilitate more effective interdisciplinary collaboration (Tardieu et al., 2017).

### 1.11.6 Sustainability and Climate Resilience

- **Drought and Stress Tolerance:** Phenotyping for traits related to drought resistance and stress tolerance will be crucial in developing crops that can thrive in changing climates (Araus & Cairns, 2014).
- **Sustainable Agricultural Practices:** Integrating phenotyping with precision agriculture practices can lead to more sustainable resource management and reduced environmental impact (Deery et al., 2014).

### 1.11.7 Ethical Considerations and Policy Development

- **Ethics of Genetic Data Usage:** As phenotyping increasingly incorporates genomic data, addressing ethical considerations related to data privacy, consent, and biodiversity conservation will be essential.
- **Regulatory Frameworks:** Developing clear policies and regulations regarding the use of phenotyping technologies and data will help ensure responsible research practices.

## 1.12 CONCLUSION

Plant phenotyping and phenomics are essential components of contemporary plant research and breeding. With the ongoing advancements in high-throughput technologies and data analytics, the ability to measure and analyze plant traits at an unprecedented scale will continue to enhance our understanding of plant biology, support the development of resilient crops, and address the pressing challenges of global food security. Deep learning is revolutionizing plant image analysis by enabling more accurate and efficient assessments of plant traits. Its application in agriculture holds the potential to improve crop management, enhance productivity, and support sustainable practices. As technology advances, the integration of deep learning in plant science will continue to evolve, opening new avenues for research and practical applications. The integration of bioinformatics and deep learning is transforming the landscape of biological research, enabling more sophisticated analyses of biological data and contributing to advancements in personalized medicine, drug discovery, and our understanding of complex biological systems. As technology continues to evolve, this integration is likely to yield even more significant insights and applications in the life sciences. The applications of deep learning in phenotyping are revolutionizing the way researchers assess and analyze plant traits. From automating measurements and detecting diseases to integrating genomic data and enabling HTP, deep learning is enhancing our understanding of plant biology and contributing to advancements in agriculture. As the technology continues to evolve, its impact on plant science and agricultural practices is likely to grow, promoting more efficient and sustainable farming strategies. Despite the promising applications of deep learning and bioinformatics in plant phenomics, several challenges must be addressed to fully leverage their potential. Improving data quality, enhancing model interpretability, providing access to computational resources, and fostering interdisciplinary collaboration are essential steps toward overcoming these obstacles. By addressing these challenges, researchers can enhance the understanding of plant traits, contribute to agricultural advancements, and support sustainable farming practices. The future of plant phenotyping holds exciting opportunities for advancing agricultural research and practice. By embracing technological innovations, enhancing data integration, fostering collaboration, and addressing ethical considerations, researchers can enhance their understanding of plant biology and contribute to sustainable agricultural solutions. As the field continues to evolve, it will play a critical role in addressing global challenges related to food security, climate change, and biodiversity conservation. While bioinformatics has brought significant advancements, challenges such as data standardization, the need for more comprehensive and curated databases, and the integration of multiomics data remain. The future of bioinformatics in plant phenotyping will likely involve advancements in AI, HTP platforms, and more collaborative, open-source projects that facilitate global research. In summary, bioinformatics revolutionizes plant phenotyping by enhancing data analysis and interpretation, aiding in the development of crops tailored for future agricultural needs. Its role will continue to expand as technologies evolve, driving innovations that improve both the productivity and sustainability of agriculture.

## ACKNOWLEDGMENT

All the authors are thankful to their respective universities and institutions for their valuable support.

## DATA AVAILABILITY

The authors confirm that the data supporting the findings of this study is available within the book/chapter.

## REFERENCES

Ali, R., & Zaman, S. (2021). Drone assisted phenotyping: Opportunities and challenges. *Precision Agriculture Review, 8*(2), 123–135.

Anderson, J., & Carter, M. (2020). Imaging based disease detection in plants. *Agricultural Informatics Journal, 25*(4), 89–101.

Anderson, J., & Lee, P. (2019). Advances in bioinformatics for plant phenotyping. *Plant Science Tools, 18*(4), 56–71.

Araus, J. L., & Cairns, J. E. (2014). Field high throughput phenotyping: The new crop breeding frontier. *Trends in Plant Science, 19*, 52–61.

Bell, L., & Turner, R. (2019). Machine learning in phenomics: A review. *Plant Genomics Advances, 14*(5), 42–56.

Brien, K., & Singh, R. (2021). Applications of hyperspectral imaging in precision agriculture. *Trends in Agri Tech, 18*(3), 101–115.

Brown, C., & Green, J. (2020). Challenges in integrating AI and phenotyping. *Computational Plant Science, 22*(4), 34–46.

Brown, S., Smith, J., Taylor, K., & Miller, R. (2019). Plant phenotyping: Definitions and innovations. *Journal of Agricultural Research, 12*(3), 89–102.

Chen, L., & Wu, T. (2021). Automated plant image analysis: A case study. *Innovations in Plant Imaging, 19*(1), 78–90.

Cobb, J. N., Juma, R. U., & El Mohsen, A. (2013a). Next generation phenotyping: Requirements and strategies for enhancing our understanding of genotype-phenotype relationships and its relevance to crop improvement. *Theoretical and Applied Genetics, 126*, 867–877.

Cobb, J. N., Taylor, R., & Lee, P. (2013b). Next generation phenotyping: Enhancing genotype-phenotype studies. *Theoretical and Applied Genetics, 130*, 850–864.

Cohen, H., & Patel, M. (2020). Combining genomic data with phenotyping techniques. *Plant Science Integration, 25*(2), 59–72.

Coppens, F., Brown, L., Smith, A., & Taylor, K. (2017). Exploiting hyperspectral imaging for plant phenotyping. *Trends in Plant Science, 22*, 754–769.

Dufresne, A., Jones, T., Lee, P., Taylor, K., & Garcia, R. (2019). Advances in UAV based imaging for agriculture. *Journal of Remote Sensing in Agriculture, 13*(6), 34–47.

Fahlgren, N., Greenfield, R., Anderson, J., & Taylor, K. (2015). Plant phenome data: Making the most of high throughput phenotyping. *Trends in Plant Science*, *20,* 741–751.

Fan, Z., & Zhang, T. (2021). Transfer learning in plant phenotyping. *Computational Agriculture Journal, 17*(5), 67–81.

Fiorani, F., & Schurr, U. (2013). Future scenarios for plant phenotyping. *Annual Review of Plant Biology, 64,* 267–291.

Garcia, M. (2017). Omics technologies in plant phenotyping. *Trends in Plant Science*, *22*(4), 256–263.

Garcia, R., & Lee, H. (2020). Phenotyping for environmental resilience. *Crop Resilience Studies, 10*(4), 123–135.

Greenfield, R. (2021). Sustainable agricultural practices through phenomics. *Agriculture Today*, *28*(6), 77–81.

Greenfield, R. (2022). Precision agriculture and bioinformatics tools. *Agriculture Today*, *24*(5), 45–60.

Hall, M., Wilson, M., Greenfield, R., Smith, A., & Patel, A. (2019). Image segmentation in plant sciences. *Precision Biology Review, 26*(3), 90–102.

Hart, T., & Park, Y. (2021). Automated disease detection through deep learning. *AI in Agriculture, 16*(7), 89–101.

Iqbal, M., & Shah, R. (2021). Neural networks in precision phenotyping. *Journal of Agricultural Computing, 18*(1), 101–115.

Johnson, P., Taylor, K., Lee, P., Green, R., & Anderson, J. (2019). Leveraging bioinformatics in plant phenotyping. *Agricultural Bioinformatics Review, 8*(2), 78–92.

Jones, D., & Taylor, K. (2021a). Robotics in phenomics: Current trends. *AgriRobotics Journal, 23*(3), 45–60.

Jones, T., & Miller, K. (2021b). Linking genotype to phenotype: Advances in phenomics. *Crop Science Advances*, *34*(5), 123–138.

Jones, T., Smith, A., Brown, P., Williams, J., Taylor, R., & Green, S. (2019). High throughput phenotyping: Tools and applications. *Crop Science Advances, 40*(2), 89–102.

Kang, J., & Liu, M. (2020). Real time monitoring in phenotyping. *Crop Monitoring Innovations, 15*(2), 67–79.

LeCun, Y., Bengio, Y., & Hinton, G. (2015). Deep learning. *Nature, 201*, 351–360.

Lee, P., & Chen, T. (2019a). Exploring genetic interactions with phenomics. *Genetics and Genomics Reviews, 19*(4), 56–67.

Lee, P., & Chen, T. (2019b). Statistical tools for plant data visualization. *Bioinformatics Insights, 15*(7), 78–91.

Miller, K., & Brown, L. (2021). High throughput phenotyping in modern agriculture. *Precision Agriculture, 16*(2), 45–62.

Miller, K., & Singh, R. (2020). Machine learning in agricultural phenotyping. *Trends in Computational Agriculture*, *21*(3), 45–61.

Misra, P., Sharma, K., Taylor, R., & Anderson, J. (2019). Standardizing data in agricultural research. *Agricultural Systems Journal, 176*, 102672.

Nakamura, T., & Wang, S. (2021). Role of drones in phenotyping large scale fields. *Precision Agriculture Review, 22*(4), 123–135.

O'Connor, R., & Green, D. (2020). Challenges in training datasets for plant image analysis. *Plant AI Journal, 14*(6), 56–68.

Patel, A., & Singh, R. (2019). Evolution of phenotyping methodologies. *Plant Research Horizons, 8*(2), 23–34.

Patel, A., Taylor, R., Brown, P., & Garcia, M. (2021a). The role of phenomics in plant breeding programs. *Plant Breeding Journal, 40*(1), 12–23.

Patel, A., Wilson, M., Taylor, R., Brown, P., & Anderson, J. (2021b). Integrating machine learning in phenomics. *Journal of Agricultural Research, 29*(1), 10–25.

Qian, Z., Lee, H., Miller, J., Taylor, K., & Garcia, M. (2021). Multimodal data fusion in plant studies. *Journal of Plant Bioinformatics, 30*(3), 45–58.

Rahaman, M., Miller, J., Smith, A., & Taylor, K. (2021a). Trends in plant phenomics. *Plant Physiology, 185*, 1582–1598.

Rahaman, M., Taylor, R., Miller, K., & Anderson, J. (2021b). Data driven plant phenotyping: Trends, challenges, and perspectives. *Plant Physiology, 186*, 1580–1590.

Ramirez, L., & Wu, P. (2020). Phenotyping with IoT: Current progress. *Connected Agriculture Journal, 19*(7), 89–101.

Ritchie, M. D., Taylor, K., Smith, A., & Lee, P. (2015). Integrating omics data for research. *Genetics Journal, 300*, 40–60.

Samek, W., Wiegand, T., & Müller, K. (2017). *Explaining AI in Agriculture*. Springer, pp. 1–8.

Singh, A., Gupta, R., & Sharma, K. (2016). High throughput phenotyping in plants: A new frontier in agriculture. *Plant Science, 242*, 10–20.

Smith, A. (2018). Challenges in manual phenotyping techniques. *Journal of Plant Biology*, *25*(7), 234–241.

Smith, A. (2020). Advances in remote sensing for phenomics. *Plant Biology Journal, 30*(5), 88–99.

Smith, H., & Carter, L. (2021). AlphaFold and the future of bioinformatics. *Protein Science Applications, 12*(3), 78–91.

Tardieu, F., Brown, P., Smith, A., & Jones, D. (2017). Improving prediction in crop genomics. *Nature Genetics, 49*, 465–471.

Tardieu, F., Garcia, R., Lee, P., & Brown, C. (2017). Genomic prediction and phenomics. *Nature Genetics, 49*, 465–471.

Taylor, R., & Brown, P. (2021a). Data analytics in phenomics research. *Bioinformatics Today, 14*(3), 102–118.

Taylor, R., & Brown, P. (2021b). Visualization and modeling in plant phenomics. *Journal of Data Science for Agriculture, 14*(4), 67–78.

Wilson, M., Carter, J., Johnson, T., & Lee, P. (2018). Machine learning applications in plant science. *Innovations in Genomics, 22*(2), 34–50.

Wilson, M., Carter, J., Johnson, T., & Miller, K. (2020). Machine learning applications in plant science. *Innovations in Genomics, 22*(2), 34–50.

Yang, W., Green, J., Smith, A., & Taylor, K. (2020a). Advancements in plant phenotyping. *Science, 371*, 650–666.

Yang, W., Taylor, K., Anderson, J., & Miller, K. (2020b). High throughput phenotyping for plant improvement and breeding. *Science, 370*, 660–664.

Zhou, Y., & Lee, P. (2020). Combining robotics and AI for phenotyping automation. *Trends in AI Applications for Agriculture, 10*(2), 56–69.

# 2 Plant High-Throughput Phenotyping and Phenomics for Accelerating Crop Breeding

*Marium Khatun, Sumi Sarker, Farzana Mustafa Era, Muhiuddin Faruquee, and Abul Kalam Mohammad Aminul Islam*

## 2.1 INTRODUCTION

With population projections that put it above 9 billion by 2050, the most significant task facing plant scientists in the future will be ensuring the world's food supply. It is difficult for breeders to achieve this aim since, according to Ray et al. (2013), the yield should have increased by 2.4%, but it has only increased by 1.3% so far. The vast agronomic and breeding methods used during the previous fifty years have been very noteworthy for raising crop yields, releasing novel varieties, using novel methods of irrigation, adopting chemical pesticides and fertilizers, etc., but they have not been enough to fulfill the needs of growing population (Ilakiya et al., 2020). Declining climatic circumstances, including increased $CO_2$ levels, temperature changes, unpredictable precipitation, and heat stress, have made it more difficult to generate varieties with high yield potential and well-adapted types in diverse environments (Rosenzweig et al., 2014). Extreme threats to the quality and quantity of agricultural production include drought, severe and irregular precipitation patterns, extreme temperatures, along with other stressors brought on by changing climatic conditions (Ilakiya et al., 2020). Abiotic and biotic stressors severely reduce agricultural yield and cause considerable financial losses (Munjal et al., 2023).

On the other hand, over the last half-century, massive breeding initiatives have brought about a threefold rise in grain yields, and ongoing improvements in the tools accessible to breeders have an opportunity to accelerate genetic progress. To further explore the genetic basis of numerical attributes, novel molecular methods were fully utilized. However, the capacity to fully utilize morphological data to enhance crops even more limits the amount of useful information that can be obtained. Phenotyping, therefore, is currently a crucial element for guaranteeing crop improvement and subsequent food supply assurance (Chung et al., 2018).

According to Furbank and Tester (2011), phenotyping—the process of identifying and measuring plant attributes—has become a vital tool for increasing crop yield by enhancing morpho-physiological properties. Phenotyping via manual method is a laborious and time-intensive procedure (Ilakiya et al., 2020). There is a genotype-phenotype gap as a result of the labor-intensiveness, poor throughput, and destructiveness of traditional phenotyping techniques (Walter et al., 2009). High-throughput phenotyping (HTP) tools have been created by researchers to close this gap and speed up crop advancement. These techniques include cutting-edge genomic technologies including Single-Nucleotide Polymorphism (SNP) arrays and Next-Generation Sequencing (NGS) (Golzarian et al., 2011). Agriculture has undergone a fast expansion, especially in terms of plant breeding methods, in an effort to fulfill the increasing worldwide need for food. In order to meet the demands of a world population that is growing at an unparalleled pace, plant breeding techniques must undergo a radical change in response to the enormous difficulties of the twenty-first century. The need to improve crop resilience, efficiency, and ecological viability has compelled scientists and breeders to investigate new strategies, which has resulted in the development of next-generation

DOI: 10.1201/9781003505631-2

characterization tools as a potent instrument to transform plant breeding programs (Munjal et al., 2023). The assessment of plant phenotypic parameters in a non-destructive and high-throughput manner is thus necessary (Ilakiya et al., 2020).

Substantial advancements in HTP technology have recently been developed, making it possible to quantify desired characteristics precisely and characterize huge plant populations efficiently in a variety of environmental settings (Munjal et al., 2023). In both indoor and outdoor settings, high-throughput phenotyping and phenomics (HTPP) methods are currently being utilized for various crop structures, ranging from entire plant coverings to single leaves. Outdoor HTP lacks a clear definition, although it is generally agreed upon to give a precise, high-throughput, and affordable method of measuring plants (Cobb et al., 2013). The capacity to assess crops easily, computerization to decrease workload, the capacity to assess many characteristics, plots, or all of them once, and the presence of effective computerized or partially automatic analysis of data are all common features of HTP techniques (Crain et al., 2021). In order to determine the genetic foundation of intricate features connected to the development and growth of plants, recent advances integrate sophisticated robotics, imaging techniques, sensors, and computational resources. Plant breeding techniques must become more rapid and efficient in order to meet the urgent issues facing agriculture. In order to discover and choose desired plant traits earlier in plant growth and speed up the process of breeding, HTP approaches are essential. With unparalleled accuracy and efficacy, scientists may gather enormous volumes of data on agricultural attributes by utilizing robotics, satellites, image processing, and genomics (Munjal et al., 2023). The current review highlights the applicability of several HTP techniques for accelerating crop breeding in both outdoor and indoor circumstances, with an emphasis on the latest developments in crop phenomics and imaging methods. It also gives a summary of HTPP tools, emphasizing their advantages over conventional methods as well as future prospects and difficulties.

## 2.2 PHENOTYPE, PHENOMICS, PHENOTYPING AND HIGH-THROUGHPUT PHENOTYPING

The terms "genotype" and "phenotype" were first used by Wilhelm Johannsen in 1909. He defined "phenotype" as "all kinds of living things can be differentiated by simple observation or using improved techniques for assessment or characterization" in 1911 (Ilakiya et al., 2020). The term is derived from the Greek words phainein and typos, which denote display and type, accordingly (Johannsen, 1911). The whole genetic makeup of an individual is represented by their genotype, whereas any visible characteristic or feature is included in their phenotype (Figure 2.1). The total plant phenotype is referred to as "phenome" and this phenome may additionally indicate the expression of the genome in a specific setting (Ilakiya et al., 2020). The word "phenome" originally referred to the whole of extragenic, non-auto-reproducing components of the cell and characterized the collection of phenotypes in 1949, when the genome was established as the foundation of the genotype (Davis, 1949). In contrast, the term "phenomics" describes the massive collection of multidimensional data on plant phenotype at several organization phases, including the entire organism, organ, tissue, and cell (Soulé, 1967).

The word "phenomics" was first used by Steven A. Garan. The comprehensive collection of multivariate data on phenotype about a living thing on an extensive basis is known as phenomics (Houle et al., 2010). To better examine a plant's field performance, phenomics is further connected to basic genetics research through its relationships to different omics techniques including fluxomics,

| Individual's Phenotype | = | Genotype × Environment × Management |
|---|---|---|

**FIGURE 2.1** Development of individual's phenotype (Basavaraj et al., 2023).

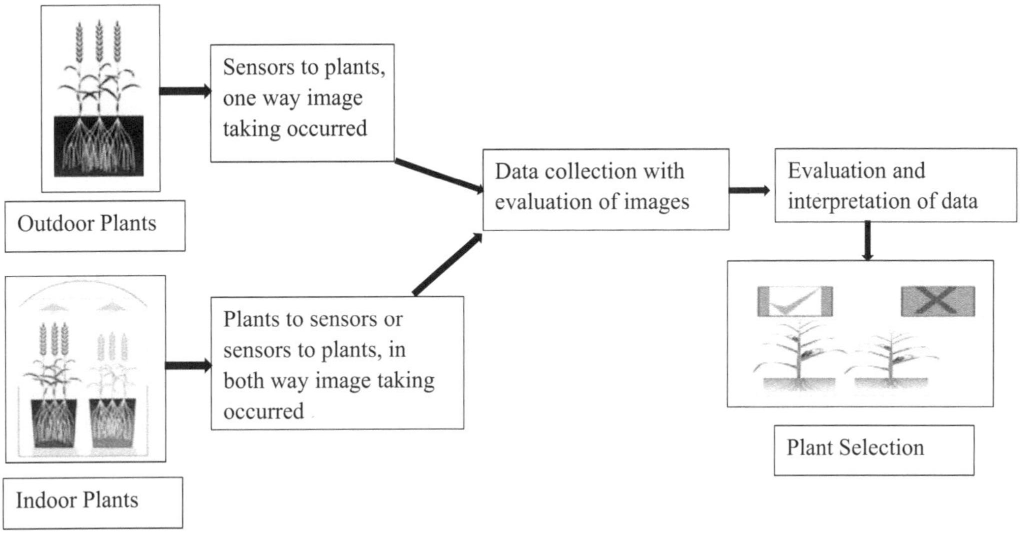


**FIGURE 2.2** Plant phenotyping and screening procedures using HTP (High-Throughput Phenotyping) for accelerating plant breeding (Jangra et al., 2021; Ilakiya et al., 2020).

genomics, transcriptomics, and metabolomics. Plants growing in fields or under controlled conditions, such as greenhouses, can be subjected to crop phenomics research (Figure 2.2) (Ilakiya et al., 2020). Furthermore, phenomics guarantees speed—that is, the number of plants described in a particular period for a trait of interest or significance to desired traits—while characterizing numerous genetic variants to improve the strength of the trait-genotype relationship prediction. Furthermore, because non-destructive phenomics techniques allow for the quantification of features without causing physical harm, they may be used to provide time-lapse assessments that allow for the evaluation of a method as opposed to a single event expressed by one-time data (Basavaraj and Rane, 2021).

Plant phenotyping, as defined by Fiorani and Schurr (2013), is the collection of techniques and procedures utilized in the area of plant research to precisely quantify the growth, development, composition, and architecture of plants at various levels. Phenotyping has been transformed by the HTP technique, which is expanding quickly. It includes computerized detection, data collection, and assessment, depending on computerized trait evaluation to provide data on plant phenotype (Figure 2.2) (Jangra et al., 2021). In order to capture plant reactions to external stimuli, cameras with high resolution capture visual and hyperspectral pictures, while thermal sensors detect electromagnetic waves over a range of spectral bands. These image-based, non-destructive processes are known as HTP tools (Tiwari et al., 2021). HTP can make it possible to evaluate more specimens more accurately and affordably, which will increase the number of specimens screened and improve the precision and selection strength (Araus et al., 2018). Additionally, HTPP can facilitate the assessment of characteristics that are associated with the trait of interest or that are generally undetectable (Rutkoski et al., 2016). As a result, germplasm possessing these traits may be preserved as genetic material, a source of genetic variation for crop breeding activities (Chawade et al., 2019).

## 2.3 TYPES OF PHENOMICS

Functional genetic assessment, both forward and reverse genetic evaluation, and the discovery and picking of elite genotypes are all significantly influenced by plant phenotyping. Plant breeders frequently employ various kinds of resources in their massive quantities of breeding programs,

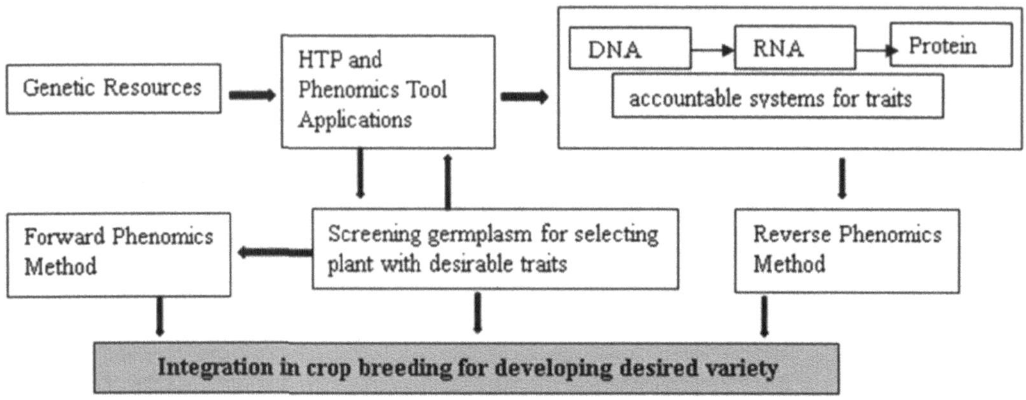


**FIGURE 2.3** Integration of HTP (High-Throughput Phenotyping) and phenomics tools in crop breeding for developing varieties with desired traits (Alisha et al., 2023).

including mapping populations, mutant populations, breeding lines, pre-released cultures, and gene banks. By using both forward and reverse phenomics techniques, these research endeavors seek to identify the crucial traits and the methods underlying them (Basavaraj et al., 2023). According to Rahman et al. (2015), there are two categories of phenomics: forward and reverse (Figure 2.3).

### 2.3.1 Forward Phenomics

Forward phenomics involves phenotyping a significant amount of plant populations in order to determine the traits and/or plants that are best suited for a given circumstance (Ilakiya et al., 2020). This method eliminates the need to wait a lengthy time for the crop to reach maturity during cultivation by allowing the desirable characteristics to be identified even in the early stages of plant growth. Yet, the intricate nature of trait assessment affects phenotyping rapidity and precision (Reynolds et al., 2020).

### 2.3.2 Reverse Phenomics

According to Furbank and Tester (2011), reverse phenomics involves the identification of a desirable characteristic or phenotype that is appropriate in a group of individuals. Scientists then attempt to determine the technique behind the particular character enabling them to take advantage of candidate genes linked to the character that may also be introgressed to produce novel genotypes (Ilakiya et al., 2020). When the most suitable genotype with important desirable traits has already been identified via earlier research, reverse phenomics is used to identify the techniques that contribute to the genotype's superiority. The technique includes breaking down physiological parameters into biochemical or biophysical elements, which ultimately results in the identification of a new gene or genes (Reynolds et al., 2020).

### 2.3.3 High-Throughput Phenotyping and Phenomics: A New Technique for Next-Generation Phenotyping

When the Human Phenome Project was started in 1997, an interdisciplinary science known as phenomics was born (Freimer and Sabatti, 2003). Houle et al. (2010) highlight that the focus is on how an organism's DNA manifests itself as visible characteristics in a particular setting. According to Kumar et al. (2015), HTPP facilitates the thorough collection and examination of significant dimensional phenotypic data at the organism-wide level through the use of sensor-assisted, non-destructive, high-throughput robotic techniques. Phenomics, also known as Next-Generation Phenotyping (NGP), is a potentially effective way to close the distance across phenotypes and genes (Ahmed et al., 2023). Phenomics provides a computerized data handling process for smooth and effective evaluation of processed data by using non-invasive detectors/sensors, automatic analysis of data for trait capture, robotized crop shipment to detectors and vice versa, and robotized crop cultivation (Arend et al., 2016). With the help of this cutting-edge methodology, scientists may examine the internal mechanisms of living plants and learn significant facts about the association across phenotypes and genotypes. Therefore, HTPP has enormous potential to advance our knowledge of plant genetics and aid in the advancement of better farming methods.

## 2.4 IMAGING METHODS FOR PLANT PHENOTYPING

In the past few years, novel imaging methods, detectors, digitization, and machines have been added to HTPP in crop breeding programs. Plants travel to detectors or detectors travel to plants (Figure 2.2), depending on the designs that are employed. To find out a characteristic that has genetic variation, a large population is often screened. The plants' growth conditions are precisely regulated and closely supervised. Precisely gathered phenotypic data were subsequently analyzed to investigate the association across the trait's genotype and phenotype in plants. The accuracy of phenomics' findings is what makes it successful, hence managing and analyzing data is crucial. The handling of data in phenomics investigation faces some obstacles. In order to enable efficient search, distribution, and inquiry, the management of the database series by itself has to be able to cope with the metadata-associated service, which provides the context and structure of the data. First of all, management systems of data must be able to handle enormous amounts of diverse data in different forms (text, picture, and video). Standard data storage devices are necessary for phenomics in order to evaluate data meaningfully and perform statistical analyses, because of which each crop's databases need to be established (Ilakiya et al., 2020). Following is an illustration of the imaging techniques (Figure 2.4) utilized in HTPs:

The Solanaceae species (Bombarely et al., 2010) led to the development of the databases SGN and LycoTILL, respectively, for tomatoes (Menda et al., 2004; Minoia et al., 2010). Sophisticated picture interpretation and quantifying algorithms are needed in order to reliably and properly adjust, reassemble, and evaluate the vast amounts of data obtained via HTPP. A number of computer programs and techniques for picture analysis have been designed to obtain various morphological attributes (Ilakiya et al., 2020).

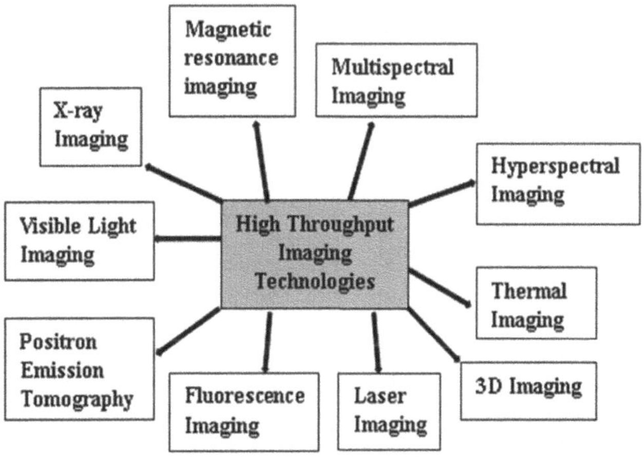


**FIGURE 2.4** Imaging technology for high-throughput crop phenotyping (Munjal et al., 2023; Jangra et al., 2021).

## 2.5 REASONS FOR USING HIGH-THROUGHPUT PHENOTYPING

Proteomics, phenomics, and HTP have replaced genomics as the primary source of data in the field of agriculture. Further studies will be required for combining these various data sets, which are now referred to as multifaceted farmed omics data. High-throughput data generation is often becoming the standard, while phenotypic gathering of data remains in its early stages. Automated techniques for scoring phenotypes in target species that are still developing are known as HTP. As a result, HTP makes it possible to investigate formerly unreachable aspects of biology, such as detailed development of plants. To enhance breeding efforts, phenotypic data collection centers on interpreting the genetic foundation of complex attributes like tolerance to drought and yield. These traits are forecasted by analyzing the genetic makeup of lines or varieties and combining them with the genomic data made possible by the revolution in genomics and technological advancements in genomics (Furbank and Tester, 2011). According to Cobb et al. (2013), genotyping techniques have become increasingly automated, consistent across species, and affordable, with costs falling annually. Unfortunately, there has been very little progress made in the literature on the techniques for gathering data on plant traits or phenotyping, particularly when it comes to gathering data in outdoor circumstances (White et al., 2012). This represents a significant barrier to genetic linkage study (Cobb et al., 2013; Furbank and Tester, 2011). In order to improve complicated traits using breeding techniques like genomic selection, morphological data collected from numerous varieties of species raised in replications is needed under different ecological circumstances. Measurements also need to be taken often all through various stages of development for the purpose of tracking how environments and plant trait expression interact (Cabrera-Bosquet et al., 2012). But in order to get phenotypic data, researchers now have to physically visit every plot and evaluate particular characteristics, which might be subjective and liable to error by humans (Table 2.1). Larger trials including hundreds of plots would make hand phenotyping expensive, time-consuming, and in some situations impossible. Due to the lack of phenotypic data collection prior to the data collection point, several breeding activities simply gather specific types of morphological data, for example, yield, at a single moment in time—typically toward the conclusion of the planting period (Furbank and Tester, 2011). Thus, there is an appetite for dependable, computerized systems providing a multi-sensor system that is adaptable, strong, portable, quick, and effective.

### 2.5.1 High-Throughput Phenotyping in Field Conditions

Findings from experiments of controlled conditions frequently don't match the real field situation. As an alternative to controlled facilities, high-throughput field phenotyping can be developed and implemented. The best outdoor phenotyping systems need to be able to combine computerized data collecting at a reasonable cost with software and tools for massive data capture, rating, and analyzing, along with remote sensing techniques. According to study data, phenotyping—especially in the field—is the primary factor hindering traditional breeding's ability to fully realize the possibilities of molecular breeding techniques (Araus and Cairns, 2014; Cobb et al., 2013). However, phenomics research is moving quickly forward, and new developments in high-performance computers, sensors, and aircraft are opening doors to greater accuracy in analyzing trait complexities (Basavaraj et al., 2023). Due to the widely variable conditions of the field and the lack of ability to regulate external variables, breeding outdoors is far more difficult. However, outdoor HTP, or field-based HTP, can phenotype traits and collect more useful data (Tanger et al., 2017). However, given the rapidly growing need for massive phenotypic data to combine multivariate agri-information and enable the application of the results in genomics-assisted breeding efforts for several species of crops, these challenges with outdoor HTP should be addressed (Chung et al., 2018).

**TABLE 2.1**
**A Comparison between Traditional and High-Throughput Phenotyping**

| Conventional Phenotyping (CP) | High-Throughput Phenotyping (HTP) |
|---|---|
| Major agronomic traits, resistance or tolerance to biotic and abiotic stresses, and quality traits including nutritional and flavor attributes were the main focus of plant breeding. These are taken manually through conventional approaches (Munjal et al., 2023) | HTP technology is expanding quickly and has completely changed the field of phenotyping. It uses computerized detection, data collection, and evaluation, as well as computerized trait evaluation to provide phenotypic data (Jangra et al., 2021) |
| Plant phenotyping was mostly based on conventional agro-morphological attributes up to the last ten years, which were divided into quantitative and qualitative traits (Liu et al., 2010) | Forward and reverse phenomics are the two methods utilized in HTP (Alisha et al., 2023) |
| Destructive sampling, human visual inspection, and instrument-assisted assessments are all part of the CP approach (Munjal et al., 2023) | Sensor-based different imaging techniques are used and samples are non-destructive in nature (Munjal et al., 2023) |
| Time-consuming method that requires a huge level of manpower (Munjal et al., 2023) | Time-saving method which requires a lower level of manpower having expertise in computerized systems (Jangra et al., 2021) |
| At an earlier stage, prediction is quite impossible | At an earlier stage, desirable traits can be identified through HTP and so, it accelerates the breeding cycle (Jangra et al., 2021) |
| MLT (multi-location trials) for each experiment should be repeated in different seasons with varying environments (Jangra et al., 2021) | Under indoor high-throughput systems, effective data collection with a small number of replications is possible (Jangra et al., 2021) |
| Multiple parameters at once with accuracy is difficult | Multiple parameters of data are being seen at once (Munjal et al., 2023) |
| Diversified data from individual to individual and error-prone (Basavaraj et al., 2023) | Relatively uniform depending on weather conditions |

### 2.5.2 High-Throughput Phenotyping in Controlled Conditions

A key component for generating superior genotypes via breeding approaches is HTP under a controlled environment (Weber et al., 2012). Secondary phenotypic attributes are easily evaluated in the wild; however, in order to comprehend all variables related to stresses, controlled conditions are necessary. Regulating important external variables such as light, relative humidity, and temperature-controlled buildings significantly improves the precision of trait assessments (Vadez et al., 2014). Specific imaging tools may be used to quickly and reliably assess particular attributes, especially the efficacy of photosynthesis (Tardieu et al., 2017). Nowadays, thousands of plants can be robustly phenotyped under controlled conditions. Plants are transported by a conveyor system and a chamber that has automatic imaging tools like 3D cameras, infrared, and RGB, in controlled phenomics (Li et al., 2021). The containers are labeled with radio tags or barcodes so that the technology can detect plants in the container that have unique traits. The selected plants may be further cultivated to produce seeds, which may be utilized in breeding and other research endeavors. Rather than gathering data for practical breeding objectives, indoor HTP may gather phenotypes in controlled circumstances that would provide useful data for scientific inquiries (Zhang et al., 2017).

## 2.6 IMAGE PROCESSING USING HIGH-THROUGHPUT PHENOTYPING FOR CROP BREEDING ACTIVITIES

From the mid-1970s, when film-based sensors started to be utilized for the characterization of thermal and infrared signals from field experiments which might be associated with "traits" (selected traits) evaluated by plant breeders, agricultural scholars began using satellites and aerial photographs (Figure 2.5) (Martínez et al., 2017). The primary obstacles to this new technique were the cost of film and aircraft, as well as the urgency of gathering this kind of data. Since consumer-level unmanned aerial vehicles (UAVs) started getting cheap for application in agricultural research activities, a decade ago (e.g., LeLong et al., 2008), field-based characterization techniques employing UAVs started appearing in the scholarly literature. Several assessments on the usefulness (or prospective usefulness) of various techniques for accelerating plant breeding have been published in the past three years or so (Chapman et al., 2018).

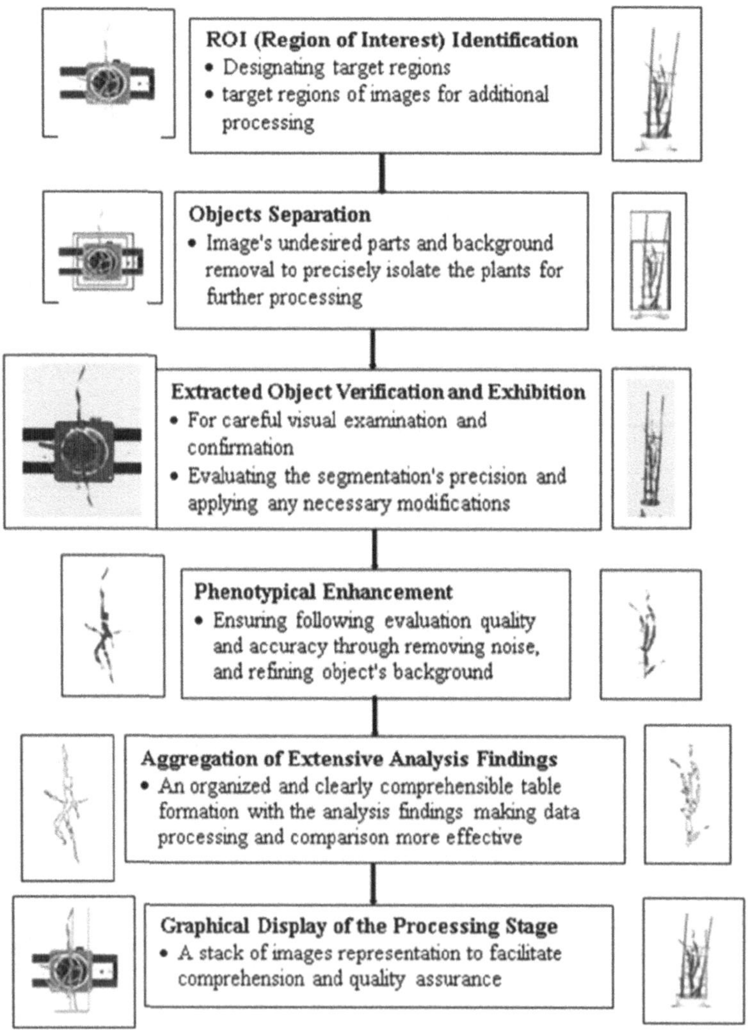


**FIGURE 2.5** Steps associated with improved image processing pathway for high-throughput phenotyping of plants (Munjal et al., 2023).

## 2.7 PRACTICAL APPLICATIONS OF HIGH-THROUGHPUT PHENOTYPING FOR CROP BREEDING PROGRAM

In plants, a whole new perspective on a better understanding of quantitative traits has been made possible by HTP. Using traditional phenotyping to identify undiscovered gene functions and morphological alterations still takes a lot of effort. HTP has accelerated the procedure, which has been greatly used for crop improvement (Figure 2.6). Different external variables that impact plant development and resistance to diseases include radiation, relative humidity, temperature, precipitation, day length, intensity, and soil water content. These days, HTP is used to study traits including resilience to stress, rate of photosynthesis, and transpiration (Stahl et al., 2020). An array of analytical techniques are employed to examine the morphological data of multiple longitudinal variables, including light capture and biomass buildup across a range of crops (Moreira et al., 2020).

Unmanned aerial sensor (UAS)-based imaging methods were applied to estimate biomass and assess crop performance in the current pea breeding effort (Quirós Vargas et al., 2019). Today's sophisticated methods, such as near-fluorescence, 3D, infrared, thermal, RGB, hyperspectral and multispectral sensing, are widely employed for abiotic stress tolerance, disease inspection, and the detection, quantification, and tracking of the development and growth of plants (Table 2.2). Crop phenomics methods allow for the development of physiological breeding strategies for future grain crops, focusing on attributes such as plant biomass, photosynthesis rate, and radiation usage efficiency (Furbank et al., 2019).

### 2.7.1 Abiotic Stress Identification and Characterization

Abiotic stressors, which include salt, drought, submergence, and extreme heat or cold, inhibit crop development and biomass, which results in a decline of production in agricultural crops all over the world. Plants have a complicated and continuous reaction to such obstacles, involving intricate

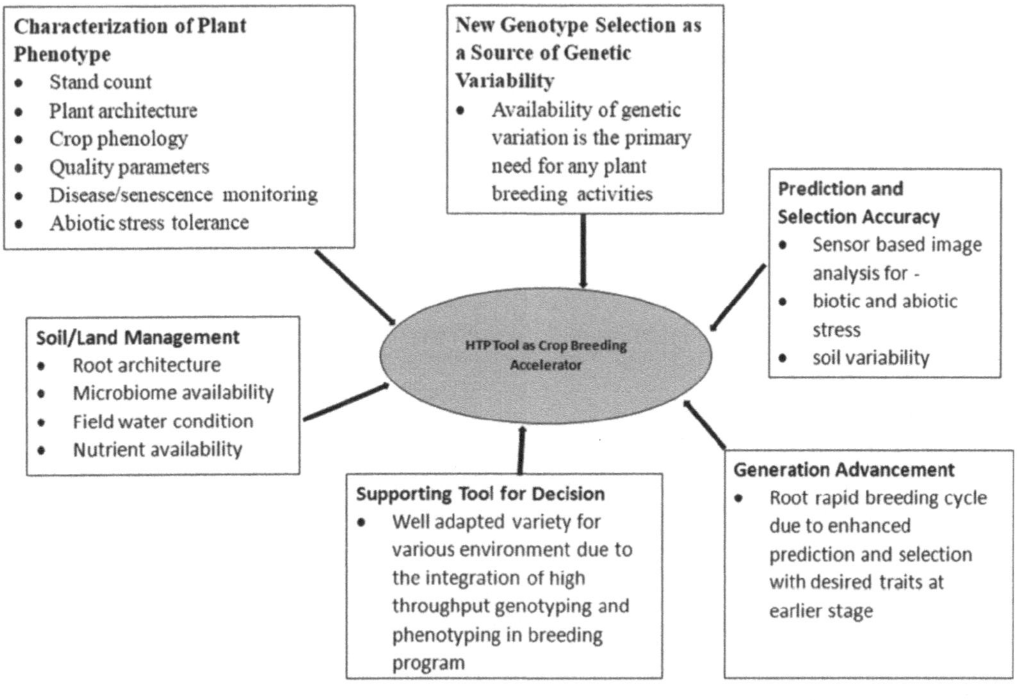


**FIGURE 2.6** Six pillars of a breeding program for accelerating genetic gain utilizing HTPP (High-Throughput Phenotyping and Phenomics) tool (Chaudhari et al., 2024; Araus et al., 2018).

**TABLE 2.2**
**Potential Applications of HTPT (High-Throughput Phenotyping Technique) for Crop Breeding Program**

| Traits Phenotyped | HTPT/Tools/Sensors | References |
|---|---|---|
| Earlier crop inspection | UAVs | Yang et al. (2020) |
| Nitrogen stress | Near-infrared and visible spectrum reflectance | Camino et al. (2018) |
| Diseases incidence | Near-ground optical sensors (RGB, thermal, hyperspectral cameras) | Behmann et al. (2018) |
| Phenology of wheat, soybeans, maize, canola | RF algorithm, multitemporal RADARSAT-2 data set, SAR polarimetric decompositions | Wang et al. (2019b) |
| Lodging of wheat | Ground-based sensor | Fitch et al. (1984) |
| Lodging of rice | Motor-powered cameras | Ogden et al. (2002) |
| RSA of maize | Shovelomics | Trachsel et al. (2011) |
| Tiller density, plant biomass, lodging, water content | Multiple detector system | Busemeyer et al. (2013) |
| Leaf growth and RSA | GrowScreen-Agar | Nagel et al. (2020) |
| Vegetative index, canopy temperature, yield | Phenocart | Crain et al. (2016) |
| Moisture content, LAI, seed health, seed constitution, transpiration rate, canopy temperature, stomatal permeability | Near-infrared sensor, thermal imagery | Jangra et al. (2021) |
| Abiotic and biotic stress | Vinobot and Vinocule | Shafekhani et al. (2017) |
| Canopy growth and development of cotton | GPhenoVision, high-resolution imagery | Jiang et al. (2018) |
| Plant height, LAI, leaf angle against the stem, plant biomass, penetration depth of light | PlantEye, four-channel multispectral camera | Kumar et al. (2023) |
| Dynamics of transpiration | FieldScan, DroughtSpotter | Kumar et al. (2023) |
| Moisture stress, loss of heat, rate of transpiration, senescence of leaves, temperature of leaves and canopies, disease and pathogen screening | Thermal infrared thermometers | Li et al. (2014) and O'Shaughnessy et al. (2011) |
| NDVI, moisture content, LAI, heat loss, rate of transpiration | Infrared thermography | Prashar et al. (2013) |
| Carbohydrates, starch, protein, and moisture content of the plant, LAI, plant disease | Thermal cameras, hyperspectral imagery | Yol et al. (2015) and Busemeyer et al. (2013) |
| Plant health, senescence, phenology, blooming time, plant height, biomass, LAI, germination percentage, growth and yield contributing factors | Visible light imagery | Sugiura et al. (2016) and Clark et al. (2011) |
| Moisture content, pigment and nutrient content, senescence, the effectiveness of photosynthesis | Multispectral imagery system | Tiwari et al. (2021) |
| Physiological analysis, moisture content, morphometric traits, disease and pest incidence | Magnetic resonance imager | Alisha et al. (2023) |
| LAI, RSA, plant height, leaf angle distributions, canopy organization, and plant structure | 3D computerized imager | Galkovskyi et al. (2012) |
| Heat and drought stress, shoot architecture, photosynthesis rate, NPQ, disease and pest status | Fluorescence sensing cameras | Tatagiba et al. (2015) and Brabandt et al. (2014) |

*Note:* LAI, leaf area index; NDVI, normalized difference vegetation index; NPQ, non-photochemical quenching; RSA, root system architecture; SAR, synthetic aperture radar; UAV, unmanned aerial vehicle.

relationships among various pathways across various regulatory aspects (Saito and Matsuda, 2010). Characterization for tolerance to such environmental stressors is sometimes difficult. It takes a lot of time and is not practicable to screen an extensive mapping population or a group of genotypes for quantitative trait locus (QTL) evaluation and association mapping of resistance to abiotic stress. However, the development of HTP methods has made it feasible to conduct stress tests on different characteristics for large populations in a nondestructive manner. Abiotic stressors, like salt and drought, are strongly linked to biochemical, physiological, and morphological alterations (Alisha et al., 2023). According to Schlemmer et al. (2005), drought, for example, directly impacts intercellular structures and cellular turgidity, which in turn impacts root and shoot structure. This makes it challenging to assess using conventional methods and requires the use of HTP technologies.

### 2.7.2 Water Stress-Associated Traits

Plant responses to water stress are thought to be complicated and varied. Plants with particular traits, such as transpiration efficacy, canopies conductivity, leaf area index (LAI), relative water content, harvest index, entire dry matter accumulation, and crop length, are more likely to withstand drought. Among legumes, drought is correlated with rapid ground cover, initial plant vigor, a lengthy and extensive root architecture, higher biomass of roots, and higher moisture content of leaves. Responses to drought entail a variety of alterations in morphology, physiology, and metabolic activities. Either floral productivity, which is significantly decreased under extreme conditions, or the survival of plants is a quality that contributes to increased production. In order to choose genotypes that are resistant to drought for upcoming breeding efforts, it might be very beneficial to analyze such responses. Under water stress, data may be gathered and changes can be tracked over time via HTP. Quick evaluation of plant development is made possible by RGB cameras, which provide precise information at quick collection speeds. Measuring nitrogen levels, biomass, and the responsiveness of leaves to water scarcity may also be done using it. Because digital photography is less expensive and does not need as much skill, it is necessary for characterization of complicated features like drought. Different methods are evaluated in *Arabidopsis thaliana*, where plant pictures are obtained with a 96.7% precision using a stationary camera capturing numerous objects at once in an automated manner (Minervini et al., 2014). In research on wine grapes, leaf conductivity, stem moisture content, and LAI were tracked using thermal and visual RGB imagery (Moller et al., 2007). Plant canopy characteristics were measured from field trials of wheat and soybean using multi-sensor platforms equipped with RGB web-based cameras and transportable spectrometers (Bai et al., 2016).

### 2.7.3 Salinity Stress-Associated Traits

One significant element influencing agricultural crop productivity is the salinity of the soil. Ion poisoning is brought on by salt, and plants develop more slowly as a result of decreased photosynthetic activity and altered physiological conditions. Plant tolerance to salinity is measured by the shoot ion-independent tolerance score. According to Läuchli and Grattan (2007), salt stress decreases the length of roots, plant biomass, LAI, and entire chlorophyll content at various stages of development. Osmotic stresses and electrical imbalance, which are brought on by a reduction in stomatal opening, variations in the activity of antioxidant enzymes, and a reduced capacity for reactive oxygen species (ROS) detoxification, also impact a number of metabolic and physiological activities. Tissues of plants subjected to high salinity levels experience a large concentration of $Na+$ and $Cl-$ ions, which cause a toxic effect on ions and an acute ionic disequilibrium. Numerous biological elements suffer damage from oxidation as a result of ROS production brought on by salinity. To produce agricultural plants that can withstand salt, physiological and genetic investigations can benefit from the application of high-throughput testing in the analysis of images. The discovered lines may be used in breeding efforts going forward thanks to these procedures. Awlia et al. (2016) discovered an alteration in rosette coloration in *A. thaliana*, whereas Al-Tamimi et al. (2016) used high-throughput non-invasive characterization in rice to quantify other variables including transpiration percentage, transpiration usage efficacy, and comparable growth rate. Salt stress-related alterations in photosynthetic efficiency are captured using chlorophyll fluorescence imagery. Certain cutting-edge techniques, like WIVAM (http://www.wivam.be) and PHENOPSIS (Granier et al., 2006), evaluate tolerance to salt in wheat (Harris et al., 2010), rice (Hairmansis et al., 2014), and barley (Meng et al., 2017). A hyperspectral scanning method was recently employed to evaluate several okra (*Abelmoschus esculentus* L.) genotypes in order to choose a crop that could withstand salt (Feng et al., 2020).

### 2.7.4 Biotic Stress Identification and Characterization

When pests and diseases harm plants, both quantity and quality of crop yield decline. Thus, it is essential to identify pest and disease incidence in plants as soon as possible. The symptoms of plant diseases, including cankers, blight, lesions, wilt, or rot, are often determined by early visual assessments before a disease is identified. Alternatively, they are predicated on observable indications of the pathogen's (*Puccinia spp.* uredospore) mycelium development. Extensive study and inquiry are necessary for visual prediction. Recently, a number of sensor-based techniques have been employed to take advantage of plants' optical qualities. Digital photography is an effective way to investigate and characterize plants during stress situations in order to comprehend the morphology of plant growth (Bock et al., 2010). The sensitivity of sugar beet lines to disease varied, and the distinction between sensitive and resistant crops was also assessed. Therefore, the presence of infection was tested using the chlorophyll fluorescence imagery approach (Gamborg and Phillips, 2013). According to Scholes and Rolfe (2009), it is possible to identify the existence of fungal infections prior to the expression of symptoms by evaluating the photosynthetic efficiency of various leaf regions affected by the infection. Yang et al. (2012) developed an NIR hyperspectral scanning method having a total precision of 92% for the identification of blast disease in rice seedlings. Using hyperspectral data, *Phaeosphaeria* leaf spots that occur in maize may be detected earlier with an 88% total precision (Adam et al., 2017). Therefore, time-lapsed and highly precise assessment to track diseases of plants might be provided via high-throughput screening.

### 2.7.5 Plant Growth-Associated Traits

In order to determine plant biomass, it is necessary to track the growth of plants, which may be assessed in terms of the growth of roots and shoots. According to Kuijken et al. (2015), roots are vital organs that need to be regulated. With the use of HTP approaches, root characteristics may be chosen in order to increase crop improvement. Agar plates, hydroponic systems, and aeroponic systems were among the methods utilized to screen an extensive number of plants for root characteristics. These methods are being used in conjunction with photographic devices, and root length, number, and thickness may be measured by image processing (Gregory et al., 2009). Utilizing non-invasive imagery methods for crop organ characterization is an effective way to evaluate physiological features. Plant growth evaluation is done using PlaRoM (plant root monitoring platform) (Yazdanbakhsh and Fisahn, 2009). The HTP approach is employed in oilseed rape and wheat to assess the root characteristics of seedlings (Makanza et al., 2018). A phenotyping platform is utilized in sorghum to evaluate the development angle of nodal roots, a crucial aspect of the root system architecture (RSA) (Joshi et al., 2017). Rhizotubes facilitate hemagglutinin production from crop roots and shoots, such as those of *Vitis vinifera*, *Medicago truncatula*, *Vulpia myuros*, *Brassica napus*, and *Pisum sativum* (Jeudy et al., 2016). Similarly, RootReader2D software was used to conduct phenotyping of root experiments in tomato, maize, *A. thaliana*, rice, and sorghum (Clark et al., 2013). These studies contribute to an improved comprehension of the interactions among root functions and their structure. A useful technique for examining rice leaf development and assessing the health of rice panicles is spectral analysis (Liu et al., 2010). Evaluating fruits and vegetables is also beneficial (Lorente et al., 2012). In order to anticipate the biomass of crops in both barley and wheat, the predicted stem area is measured (Golzarian et al., 2011). HTP has recently been employed to research rapeseed plant growth and ultimate yield (Li et al., 2020).

### 2.7.6 Physiological Traits

C3 and C4 are the two main photosynthetic processes found in plants that perform photosynthesis. Because C4 plants are better at photosynthesizing and concentrating carbon dioxide inside their leaves than C3 plants, scientists are interested in substituting crop species' C3 route with the more effective C4 system (Caemmerer et al., 2012). However, Rubisco's inefficiency is the main factor restricting photosynthetic efficiency. The Rubisco efficacy varies across plants. Researchers are looking through numerous germplasms of wheat to find a genotype with a more effective Rubisco and faster rates of photosynthetic ability by using phenomics which may thrive effectively under drought, salt, and nutrient shortage conditions (Baker, 2008; Tackenberg, 2007).

### 2.7.7 Qualitative Traits

Several biological molecules, including proteins, carbohydrates, flavonoids, and various fatty acids, make up grains. Also, the most important characteristic for millers and farmers is seed firmness or structure. The traditional approach of measuring seed firmness is damaging and laborious. There is a great demand for quick and nondestructive measurement techniques. It is well recognized that *Aspergillus* and *Fusarium*, fungi that produce aflatoxin and are dangerous to both humans and animals, have an impact on maize seeds. Enzyme-linked immunosorbent assay (ELISA) and reverse transcriptase–polymerase chain reaction (RT-PCR) are the traditional methods used to quantify mycotoxin; however, both methods are time-consuming and only analyze just a few samples at once (Fox and O'Hare, 2017). The most crucial step of the product development procedure in the starch sector is measuring starch architecture. HTP techniques can significantly contribute to reducing costs and speeding up the procedure without sacrificing quality (Basavaraj et al., 2023).

### 2.7.8 Associated Traits with Nutrient Deficiencies

The availability of nutrients in the soil is a crucial environmental component that influences the overall development of plants. When there are harmful minerals present, the availability of nutrients to plants is reduced, which hinders their development and growth and causes limited growth of roots when there is a shortage of water. Barley and wheat plants growing in boron-deficient soil were studied using RGB imagery (Hayes et al., 2013; Schnurbusch et al., 2010). Legumes with deficiencies in nitrogen, phosphorus, and potassium may be found via chlorophyll fluorescence imagery (Donnini et al., 2013). Plants living in deficient soils go through specific physiological alterations. For instance, when they are stressed by dryness, their stomata shut and their photosynthesis zone decreases. Thus, to investigate the leaf alignment, canopy temperature, and structure of plants during stress, several kinds of approaches were used, like RGB imagery with infrared thermographs (Leinonen and Jones, 2004). For the purpose of measuring damage to plants, the findings from these pictures' elucidations are therefore more helpful in the early diagnosis of nutrient deficiencies and toxins. Using hyperspectral pictures of crops, genotypic variations for several micronutrient deficits were evaluated (Ji-Yong et al., 2012).

### 2.7.9 Yield and Yield-Associated Traits

Grain yield is considered a useful feature for plant breeders of crops such as maize. The genetic composition of an individual, methods of management, and growing circumstances all affect this trait's expression (Baye et al., 2011). In addition to grain production, a group of constituent attributes referred to as secondary attributes are essential to the performance of crops. G×E×M also affects these plant traits, adding another level of complexity to breeding for the desired surroundings (Reynolds and Langridge, 2016). Yield and its associated parameters, such as crop biomass and height, are conventionally evaluated by hand, a time-intensive, laborious, and error-prone process. For example, when maize reaches maturity, its height is determined by hand, although the plants are usually tall at that point. When many plants need to be assessed, it turns difficult to ensure accuracy in this way, necessitating the use of HTP techniques (Wang et al., 2019a). It is essential to estimate plant growth rate rather than biomass at any particular time in order to comprehend how plants respond to external stimuli throughout the period. According to the traditional method, this activity is completed by using a destructive strategy to periodically harvest plants at various phases of crop development. There is much potential for HTP since any phenotyping method permitting non-destructive techniques can be very beneficial. Measuring the grain yield elements per cob in a crop like maize is time-consuming and laborious. Additionally, this expands the potential applications of HTP for raising agricultural yield (Kienbaum et al., 2021).

### 2.7.10 Post-Harvest Traits Characterization

Imagery and digital sensor technology are used by the seed and milling sectors to regularly evaluate quality characteristics. Gegas et al. (2010) analyzed the genetic makeup of grain using 2D imaging in conjunction with an attribute retrieval phenotyping process in wheat varieties. The majority of grain crops may use this protocol and approach, and with further refinement, it can be expanded to include any portion of the plant. An integrative computerized system was constructed and used to characterize the spikelets of rice, and it is currently feasible to assess 1,440 plants every day with above 95% accuracy (Duan et al., 2011). Additionally, candidate genes linked to tassel (Gage et al., 2017), stalks (Mazaheri et al., 2019), and vascular bundle and rind characteristics in maize kernels (Miller et al., 2017) were found. According to Komyshev et al. (2017), the SeedCounter smartphone app, which is based on Android, accurately calculates grain number and size. Furthermore, over time, instruments for recording morphology and other intricacies of seeds and associated characteristics were developed. "PhenoSeeder" collects 3D measurements of individual seed, the parameters obtained have a correlation with features associated with vigor in the initial stages of seedling development (Jahnke et al., 2009). While Singh et al. (2010) utilized hyperspectral imaging to identify insect-damaged wheat grains and determine the protein content of the grain, Wallays et al. (2009) used the technology to assess clean grains. Combining genomic techniques with huge phenotyping data sets is crucial to understanding the genetic makeup of important agronomic parameters. Utilizing such integrated methods, for example, the amino acid contents of rice grains (Sun et al., 2019) and the panicle structure of sorghum (Zhou et al., 2019) were studied. In order to examine maize ears, Warman and Fowler (2019) created a high-throughput image-based characterizing method. Green fluorescent protein and anthocyanin seed markers are easily distinguished by this approach. In comparison to conventional methods, recently developed phenomics technologies promise faster and more accurate phenotyping of plant responses. Nowadays, many of these techniques are essential components of phenomics systems that range from controlled to field settings.

## 2.8 GENETIC MATERIAL CHARACTERIZATION

Genetic material characterization, the use of various phenotyping methods to estimate a trait's true heritability through the discovery of novel QTLs, and genomic selection for complicated attributes are all made possible by an HTP technique. Another trustworthy technique for examining distinct qualities in genetic material is chlorophyll fluorescence phenomics. According to Jangra et al. (2021), in genetic mapping, the imagery method offers physiological data appropriate for quantitative trait mapping. Additionally, a number of applications for mapping QTLs for the leaf region were suggested (Baker and Rosenqvist, 2004). The rice genomic data that regulates complicated attributes may be accessed by integrating genome-wide association studies (GWAS) with HTP (Yang et al., 2014). Prado et al. (2018) used a characterization system to select QTLs linked to biomass buildup and exchange of gases in recent maize research. According to Feng et al. (2017), hyperspectral imagery in the near-infrared region may be utilized to show the variations among non-GM ancestors and genetically modified corn with about 100% forecast and computation precision. Further applications of hyperspectral imagery include genotype selection in Chinese cabbage (Cheng et al., 2014), rice (Kong et al., 2014), maize (Zhang et al., 2012), and several types of black beans (Zhang et al., 2014).

## 2.9 TAKING ADVANTAGE OF GENOMICS POTENTIAL

Morphological and physiological data collected utilizing the latest computerized phenotyping tools accessible worldwide can be linked to the genetic resources of a plant species. To fully understand the plant phenome in its entirety across a range of surroundings, HTP techniques collect accurate and exact data for evaluation. Thus, similar to genomics systems, characterization systems generate databases that gather phenotypic responses to the surroundings that encompass a range of plant characteristics and factors, such as TRY (the plant trait database) or the database of Plant Meta-phenomics (Kumar et al., 2023). Along with additional "omics" data including transcriptomic, proteomic, and metabolomic data, and the global genomic databases TAIR, TIGR, and NCBI, such phenotyping databases are becoming more important in comprehending the genetic basis of intricate attributes. Accurate and effective, HTP is essential to fine-tuning traits, irrespective of the genetic method used to induce allelic recombination or measure diversity using resequencing technology. Allelic variability may be employed to focus on the associated genes and discover the role of a particular gene or genes for the development and growth of agricultural plants (Kumar et al., 2015). Phenomics may be utilized to obtain this information.

## 2.10 FUTURE PROSPECTS AND CHALLENGES

The fast advancement of a wide range of tools, such as robotics, sensors, data extraction, image processing, and remote control, has benefited HTPP tools over the past few years. Plot-level assessments can now be made in a matter of seconds because of HTPP, which previously posed the biggest obstacle to phenotyping. The massive quantity of data management, however, is the barrier that can prevent such systems from being fully implemented (Cobb et al., 2013). The programming utilized in a manner to include environmental, morphological, genotypic, and managemental information is one of the primary limiting elements that have to be addressed, along with handling of data and the application of bioinformatics to analyze the data related to phenotype (Maes and Steppe, 2012). More sophisticated analytic methods than only standard statistical tools are needed (White et al., 2012). In regard to plant efficiency, scientists are unable to physically understand the meaning of any of the values acquired by HTPPs (Araus and Cairns, 2014).

Clearly, there is a wide range of difficulties for plant phenomics over the next five to ten years. Most importantly, in the field of phenomics, a significant shift into the large-scale data era—which is distinguished by its multiscale nature, multiple dimensions, and high-throughput capacity—is taking place. Researchers investigate a range of HTPP methodologies that deal with plant physiological data, structure, and morphology. These methodologies display three important multi-characteristics: multi-scale (that includes plant physiological data, its structure, and morphology at different levels, from cells to entire plant levels); multi-level (that includes conventional small-scale to medium along with enormous-scale omics data); and multi-domain (that includes genomics, phenomics, and other essential fields). Understanding the limits of single data of phenotype in fulfilling the requirements of analysis of association in the nascent "omics" age, scientists also recognize that systematic and complete phenomics data will serve as the foundation for subsequent studies (Munjal et al., 2023).

## 2.11 CONCLUSION

In challenging field conditions, HTPP is a novel and efficient tool that allows for the acquisition of dense and deep phenotypic data of plants that are not possible with traditional phenotyping techniques. Phenotyping tools were developed to monitor the traits of crops in a variety of climate conditions. These tools can be installed in indoor or outdoor environmental conditions. Sophisticated phenotyping techniques have been developed by multiple companies worldwide for agricultural trait-specific applications. These characterization tools, which fuse enormous-scale genomic analysis with the HTPP technique, have not only significantly increased our understanding of the constantly evolving growth process of plants but have also given breeders a novel way for maximizing the architecture of plants for the purpose of breeding approach. HTPP technique, which is based on recently developed phenotyping systems, detectors, processing of images, and data management methods of data, will ultimately enhance the precision of evaluation of the plant traits measurement and further speed up the effectiveness of novel trait recognition in plant breeding programs.

## REFERENCES


Adam, E., Deng, H., Odindi, J., Abdel-Rahman, E. M., & Mutanga, O. (2017). Detecting the early stage of phaeosphaeria leaf spot infestations in maize crop using in situ hyperspectral data and guided regularized random forest algorithm. *Journal of Spectroscopy*, *2017*(1), 6961387.

Ahmed, H. G. M. D., Zeng, Y., Fiaz, S., & Rashid, A. R. (2023). Applications of high-throughput phenotypic phenomics. In *Sustainable Agriculture in the Era of the OMICs Revolution* (pp. 119–134). Cham: Springer International Publishing.

Alisha, J., Kumari, A., & Munjal, R. (2023). Unleashing the potentials of high-throughput phenotyping for accelerating crop breeding. *The Pharma Innovation Journal*, *12*(7S), 736-744.

Araus, J. L., & Cairns, J. E. (2014). Field high-throughput phenotyping: the new crop breeding frontier. *Trends in Plant Science*, *19*(1), 52–61.

Araus, J. L., Kefauver, S. C., Zaman-Allah, M., Olsen, M. S., & Cairns, J. E. (2018). Translating high-throughput phenotyping into genetic gain. *Trends in Plant Science*, *23*(5), 451–466.

Arend, D., Junker, A., Scholz, U., Schüler, D., Wylie, J., & Lange, M. (2016). PGP repository: a plant phenomics and genomics data publication infrastructure. *Database*, *2016*, baw033.

Awlia, M., Nigro, A., Fajkus, J., Schmoeckel, S. M., Negrão, S., Santelia, D., … & Panzarová, K. (2016). High-throughput non-destructive phenotyping of traits that contribute to salinity tolerance in Arabidopsis thaliana. *Frontiers in Plant Science*, *7*, 1414.

Bai, G., Ge, Y., Hussain, W., Baenziger, P. S., & Graef, G. (2016). A multi-sensor system for high throughput field phenotyping in soybean and wheat breeding. *Computers and Electronics in Agriculture*, *128*, 181–192.

Baker, N. R. (2008). Chlorophyll fluorescence: a probe of photosynthesis in vivo. *Annual Review of Plant Biology*, *59*(1), 89–113.

Baker, N. R., & Rosenqvist, E. (2004). Applications of chlorophyll fluorescence can improve crop production strategies: an examination of future possibilities. *Journal of Experimental Botany*, *55*(403), 1607–1621.

Basavaraj, P. S., & Rane, J. (2020). Avenues to realize potential of phenomics to accelerate crop breeding for heat tolerance. *Plant Physiology Reports*, *25*(4), 594–610.

Basavaraj, P. S., Rane, J., Prathibha, M. D., Boraiah, K. M., & Kumar, M. (2023). Advances in high-throughput phenotyping of maize (Zea Mays L.) for climate resilience. In S. H. Wani, Z. A. Dar, & G. P. Singh (Eds.), *Maize Improvement: Current Advances in Yield, Quality, and Stress Tolerance under Changing Climatic Scenarios* (pp. 259–288). Cham: Springer International Publishing.

Baye, T. M., Abebe, T., & Wilke, R. A. (2011). Genotype–environment interactions and their translational implications. *Personalized Medicine*, *8*(1), 59–70.

Behmann, J., Acebron, K., Emin, D., Bennertz, S., Matsubara, S., Thomas, S., … & Rascher, U. (2018). Specim IQ: evaluation of a new, miniaturized handheld hyperspectral camera and its application for plant phenotyping and disease detection. *Sensors*, *18*(2), 441.

Bock, C. H., Poole, G. H., Parker, P. E., & Gottwald, T. R. (2010). Plant disease severity estimated visually, by digital photography and image analysis, and by hyperspectral imaging. *Critical Reviews in Plant Sciences*, *29*(2), 59–107.

Bombarely, A., Menda, N., Tecle, I. Y., Buels, R. M., Strickler, S., Fischer-York, T., … & Mueller, L. A. (2010). The Sol Genomics Network (solgenomics. net): growing tomatoes using Perl. *Nucleic Acids Research*, *39*(suppl_1), D1149–D1155.

Brabandt, H., Bauriegel, E., Gärber, U., & Herppich, W. B. (2014). ΦPSII and NPQ to evaluate Bremia lactucae-infection in susceptible and resistant lettuce cultivars. *Scientia Horticulturae*, *180*, 123–129.

Busemeyer, L., Mentrup, D., Möller, K., Wunder, E., Alheit, K., Hahn, V., … & Ruckelshausen, A. (2013). BreedVision—a multi-sensor platform for non-destructive field-based phenotyping in plant breeding. *Sensors*, *13*(3), 2830–2847.

Cabrera-Bosquet, L., Crossa, J., von Zitzewitz, J., Serret, M. D., & Luis Araus, J. (2012). High-throughput phenotyping and genomic selection: the frontiers of crop breeding converge F. *Journal of Integrative Plant Biology*, *54*(5), 312–320.

Caemmerer, S. V., Quick, W. P., & Furbank, R. T. (2012). The development of C4 rice: current progress and future challenges. *Science*, *336*(6089), 1671–1672.

Camino, C., González-Dugo, V., Hernández, P., Sillero, J. C., & Zarco-Tejada, P. J. (2018). Improved nitrogen retrievals with airborne-derived fluorescence and plant traits quantified from VNIR-SWIR hyperspectral imagery in the context of precision agriculture. *International Journal of Applied Earth Observation and Geoinformation*, *70*, 105–117.

Chapman, S. C., Zheng, B., Potgieter, A. B., Guo, W., Baret, F., Liu, S., … & Jordan, D. R. (2018). Visible, near infrared, and thermal spectral radiance on-board UAVs for high-throughput phenotyping of plant breeding trials. In P. S. Thenkabail, J. G. Lyon, & A. Huete (Eds.), *Biophysical and Biochemical Characterization and Plant Species Studies* (pp. 275–299). Boca Raton: CRC Press.

Chaudhari, G. R., Shekhvat, N., Chavda, D., Ragi, S., & Kumar, S. (2024). Recent advances in high throughput phenotyping for crop improvement. In G. R. Chaudhari, K. Prakash, & S. R. Patel (Eds.), *Modern Plant Breeding* (pp. 144–164). New Delhi: Brillion Publishing.

Chawade, A., van Ham, J., Blomquist, H., Bagge, O., Alexandersson, E., & Ortiz, R. (2019). High-throughput field-phenotyping tools for plant breeding and precision agriculture. *Agronomy*, *9*(5), 258.

Cheng, S. X., Kong, W. W., Zhang, C., Liu, F., & He, Y. (2014). Variety recognition of Chinese cabbage seeds by hyperspectral imaging combined with machine learning. *Spectroscopy and Spectral Analysis*, *34*(9), 2519–2522.

Chung, Y. S., Kim, K. S., & Kim, C. (2018). Perspectives on high throughput phenotyping in developing countries. *Korean Journal of Agricultural Science*, *45*(3), 317–323.

Clark, R. T., Famoso, A. N., Zhao, K., Shaff, J. E., Craft, E. J., Bustamante, C. D., … & Kochian, L. V. (2013). High-throughput two-dimensional root system phenotyping platform facilitates genetic analysis of root growth and development. *Plant, Cell & Environment*, *36*(2), 454–466.

Clark, R. T., MacCurdy, R. B., Jung, J. K., Shaff, J. E., McCouch, S. R., Aneshansley, D. J., & Kochian, L. V. (2011). Three-dimensional root phenotyping with a novel imaging and software platform. *Plant Physiology*, *156*(2), 455–465.

Cobb, J. N., DeClerck, G., Greenberg, A., Clark, R., & McCouch, S. (2013). Next-generation phenotyping: requirements and strategies for enhancing our understanding of genotype–phenotype relationships and its relevance to crop improvement. *Theoretical and Applied Genetics*, *126*, 867–887.

Crain, J., Wang, X., Lucas, M., & Poland, J. (2021). Experiences of applying field-based high-throughput phenotyping for wheat breeding. In J. Zhou & H. T. Nguyen (Eds.), *High-Throughput Crop Phenotyping* (pp. 71–99). Springer, Cham, Switzerland.

Crain, J. L., Wei, Y., Barker III, J., Thompson, S. M., Alderman, P. D., Reynolds, M., … & Poland, J. (2016). Development and deployment of a portable field phenotyping platform. *Crop Science*, *56*(3), 965–975.

Davis, B. D. (1949). The isolation of biochemically deficient mutants of bacteria by means of penicillin. *Proceedings of the National Academy of Sciences, 35*(1), 1–10.

Donnini, S., Guidi, L., Degl'Innocenti, E., & Zocchi, G. (2013). Image changes in chlorophyll fluorescence of cucumber leaves in response to iron deficiency and resupply. *Journal of Plant Nutrition and Soil Science*, *176*(5), 734–742.

Duan, L., Yang, W., Huang, C., & Liu, Q. (2011). A novel machine-vision-based facility for the automatic evaluation of yield-related traits in rice. *Plant Methods*, *7*, 1–13.

Feng, X., Zhan, Y., Wang, Q., Yang, X., Yu, C., Wang, H., … & He, Y. (2020). Hyperspectral imaging combined with machine learning as a tool to obtain high-throughput plant salt-stress phenotyping. *The Plant Journal*, *101*(6), 1448–1461.

Feng, X., Zhao, Y., Zhang, C., Cheng, P., & He, Y. (2017). Discrimination of transgenic maize kernel using NIR hyperspectral imaging and multivariate data analysis. *Sensors*, *17*(8), 1894.

Fiorani, F., and Schurr, U. (2013). Future scenarios for plant phenotyping. *Annual Review of Plant Biology*, *64*, 267–291.

Fitch, B. W., Walraven, R. L., & Bradley, D. E. (1984). Polarization of light reflected from grain crops during the heading growth stage. *Remote Sensing of Environment*, 15(3), 263–268.

Fox, G. P., & O'Hare, T. J. (2017). *Analysing Maize Grain Quality* (pp. 237–260). Cambridge: Burleigh Dodds Science Publishing.

Freimer, N., & Sabatti, C. (2003). The human phenome project. *Nature Genetics*, *34*(1), 15–21.

Furbank, R. T., Jimenez-Berni, J. A., George-Jaeggli, B., Potgieter, A. B., & Deery, D. M. (2019). Field crop phenomics: enabling breeding for radiation use efficiency and biomass in cereal crops. *New Phytologist*, *223*(4), 1714–1727.

Furbank, R. T., & Tester, M. (2011). Phenomics–technologies to relieve the phenotyping bottleneck. *Trends in Plant Science*, *16*(12), 635–644.

Gage, J. L., Miller, N. D., Spalding, E. P., Kaeppler, S. M., & de Leon, N. (2017). TIPS: a system for automated image-based phenotyping of maize tassels. *Plant Methods*, *13*, 1–12.

Galkovskyi, T., Mileyko, Y., Bucksch, A., Moore, B., Symonova, O., Price, C. A., … & Weitz, J. S. (2012). GiA Roots: software for the high throughput analysis of plant root system architecture. *BMC Plant Biology*, *12*, 1–12.

Gamborg, O., & Phillips, G. C. (Eds.). (2013). *Plant Cell, Tissue and Organ Culture: Fundamental Methods*. Germany: Springer Science & Business Media.

Gegas, V. C., Nazari, A., Griffiths, S., Simmonds, J., Fish, L., Orford, S., … & Snape, J. W. (2010). A genetic framework for grain size and shape variation in wheat. *The Plant Cell*, *22*(4), 1046–1056.

Golzarian, M. R., Frick, R. A., Rajendran, K., Berger, B., Roy, S., Tester, M., & Lun, D. S. (2011). Accurate inference of shoot biomass from high-throughput images of cereal plants. *Plant Methods*, *7*, 1–11.

Granier, C., Aguirrezabal, L., Chenu, K., Cookson, S. J., Dauzat, M., Hamard, P., … & Tardieu, F. (2006). PHENOPSIS, an automated platform for reproducible phenotyping of plant responses to soil water deficit in Arabidopsis thaliana permitted the identification of an accession with low sensitivity to soil water deficit. *New Phytologist*, *169*(3), 623–635.

Gregory, P. J., Bengough, A. G., Grinev, D., Schmidt, S., Thomas, W. B. T., Wojciechowski, T., & Young, I. M. (2009). Root phenomics of crops: opportunities and challenges. *Functional Plant Biology*, *36*(11), 922–929.

Hairmansis, A., Berger, B., Tester, M., & Roy, S. J. (2014). Image-based phenotyping for non-destructive screening of different salinity tolerance traits in rice. *Rice*, *7*, 1–10.

Harris, B. N., Sadras, V. O., & Tester, M. (2010). A water-centred framework to assess the effects of salinity on the growth and yield of wheat and barley. *Plant and Soil*, *336*, 377–389.

Hayes, J. E., Pallotta, M., Baumann, U., Berger, B., Langridge, P., & Sutton, T. (2013). Germanium as a tool to dissect boron toxicity effects in barley and wheat. *Functional Plant Biology*, *40*(6), 618–627.

Houle, D., Govindaraju, D. R., & Omholt, S. (2010). Phenomics: the next challenge. *Nature Reviews Genetics*, *11*(12), 855–866.

Ilakiya, T., Parameswari, E., Davamani, V., Swetha, D., & Prakash, E. (2020). High-throughput crop phenotyping in vegetable crops. *Pharma Innovation*, *9*, 184–191.

Jahnke, S., Menzel, M. I., Van Dusschoten, D., Roeb, G. W., Bühler, J., Minwuyelet, S., … & Schurr, U. (2009). Combined MRI–PET dissects dynamic changes in plant structures and functions. *The Plant Journal*, *59*(4), 634–644.

Jangra, S., Chaudhary, V., Yadav, R. C., Neelam R., & Yadav, N. R. (2021). High-throughput phenotyping: A platform to accelerate crop improvement. Phenomics, 1(2), 31–53.

Jeudy, C., Adrian, M., Baussard, C., Bernard, C., Bernaud, E., Bourion, V., … & Salon, C. (2016). RhizoTubes as a new tool for high throughput imaging of plant root development and architecture: test, comparison with pot grown plants and validation. *Plant Methods*, *12*, 1–18.

Jiang, Y., Li, C., Robertson, J. S., Sun, S., Xu, R., & Paterson, A. H. (2018). GPhenoVision: a ground mobile system with multi-modal imaging for field-based high throughput phenotyping of cotton. *Scientific Reports*, *8*(1), 1213.

Ji-Yong, S., Xiao-Bo, Z., Jie-Wen, Z., Kai-Liang, W., Zheng-Wei, C., Xiao-Wei, H., … & Holmes, M. (2012). Nondestructive diagnostics of nitrogen deficiency by cucumber leaf chlorophyll distribution map based on near infrared hyperspectral imaging. *Scientia Horticulturae*, *138*, 190–197.

Joshi, D. C., Singh, V., Hunt, C., Mace, E., van Oosterom, E., Sulman, R., … & Hammer, G. (2017). Development of a phenotyping platform for high throughput screening of nodal root angle in sorghum. *Plant Methods*, *13*, 1–12.

Kienbaum, L., Correa Abondano, M., Blas, R., & Schmid, K. (2021). DeepCob: precise and high-throughput analysis of maize cob geometry using deep learning with an application in genebank phenomics. *Plant Methods*, *17*(1), 91.

Komyshev, E., Genaev, M., & Afonnikov, D. (2017). Evaluation of the SeedCounter, a mobile application for grain phenotyping. *Frontiers in Plant Science*, *7*, 1990.

Kong, W., Zhang, C., Liu, F., Nie, P., & He, Y. (2013). Rice seed cultivar identification using near-infrared hyperspectral imaging and multivariate data analysis. *Sensors*, *13*(7), 8916–8927.

Kuijken, R. C., van Eeuwijk, F. A., Marcelis, L. F., & Bouwmeester, H. J. (2015). Root phenotyping: from component trait in the lab to breeding. *Journal of Experimental Botany*, *66*(18), 5389–5401.

Kumar, B. S., Jeannie, S., Doppalapudi, V. R., & Kumar, A. A. (2023). Revolutionizing crop phenotyping: breakthroughs in high throughput methods and their multifaceted implementations. In A. K. Gore, M. Nagora, K. Lal, K. Maniratnam, & A. Mangara (Eds.), *Advances in Agronomy* (Vol. 1). Gurugram: Bigfoot06 Publications.

Kumar, J., Pratap, A., & Kumar, S. (2015). Plant phenomics: an overview. In J. Kumar, A. Pratap, & S. Kumar (Eds.), *Phenomics in Crop Plants: Trends, Options and Limitations* (pp. 1–10). New Delhi: Springer.

Läuchli, A., & Grattan, S. R. (2007). Plant growth and development under salinity stress. In M. A. Jenks, P. M. Hasegawa, & S. M. Jain (Eds.), *Advances in Molecular Breeding toward Drought and Salt Tolerant Crops* (pp. 1–32). The Netherlands: Springer.

Leinonen, I., & Jones, H. G. (2004). Combining thermal and visible imagery for estimating canopy temperature and identifying plant stress. *Journal of Experimental Botany*, *55*(401), 1423–1431.

Lelong, C. C., Burger, P., Jubelin, G., Roux, B., Labbé, S., & Baret, F. (2008). Assessment of unmanned aerial vehicles imagery for quantitative monitoring of wheat crop in small plots. *Sensors*, *8*(5), 3557–3585.

Li, H., Feng, H., Guo, C., Yang, S., Huang, W., Xiong, X., … & Yang, W. (2020). High-throughput phenotyping accelerates the dissection of the dynamic genetic architecture of plant growth and yield improvement in rapeseed. *Plant Biotechnology Journal*, *18*(11), 2345–2353.

Li, L., Zhang, Q., & Huang, D. (2014). A review of imaging techniques for plant phenotyping. *Sensors*, 14(11), 20078–20111.

Li, Y., Wen, W., Guo, X., Yu, Z., Gu, S., Yan, H., & Zhao, C. (2021). High-throughput phenotyping analysis of maize at the seedling stage using end-to-end segmentation network. *PLoS One*, *16*(1), e0241528.

Liu, Z. Y., Shi, J. J., Zhang, L. W., & Huang, J. F. (2010). Discrimination of rice panicles by hyperspectral reflectance data based on principal component analysis and support vector classification. *Journal of Zhejiang University Science B*, *11*(1), 71–78.

Lorente, D., Aleixos, N., Gómez-Sanchis, J. U. A. N., Cubero, S., García-Navarrete, O. L., & Blasco, J. (2012). Recent advances and applications of hyperspectral imaging for fruit and vegetable quality assessment. *Food and Bioprocess Technology*, *5*, 1121–1142.

Maes, W. H., & Steppe, K. (2012). Estimating evapotranspiration and drought stress with ground-based thermal remote sensing in agriculture: a review. *Journal of Experimental Botany*, *63*(13), 4671–4712.

Makanza, R., Zaman-Allah, M., Cairns, J. E., Magorokosho, C., Tarekegne, A., Olsen, M., & Prasanna, B. M. (2018). High-throughput phenotyping of canopy cover and senescence in maize field trials using aerial digital canopy imaging. *Remote Sensing*, *10*(2), 330.

Martínez, J., Egea, G., Agüera, J., & Pérez-Ruiz, M. (2017). A cost-effective canopy temperature measurement system for precision agriculture: a case study on sugar beet. *Precision Agriculture*, *18*, 95–110.

Mazaheri, M., Heckwolf, M., Vaillancourt, B., Gage, J. L., Burdo, B., Heckwolf, S., … & Kaeppler, S. M. (2019). Genome-wide association analysis of stalk biomass and anatomical traits in maize. *BMC Plant Biology*, *19*, 1–17.

Menda, N., Semel, Y., Peled, D., Eshed, Y., & Zamir, D. (2004). In silico screening of a saturated mutation library of tomato. *The Plant Journal*, *38*(5), 861–872.

Meng, R., Saade, S., Kurtek, S., Berger, B., Brien, C., Pillen, K., … & Sun, Y. (2017). Growth curve registration for evaluating salinity tolerance in barley. *Plant Methods*, *13*, 1–9.

Miller, N. D., Haase, N. J., Lee, J., Kaeppler, S. M., de Leon, N., & Spalding, E. P. (2017). A robust, high-throughput method for computing maize ear, cob, and kernel attributes automatically from images. *The Plant Journal*, *89*(1), 169–178.

Minervini, M., Abdelsamea, M. M., & Tsaftaris, S. A. (2014). Image-based plant phenotyping with incremental learning and active contours. *Ecological Informatics*, *23*, 35–48.

Minoia, S., Petrozza, A., D'Onofrio, O., Piron, F., Mosca, G., Sozio, G., … & Carriero, F. (2010). A new mutant genetic resource for tomato crop improvement by TILLING technology. *BMC Research Notes*, *3*, 1–8.

Moller, M., Alchanatis, V., Cohen, Y., Meron, M., Tsipris, J., Naor, A., … & Cohen, S. (2007). Use of thermal and visible imagery for estimating crop water status of irrigated grapevine. *Journal of Experimental Botany*, *58*(4), 827–838.

Moreira, F. F., Oliveira, H. R., Volenec, J. J., Rainey, K. M., & Brito, L. F. (2020). Integrating high-throughput phenotyping and statistical genomic methods to genetically improve longitudinal traits in crops. *Frontiers in Plant Science*, *11*, 681.

Munjal, R., Beniwal, J., Dhundwal, A., Goyal, A., Kumari, A., & Behl, R. K. (2023). Accelerating crop breeding in the 21st century: a comprehensive review of next generation phenotyping techniques and strategies. *Ekin Journal of Crop Breeding and Genetics*, *9*(2), 160–171.

Ogden, R. T., Miller, C. E., Takezawa, K., & Ninomiya, S. (2002). Functional regression in crop lodging assessment with digital images. *Journal of Agricultural, Biological, and Environmental Statistics*, *7*, 389–402.

O'Shaughnessy, S. A., Hebel, M. A., Evett, S. R., & Colaizzi, P. D. (2011). Evaluation of a wireless infrared thermometer with a narrow field of view. *Computers and Electronics in Agriculture*, *76*(1), 59–68.

Prado, S. A., Cabrera-Bosquet, L., Grau, A., Coupel-Ledru, A., Millet, E. J., Welcker, C., & Tardieu, F. (2018). Phenomics allows identification of genomic regions affecting maize stomatal conductance with conditional effects of water deficit and evaporative demand. *Plant, Cell & Environment*, *41*(2), 314–326.

Prashar, A., Yildiz, J., McNicol, J. W., Bryan, G. J., & Jones, H. G. (2013). Infra-red thermography for high throughput field phenotyping in Solanum tuberosum. *PLoS One*, *8*(6), e65816.

Quirós Vargas, J. J., Zhang, C., Smitchger, J. A., McGee, R. J., & Sankaran, S. (2019). Phenotyping of plant biomass and performance traits using remote sensing techniques in pea (Pisum sativum, L.). *Sensors*, *19*(9), 2031.

Rahman, H., Ramanathan, V., Jagadeeshselvam, N., Ramasamy, S., Rajendran, S., Ramachandran, M., … & Muthurajan, R. (2015). Phenomics: technologies and applications in plant and agriculture. In D. Barh, M. S. Khan, & E. Davies (Eds.), *PlantOmics: The Omics of Plant Science* (pp. 385–411). New Delhi: Springer.

Ray, D. K., Mueller, N. D., West, P. C., & Foley, J. A. (2013). Yield trends are insufficient to double global crop production by 2050. *PLoS One*, *8*(6), e66428.

Reynolds, M., Chapman, S., Crespo-Herrera, L., Molero, G., Mondal, S., Pequeno, D. N., … & Sukumaran, S. (2020). Breeder friendly phenotyping. *Plant Science*, *295*, 110396.

Reynolds, M., & Langridge, P. (2016). Physiological breeding. *Current Opinion in Plant Biology*, *31*, 162–171.

Rosenzweig, C., Elliott, J., Deryng, D., Ruane, A. C., Müller, C., Arneth, A., Boote, K. J., Folberth, C., Glotter, M., Khabarov, N., Neumann, K., Piontek, F., Pugh, T. A. M., Schmid, E., Stehfest, E., Yang, H., Jones, J. W., & Jones, J. W. (2014). Assessing agricultural risks of climate change in the 21st century in a global gridded crop model intercomparison. *Proceedings of the National Academy of Sciences*, *111*(9), 3268–3273.

Rutkoski, J., Poland, J., Mondal, S., Autrique, E., Pérez, L. G., Crossa, J., … & Singh, R. (2016). Canopy temperature and vegetation indices from high-throughput phenotyping improve accuracy of pedigree and genomic selection for grain yield in wheat. *G3: Genes, Genomes, Genetics*, *6*(9), 2799–2808.

Saito, K., & Matsuda, F. (2010). Metabolomics for functional genomics, systems biology, and biotechnology. *Annual Review of Plant Biology*, *61*(1), 463–489.

Schlemmer, M. R., Francis, D. D., Shanahan, J. F., & Schepers, J. S. (2005). Remotely measuring chlorophyll content in corn leaves with differing nitrogen levels and relative water content. *Agronomy Journal*, *97*(1), 106–112.

Schnurbusch, T., Hayes, J., & Sutton, T. (2010). Boron toxicity tolerance in wheat and barley: Australian perspectives. *Breeding Science*, *60*(4), 297–304.

Scholes, J. D., & Rolfe, S. A. (2009). Chlorophyll fluorescence imaging as tool for understanding the impact of fungal diseases on plant performance: a phenomics perspective. *Functional Plant Biology*, *36*(11), 880–892.

Singh, C. B., Jayas, D. S., Paliwal, J., & White, N. D. (2010). Identification of insect-damaged wheat kernels using short-wave near-infrared hyperspectral and digital colour imaging. *Computers and Electronics in Agriculture*, *73*(2), 118–125.

Soulé, M. (1967). Phenetics of natural populations I. Phenetic relationships of insular populations of the side-blotched lizard. *Evolution*, *21*(3), 584–591.

Stahl, A., Wittkop, B., & Snowdon, R. J. (2020). High-resolution digital phenotyping of water uptake and transpiration efficiency. *Trends in Plant Science*, *25*(5), 429–433.

Sugiura, R., Tsuda, S., Tamiya, S., Itoh, A., Nishiwaki, K., Murakami, N., … & Nuske, S. (2016). Field phenotyping system for the assessment of potato late blight resistance using RGB imagery from an unmanned aerial vehicle. *Biosystems Engineering*, *148*, 1–10.

Sun, D., Cen, H., Weng, H., Wan, L., Abdalla, A., El-Manawy, A. I., … & He, Y. (2019). Using hyperspectral analysis as a potential high throughput phenotyping tool in GWAS for protein content of rice quality. *Plant Methods*, *15*, 1–16.

Tackenberg, O. (2007). A new method for non-destructive measurement of biomass, growth rates, vertical biomass distribution and dry matter content based on digital image analysis. *Annals of Botany*, *99*(4), 777–783.

Tanger, P., Klassen, S., Mojica, J. P., Lovell, J. T., Moyers, B. T., Baraoidan, M., ... & McKay, J. K. (2017). Field-based high throughput phenotyping rapidly identifies genomic regions controlling yield components in rice. *Scientific Reports*, *7*(1), 42839.

Tardieu, F., Cabrera-Bosquet, L., Pridmore, T., & Bennett, M. (2017). Plant phenomics, from sensors to knowledge. *Current Biology*, *27*(15), R770–R783.

Tatagiba, S. D., DaMatta, F. M., & Rodrigues, F. Á. (2015). Leaf gas exchange and chlorophyll a fluorescence imaging of rice leaves infected with Monographella albescens. *Phytopathology*, *105*(2), 180–188.

Tiwari, J. K., Changan, S. S., Buckseth, T., Singh, R. K., Singh, B., Luthra, S. K., ... & Kumar, M. (2021). In J. Zhou & H. T. Nguyen (Eds.), *High-Throughput Phenotyping in Potato Breeding: High-Throughput Crop Phenotyping* (pp. 165–182). Springer, Cham, Switzerland.

Trachsel, S., Kaeppler, S. M., Brown, K. M., & Lynch, J. P. (2011). Shovelomics: high throughput phenotyping of maize (Zea mays L.) root architecture in the field. *Plant and Soil*, *341*, 75–87.

Vadez, V., Kholova, J., Medina, S., Kakkera, A., & Anderberg, H. (2014). Transpiration efficiency: new insights into an old story. *Journal of Experimental Botany*, *65*(21), 6141–6153.

Wallays, C., Missotten, B., De Baerdemaeker, J., & Saeys, W. (2009). Hyperspectral waveband selection for on-line measurement of grain cleanness. *Biosystems Engineering*, *104*(1), 1–7.

Walter, A., Silk, W. K., & Schurr, U. (2009). Environmental effects on spatial and temporal patterns of leaf and root growth. *Annual Review of Plant Biology*, *60*(1), 279–304.

Wang, J., Badenhorst, P., Phelan, A., Pembleton, L., Shi, F., Cogan, N., ... & Smith, K. (2019a). Using sensors and unmanned aircraft systems for high-throughput phenotyping of biomass in perennial ryegrass breeding trials. *Frontiers in Plant Science*, *10*, 1381.

Wang, J., Dai, Q., Shang, J., Jin, X., Sun, Q., Zhou, G., & Dai, Q. (2019b). Field-scale rice yield estimation using sentinel-1A synthetic aperture radar (SAR) data in coastal saline region of Jiangsu Province, China. *Remote Sensing*, *11*(19), 2274.

Warman, C., & Fowler, J. E. (2019). Custom built scanner and simple image processing pipeline enables low-cost, high-throughput phenotyping of maize ears. *Biorxiv*, 780650.

Weber, V. S., Melchinger, A. E., Magorokosho, C., Makumbi, D., Bänziger, M., & Atlin, G. N. (2012). Efficiency of managed-stress screening of elite maize hybrids under drought and low nitrogen for yield under rainfed conditions in Southern Africa. *Crop Science*, *52*(3), 1011–1020.

White, J. W., Andrade-Sanchez, P., Gore, M. A., Bronson, K. F., Coffelt, T. A., Conley, M. M., ... & Wang, G. (2012). Field-based phenomics for plant genetics research. *Field Crops Research*, *133*, 101–112.

Yang, W., Feng, H., Zhang, X., Zhang, J., Doonan, J. H., Batchelor, W. D., ... & Yan, J. (2020). Crop phenomics and high-throughput phenotyping: past decades, current challenges, and future perspectives. *Molecular Plant*, *13*(2), 187–214.

Yang, W., Guo, Z., Huang, C., Duan, L., Chen, G., Jiang, N., ... & Xiong, L. (2014). Combining high-throughput phenotyping and genome-wide association studies to reveal natural genetic variation in rice. *Nature Communications*, *5*(1), 5087.

Yang, Y., Chai, R., & He, Y. (2012). Early detection of rice blast (Pyricularia) at seedling stage in Nipponbare rice variety using near-infrared hyper-spectral image. *African Journal of Biotechnology*, *11*(26), 6809–6817.

Yazdanbakhsh, N., & Fisahn, J. (2009). High throughput phenotyping of root growth dynamics, lateral root formation, root architecture and root hair development enabled by PlaRoM. *Functional Plant Biology*, *36*(11), 938–946.

Yol, E., Toker, C., & Uzun, B. (2015). Traits for phenotyping. In J. Kumar, A. Pratap, & S. Kumar (Eds.), *Phenomics in Crop Plants: Trends, Options and Limitations* (pp. 11–26). India: Springer.

Zhang, C., Liu, F., Zhang, H. L., Kong, W. W., & He, Y. (2014). Identification of varieties of black bean using ground based hyperspectral imaging. *Spectroscopy and Spectral Analysis*, *34*(3), 746–750.

Zhang, X., Huang, C., Wu, D., Qiao, F., Li, W., Duan, L., ... & Yan, J. (2017). High-throughput phenotyping and QTL mapping reveals the genetic architecture of maize plant growth. *Plant Physiology*, *173*(3), 1554–1564.

Zhang, X., Liu, F., He, Y., & Li, X. (2012). Application of hyperspectral imaging and chemometric calibrations for variety discrimination of maize seeds. *Sensors*, *12*(12), 17234–17246.

Zhou, Y., Srinivasan, S., Mirnezami, S. V., Kusmec, A., Fu, Q., Attigala, L., ... & Schnable, P. S. (2019). Semiautomated feature extraction from RGB images for sorghum panicle architecture GWAS. *Plant Physiology*, *179*(1), 24–37.

# 3 Remote Sensing Technologies for High-Throughput In-Field Crop Phenotyping

*Tasfia Binte Anam Priya, Falguni Akter Shraboni, M. Kamal Hossain, Stephen P. Klassen, Suresh Kadaru, and Abul Kalam Mohammad Aminul Islam*

## 3.1 INTRODUCTION

According to a recent study, the estimated global population for the year 2060 is 9.3 billion (Goujon, 2019). The current food crisis highlights the need for new strategies and methods to increase global food supply (Wang et al., 2018). A potential solution to future food supply issues might be the use of crop monitoring techniques along with efficient and effective technology, to boost food production in the face of space limitations (Ma et al., 2022). Plant phenotyping is defined by Fiorani and Schurr (2013) as a collection of procedures and techniques used to evaluate the structure, composition, and growth rates of plants at different dimensions. Plant phenotyping aims to analyze plant traits in order to determine the components that affect them, which may help plants produce better yield under restricted growth environments (Fiorani & Schurr, 2013).

Conventional approaches for plant phenotyping include manual and destructive sampling methods, which can be rather expensive, labor-intensive, and time-consuming. Furthermore, there are several phases in manual sampling and analysis procedures that require human intervention, increasing the chances of inaccuracy at each stage (Atefi et al., 2021). The precise and quick acquisition of phenotypic data is essential for examining the relationship between genomic and phenotypic information. The selection of desirable genotypes, acceleration of genetic gains, and identification of genetic variants providing maximum outcomes in a variety of dynamic target environments are all dependent on the ability to phenotype several plant species for numerous traits across multiple environments (Costa et al., 2019). Crop monitoring and the accurate and reliable measurement of important agronomical traits are still in the early phases of development, making plant phenotyping the bottleneck (Mahlein, 2016). High-throughput phenotyping emerges as an essential tool to respond to this bottleneck since it uses remote sensing technologies, which provide structured, non-destructive methods of delivering frequent information on a local to global scale (Weiss et al., 2020).

### 3.1.1 High-Throughput Crop Phenotyping

Now that high-throughput field phenotyping (HTFP) technologies have been developed, a new approach to data processing and sensing has emerged with the goal of increasing throughput. Due to advancements in modern remote sensing technologies, scientists have promptly designed non-invasive, high-throughput plant phenotyping approaches (Rebetzke et al., 2019). Field phenotyping can combine measurements of different phenotypes (Jimenez-Berni et al., 2018), which may not be visible to the human eye. By using specialized remote sensing techniques, high-throughput plant phenotyping studies strive to accomplish rapid and accurate methods for monitoring crops on a large


DOI: 10.1201/9781003505631-3

scale with little influence on the environment (Ma et al., 2022). Mainly three types of phenotyping can be noted based on the plant part: High-throughput shoot phenotyping, High-throughput root phenotyping, and High-throughput seed phenotyping, which have been discussed below.

#### 3.1.1.1 High-Throughput Shoot Phenotyping

Shoot phenotyping is a crucial aspect of plant biology and agricultural research. It involves the detailed characterization and measurement of various traits and features of plant shoots, including their morphology, growth patterns, physiological properties, and responses to environmental stimuli. Phenotyping of plants can be done in both natural and controlled environments and has been simplified with the advancement of several complex and high-throughput remote sensing technologies recently.

It is possible to monitor plant shoots using a number of phenotyping platforms under controlled conditions, such as PHENOPSIS, which studied how *Arabidopsis* responded to water scarcity (Granier et al., 2006). GROWSCREEN was developed to measure the relative growth rate and total leaf area of plants under varying lighting conditions (Walter et al., 2007). The GROWSCREEN FLUORO imaging system could phenotype leaf development and chlorophyll fluorescence in *Nicotiana tabacum* and *Arabidopsis thaliana* simultaneously (Jansen et al., 2009). Another phenotyping platform called Phenoscope was developed to collect data on water content and shoot development from 735 pots (Tisné et al., 2013). Phenovator was developed to collect data on photosynthesis, and plant development features, as well as for predicting accumulation of leaf area (Flood et al., 2016).

There are numerous platforms developed for shoot phenotyping in controlled environments, but the main focus of this review is field-based crop phenotyping. High-throughput shoot phenotyping for automatic detection of single plants in wheat may be achieved in the field using drone-based RGB imagery combined with sophisticated image processing algorithms (Jin et al., 2017a). Currently, there aren't any non-destructive techniques that are both widely used and accurate enough to calculate total canopy biomass. Using active light detection and ranging (LiDAR) devices or stereo imagery, researchers can evaluate a three-dimensional (3D) map of the crop's visible outer canopy and assess its structure. Automated leaf identification is one example of how stereo imaging may be used to separate plant components using color information (Müller-Linow et al., 2015). The efficiency of photosynthetic light responses may be measured using fluorescence-based approaches, which have recently gained popularity (Keller et al., 2019). Disease detection is difficult for automated field phenotyping but important for many breeding programs. Recent promising research shows that a combination of visible and hyperspectral imagery along with machine learning may detect *Cercospora* infection in wheat (Thomas et al., 2018). Weed detection is also essential to prevent yield loss in field conditions. Huang et al. (2018a) detected weeds in a rice field with higher accuracy. A recent study employed machine learning to identify reproductive organs (Zhou et al., 2018). Researchers have also utilized unmanned aerial vehicle (UAV) data to identify lodging resistance in rice (Yang et al., 2017a).

#### 3.1.1.2 High-Throughput Root Phenotyping

Phenotyping roots in the field can be extremely challenging (Wasson et al., 2012). So far, remote sensing technologies for root phenotyping have mostly been tested in controlled environments, making it difficult to apply the findings in the field (Atkinson et al., 2019). One of the most significant challenges in creating platforms that can directly evaluate root characteristics in the field without affecting root activity is the underground character of roots (Araus et al., 2022).

Root phenotyping in controlled environments has been made possible by the availability of several advanced non-destructive techniques for root examination. X-ray computed tomography stands out among these technologies as the most effective. It is crucial for understanding root architectural characteristics and their role in water and nutrient absorption (Pfeifer et al., 2015). Root development in opaque soil may be seen and measured in three dimensions (3D) using Magnetic Resonance

Imaging (MRI) and X-ray Computed Tomography (CT). The root structural traits (RSA) of 12 different rice genotypes were estimated using an automated platform for non-invasive imaging and analysis (Iyer-Pascuzzi et al., 2010). The RootViz application was used in wheat (*Triticum aestivum*) by Tracy et al. (2012) for visualization of root structure throughout the early stages of plant growth and development.

Many 2D imaging programs have been developed as an alternative to 3D platforms for root phenotyping (Atkinson et al., 2019). Currently, a few software has been created for root phenotyping in controlled conditions, ranging from manual programs like DART (Data Analysis of Root Tracings), which was used in roots of tomato plants (Le Bot et al., 2010), RootNav tested on rice, *Arabidopsis*, and wheat cultivars (Pound et al., 2013), and ImageJ for wheat genotypes (Slack et al., 2018). Novel imaging-based tools such as RootTrace, RootReader3D, and RootAnalyzer have also been developed (Chopin et al., 2015; Clark et al., 2013; French et al., 2009). Novel approaches have been used for documenting the growth and architecture of maize root systems such as "Rhizoslides" (Le Marié et al., 2014), RhizoTubes (Jeudy et al., 2016), RADIX (Le Marié et al., 2016), etc.

Remote sensing techniques to evaluate root characteristics in field conditions are still limited. Most of these are usually time-consuming, difficult, and costly, affected by soil characteristics, and need to be integrated with other disciplines to optimize information extraction (Araus et al., 2022). Root crown phenotyping, or "shovelomics," is a method for continuously assessing root characteristics manually by excavating the top 15–30 cm of the rhizosphere and analyzing its morphological and architectural features (Wasson et al., 2020). Ground-based geophysical platforms were utilized to profile soil and roots using electrical resistance tomography (ERT) in sorghum and maize in the field (Srayeddin & Doussan, 2009). Additionally, Whalley et al. (2017) used EMI sensors to infer root activities in wheat in field settings by tracking changes in soil water.

Alternative methods of indirectly evaluating root function have been suggested, which involve measuring crop above-ground traits like canopy temperature and stable carbon isotope structure (Tracy et al., 2020). The assessment of shoot proxy traits also allows for the indirect assessment of root performance. However, problems can occur depending on indirect measurement of root growth in field conditions (Wasson et al., 2012). To overcome these limitations, remote sensing technologies should be implemented to further study root traits in field conditions.

#### 3.1.1.3 High-Throughput Seed Phenotyping

Seed phenotyping mostly involves laboratory-based image acquisition and analysis, therefore, finding instances of field-based seed phenotyping is difficult. SmartGrain software was created for automated shape parameter calculations, including length, breadth, area, and perimeter length in several crop seeds including *Arabidopsis* and rice (Tanabata et al., 2012). PhenoSeeder is another tool for the physical handling and phenotyping of seeds of varying sizes (Jahnke et al., 2016). SeedGerm was developed as an affordable phenotyping tool for computerized imaging and phenotypic analysis of germination in crop seeds (Colmer et al., 2020). A portable imaging system (BELT), combined with phenoSEED, software for image acquisition and analysis, was created for the optical analysis of small seeds such as lentils (Halcro et al., 2020).

A novel low-cost, portable device that can do high-throughput phenotyping of crop seeds in field conditions was recently developed, with a phenotypic parameter measurement accuracy of above 95%, indicating that this instrument may be used for precise assessment of crop seed parameters in future research (Zhihong et al., 2016). Some of the remote sensing platforms used for different types of high-throughput phenotyping in both controlled and field conditions have been summarized in Table 3.1.

#### 3.1.1.4 Importance of HTP in Integration of Phenomics and Genomics

The best way to describe crop phenomics is as an interdisciplinary field that studies the effects of crop growth on organism-wide multidimensional phenotypes using high-throughput accurate accumulation and analysis (Yang et al., 2020). The research bottleneck in plant science has shifted from genomics to phenomics as a consequence of unprecedented advances in DNA sequencing

**TABLE 3.1**
**Overview of Remote Sensing Platforms for Different Types of High-Throughput Phenotyping**

| Types of HTP | Environment | Remote Sensing Platforms | Crops | Measured Traits | References |
|---|---|---|---|---|---|
| High-throughput Shoot Phenotyping | Controlled | PHENOPSIS, GROWSCREEN, GROWSCREEN FLUORO, Phenoscope, Phenovator | *Nicotiana tabacum*, *Arabidopsis thaliana* and other plant species | Total leaf area, chlorophyll fluorescence, predicted leaf area accumulation, shoot development, relative growth rate, performance in response to water shortage, salt stress | Flood et al. (2016), Granier et al. (2006), Jansen et al. (2009), Tisné et al. (2013), and Walter et al. (2007) |
| | Field | UAVs, Satellites, RGB sensors, LiDARs, Hyperspectral and Multispectral Sensors, Stereo Imagery, Fluorescence Imagery | Wheat, barley, rice, soybean, sugar beet, spinach | Canopy biomass, automated leaf identification, chlorophyll fluorescence, disease and weed detection, reproductive organ detection, lodging resistance | Huang et al., (2018a), Jin et al. (2017a), Keller et al. (2019), Müller-Linow et al. (2015), Thomas et al. (2018), Yang et al. (2017a), and Zhou et al. (2018) |
| High-throughput Root Phenotyping | Controlled | X-ray Computed Tomography (CT), Magnetic Resonance Imaging (MRI), Software applications e.g. DART, ImageJ, RootAnalyzer, RootTrace, RootNav, RootReader3D, RhizoTubes, Rhizoslides, RADIX | Rice, barley, wheat, canola, tomato, maize, bean, pea, grapevine, *Arabidopsis*, rapeseed, cowpea | 2D and 3D RSA (Root System Architecture) characteristics such as Root diameter, root thickness, root growth and development, root biomass | Chopin et al. (2015), Clark et al. (2013), French et al. (2009), Iyer-Pascuzzi et al. (2010), Jeudy et al. (2016), Le Bot et al. (2010), Le Marié et al. (2014, 2016), Pfeifer et al. (2015), Pound et al. (2013), Slack et al. (2018), and Tracy et al. (2012) |
| | Field | Shovelomics, Electrical Resistance Tomography (ERT), EMI Sensors, Geophysical Platforms | Wheat, maize, sorghum | Root architecture, soil and root profile, root activities due to changes in soil water | Srayeddin and Doussan (2009), Wasson et al. (2020), and Whalley et al. (2017) |
| High-throughput Seed Phenotyping | Controlled | Imaging and Image analysis devices, Softwares, e.g. SmartGrain, SeedGerm, BELT, PhenoSEED, Phenoseeder | Rice, wheat, bean, *Arabidopsis thaliana*, rapeseed, barley | Coleoptile length, seed color, shape and size, seed density, automated seed imaging, seed germination analysis | Colmer et al. (2020), Halcro et al. (2020), Jahnke et al. (2016), and Tanabata et al. (2012) |
| | Field | Smartphone based portable instrument | Not specified | seed length, seed width, area of seed, total number of seeds | Zhihong et al. (2016) |

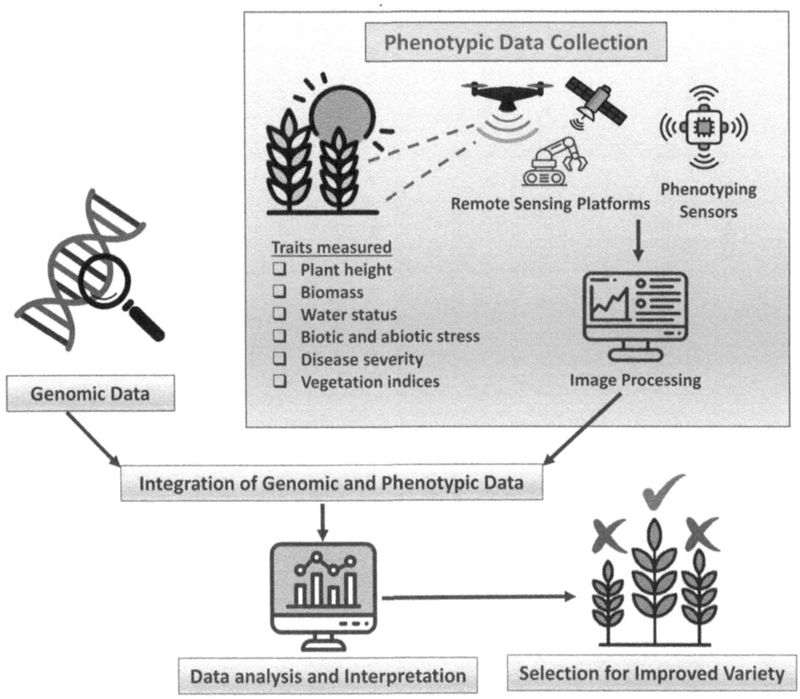


**FIGURE 3.1** Integration of genomics and phenomics for high-throughput phenotyping of crops.

technologies, which have greatly advanced our knowledge and ability to understand crop genomes (Edwards et al., 2013). However, there is still a lack of understanding in plant research when it comes to connecting genomic and phenotypic data. Researchers have been working to create non-invasive, high-throughput phenotyping technologies that can improve crops by collecting data on traits and linking it to genomic information (Araus & Cairns, 2014).

The development of high-throughput phenomics over the last decade has greatly enhanced the efficiency and accuracy of phenotyping, using several methods such as genome-wide association studies (GWAS), QTL mapping, and candidate gene association studies. In an experiment by Yang et al. (2014), 141 associated loci were found using the rice automatic phenotyping platform (RAP) in conjunction with genome wide association studies (GWAS). Zhang et al. (2017) also found 988 QTLs in maize by integrating the high-density linkage map with the high-throughput phenotyping platform, which helped to study QTL distribution over the whole genome. Screening thousands of lines for specific traits rapidly is becoming more relevant as functional genomics and molecular breeding advance at a fast pace. A novel outlook on gene mapping, plant genomics, and breeding research tools can be made possible by integrating remote sensing and high-throughput phenotyping with large-scale GWAS or QTL analysis (Figure 3.1). A new era in intelligent crop breeding may begin when phenomics data is improved and aggregated with different genome-specific technologies (Fan et al., 2021). This review seeks to highlight the use of different remote sensing techniques that are currently regarded as valuable tools for high-throughput crop phenotyping, particularly in field conditions, and the challenges and prospects for this developing technology.

## 3.2 HIGH-THROUGHPUT PHENOTYPING PLATFORMS

Research using plant phenotyping platforms has expanded greatly due to advancements in remote sensing technologies. A high-throughput plant phenotyping platform can screen more samples with more precision and at a lower cost, which improves the selection intensity and accuracy (Von Mogel, 2013).

A number of plant phenotyping systems that include various sensors have been launched or developed by various research institutes. There are two main types of high-throughput phenotyping platforms now in use: ground-based phenotyping platforms and aerial-based phenotyping platforms (Qiu et al., 2018).

### 3.2.1 Ground-Based Phenotyping Platforms

Conveyors, trucks, or robots equipped with a range of sensors mounted at varying heights and locations mainly constitute ground-based phenotyping platforms for measuring several plant phenotyping traits (Enciso et al., 2017). The ground-based plant phenotyping systems have the potential to enhance work efficiency while decreasing the need for human labor (Qiu et al., 2018). Ground systems typically have more sensors and equipment, so they can carry a heavier payload. Additionally, ground systems have more viewing angles and higher resolution data, which allows them to measure phenotypic traits at deeper layers. Furthermore, ground platforms are less affected by wind, thus they are able to follow a straight path and conduct more uniform longitudinal scanning (Xu & Li, 2022).

However, there are some restrictions on the ground-based plant phenotyping systems. Covering a field often requires a considerable amount of time. Consequently, ground-based phenotyping devices cannot quickly and accurately measure plant phenotypes on a broad scale (Ma et al., 2022).

#### 3.2.1.1 Tractor Based Systems

The USDA Maricopa Agricultural Centre in Arizona was one of the first institutes to design a tractor-based system that could assess the temperature and height of eight rows of canopies simultaneously using a combination of infrared thermometers and ultrasonic sensors (Andrade-Sanchez et al., 2014). Another such system is BreedVision, which integrates a number of sensors, such as an RGB camera, a laser distance sensor, a light curtain sensor, or a 3D time of flight camera with a tractor (Busemeyer et al., 2013). Another example is the GPhenoVision, a multisensor tractor system that measures the morphological features of cotton (Jiang et al., 2018). In order to estimate plant biomass and nitrogen uptake, the PhenoTrac 4 tractor was tested in barley and wheat breeding nurseries (Barmeier & Schmidhalter, 2016). In another study, a tractor was equipped with a LiDAR scanner and an RTK-GPS to construct three-dimensional models of cotton plants, which were then used to estimate the plant's volume, canopy area, and height (Sun et al., 2018).

Despite the tractor-based system's large payload capacity, the tractor's weight might become a problem as it can compress the soil and stunt the crop's development. Furthermore, the vibrations from the tractor can harm the sensors and diminish the data quality, which would be a major problem for the data-gathering system. Conditions of the field's soil may restrict the tractor's work efficiency (Xu & Li, 2022).

#### 3.2.1.2 Pushcart and Motorized Cart

Pushcarts and motorized carts were created as an alternative to tractors to address the problem of soil compaction. The majority of cart-type systems include metal frames with bicycle wheels, rendering them to be lightweight, inexpensive, and easy to install sensors (Xu & Li, 2022). Field trials using wheat breeding lines were conducted using Phenocart to analyze the relationship between canopy temperature, plant health, and grain production (Crain et al., 2016). A proximal sensing cart equipped with digital cameras, infrared thermometers, and a GPS receiver, was employed in wheat, barley, and cotton breeding nurseries (White & Conley, 2013). A hydraulic drive system called Phenomobile was utilized in a wheat breeding trial to measure plant stress, canopy temperature, number of spikes in the canopy, and estimated canopy height (Deery et al., 2014). Compared to tractors, the pushcart's location is more easily controlled, allowing it to stop and scan the field at any position (Meacham-Hensold et al., 2020). The main problem with pushcarts and motorized carts is that they are impractical and wasteful in big fields since they need human strength and control (Xu & Li, 2022).

#### 3.2.1.3 Gantry and Cable-Driven System

Gantry-based systems can travel linearly on parallel tracks and employ a gantry to transport sensors, allowing both horizontal and vertical movement of the camera. A popular gantry-based system is Field Scanalyzer, which can carry the chlorophyll fluorescence imager and other heavy sensors due to its large payload (Virlet et al., 2017). The gantry system PhénoField®, run by the French applied research institution ARVALIS, was mainly used for wheat breeding (Beauchêne et al., 2019).

Similar to the gantry-based system, the cable-driven system is permanently installed on-site and uses towers positioned throughout the field to hold sensors hung from cables. The NU-Spidercam from the University of Nebraska-Lincoln (Bai et al., 2019) and the field phenotyping platform (FIP) at the Swiss Federal Institute of Technology in Zurich (Kirchgessner et al., 2017) are two examples of representative systems.

Due to their autonomy, gantry-based and cable-driven platforms may thoroughly examine the field several times throughout the growth season without contacting any soil or plants, which is a major benefit. The ability to precisely control the sensors for crop scanning is an additional benefit. Consequences include but are not limited to, expensive costs, maintenance, and a fixed experimental location with limited field coverage. Systems typically scan the canopy's top view; they can't provide data from below the canopy without scanning from the sides (Xu & Li, 2022)

#### 3.2.1.4 Phenotyping Robots

To solve the labor shortage in agriculture, the development of phenotyping robots has evolved dramatically during the last ten years. The agricultural robot is an excellent tool to manage vast farms with less human labor through automation, which also makes it an ideal remote sensing technology for high-throughput field phenotyping. As a result, the development of phenotyping robots to replace other ground phenotyping systems is becoming popular (Xu & Li, 2022). Robotic phenotyping may make it possible to effectively track how plant features develop over time in both experimental and natural environments. The primary advantage of robots over tractor-based systems is that they are entirely autonomous and don't require skilled operators to do farming tasks (White et al., 2012).

A variety of agricultural tasks have been replaced by autonomous robots, including harvesting in sweet pepper (Arad et al., 2020), spraying in wheat (Adamides et al., 2017), pest and weed control in tomato and lettuce (Raja et al., 2020), and pruning in apples (Zahid et al., 2020). Since autonomous robotic systems have the potential to greatly enhance the speed, coverage, accuracy, and economy of plant trait assessments, they are thought of as essential parts of high-throughput plant phenotyping (Atefi et al., 2021). Based on their means of motion, the current generation of phenotyping robots may be divided into three groups: wheeled, tracked, and wheel legged robots (Xu & Li, 2022)

##### *3.2.1.4.1 Wheeled Robot*

The most prevalent phenotyping robot is the wheeled robot. In general, there are two types of wheeled robots: those that have complete mobility (like omnidirectional robots) and those that have locally limited mobility (like skid steering and differential drive robots). For high-throughput field phenotyping, commercial robotic platforms like Clearpath's Jackal and Vinoculer are popular choices for skid steering robots (Sampaio et al., 2021; Shafiekhani et al., 2017). RGB images from TerraSentia were utilized for identification and counting of the maize stand (Zhang et al., 2020b). Phenobot 1.0 is a two-wheel drive and two-wheel steering (2WD2WS) robotic platform combined with stereo cameras, navigation, and an auto steering system for image acquisition, which was used to test algorithms for the 3D image construction and prediction of stem diameter and plant height in sorghum (Salas Fernandez et al., 2017). Some exemplary four-wheel drive and four-wheel steering (4WD4WS) robots are the Thorvald II (Grimstad & From, 2017) and Ladybird (Underwood et al., 2017).

#### *3.2.1.4.2 Tracked Robot*

Tracked robots have a greater capacity to adapt to different types of terrain than wheeled robots because of their increased surface area in contact with the ground. Because of the low ground pressure, it can operate in muddy fields and other tough terrains where wheeled robots are unable to. A tracked robot called TERRA-MEPP was created to phenotype sorghum (Young et al., 2019). Phenomobile V2 is another heavy-duty tracked robot with high-throughput used under difficult soil conditions (Baret et al., 2022). LT2 from SuperDroid Robots was also developed for non-invasive high-throughput phenotyping in fields (Gao et al., 2018).

#### *3.2.1.4.3 Wheel Legged Robot*

Wheel-legged robots provide both the high terrain flexibility of the legged robot and the speed of the wheeled robot. Wheel-legged robots, like the BoniRob, may achieve great speed and change their dimensions to accommodate a variety of field layouts (Ruckelshausen et al., 2009). Image datasets from BoniRob were used to identify and locate crop leaves and classify weeds as well as plant categorization and mapping (Chebrolu et al., 2017; Weyler et al., 2021). The disadvantage of the wheel-legged robot is that it is less durable than a wheeled robot due to its increased complexity (Xu & Li, 2022). A summary of the ground-based phenotyping platforms is highlighted in Table 3.2.

## 3.2.2 Aerial-Based Plant Phenotyping Platforms

Since aerial crop phenotyping platforms can overcome the restrictions of certain ground-based platforms, their use is on the rise. In a matter of minutes, plant phenotyping characteristics may be measured from several plots using aerial-based platforms (Ma et al., 2022). When compared to ground-based platforms, aerial platforms have the advantage of quick measurements of large-scale plants. Therefore, data obtained from aerial media may accurately represent plant conditions on a broad scale without any time lag (Li et al., 2014). The payload capacity of airborne plant phenotyping platforms is limited, however, thus it is not possible to take several measurements of plant phenotyping using only one platform (Sankaran et al., 2015). Satellites and Unmanned aerial vehicles (UAVs) are the most commonly used aerial platforms nowadays.

### 3.2.2.1 Satellites

Data mining techniques and the gathering of satellite images for use in remote sensing applications have been around for over 50 years (Zhang et al., 2020a). Among this technology's many advantages is its ability to acquire data from a larger area in a systematic and comparable approach, without the introduction of sample bias, and without causing any damage (Liaghat, 2010). Tattaris et al. (2016) used the Digital Globe WorldView-2 satellite to assess germplasm in a wheat breeding experiment. In agriculture, the use of satellite-based remote sensing has been documented in the last few years for a number of purposes, such as estimation of plant height, leaf area index (LAI), crop growth and phenology, crop type identification, yield prediction, nutrition and disease management, and other agricultural practices (Jin et al., 2017b; Kayad et al., 2016; Yuan et al., 2016). The efficacy of these sensing methods has been acknowledged with the improvement of resolutions in modern sensors, the speed of data acquisition, accessibility, and availability of open data for some sources over several years (Sibanda et al., 2017; Wei et al., 2017). With satellite-based remote sensing technologies, data may be quickly acquired, leading to early agronomic management choices that can pinpoint where to intervene to avoid output losses (Zhang et al., 2020a). The weather, imaging frequency, resolution, imaging expenses, and access time to acquired images are the main constraints of satellite imaging. Higher resolution in satellite imagery is required for plant breeding studies due to the often tiny plot sizes (Chawade et al., 2019).

**TABLE 3.2**
**Summary of Ground-Based High-Throughput Phenotyping Platforms**

| Name of the Platform | Sensors Used | Crops | Crop Parameters | References |
|---|---|---|---|---|
| | | **Tractor Based Systems** | | |
| LeeAgra 3434 DL | The Multispectral Crop Canopy Sensor, Infrared Radiometer Sensor, Sonar Proximity Sensor | Cotton | Plant canopy height, reflectance, temperature | Andrade-Sanchez et al. (2014) |
| BreedVision | RGB Camera, Hyperspectral Imaging Sensors, 3D Time of Flight (ToF) Cameras | Triticale (*Triticosecale wittmack* L.) | Plant moisture content, plant height, dry biomass yield, tiller density | Busemeyer et al. (2013) |
| GPhenoVision | Hyperspectral Camera, Thermal Camera, RGB-D Camera | Cotton | Canopy temperature, height, width, leaf volume, projected leaf area | Jiang et al. (2018) |
| PhenoTrac 4 | Multispectral and Hyperspectral Sensors | Wheat, barley | Plant biomass, plant nitrogen uptake, dry weight of plant organs | Barmeier and Schmidhalter (2016) |
| Spider DL | LiDAR sensor | Cotton | Canopy height, projected canopy area, plant volume | Sun et al. (2018) |
| | | **Pushcart and Motorized Cart** | | |
| Phenocart | Infrared Thermometer (IRT), RGB and Near Infrared Cameras | Wheat | Normalized difference vegetation index (NDVI), canopy temperature, plant biomass, relative growth rates | Crain et al. (2016) |
| Professor | Multiple Proximal Sensors and Cameras | Wheat, maize, cotton | Different plant traits under field conditions | Thompson et al. (2018) |
| Proximal Sensing Cart | RGB Cameras, Infrared Thermometers, Ultrasonic Proximity Sensors | Wheat, barley, *camelina*, cotton | Responses to drought, heat stress, and nutritional deficiencies, crop growth and development | White and Conley (2013) |
| Phenomobile | LiDAR Sensors, RGB Cameras, Thermal Infrared Cameras, Spectroradiometer, Hyperspectral Camera | Wheat, *Brassica napus*, maize, Japanese millet (*Echinochloa esculenta*) | Canopy structural parameters, height, identifying individual plant organs, leaf area and volume, canopy temperature, vegetation indices | Deery et al. (2014) |
| | | **Gantry and Cable Driven System** | | |
| Field Scanalyzer | RGB, Chlorophyll Fluorescence and Thermal Infrared Cameras, Hyperspectral Sensors | Wheat | Plant height, spacing, biomass, Canopy temperature, Spectral indices | Virlet et al. (2017) |
| PhénoField® | RGB Camera, Spectroradiometer, LiDAR Sensor | Wheat | Plant height, average leaf angle, chlorophyll index, green area index, | Beauchêne et al. (2019) |
| Field Phenotyping Platform (FIP) | RGB Camera, Multispectral Camera, Thermal Camera, NIR Camera | Wheat | Plant height, spectral reflectance, canopy temperature, ground cover | Kirchgessner et al. (2017) |
| NU-Spidercam | LiDAR, Thermal Infrared Camera, Multispectral Camera, Visible Near Infrared Spectrometer | Maize, Soybean | Vegetation coverage, canopy and soil NDVI, soil and canopy temperature, canopy reflectance, canopy height | Bai et al. (2019) |

(*Continued*)

**TABLE 3.2 (*Continued*)**
**Summary of Ground-Based High-Throughput Phenotyping Platforms**

| Name of the Platform | Sensors Used | Crops | Crop Parameters | References |
|---|---|---|---|---|
| | | **Wheeled Robots** | | |
| Jackal robot | RGB-D Camera, Stereo Depth Camera | Maize | Temperature, humidity, luminosity, 3D reconstruction of plants | Sampaio et al. (2021) |
| Vinoculer<br>Vinobot | RGB and Thermal Cameras<br>LiDAR, Stereo Camera, Environmental Sensors | Maize, sorghum | 3D Models of the plants, plant height, leaf area index, air temperature | Shafiekhani et al. (2017) |
| Phenobot 1.0 | Stereo and RGB cameras | Sorghum | Plant architecture e.g. stem diameter, plant height | Salas Fernandez et al. (2017) |
| TerraSentia | RGB Camera, LiDAR, Multispectral Camera, Hyperspectral Camera, RGB-D Camera | Maize | Estimating plant population | Zhang et al. (2020b) |
| Ladybird | RGB Camera, LiDAR, Thermal Camera, Hyperspectral Camera, Stereo Camera, | Wheat, barley, lentil, bean, field pea, chickpea | Plant height, NDVI, crop row detection, canopy closure, weed detection | Underwood et al. (2017) |
| Thorvald II | RGB Camera, Infrared Camera | Wheat | Different phenotypical measurements | Grimstad and From (2017) |
| | | **Tracked Robots** | | |
| TERRA-MEPP | Time of Flight (ToF) Depth Sensors, Stereo Cameras | Sorghum | Plant height, stem width | Young et al. (2019) |
| Phenomobile V2 | LiDAR, RGB Cameras and Multispectral Cameras | Tall crops such as sunflower, maize, wheat, rapeseed or peas. | Average Leaf Angle (ALA), Fraction of light intercepted (FIPAR), Green area index (GAI) | Baret et al. (2022) |
| Superdroid LT2 | RGB Camera | Soybean, maize | Canopy phenotypic data acquisition | Gao et al. (2018) |
| | | **Wheel Legged Robots** | | |
| BoniRob | 3D Time of Flight Cameras, Multispectral Camera, RGB-D Sensor, LiDAR, Hyperspectral and Depth Sensors | Maize, wheat, sugar beet | Identifying and locating crop leaves, weed classification, plant categorization, mapping | Chebrolu et al. (2017) and Weyler et al. (2021) |

*Source:* Modified from Xu and Li (2022).

#### 3.2.2.2 Unmanned Aerial Vehicles (UAVs)

A new remote sensing technology that is quickly becoming popular among both the general public and the scientific community is Unmanned Aerial Vehicles (UAVs). UAVs present a flexible platform for remote sensing by providing aerial images with high spatial and temporal resolutions (Holman et al., 2016). UAVs can be mounted with various sensors to measure various plant phenotyping parameters (Ma et al., 2022). Unmanned Aerial Vehicles (UAV) have evolved into an innovative, inexpensive platform for remote sensing owing to the downsizing, increased precision, and decreased cost of several technologies. Unmanned Aerial Vehicles (UAVs) surpass the constraints in resolution and cost that often hinder the utilization of satellite and airborne remote sensing (Holman et al., 2016). Moreover, UAVs can swiftly gather data from the whole study site while reducing the impact of environmental factors like wind, clouds, and sun angle (Xie & Yang, 2020). The most commonly used forms of UAVs include fixed-wing gliders, parachutes, hexacopters, octocopters, rotocopters, etc.

In order to evaluate crop phenotypic traits such as crop height, vigor, and growth in plant breeding as markers of agricultural performance, a few research have been conducted employing UAV-based remote sensing. Image classification based on high-resolution RGB images is often used to get crop metrics such as vegetation coverage, biomass, and crop height (Li et al., 2016). Wan et al. (2020) developed an effective technique for calculating yield in rice using UAV-based multispectral and RGB images. A fixed-wing glider fitted with a multispectral camera was piloted at a height of 150 meters to assess variation due to low nitrogen status in maize germplasms (Zaman-Allah et al., 2015). Tattaris et al. (2016) assessed the biomass and yield of wheat trials using an octocopter equipped with multispectral and thermal sensors. In order to estimate vegetation indices in wheat experiments, a rotocopter equipped with a multispectral camera was utilized (Khan et al., 2018). Crop lodging in wheat, canopy temperature in sugarcane, and ground cover in sorghum were all investigated using a phenocopter equipped with different sensors (Chapman et al., 2014). The height of sugarcane was measured using a fixed-wing glider equipped with an RGB camera (De Souza et al., 2017). Xu et al. (2018) assessed cotton trials using an octocopter equipped with an RGB camera. In one study, crop biomass and crop yield were predicted using a UAV equipped with RGB and hyperspectral cameras for obtaining images of the potato crop canopy (Li et al., 2020). UAVs have been utilized for chlorophyll fluorescence, monitoring of drought, and water status in citrus fruits (Berni et al., 2012). Table 3.3 provides an overview of the aerial-based high-throughput phenotyping platforms.

## 3.3 HIGH-THROUGHPUT PLANT PHENOTYPING SENSORS

Many sensors are now essential components for measuring plant phenotyping characteristics since they can provide quick, non-invasive plant information (Fiorani & Schurr, 2013). Research indicates that both new and traditional sensors are integrated with remote sensing platforms for measuring plant phenotyping parameters, including visible spectrum image sensors, thermal image sensors, infrared image sensors, multispectral image sensors, hyperspectral image sensors, fluorescence sensors, and different 3D imaging sensors such as Light detection and ranging (LiDAR), depth image sensors etc. The many sensor types that are often used for plant phenotyping are covered in the subsections that follow.

### 3.3.1 Visible Spectrum Image Sensors

The most popular and significant sensing technology for crop phenotyping is the visible light image sensor, mainly the RGB sensor, which has the advantages of cheap cost, high efficiency, and easy advertising (Fan et al., 2021). Furthermore, each year has seen significant improvements in the technical specifications of visible spectrum image sensors, including image quality, sensitivity to light, and focus capacity (Mahlein, 2016). Plant physiological and textural characteristics, biotic stress,

**TABLE 3.3**
**Summary of Aerial Based High-Throughput Phenotyping Platforms**

| Name of the Platform | Sensors Used | Crops | Applications | References |
|---|---|---|---|---|
| **Satellites** | | | | |
| Digital Globe WorldView-2 (WV-2) | Multispectral Sensor | Wheat | Estimation of biomass from vegetation index (NDVI) and canopy temperature (CT) | Tattaris et al. (2016) |
| Landsat-8 | Different Spectral Cameras | Alfalfa | Developing a remote sensing-based yield prediction system | Kayad et al. (2016) |
| SPOT-6 | Multispectral Camera | Wheat | Monitoring of crop diseases | Yuan et al. (2016) |
| WorldView-3 (WV-3) | Multispectral Sensor | Grassland | Predicting the above-ground biomass | Sibanda et al. (2017) |
| WorldView-2, WorldView-3, Pleiades-1A, and SPOT-6 | RGB and Multispectral Camera | Oilseed rape | Estimating leaf area index | Wei et al. (2017) |
| **UAVs** | | | | |
| A multiple-rotor wing UAV | RGB Camera | Maize | Estimating canopy height and aboveground biomass | Li et al. (2016) |
| Octorotor UAV | RGB and Multispectral Camera | Rice | Calculating grain yield | Wan et al. (2020) |
| Fixed wing glider | Multispectral Camera | Wheat | Phenotyping for low nitrogen stress tolerance | Zaman-Allah et al. (2015) |
| AscTec Falcon 8 UAV | Thermal and Multispectral Sensors | Wheat | Predicting yield and biomass | Tattaris et al. (2016) |
| 3DR Solo drone | RGB and Multispectral Sensors | Wheat | Estimation of vegetation indices | Khan et al. (2018) |
| Phenocopter | RGB and Thermal Camera | Sorghum, wheat, cotton, sugarcane | Estimation of canopy temperature, crop lodging, and ground cover | Chapman et al. (2014) |
| EBee Ag drone | RGB Camera | Sugarcane | Estimation of plant height | De Souza et al. (2017) |
| Octocopter (s1000+, DJI) | RGB Camera | Cotton | Detection and estimating the number of flowers | Xu et al. (2018) |
| UAV helicopter platform | Thermal and Multispectral Sensors | Peach orchard, corn, olive | Estimation of biophysical parameters such as water stress detection, chlorophyll content, leaf area index, and Photochemical Reflectance Index (PRI) | Berni et al. (2012) |

and health status may all be shown by a visible spectrum image sensor, which generates an RGB image gathering the visible radiations of objects (Qiu et al., 2018). Time and money can be saved by automating the data capture of plant phenotyping characteristics, eliminating the need for farmers and researchers to visit the fields for plant monitoring and measurement (Fan et al., 2021).

The biomass and height of barley plants were measured using a UAV equipped with an RGB camera (Bendig et al., 2015). Klodt et al. (2015) calculated the leaf surface area of grapevine along with fruit-to-leaf ratios using a typically visible spectrum image sensor. Different spectra may provide diverse information on plant phenotyping, including temperature and water content (Fahlgren et al., 2015a). Korobiichuk et al. (2018) investigated a technique for measuring corn's nitrogen status using the visible spectrum image sensor without the need for extra artificial lighting. Sugiura et al. (2016) developed a protocol to evaluate late blight severity in potatoes using RGB imagery. In agricultural production, it may be employed to track lodging percentage, canopy coverage, flowering dynamics, and crop emergence rate (Hall et al., 2016). A major drawback in visible spectrum-based plant phenotyping is differentiating between similar colored plants and backgrounds (Ma et al., 2022).

### 3.3.2 Infrared Spectrum Image Sensors

The near-infrared (NIR) image sensor's spectral sensitivity allows it to detect both shortwave and near-infrared wavelengths, which is helpful for determining the water content of leaves (Fahlgren et al., 2015b). The plant water content of barley was determined by using an NIR imaging sensor (Chen et al., 2015). Instead of visible light, near-infrared and shortwave infrared are considerably more effective at identifying Potato virus Y (PVY) in potatoes (Griffel et al., 2018). Briglia et al. (2019) proposed the idea that grapevines' morphological analysis and responses to drought may be studied using RGB and NIR image analysis. A limitation of infrared spectrum sensors is that information on other wavelengths may be lost due to the NIR image sensor's restriction to certain bands (Ma et al., 2022).

### 3.3.3 Thermal Infrared Image Sensors

Temperature mapping of a plant may be obtained by thermal infrared imaging. The primary benefits of thermal imaging over other types of sensors are its non-invasiveness, ease of use, and superior accuracy. Potential uses of thermal image-based remote sensing in phenotyping include prediction of water stress, identifying plant diseases and pathogens, measuring fruit maturity, and detecting injury to crops (Costa et al., 2019; Ma et al., 2022). Matos et al. (2014) used thermal image sensors to investigate the effects of temperature and light fluctuations on plant growth rate. Using thermal infrared imaging, Crusiol et al. (2020) determined the temperature and water status of soybean plants, demonstrating the connection between soil water availability and plant thermal infrared behavior. However, the ambient circumstances may readily alter the thermal infrared image sensor, making it difficult to distinguish small temperature differences (Sankaran et al., 2015).

### 3.3.4 Multispectral Image Sensors

The multispectral image sensor has a few distinct broad spectral bands which can detect light, making it capable of providing more precise spectrum data (Ma et al., 2022). The multispectral camera has several applications in agriculture, including crop type identification, growth monitoring and analysis, yield forecast, and computation of vegetation-covered areas (Fan et al., 2021). A variety of phenotypic traits related to plants can be measured including water stress, plant stress, disease detection, and vegetation indices, which are based on data detected in the visible and near-infrared spectra (Kitić et al., 2019).

Andrade-Sanchez et al. (2014) measured the cotton canopy reflectance using a multispectral image sensor. An unmanned aerial system fitted with sensors was used by (Zaman-Allah et al., 2015) to get multispectral imagery of the performance of maize in low nitrogen conditions. Kumar et al. (2020) obtained the near-infrared and RGB images using a multispectral imaging sensor installed in a UAV, which were then utilized to create the vegetative index for monitoring the health of maize crops. The major limitation of multispectral image sensors is their limitation to a few spectral bands (Ma et al., 2022).

### 3.3.5 Hyperspectral Image Sensors

Compared to other types of sensors, hyperspectral image sensors provide a possible way for more precise and thorough plant phenotyping owing to their hundreds of minute spectral bands (Ma et al., 2022). As it can provide insightful information about plant parameters including plant stress, chlorophyll and water status, the hyperspectral image sensor is becoming a common remote sensing technique for high-throughput plant phenotyping (Moghimi et al., 2018). Hyperspectral imaging may be used to detect water levels, disease severity, and other physiological features by simultaneously providing an abundance of spectral and spatial information (Yang et al., 2020).

Backhaus et al. (2011) described a hyperspectral imaging approach for classifying the nutrition level of crops. To identify sugar beet diseases, Mahlein et al. (2013) created unique spectral disease indicators, or SDIs. In order to screen drought-tolerant maize varieties, hyperspectral images were used to identify plant response to the drought (Asaari et al., 2018). Hyperspectral images were also analyzed to assess the cold damage to maize seedlings using a convolutional neural network (CNN) model (Yang et al., 2019). However, Hyperspectral data processing can get complex, and the hyperspectral sensor is expensive (Mahlein et al., 2018).

### 3.3.6 Fluorescence Sensors

By sensing fluorescence, fluorescence sensors are able to quantify chlorophyll content in plants (Ma et al., 2022). Fluorescence imaging can detect physiological signals including photosynthetic activity and reactive oxygen species, when equipped with particular lighting (Fichman et al., 2019). Consequently, plant stress and disease detection may be accomplished using the fluorescence sensor (Li et al., 2014). The LemnaTec 3D Scanalyzer system was equipped with a fluorescence sensor to identify fluorescent images in rice, which can be used to identify plant diseases such as necrosis, senescence, and chlorosis (Hairmansis et al., 2014). The LemnaTec system's fluorescence sensor was also used by Chen et al. (2015) to detect signals of chlorophyll fluorescence. Fluorescence imaging is less susceptible to environmental influences than visible light imaging, and its singular imaging channel helps scientists carry out quantitative threshold analysis (Fan et al., 2021).

Nevertheless, the view restriction and the need for intense light are the fluorescence sensor's drawbacks (Qiu et al., 2018). Furthermore, the majority of fluorescent images only show individual leaves or model crop seedlings. Large-scale plant phenotyping might potentially benefit from the use of more reliable data processing tools (Li et al., 2014). A newly emerged phenotyping tool, solar-induced fluorescence (SIF) has an advantage over active fluorescence technologies for broad fields, aerial surveys, and satellites since it passively monitors fluorescence emission in atmospheric absorption bands (Pinto et al., 2016).

### 3.3.7 3D Mapping Sensors

Currently, the most common imaging sensor technologies used for 3D mapping of plants are depth image sensors, light detection and ranging (LiDAR) sensors, stereo vision, time of flight cameras, etc.

#### 3.3.7.1 Light Detection and Ranging (LiDAR)

Recent years have seen a rise in the usage of LiDAR sensors for phenotypic analysis through the acquisition of three-dimensional point clouds of plants (Fan et al., 2021). With the use of laser photoelectric detection as a light source, LiDAR is an active remote sensing tool that may collect more precise three-dimensional morphological and spatial data on plants (Jin et al., 2018). Its benefits include low altitude performance detection, excellent anti-interference ability, and high resolution. In order to extract features related to leaf width, leaf length, and plant height, it may get high-throughput phenotypic data of a single plant or a population from multiple sources.

The vertical and horizontal vegetation canopy structure can be precisely characterized by LiDAR imagery, which also allows for rapid analysis (Fan et al., 2021). Because of their high resolution and long-range, numerous LiDAR sensors might become a viable solution for the light sensitivity issue in other sensors (Zhao et al., 2019).

LiDAR is often implemented for 3D reconstruction of plants and trait estimation in high-throughput plant phenotyping research (Guo et al., 2018). A ground vehicle equipped with LiDAR was used by Walter et al. (2019) to measure the canopy height and biomass of wheat in fields. A LiDAR was installed on a ground-based phenotyping platform to assess the above-ground biomass, ground cover, and canopy height of maize (Jimenez-Berni et al., 2018). LiDAR was used in another study to analyze the development of cotton plants and assess their volume, canopy area, and plant height (Sun et al., 2018).

Even though LiDAR offers a way to precisely measure specific plant features, there are some issues that need to be resolved. According to Sankaran et al. (2015), one drawback of LiDAR is its extreme sensitivity to even minute changes in route length. Furthermore, the cost of extremely precise LiDAR is quite costly (Qiu et al., 2018).

#### 3.3.7.2 Depth Image Sensors

Images with real-time depth information may be obtained using a depth image sensor. Currently, the RGB-D image sensor is the most widely used depth image sensor comprised of infrared, depth, and digital image emitters (Ma et al., 2022). RGB-D image sensors are extensively used in many different applications, including object identification, 3D reconstruction, and remote control since they can provide both RGB and depth images of an object at the same time (Xia et al., 2015). In order to segment plant leaves, Seiffert et al. (2010) used a Kinect RGB-D image sensor to take RGB and depth images of the plants, which may be utilized for stress and disease resistance. Vulnerability to visual interference, limited sensing distance, and low resolution are the drawbacks of depth image sensors (Kadambi et al., 2014).

#### 3.3.7.3 Stereo Vision Sensors

Motion-based structural methods for 3D data are used in stereo vision with two or more cameras (Campbell et al., 2018). In situ applications of stereo vision have been effective; for instance, stereo vision was used for wilt detection (Mizuno et al., 2007). It has also proved its effectiveness in field operations for imaging at greater scales, as shown by the use of aerial stereo imagery for estimating plant growth (Rovira-Más et al., 2005). Jin and Tang (2009) proved the application of stereo sensors for sensing plant spacing in maize in the field and greenhouse simultaneously. PhenoStereo, a portable stereo imaging module, was created to measure the stem diameter in sorghum, for assessing the stalk strength and biomass potential in the field using an enhanced 3D modeling procedure (Xiang et al., 2020).

Apart from texture, sunlight has a significant role in determining the functioning of stereo vision. Strong sunlight and the stereo-matching procedure restrict the application's breadth or scale and decrease the effectiveness of stereo vision (Campbell et al., 2018).

#### 3.3.7.4 Time of Flight (ToF) Sensors

Time of Flight cameras typically provide low-resolution depth images using near-infrared emitters. Throughout the last several years, their resolution has been progressively improving. Their ability to generate images at high frame rates and depth makes them ideal for real-time applications. However, there are limited applications of ToF cameras for plant phenotyping compared to stereo vision and LiDAR due to being more expensive than traditional 3D imaging systems (Li et al., 2014).

Song et al. (2011) fused stereo images and ToF images for measuring plant leaf area in a greenhouse. In maize fields, a ToF camera was utilized to assess the inter-plant spacing (Nakarmi & Tang, 2012). Low resolution and sensitivity to sunlight or external light have emerged as the main obstacles to ToF applications (Kazmi et al., 2014). Table 3.4 highlights some applications of high-throughput phenotyping sensors.

**TABLE 3.4**
**Overview of Some Applications of High-Throughput Phenotyping Sensors**

| Type of Sensor | Crops | Applications | Advantages | Limitations | References |
|---|---|---|---|---|---|
| Visible Spectrum Image Sensors | *Arabidopsis*, wheat, maize, potato, rice, barley, grapevine, foxtail millet | Measuring plant biomass, height, leaf surface area, temperature and water content, nitrogen status, detection of diseases, lodging percentage, canopy coverage, vegetation indices | Cheap cost, high efficiency, and easy advertising | Difficult to differentiate between similar backgrounds, limited spectral bands, affected by light | Bendig et al. (2015), Hall et al. (2016), Klodt et al. (2015), Korobiichuk et al. (2018), Mahlein (2016), Sugiura et al. (2016), and Fahlgren et al. (2015a) |
| Infrared Spectrum Image Sensors | Wheat, maize, potato, grapevine, barley, lily, foxtail millet | Estimation of plant water content, response to drought, disease identification | Easy handling, non-destructive | Restriction to certain bands, affected by weather conditions | Briglia et al. (2019), Chen et al. (2015), Fahlgren et al. (2015b), and Griffel et al. (2018) |
| Thermal Infrared Image Sensors | Wheat, maize, rice, cotton, soybean, barley, *Brachypodium distachyon* | Measuring temperature and water status, crop growth rate, prediction of water stress, fruit maturity, identification of plant diseases and pathogens | Non-invasiveness, ease of use, and superior accuracy | Difficult to distinguish small temperature differences, affected by weather and environment | Costa et al. (2019), Crusiol et al. (2020), and Matos et al. (2014) |
| Multispectral Image Sensors | Wheat, canola, maize, cotton, tomato | Crop type identification, growth monitoring and analysis, yield forecast, canopy reflectance, vegetation indices, water stress, plant stress, disease detection | More precise data acquisition, a wide range of spectral bands | Less range of spectral bands compared to hyperspectral sensors, and difficulty in data management | Andrade-Sanchez et al. (2014), Kitić et al. (2019), Zaman-Allah et al. (2015), and Kumar et al. (2020) |
| Hyperspectral Image Sensors | Wheat, tobacco, rice, maize, sugar beet | Detection of plant stress, chlorophyll, water status, disease severity, classifying crop's nutrition level, identification of plant response to drought, cold damage | More precise and thorough plant phenotyping | Complex data processing, expensive | Asaari et al. (2018), Backhaus et al. (2011), Mahlein et al. (2013, 2018), and Yang et al. (2019) |

(*Continued*)

**TABLE 3.4 (*Continued*)**
**Overview of Some Applications of High-Throughput Phenotyping Sensors**

| Type of Sensor | Crops | Applications | Advantages | Limitations | References |
|---|---|---|---|---|---|
| Fluorescence Sensors | *Arabidopsis*, rice, maize, wheat, barley, bean, apple | Quantifying chlorophyll content, detection of photosynthetic activity, reactive oxygen species, plant stress, disease | Less susceptible to environmental influences | View restriction, intense light required, only individual plants visible | Chen et al. (2015), Hairmansis et al. (2014), Li et al. (2014), and Qiu et al. (2018) |
| Light Detection and Ranging (LiDAR) | Maize, wheat, cotton, rye, barley | Estimating canopy height, ground cover, biomass, volume, canopy area, 3D reconstruction of plants and trait estimation | High resolution, low-altitude performance detection, high accuracy | Extreme sensitivity to even minute changes, high cost | Guo et al. (2018), Jimenez-Berni et al. (2018), Jin et al. (2018), Sun et al. (2018), and Walter et al. (2019) |
| Depth Image Sensors | Wheat, sugar beet, tobacco | Object identification, 3D reconstruction of plants, detection of stress and disease | Real time depth information, relatively less expensive | Vulnerability to visual interference, limited sensing distance, low resolution | Kadambi et al. (2014), Seiffert et al. (2010), and Xia et al. (2015) |
| Stereo Vision Sensors | Maize, sorghum, soybean | 3D plant modelling, wilt detection, determination of plant spacing, estimating stem diameter, plant biomass, plant growth | Real time data acquisition, high accuracy | Affected by sunlight, complex data interpretation | Jin and Tang (2009), Mizuno et al. (2007), Rovira-Más et al. (2005), and Xiang et al. (2020) |
| Time of Flight (ToF) Sensors | Maize, pepper | Measuring plant leaf area, assessment of inter plant spacing | Ability to generate images at high frame rates and depth | Expensive, low resolution, sensitivity to sunlight | Kazmi et al. (2014), Nakarmi and Tang (2012), and Song et al. (2011) |

*Source*: Modified from Fan et al. (2021), Jangra et al. (2021), and Ma et al. (2022)

## 3.4 IMAGE PROCESSING AND DATA ANALYSIS

It is difficult to come to a standard image analysis procedure since the formats of imaging data change are greatly based on the type of imaging sensors (Yang et al., 2020). In order to examine the phenotypic traits of plants, images are examined and interpreted using different imaging techniques (Zhang & Zhang, 2018). A general overview of image analysis workflow includes (Figure 3.2): (1) Image preprocessing to reduce noise, increase contrast, cut or combine raw images, and alter space so the image may be easily handled in the next phase; (2) Image segmentation using techniques like thresholding and structural processing to separate the desired items from the background; (3) Feature extraction to obtain the unedited features in accordance with the experiment objectives (4) To filter and investigate the important characteristics, feature preprocessing and trait selection should be used once huge amounts of raw features have been extracted; (5) Data mining, such as using deep learning (e.g., recurrent neural networks [RNNs], convolutional neural networks [CNNs] or deep neural networks [DNNs]) to investigate additional spatial, temporal, and geographical information, or utilizing mechanistic models to enable dynamic growth (Chen et al., 2015). After that, data analysis is used to extract the desired plant attributes from these features. In general, the most important preprocessing steps are image segmentation followed by feature extraction (Zhang & Zhang, 2018).

### 3.4.1 Machine Learning and Deep Learning for Plant Phenotypic Data Analysis

Since remote sensing methods are being widely utilized and sensor data is most often processed and analyzed as images, image analysis has become an integral part of high-throughput plant phenotyping (Araus et al., 2022). With the development of machine learning and deep learning, an entirely new approach to analyzing plant phenotypic data has emerged (Fan et al., 2021).

#### 3.4.1.1 Machine Learning

When it came to analyzing plant imagery using conventional approaches like segmentation, the main problems were the difficulty of extracting features and the difficulty in establishing and properly modeling the link between input and output. Machine learning is based on the principle of

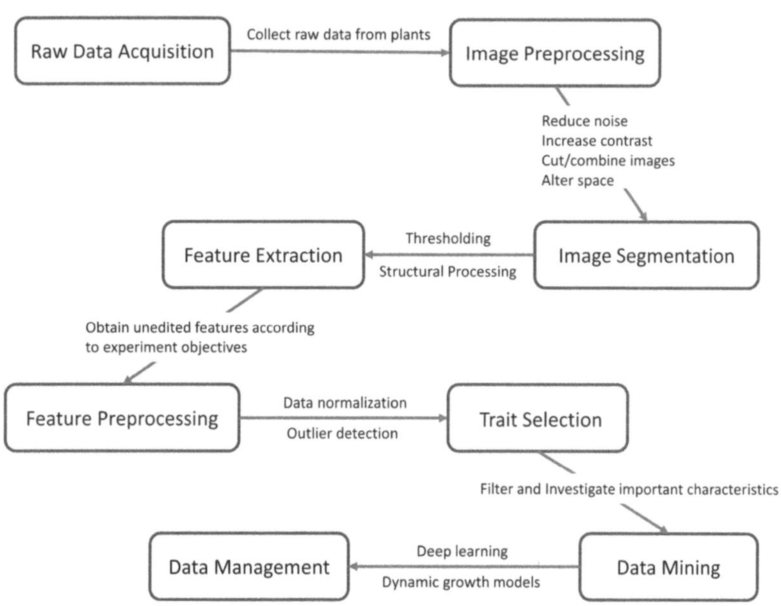


**FIGURE 3.2** A workflow of data analysis for high-throughput phenotyping.

training models using massive amounts of data or experience, and then using those models to either make predictions or judgments (Fan et al., 2021). Machine learning uses a number of statistical and probabilistic methods to examine large datasets on crop phenotyping and environmental factors in order to identify and forecast new trends and patterns (Jangra et al., 2021). One sort of machine learning that is often used to address categorization issues is support vector machine (SVM) (Fan et al., 2021). Machine learning systems were developed for early detection and differentiation of disease symptoms in cereals and vegetables using support vector machine (SVM) along with different imaging systems (Odilbekov et al., 2018; Raza et al., 2015). While machine learning techniques have many benefits, there are still certain constraints when it comes to processing and curating images, choosing appropriate methods, and choosing models (Jangra et al., 2021).

#### 3.4.1.2 Deep Learning

Deep learning has been revolutionizing machine learning in recent years by dominating certain tasks. The primary method of using deep learning for plant phenotyping analysis is collecting data from a range of sensors at the population or individual plant scale, processing it via human-machine interaction, and then applying deep learning technology training (Fan et al., 2021). According to Kamilaris and Prenafeta-Boldú (2018), deep learning has surpassed all other methods in the field of data analysis and image processing. There can be three different models for deep learning: Recurrent Neural Network (RNN), Convolutional Neural Network (CNN), and Neuronal Neural Network (NNN) (Fan et al., 2021). Convolutional neural networks (CNNs) have been shown to be the most effective deep learning architecture when it comes to image identification, segmentation, and classification (Gao et al., 2020).

Deep learning is currently the subject of several reports that aim to improve plant phenotyping using images captured from various scales. This information is primarily utilized in deep learning applications, for the study of leaf morphological traits (Dobrescu et al., 2020), in addition to counting plant vegetative and reproductive organs (Afonso et al., 2020; Zhou et al., 2019). Moreover, deep learning was used to assess the flowering stage in wheat (Wang et al., 2019), detect water and nutrient use efficiency (Takahashi & Pradal, 2021), estimate yield in maize (Khaki et al., 2021), or detect biotic and abiotic stresses (Gao et al., 2020). These methods have the potential to greatly benefit plant breeding and decision-making by providing accurate, efficient, and timely predictions of plant phenotypic data across scales and regions. They can also shed light on the evolution and regional differentiation of plant phenotypic traits (Fan et al., 2021).

## 3.5 UNLOCKING THE POTENTIAL OF BIG DATA IN HIGH-THROUGHPUT FIELD PHENOTYPING

The rapid advancement of remote sensing technology, particularly the use of novel sensors with enhanced resolutions, has resulted in a significant surge in the quantity and complexity of remote sensing data. As a research paradigm, data-intensive knowledge discovery is replacing theoretical, experimental, and computational approaches in agricultural science in this age of big data (Fan et al., 2021). High-throughput phenotyping now depends on the collection, handling, storing, analyzing, and visualizing of large amounts of data from agricultural remote sensing (Huang et al., 2018b). Big data may be defined as "large, diverse, complex, longitudinal, and/or distributed data sets generated from click streams, email, instruments, Internet transactions, satellites, sensors, video, and/or all other digital sources available today and in the future" (Coble et al., 2016). Data science and big data technologies are increasingly being included in high-throughput phenotyping techniques for the purpose of rapid data analysis and decision-making (Wolfert et al., 2017).

An American agricultural big data company is currently utilizing the findings of Stanford University's Sustainability and Artificial Intelligence Laboratory to accurately estimate soybean yields using deep learning and remote sensing data (Busemeyer et al., 2013). To monitor changes in yield, growth, acreage, drought, and other agro information for China's most important crops, the China Agriculture Remote Sensing Monitoring System (CHARMS) was first created by the Chinese Ministry of Agriculture (MOA). It consisted of six modules and a database system (Chen et al., 2011). Rathore et al. (2015) introduced a framework for real-time big data analytics used in satellite remote sensing applications. Researchers at the Institute of Remote Sensing and Digital Earth (CAS) of the Chinese Academy of Sciences developed a unique system for managing remote sensing data called FLFL. With 15 levels and five layers, it allows robust support for the swift creation, storing, retrieving, and provision of large amounts of remote sensing data (Huang et al., 2018b).

One of the biggest drawbacks of big data is the absence of standardization and data sharing across various groups. The lack of data infrastructure and resources to handle large datasets is another barrier (Bolger et al., 2019). Fan et al. (2014) noted that unforeseen endogeneity, inaccurate correlations, noise buildup, and heterogeneity are characteristics of big data. To overcome these challenges, integrated resources, collaborative efforts, and bioinformatics technology should contribute to functional genomics and breeding initiatives in crops (Yang et al., 2020).

## 3.6 ENHANCING PLANT PHENOTYPING WITH IOT (INTERNET OF THINGS)

With the use of a wide range of remote sensing technologies and sensor devices, agricultural IoT collects data on plant ontology, agricultural production, and product circulation through mobile communication wireless networks, wireless sensor networks, and the Internet. A variety of high-throughput technologies, including integrated image sensors, automated industrial equipment, and modern imaging methodologies, have made it possible to acquire phenotypic traits of plants in various growth environments (Fan et al., 2021). In addition to being essential for agricultural informatization and intelligence, IoT is a crucial technological support for modern farming that prioritizes efficiency, accuracy, and environmental preservation (Biswas et al., 2018). Agricultural IoT technology has made tremendous strides in non-destructive imaging, spectral analysis, environmental sensing, and robotics as a result of continuous collaborative research (Fan et al., 2021).

In the realm of agriculture, the Internet of Things has enormous potential. One example of IoT based field phenotyping tool is CropQuant, which combines open-source software with affordable hardware to provide sophisticated analytical answers for measuring several growth characteristics in wheat (Zhou et al., 2017). Another IoT-based model AgriPrediction, integrates a wireless network system that operates over short and medium distances, with a prediction engine and a warning system to foresee possible agricultural dysfunctions and provide economically feasible solutions to farmers (Dos Santos et al., 2019). Khattab et al. (2019) developed a monitoring system for high-throughput phenotyping and disease forecasting that is IoT based and employs Wireless Sensor Network technology. Conversely, a field study was conducted in Sao Paulo, Brazil to evaluate a proposed real-time, in situ agricultural Internet of Things device that can track soil and environmental conditions (Maia et al., 2017).

## 3.7 ADVANCES IN PORTABLE PHENOTYPING: TRANSFORMING AGRICULTURE WITH SMARTPHONES

A potentially game-changing innovation that could hasten the process of phenotyping is next-generation portable phenotyping instruments. Applications with growing usefulness have been made possible by the fast advances of mobile devices with high-resolution RGB cameras and sophisticated processing. Traditional remote sensing equipment has also seen improvements in connection and mobility due to the incorporation of smartphone technology into other customized portable devices, which has expanded the variety of optical and other sensors (Yang et al., 2020).

**TABLE 3.5**
**Portable Phenotyping Platforms for In-Field Crop Phenotyping**

| Name of the Platform | Crop | Applications | References |
|---|---|---|---|
| PocketPlant3D | Maize | Assess the structure of maize canopies | Confalonieri et al. (2017) |
| VitisBerry | Grapevine | Measure the quantity of grapevine berries | Aquino et al. (2018) |
| Smartphone app | Sugar beet | Identify and differentiate between a number of leaf diseases | Hallau et al. (2018) |
| PhenoApp | Cereals, grapevine, maize, potato, rapeseed and rice | Digitally collect phenotypic data | Röckel et al. (2022) |

PocketPlant3D is a smartphone-based program that can evaluate the structure of maize canopies (Confalonieri et al., 2017); scanning all the leaves to obtain the whole leaf architecture. It is possible to measure the quantity of grapevine berries using the Android app vitisBerry (Aquino et al., 2018). A smartphone app was introduced by (Hallau et al., 2018) that could identify and differentiate between a number of leaf diseases in sugar beets, such as *Cercospora* leaf spot, beetroot rust, and bacterial blight. PhenoApp is a free and open-source software application for android devices that allows users to digitally collect phenotypic data in both fields and greenhouses (Röckel et al., 2022).

Although studies conducted in this sector are still limited, integrating these smartphone applications with other emerging technologies such as artificial intelligence, big data, IoT, etc. could change the scenario of current high-throughput phenotyping techniques. By allowing users to capture and analyze data directly from their smartphones, it offers a cost-effective and accessible way to gather valuable information about crops, soil, and environmental conditions. Smartphone applications are ubiquitous and easy to use, making this portable phenotyping technology available to a diverse array of users, including small-scale farmers and scientists in remote locations. However, it's crucial to address challenges such as data privacy, interoperability, and ensuring the technology is accessible to all stakeholders. Some novel smartphone-based phenotyping platforms have been summarized in Table 3.5.

## 3.8 CHALLENGES AND FUTURE PROSPECTS

### 3.8.1 Challenges

High-throughput crop phenotyping in controlled environments, as well as fields, is now possible thanks to the quick development of phenotyping systems, sensors, images, and data analysis for a wide range of scales (Song et al., 2021). There is no doubt that there are still numerous issues that need to be resolved in order to estimate most crop phenotypic features more accurately, even with the advancements in remote sensing technologies and platforms.

#### 3.8.1.1 Bottleneck of Root Phenotyping

Root phenotyping is evolving quickly to better understand important performance characteristics in controlled environments. However, non-invasive root phenotyping on site is currently unattainable, and much progress is required to meet this crucial goal. Although above-ground field-based high-throughput phenotyping has made considerable progress, below-ground access is still scarce; which needs to be resolved. Regarding this situation, further research is needed on conducting in-field methods of seed and root phenotyping.

#### 3.8.1.2 Restrictions of Phenotyping Platforms

High-resolution monitoring may be accomplished with stationary phenotyping devices, however the procedure is costly and limited in scope. Although movable phenotyping platforms are the best option for ground-based phenotyping, it might be difficult to get stable sensor data when moving, and there isn't a single movable phenotyping platform that is suitable for growing crops in all environments, especially in paddy fields. Unmanned Aerial Vehicles (UAV) provide flexibility, ease of use, minimal hardware costs, and superior efficiency when compared to ground-based platforms. However, their flight duration is limited, and they can only gather crop canopy data at a restricted resolution. Although portable equipment has demonstrated remarkable mobility, their use in high-throughput phenotyping is limited since they can only gather data at fixed sites (Song et al., 2021).

#### 3.8.1.3 Complexity of Image Processing

The difficulty of image processing for hyperspectral and LiDAR data makes it essential to create software that can analyze images quickly and accurately in order to increase data processing accuracy and efficiency. For data processing and analysis to be accurate and completed quickly, data processing software efficiency needs to be increased (Yang et al., 2017b).

#### 3.8.1.4 Cost Ineffectiveness and Lack of Utilization

Despite the immense potential of remote sensing and high-throughput phenotyping, very few have practically used these strategies or demonstrated any outcomes in breeding crops (Deery & Jones, 2021). For many research groups, the cost of purchasing, running, and maintaining data processing infrastructure and remote sensing equipment might be exorbitant, which limits the use of these technologies for phenotyping applications in environments with limited resources.

### 3.8.2 Future Prospects

However, recent advancements, such as the creation of large-scale networks that provide access to modern techniques and improved technology, the widespread industry participation in the dissemination of new technologies, the field-based phenotyping of root traits, and concepts of community-driven integrated data management demonstrate how future needs could be satisfied (Watt et al., 2020)

#### 3.8.2.1 Integrating Multiple Phenotyping Platforms

In the future, flexible crop phenotyping is probably possible if the benefits of both ground and aerial platforms are combined. The benefits of both the efficiency and accuracy of UAV and ground-based phenotyping platform detection may be realized in the creation of intelligent phenotyping robots. Flexible mobility over various plots is possible with phenotyping robots that have power supply systems and autonomous navigation. Phenotyping robots are capable of concurrently gathering information about the crop growth environment and high-resolution data related to the plant canopy.

#### 3.8.2.2 Integration of Multiple Sensors and Sensor Improvements

The limitations of obtaining phenotypic data from a single sensor, coupled with the field environment's tendency to alter data quality, make it challenging to satisfy the objectives of crop breeding in the modern era. However, data combined from multiple sources can provide an explanation for novel biological occurrences. For example, a range of phenotypic data is often combined for efficient and precise analysis in the crop abiotic and biotic stress resistance screening procedure (Fan et al., 2021). The challenge for the future is the creation of innovative sensors for the detection of plant signal molecules, as well as the simultaneous capture of numerous agricultural attributes through the integration of multiple sensors.

#### 3.8.2.3 Popularization and Application of Data Science and Novel Algorithms

There is an urgent need to accelerate the implementation of data science and other relevant technology in agriculture, particularly in breeding crops. Since deep learning often requires massive volumes of data to train models, the lack of data set sharing in this field makes data collection a tedious and difficult process. The growing application of all imaging technologies for plant phenotyping has made "big data" analytics a crucial problem requiring immediate attention. A wide variety of standardization objectives, from the level of producing high-quality data to their integration in vast, varied multi-omics data landscapes, are the focus of ongoing community efforts in this direction (Watt et al., 2020). Combining data science and remote sensing can bring new technical possibilities for improving throughput and accuracy in all facets of non-invasive plant phenotyping. There are now more options available compared to complex and costly phenotyping techniques due to the growing availability of reliable and inexpensive approaches linked to mobile computing devices, smartphones, and open computer architectures. Farmers and researchers will be able to conduct timely interventions, such as fertilization, precision irrigation, and pest control, based on insights obtained from remote sensing data thanks to real-time monitoring capabilities combined with comprehensive decision support systems.

#### 3.8.2.4 Multidisciplinary Collaboration

Last but not least, effective plant phenotyping requires combining the knowledge of a diverse group of engineers, physicists, mathematicians, and plant biologists. Researchers need to have access to platforms of varying scales. It is necessary to incorporate experiences from fields and growing facilities. In order to keep up with the rapid advancement of genetic research, efficient and regular communication of research findings will aid in the rapid removal of the phenotyping bottleneck (Zhang & Zhang, 2018). Accelerating scientific discoveries and technical advancements, open data initiatives, and collaborative platforms will stimulate data sharing, interoperability, and reproducibility throughout the remote sensing and phenotyping sectors.

## 3.9 CONCLUSION

Remote sensing technologies have significantly transformed field-based high-throughput crop phenotyping by equipping researchers with powerful tools to enable rapid, accurate, and non-destructive screening of plants. Recent advancements in data analysis methods, robots, and sensors further enhanced efficiency in remote sensing technologies and high-throughput phenotyping. These technologies can not only improve the understanding of genotype-phenotype relationships but also promote sustainable agricultural practices such as efficient crop monitoring, accurate yield estimation, and early disease detection. The continuous advancements in remote sensing hold great promise for further enhancing crop phenotyping processes and supporting global food security initiatives. In conclusion, embracing remote sensing technologies will be crucial for advancing agricultural practices, improving crop productivity, and ensuring food security in the midst of changing environmental conditions.

## REFERENCES

Adamides, G., Katsanos, C., Parmet, Y., Christou, G., Xenos, M., Hadzilacos, T., & Edan, Y. (2017). HRI usability evaluation of interaction modes for a teleoperated agricultural robotic sprayer. *Applied Ergonomics*, *62*, 237–246. https://doi.org/10.1016/j.apergo.2017.03.008

Afonso, M., Fonteijn, H., Fiorentin, F. S., Lensink, D., Mooij, M., Faber, N., Polder, G., & Wehrens, R. (2020). Tomato fruit detection and counting in greenhouses using deep learning. *Frontiers in Plant Science*, *11*, 571299. https://doi.org/10.3389/fpls.2020.571299

Andrade-Sanchez, P., Gore, M. A., Heun, J. T., Thorp, K. R., Carmo-Silva, A. E., French, A. N., Salvucci, M. E., & White, J. W. (2014). Development and evaluation of a field-based high-throughput phenotyping platform. *Functional Plant Biology*, *41*(1), 68. https://doi.org/10.1071/FP13126

Aquino, A., Barrio, I., Diago, M. P., Millan, B., & Tardaguila, J. (2018). vitisBerry: An Android-smartphone application to early evaluate the number of grapevine berries by means of image analysis. *Computers and Electronics in Agriculture*, *148*, 19–28. https://doi.org/10.1016/j.compag.2018.02.021

Arad, B., Balendonck, J., Barth, R., Ben-Shahar, O., Edan, Y., Hellström, T., Hemming, J., Kurtser, P., Ringdahl, O., Tielen, T., & Van Tuijl, B. (2020). Development of a sweet pepper harvesting robot. *Journal of Field Robotics*, *37*(6), 1027–1039. https://doi.org/10.1002/rob.21937

Araus, J. L., & Cairns, J. E. (2014). Field high-throughput phenotyping: The new crop breeding frontier. *Trends in Plant Science*, *19*(1), 52–61. https://doi.org/10.1016/j.tplants.2013.09.008

Araus, J. L., Kefauver, S. C., Vergara-Díaz, O., Gracia-Romero, A., Rezzouk, F. Z., Segarra, J., Buchaillot, M. L., Chang-Espino, M., Vatter, T., Sanchez-Bragado, R., Fernandez-Gallego, J. A., Serret, M. D., & Bort, J. (2022). Crop phenotyping in a context of global change: What to measure and how to do it. *Journal of Integrative Plant Biology*, *64*(2), 592–618. https://doi.org/10.1111/jipb.13191

Asaari, M. S. M., Mertens, S., Dhondt, S., Wuyts, N., & Scheunders, P. (2018). Detection of plant responses to drought using close-range hyperspectral imaging in a high-throughput phenotyping platform. In *2018 9th Workshop on Hyperspectral Image and Signal Processing: Evolution in Remote Sensing (WHISPERS)*, pp. 1–5. https://doi.org/10.1109/WHISPERS.2018.8747228

Atefi, A., Ge, Y., Pitla, S., & Schnable, J. (2021). Robotic technologies for high-throughput plant phenotyping: Contemporary reviews and future perspectives. *Frontiers in Plant Science*, *12*, 611940. https://doi.org/10.3389/fpls.2021.611940

Atkinson, J. A., Pound, M. P., Bennett, M. J., & Wells, D. M. (2019). Uncovering the hidden half of plants using new advances in root phenotyping. *Current Opinion in Biotechnology*, *55*, 1–8. https://doi.org/10.1016/j.copbio.2018.06.002

Backhaus, A., Bollenbeck, F., & Seiffert, U. (2011). Robust classification of the nutrition state in crop plants by hyperspectral imaging and artificial neural networks. In *2011 3rd Workshop on Hyperspectral Image and Signal Processing: Evolution in Remote Sensing (WHISPERS)*, pp. 1–4. https://doi.org/10.1109/WHISPERS.2011.6080898

Bai, G., Ge, Y., Scoby, D., Leavitt, B., Stoerger, V., Kirchgessner, N., Irmak, S., Graef, G., Schnable, J., & Awada, T. (2019). NU-Spidercam: A large-scale, cable-driven, integrated sensing and robotic system for advanced phenotyping, remote sensing, and agronomic research. *Computers and Electronics in Agriculture*, *160*, 71–81. https://doi.org/10.1016/j.compag.2019.03.009

Baret, F., De Solan, B., Thomas, S., Burger, P., Liu, S., Comar, A., Rousset, C., Vanhove, R., Regnier, R., Terreni, J., & Combe, L. (2022). Phenomobile: A fully automatic robot for high-throughput field phenotyping of a large range of crops with active measurements. *IAMPS - Image Analysis Methods in the Plant Sciences*, Lyon, France. https://hal.inrae.fr/hal-03646863v1.

Barmeier, G., & Schmidhalter, U. (2016). High-throughput phenotyping of wheat and barley plants grown in single or few rows in small plots using active and passive spectral proximal sensing. *Sensors*, *16*(11), 1860. https://doi.org/10.3390/s16111860

Beauchêne, K., Leroy, F., Fournier, A., Huet, C., Bonnefoy, M., Lorgeou, J., De Solan, B., Piquemal, B., Thomas, S., & Cohan, J. P. (2019). Management and characterization of abiotic stress via PhénoField®, a high-throughput field phenotyping platform. *Frontiers in Plant Science*, *10*, 904. https://doi.org/10.3389/fpls.2019.00904

Bendig, J., Yu, K., Aasen, H., Bolten, A., Bennertz, S., Broscheit, J., Gnyp, M. L., & Bareth, G. (2015). Combining UAV-based plant height from crop surface models, visible, and near infrared vegetation indices for biomass monitoring in barley. *International Journal of Applied Earth Observation and Geoinformation*, *39*, 79–87. https://doi.org/10.1016/j.jag.2015.02.012

Berni, J. A. J., Zarco-Tejada, P. J., González-Dugo, V., & Fereres, E. (2012). Remote sensing of thermal water stress indicators in peach. *Acta Horticulturae, 962*, 325–331. https://doi.org/10.17660/ActaHortic.2012.962.45

Biswas, S. K., Devi, D., & Chakraborty, M. (2018). A Hybrid Case Based Reasoning Model for Classification in Internet of Things (IoT). *Environment: Journal of Organizational and End User Computing*, *30*(4), 104–122. https://doi.org/10.4018/JOEUC.2018100107

Bolger, A. M., Poorter, H., Dumschott, K., Bolger, M. E., Arend, D., Osorio, S., Gundlach, H., Mayer, K. F. X., Lange, M., Scholz, U., & Usadel, B. (2019). Computational aspects underlying genome to phenome analysis in plants. *The Plant Journal*, *97*(1), 182–198. https://doi.org/10.1111/tpj.14179

Briglia, N., Montanaro, G., Petrozza, A., Summerer, S., Cellini, F., & Nuzzo, V. (2019). Drought phenotyping in Vitis vinifera using RGB and NIR imaging. *Scientia Horticulturae*, *256*, 108555. https://doi.org/10.1016/j.scienta.2019.108555

Busemeyer, L., Mentrup, D., Möller, K., Wunder, E., Alheit, K., Hahn, V., Maurer, H., Reif, J., Würschum, T., Müller, J., Rahe, F., & Ruckelshausen, A. (2013). BreedVision—A multi-sensor platform for non-destructive field-based phenotyping in plant breeding. *Sensors*, *13*(3), 2830–2847. https://doi.org/10.3390/s130302830

Campbell, Z. C., Acosta-Gamboa, L. M., Nepal, N., & Lorence, A. (2018). Engineering plants for tomorrow: How high-throughput phenotyping is contributing to the development of better crops. *Phytochemistry Reviews*, *17*(6), 1329–1343. https://doi.org/10.1007/s11101-018-9585-x

Chapman, S., Merz, T., Chan, A., Jackway, P., Hrabar, S., Dreccer, M., Holland, E., Zheng, B., Ling, T., & Jimenez-Berni, J. (2014). Pheno-Copter: A low-altitude, autonomous remote-sensing robotic helicopter for high-throughput field-based phenotyping. *Agronomy*, *4*(2), 279–301. https://doi.org/10.3390/agronomy4020279

Chawade, A., Van Ham, J., Blomquist, H., Bagge, O., Alexandersson, E., & Ortiz, R. (2019). High-throughput field-phenotyping tools for plant breeding and precision agriculture. *Agronomy*, *9*(5), 258. https://doi.org/10.3390/agronomy9050258

Chebrolu, N., Lottes, P., Schaefer, A., Winterhalter, W., Burgard, W., & Stachniss, C. (2017). Agricultural robot dataset for plant classification, localization and mapping on sugar beet fields. *The International Journal of Robotics Research*, *36*(10), 1045–1052. https://doi.org/10.1177/0278364917720510

Chen, D., Neumann, K., Friedel, S., Kilian, B., Chen, M., Altmann, T., & Klukas, C. (2015). dissecting the phenotypic components of crop plant growth and drought responses based on high-throughput image analysis. *The Plant Cell*, *26*(12), 4636–4655. https://doi.org/10.1105/tpc.114.129601

Chen, Z., Zhou, Q., Liu, J., Wang, L., Ren, J., Huang, Q., Deng, H., Zhang, L., & Li, D. (2011). Charms—China agricultural remote sensing monitoring system. In *2011 IEEE International Geoscience and Remote Sensing Symposium*, pp. 3530–3533. https://doi.org/10.1109/IGARSS.2011.6049983

Chopin, J., Laga, H., Huang, C. Y., Heuer, S., & Miklavcic, S. J. (2015). RootAnalyzer: A Cross-Section Image Analysis Tool for Automated Characterization of Root Cells and Tissues. *PLOS ONE*, *10*(9), e0137655. https://doi.org/10.1371/journal.pone.0137655

Clark, R. T., Famoso, A. N., Zhao, K., Shaff, J. E., Craft, E. J., Bustamante, C. D., Mccouch, S. R., Aneshansley, D. J., & Kochian, L. V. (2013). High-throughput two-dimensional root system phenotyping platform facilitates genetic analysis of root growth and development. *Plant, Cell & Environment*, *36*(2), 454–466. https://doi.org/10.1111/j.1365-3040.2012.02587.x

Coble, K., Griffin, T., Ahearn, M., Ferrell, S., McFadden, J., Sonka, S., Fulton, J., Coble, K., Griffin, T., Ahearn, M., Ferrell, S., McFadden, J., Sonka, S., & Fulton, J. (2016). *Advancing U.S. Agricultural Competitiveness with Big Data and Agricultural Economic Market Information, Analysis, and Research.* https://doi.org/10.22004/AG.ECON.249847

Colmer, J., O'Neill, C. M., Wells, R., Bostrom, A., Reynolds, D., Websdale, D., Shiralagi, G., Lu, W., Lou, Q., Le Cornu, T., Ball, J., Renema, J., Flores Andaluz, G., Benjamins, R., Penfield, S., & Zhou, J. (2020). SeedGerm: A cost-effective phenotyping platform for automated seed imaging and machine-learning based phenotypic analysis of crop seed germination. *New Phytologist*, *228*(2), 778–793. https://doi.org/10.1111/nph.16736

Confalonieri, R., Paleari, L., Foi, M., Movedi, E., Vesely, F. M., Thoelke, W., Agape, C., Borlini, G., Ferri, I., Massara, F., Motta, R., Ravasi, R. A., Tartarini, S., Zoppolato, C., Baia, L. M., Brumana, A., Colombo, D., Curatolo, A., Fauda, V., … Rossini, L. (2017). PocketPlant3D: Analysing canopy structure using a smartphone. *Biosystems Engineering*, *164*, 1–12. https://doi.org/10.1016/j.biosystemseng.2017.09.014

Costa, C., Schurr, U., Loreto, F., Menesatti, P., & Carpentier, S. (2019). Plant phenotyping research trends, a science mapping approach. *Frontiers in Plant Science*, *9*, 1933. https://doi.org/10.3389/fpls.2018.01933

Crain, J. L., Wei, Y., Barker, J., Thompson, S. M., Alderman, P. D., Reynolds, M., Zhang, N., & Poland, J. (2016). Development and deployment of a portable field phenotyping platform. *Crop Science*, *56*(3), 965–975. https://doi.org/10.2135/cropsci2015.05.0290

Deery, D., Jimenez-Berni, J., Jones, H., Sirault, X., & Furbank, R. (2014). Proximal remote sensing buggies and potential applications for field-based phenotyping. *Agronomy*, *4*(3), 349–379. https://doi.org/10.3390/agronomy4030349

Deery, D. M., & Jones, H. G. (2021). Field phenomics: Will it enable crop improvement? *Plant Phenomics*, *2021*, 2021/9871989. https://doi.org/10.34133/2021/9871989

DeSouza, C. H. W., Lamparelli, R. A. C., Rocha, J. V., & Magalhães, P. S. G. (2017). Height estimation of sugarcane using an unmanned aerial system (UAS) based on structure from motion (SfM) point clouds. *International Journal of Remote Sensing*, *38*(8–10), 2218–2230. https://doi.org/10.1080/01431161.2017.1285082

Dobrescu, A., Giuffrida, M. V., & Tsaftaris, S. A. (2020). Doing more with less: A multitask deep learning approach in plant phenotyping. *Frontiers in Plant Science*, *11*, 141. https://doi.org/10.3389/fpls.2020.00141

DosSantos, U. J. L., Pessin, G., Da Costa, C. A., & Da Rosa Righi, R. (2019). AgriPrediction: A proactive internet of things model to anticipate problems and improve production in agricultural crops. *Computers and Electronics in Agriculture*, *161*, 202–213. https://doi.org/10.1016/j.compag.2018.10.010

Edwards, D., Batley, J., & Snowdon, R. J. (2013). Accessing complex crop genomes with next-generation sequencing. *Theoretical and Applied Genetics*, *126*(1), 1–11. https://doi.org/10.1007/s00122-012-1964-x

Enciso, J., Maeda, M., Landivar, J., Jung, J., & Chang, A. (2017). A ground based platform for high throughput phenotyping. *Computers and Electronics in Agriculture*, *141*, 286–291. https://doi.org/10.1016/j.compag.2017.08.006

Fahlgren, N., Feldman, M., Gehan, M. A., Wilson, M. S., Shyu, C., Bryant, D. W., Hill, S. T., McEntee, C. J., Warnasooriya, S. N., Kumar, I., Ficor, T., Turnipseed, S., Gilbert, K. B., Brutnell, T. P., Carrington, J. C., Mockler, T. C., & Baxter, I. (2015b). A versatile phenotyping system and analytics platform reveals diverse temporal responses to water availability in Setaria. *Molecular Plant*, *8*(10), 1520–1535. https://doi.org/10.1016/j.molp.2015.06.005

Fahlgren, N., Gehan, M. A., & Baxter, I. (2015a). Lights, camera, action: High-throughput plant phenotyping is ready for a close-up. *Current Opinion in Plant Biology*, *24*, 93–99. https://doi.org/10.1016/j.pbi.2015.02.006

Fan, J., Han, F., & Liu, H. (2014). Challenges of big data analysis. *National Science Review*, *1*(2), 293–314. https://doi.org/10.1093/nsr/nwt032

Fan, J., Zhang, Y., Wen, W., Gu, S., Lu, X., & Guo, X. (2021). The future of Internet of Things in agriculture: Plant high-throughput phenotypic platform. *Journal of Cleaner Production*, *280*, 123651. https://doi.org/10.1016/j.jclepro.2020.123651

Fichman, Y., Miller, G., & Mittler, R. (2019). Whole-plant live imaging of reactive oxygen species. *Molecular Plant*, *12*(9), 1203–1210. https://doi.org/10.1016/j.molp.2019.06.003

Fiorani, F., & Schurr, U. (2013). Future scenarios for plant phenotyping. *Annual Review of Plant Biology*, *64*(1), 267–291. https://doi.org/10.1146/annurev-arplant-050312-120137

Flood, P. J., Kruijer, W., Schnabel, S. K., Van Der Schoor, R., Jalink, H., Snel, J. F. H., Harbinson, J., & Aarts, M. G. M. (2016). Phenomics for photosynthesis, growth and reflectance in Arabidopsis thaliana reveals circadian and long-term fluctuations in heritability. *Plant Methods*, *12*(1), 14. https://doi.org/10.1186/s13007-016-0113-y

French, A., Ubeda-Tomás, S., Holman, T. J., Bennett, M. J., & Pridmore, T. (2009). High-throughput quantification of root growth using a novel image-analysis tool. *Plant Physiology*, *150*(4), 1784–1795. https://doi.org/10.1104/pp.109.140558

Gao, T., Emadi, H., Saha, H., Zhang, J., Lofquist, A., Singh, A., Ganapathysubramanian, B., Sarkar, S., Singh, A., & Bhattacharya, S. (2018). A novel multirobot system for plant phenotyping. *Robotics*, *7*(4), 61. https://doi.org/10.3390/robotics7040061

Gao, Z., Luo, Z., Zhang, W., Lv, Z., & Xu, Y. (2020). Deep learning application in plant stress imaging: A review. *AgriEngineering*, *2*(3), 430–446. https://doi.org/10.3390/agriengineering2030029

Goujon, A. (2019). Human population growth. In D. Pimentel & M. Pimentel (Eds.), *Encyclopedia of Ecology* (pp. 344–351). Elsevier. https://doi.org/10.1016/B978-0-12-409548-9.10755-9

Granier, C., Aguirrezabal, L., Chenu, K., Cookson, S. J., Dauzat, M., Hamard, P., Thioux, J., Rolland, G., Bouchier-Combaud, S., Lebaudy, A., Muller, B., Simonneau, T., & Tardieu, F. (2006). PHENOPSIS, an automated platform for reproducible phenotyping of plant responses to soil water deficit in *Arabidopsis thaliana* permitted the identification of an accession with low sensitivity to soil water deficit. *New Phytologist*, *169*(3), 623–635. https://doi.org/10.1111/j.1469-8137.2005.01609.x

Griffel, L. M., Delparte, D., & Edwards, J. (2018). Using Support Vector Machines classification to differentiate spectral signatures of potato plants infected with Potato Virus Y. *Computers and Electronics in Agriculture*, *153*, 318–324. https://doi.org/10.1016/j.compag.2018.08.027

Grimstad, L., & From, P. (2017). The Thorvald II agricultural robotic system. *Robotics*, *6*(4), 24. https://doi.org/10.3390/robotics6040024

Guo, Q., Wu, F., Pang, S., Zhao, X., Chen, L., Liu, J., Xue, B., Xu, G., Li, L., Jing, H., & Chu, C. (2018). Crop 3D—A LiDAR based platform for 3D high-throughput crop phenotyping. *Science China Life Sciences*, *61*(3), 328–339. https://doi.org/10.1007/s11427-017-9056-0

Hairmansis, A., Berger, B., Tester, M., & Roy, S. J. (2014). Image-based phenotyping for non-destructive screening of different salinity tolerance traits in rice. *Rice*, *7*(1), 16. https://doi.org/10.1186/s12284-014-0016-3

Halcro, K., McNabb, K., Lockinger, A., Socquet-Juglard, D., Bett, K. E., & Noble, S. D. (2020). The BELT and phenoSEED platforms: Shape and colour phenotyping of seed samples. *Plant Methods*, *16*(1), 49. https://doi.org/10.1186/s13007-020-00591-8

Hall, H. C., Fakhrzadeh, A., Luengo Hendriks, C. L., & Fischer, U. (2016). Precision automation of cell type classification and sub-cellular fluorescence quantification from laser scanning confocal images. *Frontiers in Plant Science*, *7*. https://doi.org/10.3389/fpls.2016.00119

Hallau, L., Neumann, M., Klatt, B., Kleinhenz, B., Klein, T., Kuhn, C., Röhrig, M., Bauckhage, C., Kersting, K., Mahlein, A. - K., Steiner, U., & Oerke, E. - C. (2018). Automated identification of sugar beet diseases using smartphones. *Plant Pathology*, *67*(2), 399–410. https://doi.org/10.1111/ppa.12741

Holman, F., Riche, A., Michalski, A., Castle, M., Wooster, M., & Hawkesford, M. (2016). High throughput field phenotyping of wheat plant height and growth rate in field plot trials using UAV based remote sensing. *Remote Sensing*, *8*(12), 1031. https://doi.org/10.3390/rs8121031

Huang, H., Deng, J., Lan, Y., Yang, A., Deng, X., & Zhang, L. (2018a). A fully convolutional network for weed mapping of unmanned aerial vehicle (UAV) imagery. *PLoS One*, *13*(4), e0196302. https://doi.org/10.1371/journal.pone.0196302

Huang, Y., Chen, Z., Yu, T., Huang, X., & Gu, X. (2018b). Agricultural remote sensing big data: Management and applications. *Journal of Integrative Agriculture*, *17*(9), 1915–1931. https://doi.org/10.1016/S2095-3119(17)61859-8

Iyer-Pascuzzi, A. S., Symonova, O., Mileyko, Y., Hao, Y., Belcher, H., Harer, J., Weitz, J. S., & Benfey, P. N. (2010). Imaging and analysis platform for automatic phenotyping and trait ranking of plant root systems. *Plant Physiology*, *152*(3), 1148–1157. https://doi.org/10.1104/pp.109.150748

Jahnke, S., Roussel, J., Hombach, T., Kochs, J., Fischbach, A., Huber, G., & Scharr, H. (2016). *pheno* seeder—A robot system for automated handling and phenotyping of individual seeds. *Plant Physiology*, *172*(3), 1358–1370. https://doi.org/10.1104/pp.16.01122

Jangra, S., Chaudhary, V., Yadav, R. C., & Yadav, N. R. (2021). High-throughput phenotyping: A platform to accelerate crop improvement. *Phenomics*, *1*(2), 31–53. https://doi.org/10.1007/s43657-020-00007-6

Jansen, M., Gilmer, F., Biskup, B., Nagel, K. A., Rascher, U., Fischbach, A., Briem, S., Dreissen, G., Tittmann, S., Braun, S., De Jaeger, I., Metzlaff, M., Schurr, U., Scharr, H., & Walter, A. (2009). Simultaneous phenotyping of leaf growth and chlorophyll fluorescence via GROWSCREEN FLUORO allows detection of stress tolerance in Arabidopsis thaliana and other rosette plants. *Functional Plant Biology*, *36*(11), 902. https://doi.org/10.1071/FP09095

Jeudy, C., Adrian, M., Baussard, C., Bernard, C., Bernaud, E., Bourion, V., Busset, H., Cabrera-Bosquet, L., Cointault, F., Han, S., Lamboeuf, M., Moreau, D., Pivato, B., Prudent, M., Trouvelot, S., Truong, H. N., Vernoud, V., Voisin, A.-S., Wipf, D., & Salon, C. (2016). RhizoTubes as a new tool for high throughput imaging of plant root development and architecture: Test, comparison with pot grown plants and validation. *Plant Methods*, *12*(1), 31. https://doi.org/10.1186/s13007-016-0131-9

Jiang, Y., Li, C., Paterson, A. H., Sun, S., Xu, R., & Robertson, J. (2018). Quantitative analysis of cotton canopy size in field conditions using a consumer-grade RGB-D camera. *Frontiers in Plant Science*, *8*, 2233. https://doi.org/10.3389/fpls.2017.02233

Jimenez -Berni, J. A., Deery, D. M., Rozas-Larraondo, P., Condon, A. (Tony) G., Rebetzke, G. J., James, R. A., Bovill, W. D., Furbank, R. T., & Sirault, X. R. R. (2018). High throughput determination of plant height, ground cover, and above-ground biomass in wheat with LiDAR. *Frontiers in Plant Science*, *9*, 237. https://doi.org/10.3389/fpls.2018.00237

Jin, J., & Tang, L. (2009). Corn plant sensing using real-time stereo vision. *Journal of Field Robotics*, *26*(6–7), 591–608. https://doi.org/10.1002/rob.20293

Jin, S., Su, Y., Gao, S., Wu, F., Hu, T., Liu, J., Li, W., Wang, D., Chen, S., Jiang, Y., Pang, S., & Guo, Q. (2018). Deep learning: Individual maize segmentation from terrestrial LiDAR data using faster R-CNN and regional growth algorithms. *Frontiers in Plant Science*, *9*, 866. https://doi.org/10.3389/fpls.2018.00866

Jin, X., Liu, S., Baret, F., Hemerlé, M., & Comar, A. (2017a). Estimates of plant density of wheat crops at emergence from very low altitude UAV imagery. *Remote Sensing of Environment*, *198*, 105–114. https://doi.org/10.1016/j.rse.2017.06.007

Jin, Z., Prasad, R., Shriver, J., & Zhuang, Q. (2017b). Crop model- and satellite imagery-based recommendation tool for variable rate N fertilizer application for the US Corn system. *Precision Agriculture*, *18*(5), 779–800. https://doi.org/10.1007/s11119-016-9488-z

Kadambi, A., Bhandari, A., & Raskar, R. (2014). 3D depth cameras in vision: Benefits and limitations of the hardware: With an emphasis on the first- and second-generation kinect models. In L. Shao, J. Han, P. Kohli, & Z. Zhang (Eds.), *Computer Vision and Machine Learning with RGB-D Sensors* (pp. 3–26). Springer International Publishing. https://doi.org/10.1007/978-3-319-08651-4_1

Kamilaris, A., & Prenafeta-Boldú, F. X. (2018). Deep learning in agriculture: A survey. *Computers and Electronics in Agriculture*, *147*, 70–90. https://doi.org/10.1016/j.compag.2018.02.016

Kayad, A. G., Al-Gaadi, K. A., Tola, E., Madugundu, R., Zeyada, A. M., & Kalaitzidis, C. (2016). Assessing the spatial variability of alfalfa yield using satellite imagery and ground-based data. *PLOS ONE*, *11*(6), e0157166. https://doi.org/10.1371/journal.pone.0157166

Kazmi, W., Foix, S., Alenyà, G., & Andersen, H. J. (2014). Indoor and outdoor depth imaging of leaves with time-of-flight and stereo vision sensors: Analysis and comparison. *ISPRS Journal of Photogrammetry and Remote Sensing*, *88*, 128–146. https://doi.org/10.1016/j.isprsjprs.2013.11.012

Keller, B., Vass, I., Matsubara, S., Paul, K., Jedmowski, C., Pieruschka, R., Nedbal, L., Rascher, U., & Muller, O. (2019). Maximum fluorescence and electron transport kinetics determined by light-induced fluorescence transients (LIFT) for photosynthesis phenotyping. *Photosynthesis Research*, *140*(2), 221–233. https://doi.org/10.1007/s11120-018-0594-9

Khaki, S., Pham, H., Han, Y., Kuhl, A., Kent, W., & Wang, L. (2021). DeepCorn: A semi-supervised deep learning method for high-throughput image-based corn kernel counting and yield estimation. *Knowledge-Based Systems*, *218*, 106874. https://doi.org/10.1016/j.knosys.2021.106874

Khan, Z., Rahimi-Eichi, V., Haefele, S., Garnett, T., & Miklavcic, S. J. (2018). Estimation of vegetation indices for high-throughput phenotyping of wheat using aerial imaging. *Plant Methods*, *14*(1), 20. https://doi.org/10.1186/s13007-018-0287-6

Khattab, A., Habib, S. E. D., Ismail, H., Zayan, S., Fahmy, Y., & Khairy, M. M. (2019). An IoT-based cognitive monitoring system for early plant disease forecast. *Computers and Electronics in Agriculture*, *166*, 105028. https://doi.org/10.1016/j.compag.2019.105028

Kirchgessner, N., Liebisch, F., Yu, K., Pfeifer, J., Friedli, M., Hund, A., & Walter, A. (2017). The ETH field phenotyping platform FIP: A cable-suspended multi-sensor system. *Functional Plant Biology*, *44*(1), 154. https://doi.org/10.1071/FP16165

Kitić, G., Tagarakis, A., Cselyuszka, N., Panić, M., Birgermajer, S., Sakulski, D., & Matović, J. (2019). A new low-cost portable multispectral optical device for precise plant status assessment. *Computers and Electronics in Agriculture*, *162*, 300–308. https://doi.org/10.1016/j.compag.2019.04.021

Klodt, M., Herzog, K., Töpfer, R., & Cremers, D. (2015). Field phenotyping of grapevine growth using dense stereo reconstruction. *BMC Bioinformatics*, *16*(1), 143. https://doi.org/10.1186/s12859-015-0560-x

Korobiichuk, I., Lysenko, V., Opryshko, O., Komarchyk, D., Pasichnyk, N., & Juś, A. (2018). Crop Monitoring for Nitrogen Nutrition Level by Digital Camera. In R. Szewczyk, C. Zieliński, & M. Kaliczyńska (Eds.), *Automation 2018* (Vol. 743, pp. 595–603). Springer International Publishing. https://doi.org/10.1007/978-3-319-77179-3_56

Kumar, A., Taparia, M., Rajalakshmi, P., Guo, W., Naik B, B., Marathi, B., & Desai, U. B. (2020). CIG based stress identification method for maize crop using UAV based remote sensing. In *2020 IEEE Sensors Applications Symposium (SAS)*, pp. 1–6. https://doi.org/10.1109/SAS48726.2020.9220016

Le Bot, J., Serra, V., Fabre, J., Draye, X., Adamowicz, S., & Pagès, L. (2010). DART: A software to analyse root system architecture and development from captured images. *Plant and Soil*, *326*(1–2), 261–273. https://doi.org/10.1007/s11104-009-0005-2

Le Marié, C., Kirchgessner, N., Flütsch, P., Pfeifer, J., Walter, A., & Hund, A. (2016). RADIX: Rhizoslide platform allowing high throughput digital image analysis of root system expansion. *Plant Methods*, *12*(1), 40. https://doi.org/10.1186/s13007-016-0140-8

Le Marié, C., Kirchgessner, N., Marschall, D., Walter, A., & Hund, A. (2014). Rhizoslides: Paper-based growth system for non-destructive, high throughput phenotyping of root development by means of image analysis. *Plant Methods*, *10*(1), 13. https://doi.org/10.1186/1746-4811-10-13

Li, L., Zhang, Q., & Huang, D. (2014). A review of imaging techniques for plant phenotyping. *Sensors*, *14*(11), 20078–20111. https://doi.org/10.3390/s141120078

Li, W., Niu, Z., Chen, H., Li, D., Wu, M., & Zhao, W. (2016). Remote estimation of canopy height and aboveground biomass of maize using high-resolution stereo images from a low-cost unmanned aerial vehicle system. *Ecological Indicators*, *67*, 637–648. https://doi.org/10.1016/j.ecolind.2016.03.036

Li, Z., Guo, R., Li, M., Chen, Y., & Li, G. (2020). A review of computer vision technologies for plant phenotyping. *Computers and Electronics in Agriculture*, *176*, 105672. https://doi.org/10.1016/j.compag.2020.105672

Liaghat. (2010). A review: The role of remote sensing in precision agriculture. *American Journal of Agricultural and Biological Sciences*, *5*(1), 50–55. https://doi.org/10.3844/ajabssp.2010.50.55

Ma, Z., Rayhana, R., Feng, K., Liu, Z., Xiao, G., Ruan, Y., & Sangha, J. S. (2022). A review on sensing technologies for high-throughput plant phenotyping. *IEEE Open Journal of Instrumentation and Measurement*, *1*, 1–21. https://doi.org/10.1109/OJIM.2022.3178468

Mahlein, A. K. (2016). Plant disease detection by imaging sensors – Parallels and specific demands for precision agriculture and plant phenotyping. *Plant Disease*, *100*(2), 241–251. https://doi.org/10.1094/PDIS-03-15-0340-FE

Mahlein, A. K., Kuska, M. T., Behmann, J., Polder, G., & Walter, A. (2018). Hyperspectral sensors and imaging technologies in phytopathology: State of the art. *Annual Review of Phytopathology*, *56*(1), 535–558. https://doi.org/10.1146/annurev-phyto-080417-050100

Mahlein, A. K., Rumpf, T., Welke, P., Dehne, H. W., Plümer, L., Steiner, U., & Oerke, E. C. (2013). Development of spectral indices for detecting and identifying plant diseases. *Remote Sensing of Environment*, *128*, 21–30. https://doi.org/10.1016/j.rse.2012.09.019

Maia, R. F., Netto, I., & Tran, A. L. H. (2017). Precision agriculture using remote monitoring systems in Brazil. In *2017 IEEE Global Humanitarian Technology Conference (GHTC)*, pp. 1–6. https://doi.org/10.1109/GHTC.2017.8239290

Matos, D. A., Cole, B. J., Whitney, I. P., MacKinnon, K. J. M., Kay, S. A., & Hazen, S. P. (2014). Daily changes in temperature, not the circadian clock, regulate growth rate in Brachypodium distachyon. *PLoS One*, *9*(6), e100072. https://doi.org/10.1371/journal.pone.0100072

Meacham-Hensold, K., Fu, P., Wu, J., Serbin, S., Montes, C. M., Ainsworth, E., Guan, K., Dracup, E., Pederson, T., Driever, S., & Bernacchi, C. (2020). Plot-level rapid screening for photosynthetic parameters using proximal hyperspectral imaging. *Journal of Experimental Botany*, *71*(7), 2312–2328. https://doi.org/10.1093/jxb/eraa068

Mizuno, S., Noda, K., Ezaki, N., Takizawa, H., & Yamamoto, S. (2007). Detection of wilt by analyzing color and stereo vision data of plant. In A. Gagalowicz & W. Philips (Eds.), *Computer Vision/Computer Graphics Collaboration Techniques* (Vol. 4418, pp. 400–411). Springer. https://doi.org/10.1007/978-3-540-71457-6_36

Moghimi, A., Yang, C., & Marchetto, P. M. (2018). Ensemble feature selection for plant phenotyping: A journey from hyperspectral to multispectral imaging. *IEEE Access*, *6*, 56870–56884. https://doi.org/10.1109/ACCESS.2018.2872801

Müller-Linow, M., Pinto-Espinosa, F., Scharr, H., & Rascher, U. (2015). The leaf angle distribution of natural plant populations: Assessing the canopy with a novel software tool. *Plant Methods*, *11*(1), 11. https://doi.org/10.1186/s13007-015-0052-z

Nakarmi, A. D., & Tang, L. (2012). Automatic inter-plant spacing sensing at early growth stages using a 3D vision sensor. *Computers and Electronics in Agriculture*, *82*, 23–31. https://doi.org/10.1016/j.compag.2011.12.011

Odilbekov, F., Armoniené, R., Henriksson, T., & Chawade, A. (2018). Proximal phenotyping and machine learning methods to identify septoria tritici blotch disease symptoms in wheat. *Frontiers in Plant Science*, *9*, 685. https://doi.org/10.3389/fpls.2018.00685

Pfeifer, J., Kirchgessner, N., Colombi, T., & Walter, A. (2015). Rapid phenotyping of crop root systems in undisturbed field soils using X-ray computed tomography. *Plant Methods*, *11*(1), 41. https://doi.org/10.1186/s13007-015-0084-4

Pinto, F., Damm, A., Schickling, A., Panigada, C., Cogliati, S., Müller-Linow, M., Ballvora, A., & Rascher, U. (2016). Sun-induced chlorophyll fluorescence from high-resolution imaging spectroscopy data to quantify spatio-temporal patterns of photosynthetic function in crop canopies. *Plant, Cell & Environment*, *39*(7), 1500–1512. https://doi.org/10.1111/pce.12710

Pound, M. P., French, A. P., Atkinson, J. A., Wells, D. M., Bennett, M. J., & Pridmore, T. (2013). RootNav: Navigating images of complex root architectures. *Plant Physiology*, *162*(4), 1802–1814. https://doi.org/10.1104/pp.113.221531

Qiu, R., Wei, S., Zhang, M., Li, H., Sun, H., Liu, G., Li, M., 1. Key Laboratory of Modern Precision Agriculture System Integration Research, Ministry of Education, China Agricultural University, Beijing 100083, China, & 2. Key Laboratory of Agricultural Information Acquisition Technology, Ministry of Agriculture, China Agricultural University, Beijing 100083, China. (2018). Sensors for measuring plant phenotyping: A review. *International Journal of Agricultural and Biological Engineering*, *11*(2), 1–17. https://doi.org/10.25165/j.ijabe.20181102.2696

Raja, R., Nguyen, T. T., Slaughter, D. C., & Fennimore, S. A. (2020). Real-time robotic weed knife control system for tomato and lettuce based on geometric appearance of plant labels. *Biosystems Engineering*, *194*, 152–164. https://doi.org/10.1016/j.biosystemseng.2020.03.022

Rathore, M. M. U., Paul, A., Ahmad, A., Chen, B. W., Huang, B., & Ji, W. (2015). Real-time big data analytical architecture for remote sensing application. *IEEE Journal of Selected Topics in Applied Earth Observations and Remote Sensing*, *8*(10), 4610–4621. https://doi.org/10.1109/JSTARS.2015.2424683

Raza, S. A., Prince, G., Clarkson, J. P., & Rajpoot, N. M. (2015). Automatic detection of diseased tomato plants using thermal and stereo visible light images. *PLoS One*, *10*(4), e0123262. https://doi.org/10.1371/journal.pone.0123262

Rebetzke, G. J., Jimenez-Berni, J., Fischer, R. A., Deery, D. M., & Smith, D. J. (2019). Review: High-throughput phenotyping to enhance the use of crop genetic resources. *Plant Science*, *282*, 40–48. https://doi.org/10.1016/j.plantsci.2018.06.017

Reynolds, M. P., Pask, A. J. D., & Mullan, D. M. (2012). *Physiological breeding I: Interdisciplinary approaches to improve crop adaptation. CIMMYT*. https://books.google.com.bd/books?id=bYdVb1JuCDoC

Röckel, F., Schreiber, T., Schüler, D., Braun, U., Krukenberg, I., Schwander, F., Peil, A., Brandt, C., Willner, E., Gransow, D., Scholz, U., Kecke, S., Maul, E., Lange, M., & Töpfer, R. (2022). PhenoApp: A mobile tool for plant phenotyping to record field and greenhouse observations. *F1000Research*, *11*, 12. https://doi.org/10.12688/f1000research.74239.2

Rovira-Más, F., Zhang, Q., & Reid, J. F. (2005). Creation of three-dimensional crop maps based on aerial stereo-images. *Biosystems Engineering*, *90*(3), 251–259. https://doi.org/10.1016/j.biosystemseng.2004.11.013

Ruckelshausen, A., Biber, P., Dorna, M., Gremmes, H., Klose, R., Linz, A., Rahe, R., Resch, R., Thiel, M., Trautz, D., & Weiss, U. (2009). BoniRob: An autonomous field robot platform for individual plant phenotyping. In E. J. Van Henten, D. Goense, & C. Lokhorst (Eds.), *Precision Agriculture'09* (pp. 841–847). Wageningen Academic. https://doi.org/10.3920/9789086866649_101

Salas Fernandez, M. G., Bao, Y., Tang, L., & Schnable, P. S. (2017). A high-throughput, field-based phenotyping technology for tall biomass crops. *Plant Physiology*, *174*(4), 2008–2022. https://doi.org/10.1104/pp.17.00707

Sampaio, G. S., Silva, L. A., & Marengoni, M. (2021). 3D reconstruction of non-rigid plants and sensor data fusion for agriculture phenotyping. *Sensors*, *21*(12), 4115. https://doi.org/10.3390/s21124115

Sankaran, S., Khot, L. R., Espinoza, C. Z., Jarolmasjed, S., Sathuvalli, V. R., Vandemark, G. J., Miklas, P. N., Carter, A. H., Pumphrey, M. O., Knowles, N. R., & Pavek, M. J. (2015). Low-altitude, high-resolution aerial imaging systems for row and field crop phenotyping: A review. *European Journal of Agronomy*, *70*, 112–123. https://doi.org/10.1016/j.eja.2015.07.004

Seiffert, U., Bollenbeck, F., Mock, H. P., & Matros, A. (2010). Clustering of crop phenotypes by means of hyperspectral signatures using artificial neural networks. In *2010 2nd Workshop on Hyperspectral Image and Signal Processing: Evolution in Remote Sensing*, pp. 1–4. https://doi.org/10.1109/WHISPERS.2010.5594947

Shafiekhani, A., Kadam, S., Fritschi, F., & DeSouza, G. (2017). Vinobot and vinoculer: Two robotic platforms for high-throughput field phenotyping. *Sensors*, *17*(12), 214. https://doi.org/10.3390/s17010214

Sibanda, M., Mutanga, O., Rouget, M., & Kumar, L. (2017). Estimating biomass of native grass grown under complex management treatments using WorldView-3 spectral derivatives. *Remote Sensing*, *9*(1), 55. https://doi.org/10.3390/rs9010055

Singh, A. K., Ganapathysubramanian, B., Sarkar, S., & Singh, A. (2018). Deep learning for plant stress phenotyping: Trends and future perspectives. *Trends in Plant Science*, *23*(10), 883–898. https://doi.org/10.1016/j.tplants.2018.07.004

Slack, S., York, L. M., Roghazai, Y., Lynch, J., Bennett, M., & Foulkes, J. (2018). Wheat shovelomics II: Revealing relationships between root crown traits and crop growth [Preprint]. *Plant Biology*. https://doi.org/10.1101/280917

Song, P., Wang, J., Guo, X., Yang, W., & Zhao, C. (2021). High-throughput phenotyping: Breaking through the bottleneck in future crop breeding. *The Crop Journal*, *9*(3), 633–645. https://doi.org/10.1016/j.cj.2021.03.015

Song, Y., Glasbey, C. A., Van Der Heijden, G. W. A. M., Polder, G., & Dieleman, J. A. (2011). Combining Stereo and Time-of-Flight Images with Application to Automatic Plant Phenotyping. In A. Heyden & F. Kahl (Eds.), *Image Analysis* (Vol. 6688, pp. 467–478). Springer. https://doi.org/10.1007/978-3-642-21227-7_44

Srayeddin, I., & Doussan, C. (2009). Estimation of the spatial variability of root water uptake of maize and sorghum at the field scale by electrical resistivity tomography. *Plant and Soil*, *319*(1–2), 185–207. https://doi.org/10.1007/s11104-008-9860-5

Sugiura, R., Tsuda, S., Tamiya, S., Itoh, A., Nishiwaki, K., Murakami, N., Shibuya, Y., Hirafuji, M., & Nuske, S. (2016). Field phenotyping system for the assessment of potato late blight resistance using RGB imagery from an unmanned aerial vehicle. *Biosystems Engineering*, *148*, 1–10. https://doi.org/10.1016/j.biosystemseng.2016.04.010

Sun, S., Li, C., Paterson, A. H., Jiang, Y., Xu, R., Robertson, J. S., Snider, J. L., & Chee, P. W. (2018). In-field high throughput phenotyping and cotton plant growth analysis using LiDAR. *Frontiers in Plant Science*, *9*, 16. https://doi.org/10.3389/fpls.2018.00016

Takahashi, H., & Pradal, C. (2021). Root phenotyping: Important and minimum information required for root modeling in crop plants. *Breeding Science*, *71*(1), 109–116. https://doi.org/10.1270/jsbbs.20126

Tanabata, T., Shibaya, T., Hori, K., Ebana, K., & Yano, M. (2012). *SmartGrain*: High-throughput phenotyping software for measuring seed shape through image analysis. *Plant Physiology*, *160*(4), 1871–1880. https://doi.org/10.1104/pp.112.205120

Tattaris, M., Reynolds, M. P., & Chapman, S. C. (2016). A direct comparison of remote sensing approaches for high-throughput phenotyping in plant breeding. *Frontiers in Plant Science*, *7*. https://doi.org/10.3389/fpls.2016.01131

Thomas, S., Behmann, J., Steier, A., Kraska, T., Muller, O., Rascher, U., & Mahlein, A. K. (2018). Quantitative assessment of disease severity and rating of barley cultivars based on hyperspectral imaging in a non-invasive, automated phenotyping platform. *Plant Methods*, *14*(1), 45. https://doi.org/10.1186/s13007-018-0313-8

Thompson, A. L., Conrad, A., Conley, M. M., Shrock, H., Taft, B., Miksch, C., Mills, T., & Dyer, J. M. (2018). Professor: A motorized field-based phenotyping cart. *HardwareX*, *4*, e00025. https://doi.org/10.1016/j.ohx.2018.e00025

Tisné, S., Serrand, Y., Bach, L., Gilbault, E., Ben Ameur, R., Balasse, H., Voisin, R., Bouchez, D., Durand-Tardif, M., Guerche, P., Chareyron, G., Da Rugna, J., Camilleri, C., & Loudet, O. (2013). Phenoscope: An automated large-scale phenotyping platform offering high spatial homogeneity. *The Plant Journal*, *74*(3), 534–544. https://doi.org/10.1111/tpj.12131

Tracy, S. R., Black, C. R., Roberts, J. A., McNeill, A., Davidson, R., Tester, M., Samec, M., Korošak, D., Sturrock, C., & Mooney, S. J. (2012). Quantifying the effect of soil compaction on three varieties of wheat (Triticum aestivum L.) using X-ray micro computed tomography (CT). *Plant and Soil*, *353*(1–2), 195–208. https://doi.org/10.1007/s11104-011-1022-5

Tracy, S. R., Nagel, K. A., Postma, J. A., Fassbender, H., Wasson, A., & Watt, M. (2020). Crop improvement from phenotyping roots: Highlights reveal expanding opportunities. *Trends in Plant Science*, *25*(1), 105–118. https://doi.org/10.1016/j.tplants.2019.10.015

Underwood, J., Wendel, A., Schofield, B., McMurray, L., & Kimber, R. (2017). Efficient in-field plant phenomics for row-crops with an autonomous ground vehicle. *Journal of Field Robotics*, *34*(6), 1061–1083. https://doi.org/10.1002/rob.21728

Virlet, N., Sabermanesh, K., Sadeghi-Tehran, P., & Hawkesford, M. J. (2017). Field scanalyzer: An automated robotic field phenotyping platform for detailed crop monitoring. *Functional Plant Biology*, *44*(1), 143. https://doi.org/10.1071/FP16163

VonMogel, K. H. (2013). Taking the *phenomics revolution* into the field. *CSA News*, *58*(3), 4–10. https://doi.org/10.2134/csa2013-58-3-1

Walter, A., Scharr, H., Gilmer, F., Zierer, R., Nagel, K. A., Ernst, M., Wiese, A., Virnich, O., Christ, M. M., Uhlig, B., Jünger, S., & Schurr, U. (2007). Dynamics of seedling growth acclimation towards altered light conditions can be quantified via GROWSCREEN: A setup and procedure designed for rapid optical phenotyping of different plant species. *New Phytologist*, *174*(2), 447–455. https://doi.org/10.1111/j.1469-8137.2007.02002.x

Walter, J. D. C., Edwards, J., McDonald, G., & Kuchel, H. (2019). Estimating biomass and canopy height with LiDAR for field crop breeding. *Frontiers in Plant Science*, *10*, 1145. https://doi.org/10.3389/fpls.2019.01145

Wan, L., Cen, H., Zhu, J., Zhang, J., Zhu, Y., Sun, D., Du, X., Zhai, L., Weng, H., Li, Y., Li, X., Bao, Y., Shou, J., & He, Y. (2020). Grain yield prediction of rice using multi-temporal UAV-based RGB and multispectral images and model transfer – A case study of small farmlands in the South of China. *Agricultural and Forest Meteorology*, *291*, 108096. https://doi.org/10.1016/j.agrformet.2020.108096

Wang, J., He, T., & Lin, Y. (2018). Changes in ecological, agricultural, and urban land space in 1984–2012 in China: Land policies and regional social-economical drivers. *Habitat International*, *71*, 1–13. https://doi.org/10.1016/j.habitatint.2017.10.010

Wang, X., Xuan, H., Evers, B., Shrestha, S., Pless, R., & Poland, J. (2019). High-throughput phenotyping with deep learning gives insight into the genetic architecture of flowering time in wheat. *GigaScience*, *8*(11), giz120. https://doi.org/10.1093/gigascience/giz120

Wasson, A. P., Nagel, K. A., Tracy, S., & Watt, M. (2020). Beyond digging: Noninvasive root and rhizosphere phenotyping. *Trends in Plant Science*, *25*(1), 119–120. https://doi.org/10.1016/j.tplants.2019.10.011

Wasson, A. P., Richards, R. A., Chatrath, R., Misra, S. C., Prasad, S. V. S., Rebetzke, G. J., Kirkegaard, J. A., Christopher, J., & Watt, M. (2012). Traits and selection strategies to improve root systems and water uptake in water-limited wheat crops. *Journal of Experimental Botany*, *63*(9), 3485–3498. https://doi.org/10.1093/jxb/ers111

Watt, M., Fiorani, F., Usadel, B., Rascher, U., Muller, O., & Schurr, U. (2020). Phenotyping: New windows into the plant for breeders. *Annual Review of Plant Biology*, *71*(1), 689–712. https://doi.org/10.1146/annurev-arplant-042916-041124

Wei, C., Huang, J., Mansaray, L., Li, Z., Liu, W., & Han, J. (2017). Estimation and mapping of winter oilseed rape LAI from high spatial resolution satellite data based on a hybrid method. *Remote Sensing*, *9*(5), 488. https://doi.org/10.3390/rs9050488

Weiss, M., Jacob, F., & Duveiller, G. (2020). Remote sensing for agricultural applications: A meta-review. *Remote Sensing of Environment*, *236*, 111402. https://doi.org/10.1016/j.rse.2019.111402

Weyler, J., Milioto, A., Falck, T., Behley, J., & Stachniss, C. (2021). Joint plant instance detection and leaf count estimation for in-field plant phenotyping. *IEEE Robotics and Automation Letters*, *6*(2), 3599–3606. https://doi.org/10.1109/LRA.2021.3060712

Whalley, W. R., Binley, A., Watts, C. W., Shanahan, P., Dodd, I. C., Ober, E. S., Ashton, R. W., Webster, C. P., White, R. P., & Hawkesford, M. J. (2017). Methods to estimate changes in soil water for phenotyping root activity in the field. *Plant and Soil*, *415*(1–2), 407–422. https://doi.org/10.1007/s11104-016-3161-1

White, J. W., Andrade-Sanchez, P., Gore, M. A., Bronson, K. F., Coffelt, T. A., Conley, M. M., Feldmann, K. A., French, A. N., Heun, J. T., Hunsaker, D. J., Jenks, M. A., Kimball, B. A., Roth, R. L., Strand, R. J., Thorp, K. R., Wall, G. W., & Wang, G. (2012). Field-based phenomics for plant genetics research. *Field Crops Research*, *133*, 101–112. https://doi.org/10.1016/j.fcr.2012.04.003

White, J. W., & Conley, M. M. (2013). A flexible, low-cost cart for proximal sensing. *Crop Science*, *53*(4), 1646–1649. https://doi.org/10.2135/cropsci2013.01.0054

Wolfert, S., Ge, L., Verdouw, C., & Bogaardt, M. J. (2017). Big data in smart farming – A review. *Agricultural Systems*, *153*, 69–80. https://doi.org/10.1016/j.agsy.2017.01.023

Xia, C., Wang, L., Chung, B. K., & Lee, J. M. (2015). In situ 3D segmentation of individual plant leaves using a RGB-D camera for agricultural automation. *Sensors*, *15*(8), 20463–20479. https://doi.org/10.3390/s150820463

Xiang, L., Tang, L., Gai, J., & Wang, L. (2020). PhenoStereo: A high-throughput stereo vision system for field-based plant phenotyping—With an application in sorghum stem diameter estimation. In *2020 ASABE Annual International Virtual Meeting,* July 13–15, 2020. https://doi.org/10.13031/aim.202001190

Xie, C., & Yang, C. (2020). A review on plant high-throughput phenotyping traits using UAV-based sensors. *Computers and Electronics in Agriculture*, *178*, 105731. https://doi.org/10.1016/j.compag.2020.105731

Xu, R., & Li, C. (2022). A review of high-throughput field phenotyping systems: Focusing on ground robots. *Plant Phenomics*, *2022*, 2022/9760269. https://doi.org/10.34133/2022/9760269

Xu, R., Li, C., Paterson, A. H., Jiang, Y., Sun, S., & Robertson, J. S. (2018). Aerial images and convolutional neural network for cotton bloom detection. *Frontiers in Plant Science*, *8*, 2235. https://doi.org/10.3389/fpls.2017.02235

Yang, G., Liu, J., Zhao, C., Li, Z., Huang, Y., Yu, H., Xu, B., Yang, X., Zhu, D., Zhang, X., Zhang, R., Feng, H., Zhao, X., Li, Z., Li, H., & Yang, H. (2017b). Unmanned aerial vehicle remote sensing for field-based crop phenotyping: Current status and perspectives. *Frontiers in Plant Science*, *8*, 1111. https://doi.org/10.3389/fpls.2017.01111

Yang, M. D., Huang, K. S., Kuo, Y. H., Tsai, H., & Lin, L. M. (2017a). Spatial and spectral hybrid image classification for rice lodging assessment through UAV imagery. *Remote Sensing*, *9*(6), 583. https://doi.org/10.3390/rs9060583

Yang, W., Feng, H., Zhang, X., Zhang, J., Doonan, J. H., Batchelor, W. D., Xiong, L., & Yan, J. (2020). Crop phenomics and high-throughput phenotyping: Past decades, current challenges, and future perspectives. *Molecular Plant*, *13*(2), 187–214. https://doi.org/10.1016/j.molp.2020.01.008

Yang, W., Guo, Z., Huang, C., Duan, L., Chen, G., Jiang, N., Fang, W., Feng, H., Xie, W., Lian, X., Wang, G., Luo, Q., Zhang, Q., Liu, Q., & Xiong, L. (2014). Combining high-throughput phenotyping and genome-wide association studies to reveal natural genetic variation in rice. *Nature Communications*, *5*(1), 5087. https://doi.org/10.1038/ncomms6087

Yang, W., Yang, C., Hao, Z., Xie, C., & Li, M. (2019). Diagnosis of plant cold damage based on hyperspectral imaging and convolutional neural network. *IEEE Access*, *7*, 118239–118248. https://doi.org/10.1109/ACCESS.2019.2936892

Young, S. N., Kayacan, E., & Peschel, J. M. (2019). Design and field evaluation of a ground robot for high-throughput phenotyping of energy sorghum. *Precision Agriculture*, *20*(4), 697–722. https://doi.org/10.1007/s11119-018-9601-6

Yuan, L., Pu, R., Zhang, J., Wang, J., & Yang, H. (2016). Using high spatial resolution satellite imagery for mapping powdery mildew at a regional scale. *Precision Agriculture*, *17*(3), 332–348. https://doi.org/10.1007/s11119-015-9421-x

Zahid, A., Mahmud, M. S., He, L., Choi, D., Heinemann, P., & Schupp, J. (2020). Development of an integrated 3R end-effector with a cartesian manipulator for pruning apple trees. *Computers and Electronics in Agriculture*, *179*, 105837. https://doi.org/10.1016/j.compag.2020.105837

Zaman-Allah, M., Vergara, O., Araus, J. L., Tarekegne, A., Magorokosho, C., Zarco-Tejada, P. J., Hornero, A., Albà, A. H., Das, B., Craufurd, P., Olsen, M., Prasanna, B. M., & Cairns, J. (2015). Unmanned aerial platform-based multi-spectral imaging for field phenotyping of maize. *Plant Methods*, *11*(1), 35. https://doi.org/10.1186/s13007-015-0078-2

Zhang, C., Marzougui, A., & Sankaran, S. (2020a). High-resolution satellite imagery applications in crop phenotyping: An overview. *Computers and Electronics in Agriculture*, *175*, 105584. https://doi.org/10.1016/j.compag.2020.105584

Zhang, X., Huang, C., Wu, D., Qiao, F., Li, W., Duan, L., Wang, K., Xiao, Y., Chen, G., Liu, Q., Xiong, L., Yang, W., & Yan, J. (2017). High-throughput phenotyping and QTL mapping reveals the genetic architecture of maize plant growth. *Plant Physiology*, *173*(3), 1554–1564. https://doi.org/10.1104/pp.16.01516

Zhang, Y., & Zhang, N. (2018). Imaging technologies for plant high-throughput phenotyping: A review. *Frontiers of Agricultural Science and Engineering*, *0*(0), 0. https://doi.org/10.15302/J-FASE-2018242

Zhang, Z., Kayacan, E., Thompson, B., & Chowdhary, G. (2020b). High precision control and deep learning-based corn stand counting algorithms for agricultural robot. *Autonomous Robots*, *44*(7), 1289–1302. https://doi.org/10.1007/s10514-020-09915-y

Zhao, F., Jiang, H., & Liu, Z. (2019). Recent development of automotive LiDAR technology, industry and trends. In X. Jiang & J.- N. Hwang (Eds.), *Eleventh International Conference on Digital Image Processing (ICDIP 2019)* (p. 178). SPIE. https://doi.org/10.1117/12.2540277

Zhihong, M., Yuhan, M., Liang, G., & Chengliang, L. (2016). Smartphone-based visual measurement and portable instrumentation for crop seed phenotyping. *IFAC-PapersOnLine*, *49*(16), 259–264. https://doi.org/10.1016/j.ifacol.2016.10.048

Zhou, C., Liang, D., Yang, X., Xu, B., & Yang, G. (2018). Recognition of wheat spike from field based phenotype platform using multi-sensor fusion and improved maximum entropy segmentation algorithms. *Remote Sensing*, *10*(2), 246. https://doi.org/10.3390/rs10020246

Zhou, C., Ye, H., Hu, J., Shi, X., Hua, S., Yue, J., Xu, Z., & Yang, G. (2019). Automated counting of rice panicle by applying deep learning model to images from unmanned aerial vehicle platform. *Sensors*, *19*(14), 3106. https://doi.org/10.3390/s19143106

Zhou, J., Reynolds, D., Cornu, T. L., Websdale, D., Orford, S., Lister, C., Gonzalez-Navarro, O., Laycock, S., Finlayson, G., Stitt, T., Clark, M. D., Bevan, M. W., & Griffiths, S. (2017). *CropQuant: An Automated and Scalable Field Phenotyping Platform for Crop Monitoring and Trait Measurements to Facilitate Breeding and Digital Agriculture*. https://doi.org/10.1101/161547

# 4 High-Throughput Field Phenotyping

## *Enhancing Crop Monitoring for Food Security*

*Amooru Harika, Seethalaxmi, Ezhumalai Sivapragasam, Sudhir Kumar, Viswanathan Chinnusamy, Mahesh Kumar, Madhurima Das, and Dhandapani Raju*

## 4.1 INTRODUCTION

By 2050, agricultural output will need to expand by 25%–70% from present levels to fulfil the rising population's demand for food, which is predicted to reach 9.7 billion people (Hunter et al., 2017). Climate change created upon with increased greenhouse gas emissions, which results in extreme temperatures and unfavorable weather, is the biggest barrier to agriculture (Praveen and Sharma, 2019). Geographical differences in the impact of global warming necessitate specialized responses for various agroecological regions of the world (Ahmed and Stepp, 2016). Due to their limited resources and ineffective farming methods, farmers in low-income countries are especially vulnerable to the effects of climate change (Maja and Ayano, 2021). Thus, to maximize crop yield while reducing the impact of agriculture on the environment, sustainable intensification of agriculture will be required through the adoption of cutting-edge techniques for cultivation, crop management and plant breeding. To achieve this goal, it is essential to develop robust cultivars with high yield potential in the environments of the target population and to offer innovative ways for automated monitoring systems to assess plant health. Tools for high-throughput field phenotyping are required in both scenarios (Chawade et al., 2019; Araus et al., 2018).

High-throughput field phenotyping (HTFP) is an emerging research area designed to address the demand for better techniques to observe and analyze plant growth and development in natural field conditions (Chawade et al., 2019; Araus et al., 2018; Holman et al., 2016). In the past, the majority of plant research has been conducted on plants in carefully regulated settings, like growth chambers or greenhouses, where environmental factors can be closely monitored (Cummings, 2008). However, studying plants in the field presents unique challenges, as environmental conditions are highly variable and difficult to control (Ninomiya, 2022). The development of HTFP has been driven by the need to overcome these challenges and improve our understanding of the growth and development of plants in the field (Ninomiya, 2022; Araus et al., 2018; Holman et al., 2016). In recent years, the field of HTFP has rapidly changed and was driven by advances in sensor technology, imaging technologies and data analytics tools. These advances have allowed researchers to collect a huge set of data on the growth and development of plants in real-time, at scale, and to analyze this data using sophisticated data analytics tools. HTFP has numerous applications in plant research and agriculture. By studying the growth and development of plants in the field, researchers can identify stress factors and diseases in crops, and develop new plant varieties that are more resistant to environmental stress (Gill et al., 2022; Singh et al., 2016). HTFP can also help to improve crop management practices, decrease the usage of water and boost agricultural sustainability.

DOI: 10.1201/9781003505631-4

High-throughput field phenotyping is an innovative methodology in plant sciences that combines modern technologies and automated systems to rapidly and non-destructively assess plant traits in field conditions (Yang et al., 2020; Shakoor et al., 2017). In the past, plant phenotyping has been dependent on labor-intensive, time-consuming manual measurements and observations, which have limited capacity to collect large amounts of data (Negrão and Julkowska, 2020). HTFP revolutionizes this process by integrating sensors, imaging technologies, and data analytics to enable high-throughput and accurate phenotypic characterization of plants in their natural conditions (Jangra et al., 2021). HTFP is important because it can speed up plant breeding and crop improvement programs by gathering quantitative and detailed information on plant characteristics like growth, morphology, stress tolerance, and yield potential (Jangra et al., 2021; Mir et al., 2019). HTFP enables breeders and scientists to select and develop superior crop varieties more efficiently. This technology-driven approach makes it possible to quickly screen large plant populations, facilitating the identification of desired traits and the development of climate-resilient, disease-resistant, and high-yielding crop varieties to address global food security challenges (Reynolds et al., 2020).

## 4.2 TRADITIONAL PHENOTYPING METHODS AND LIMITATIONS

### 4.2.1 Overview of Traditional Phenotyping Approaches

Traditional phenotyping methods in plant sciences have relied on manual observations, measurements, and subjective assessments to characterize plant traits (Xiao et al., 2022; Cobb et al., 2013). All those methods often involve labor-intensive tasks, such as visual scoring, manual measurements with rulers or calipers, and destructive sampling. Researchers would visit the field or greenhouse, examine plants individually, and record data based on their visual judgment or direct measurements (Boumans and Leonelli, 2020; Soltani, 2012). These traditional approaches have provided valuable insights into plant phenotypes but are limited in their ability to handle large-scale experiments and capture high-resolution data.

### 4.2.2 Limitations and Constraints in Traditional Phenotyping

Traditional phenotyping approaches suffer from several limitations and constraints, which have driven the development of HTFP methods (Li et al., 2021; Jiang and Li, 2020; Williams et al., 2018). Some of the key limitations include:

a. **Low Throughput and Scalability**: Traditional methods consume a lot of time and are not appropriate for wider phenotyping experiments. Assessing a large number of plants individually becomes impractical and inefficient.
b. **Subjectivity and Human Error**: Manual observations and visual scoring are subjective and prone to human error and bias. Different observers may interpret and score traits differently, leading to inconsistencies and reduced reliability of data.
c. **Destructive Sampling**: Many traditional phenotyping methods involve destructive sampling, where plants are harvested or damaged to measure traits. This limits the ability to track plant development and traits over time and restricts the possibility of repeated measurements.
d. **Limited Resolution and Precision**: Manual measurements with rulers or calipers have limitations in capturing fine-scale details and subtle variations in plant traits. This can result in reduced accuracy and precision in phenotypic data.
e. **Inability to Capture Dynamic Phenotypes**: Traditional methods often provide static snapshots of plant traits at specific time points. They fail to capture dynamic changes and responses of plants to environmental factors over time.

f. **Limited Phenotypic Trait Coverage**: Traditional phenotyping methods may focus on a few select traits due to time and resource constraints, overlooking the comprehensive assessment of multiple traits and their interactions.

These limitations have highlighted the need for advanced, high-throughput field phenotyping methods that can overcome these constraints and provide more accurate, efficient, and comprehensive phenotypic data for plant research and crop improvement programs.

## 4.3 HIGH-THROUGHPUT FIELD PHENOTYPING TECHNOLOGIES

### 4.3.1 Advanced Imaging Systems for Phenotypic Data Acquisition

HFTP requires advanced imaging systems by capturing detailed and high-resolution phenotypic data (Yang et al., 2020; Chawade et al., 2019; Araus et al., 2018). Technologies such as hyperspectral imaging (Sarić et al., 2022), thermal imaging, 3D imaging (Liu et al., 2020), and multispectral imaging (ElMasry et al., 2019) enable non-destructive and rapid assessment of various plant traits. These imaging systems can capture spatial and spectral information, allowing researchers to measure characteristics linked to plant growth, canopy structure, leaf area, chlorophyll content, stress response, and disease symptoms (Li et al., 2014). Utilizing imaging systems allows for the capture of dynamic phenotypic changes over time and improves objectivity and data accuracy.

### 4.3.2 Remote Sensing Technologies for Large-Scale Data Collection

Remote sensing technologies, such as satellites, drones, and unmanned aerial vehicles (UAVs), provide the ability to collect phenotypic data over large-scale agricultural fields (Xie and Yang, 2020). Using these technologies, researchers can track plant health, growth habits and stress responses from a bird's-eye view. With the help of multispectral or hyperspectral sensors, remote sensing platforms can collect data at various wavelengths, giving important details about pest and disease outbreaks, crop vigor, nutrient status and water stress. (Kumar et al., 2022). The large-scale and continuous data collection capability of remote sensing technologies contributes to efficient phenotyping and enables the identification of spatial variability within fields.

### 4.3.3 Automation and Robotics in Field Phenotyping

Automation and robotics have revolutionized high-throughput field phenotyping by reducing labor-intensive tasks and improving data collection efficiency (Virlet et al., 2016). Plant imaging, sample collection, and precise measurement are among the tasks that robotic platforms with sensors, cameras and robotic arms can carry out (Atefi et al., 2021). Automated systems can move through field plots, capture data from multiple plants simultaneously, and operate 24/7. This not only increases the throughput of phenotyping experiments but also enhances data accuracy, consistency, and repeatability. Automation and robotics enable researchers to scale up their experiments and handle large plant populations efficiently.

### 4.3.4 Integration of Sensor Networks for Data Acquisition

In HTFP, sensor networks are being used more and more to gather data in real-time on environmental factors and plant responses (Bai et al., 2016). These networks consist of various sensors, such as soil moisture sensors, weather stations, temperature sensors and nutrient sensors, strategically placed in the field. The data from these sensors are collected continuously, providing valuable information on the plant's microenvironment and enabling researchers to understand the interaction between plants and their surroundings. An enhanced comprehensive understanding of plant phenotypes and their responses to changing environmental conditions is achieved through the integration of sensor networks with imaging systems and automated data collection platforms. By leveraging these high-throughput field phenotyping technologies, researchers can capture large-scale, high-resolution, and dynamic phenotypic data efficiently. These technologies enable the assessment of multiple traits simultaneously, provide insights into plant-environment interactions, and accelerate the breeding and improvement of crop varieties with desirable traits, contributing to sustainable and resilient agriculture.

## 4.4 DATA PROCESSING AND ANALYSIS

### 4.4.1 Data Analytics and Machine Learning in Field Phenotyping

Large amounts of phenotypic data produced by high-throughput field phenotyping are processed and analyzed using data analytics and machine learning techniques (Arya et al., 2022; Singh et al., 2016). These methods enable researchers to extract meaningful patterns, correlations, and insights from complex datasets. Based on the gathered data, machine learning algorithms can be trained to recognize and categorize plant traits, spot anomalies and forecast phenotypic traits. These techniques aid in automating data analysis, reducing human bias, and enhancing the efficiency and accuracy of phenotype prediction.

### 4.4.2 Feature Extraction and Trait Prediction Algorithms

Feature extraction algorithms are used to identify and extract relevant features from the raw phenotypic data (Sharma et al., 2019). These features may include plant height, leaf area, color indices, texture patterns, or other quantitative measurements. Trait prediction algorithms, such as regression models or deep learning architectures, are then employed to estimate or predict specific phenotypic traits based on the extracted features. These algorithms enable researchers to quantify and characterize plant traits objectively and accurately.

### 4.4.3 Standardization and Sharing of Phenotypic Data

Standardization and sharing of phenotypic data are essential for promoting collaboration, enabling comparisons across different studies, and maximizing the utility of collected data (Becker et al., 2014). Standardized protocols for data collection, metadata annotation, and data formatting ensure consistency and interoperability. Open data initiatives and data repositories facilitate the sharing and accessibility of phenotypic data, allowing researchers worldwide to leverage existing datasets for their analyses and validations. Standardization and data-sharing efforts contribute to building comprehensive databases that support broader scientific advancements and accelerate crop improvement efforts.

### 4.4.4 Challenges and Opportunities in Data Processing and Analysis

Data processing and analysis in high-throughput field phenotyping present several challenges and opportunities (Smith et al., 2021; Mir et al., 2019):

a. **Data Integration and Fusion**: Robust data integration techniques are necessary when integrating data from various sources, including genetic information, sensors, and imaging systems. The fusion of different data modalities enhances the accuracy and reliability of phenotype predictions.

b. **Data Quality Control**: Ensuring data quality is crucial, considering the scale and complexity of high-throughput field phenotyping. In order to detect and address errors or inconsistencies in the gathered data, data cleaning methods, outlier detection, and quality control procedures must be undertaken.
c. **Computational Infrastructure and Resources**: Significant infrastructure and computational resources are needed for the processing and analysis of large-scale phenotypic datasets. High-performance computing capabilities, cloud computing, and distributed computing frameworks can address the computational demands and enable efficient data processing.
d. **Ethical Considerations and Data Privacy**: As phenotypic data becomes more accessible and shared, ethical considerations regarding data privacy, intellectual property rights, and informed consent need to be addressed to ensure responsible data usage.
e. **Integration of Multi-omics Data**: Integrating phenotypic data with genomic, transcriptomic, and metabolomic data provides a comprehensive understanding of plant traits. Developing methodologies for data integration and leveraging multi-omics approaches can uncover new insights into plant phenotypes and their genetic basis.

The advancements in data processing and analysis techniques offer tremendous opportunities for extracting meaningful information from high-throughput field phenotyping data. By addressing the challenges and leveraging these opportunities, researchers can unlock the full potential of phenotypic data and accelerate crop improvement efforts, leading to more sustainable and resilient agricultural practices.

## 4.5 APPLICATIONS OF HIGH-THROUGHPUT FIELD PHENOTYPING

### 4.5.1 Crop Improvement and Breeding Programs

Crop improvement and breeding programs greatly benefit from high-throughput field phenotyping (Jangra et al., 2021; Mir et al., 2019). Researchers can identify and choose desirable traits, such as yield potential, disease resistance, abiotic stress tolerance and nutrient use efficiency through rapid and accurate evaluation of plant traits on a large scale. According to Rebetzke et al. (2019), high-throughput phenotyping facilitates the screening of a wide variety of germplasm, the identification of superior genotypes, and the development of improved crop varieties through genetic manipulation and selective breeding. High-throughput field phenotyping provides phenotypic data that guide breeding strategies for crop improvement and helps to understand the relationships between genotype and phenotype.

### 4.5.2 Genotype-Phenotype Association Studies

High-throughput field phenotyping facilitates genotype-phenotype association studies, which aim to understand the genetic basis of specific plant traits (D'Agostino and Tripodi, 2017). Researchers can find genetic markers linked to desired traits by combining genomic information with large-scale phenotypic data collection. With the use of this data, genes or quantitative trait loci (QTLs) causing particular phenotypic variations can be identified. Studies on the relationship between genotype and phenotype shed light on the underlying genetic processes controlling complex traits and provide support for marker-assisted selection in breeding initiatives.

### 4.5.3 Disease and Stress Detection

High-throughput field phenotyping enables early detection and monitoring of plant diseases and stresses. By employing imaging systems, remote sensing technologies, and sensor networks, researchers can identify visual symptoms, physiological changes, and stress indicators in plants

(Shakoor et al., 2017). These technologies allow for the rapid detection and quantification of plant health and stress-related characteristics such as water status, leaf chlorophyll content, temperature anomalies and canopy structure. Timely detection of diseases and stresses enables effective management strategies, including targeted interventions, breeding for resistance and optimization of agronomic practices.

### 4.5.4 Precision Agriculture and Resource Management

High-throughput field phenotyping supports precision agriculture by providing valuable insights for optimizing resource management practices. Phenotypic data, combined with geospatial information, weather data, and soil characteristics, can help identify spatial variations in plant performance and resource needs within a field (Xu, 2016). With the use of this information, irrigation, fertilization, and crop protection can be managed specifically for each site, improving crop productivity and minimizing the impact on the environment and resource consumption. HTFP aids in the creation of precision agriculture decision support systems, giving farmers access to data-driven tactics for both commercially and sustainably producing crops.

## 4.6 CHALLENGES AND LIMITATIONS

### 4.6.1 Data Quality and Standardization Issues

In high-throughput field phenotyping, maintaining data quality and resolving standardization issues are two of the biggest challenges (Yang et al., 2020; Mir et al., 2019). The collection of large volumes of data from diverse sources can introduce variability and inconsistencies. Standardization of protocols for data collection, annotation, and formatting is essential to ensure consistency and interoperability across different studies and platforms. Quality control measures, data cleaning techniques, and validation procedures need to be implemented to address data quality issues and enhance the reliability and accuracy of phenotypic data.

### 4.6.2 Scalability and Cost-Effectiveness

High-throughput field phenotyping often requires significant investments in infrastructure, equipment, and skilled personnel. Scaling up phenotyping experiments to large field plots or multiple locations can be challenging due to the associated costs and logistical constraints (Smith et al., 2021; Van Eeuwijk et al., 2019). The need for advanced imaging systems, sensor networks, and automation technologies adds to the financial burden. Additionally, data processing and analysis of large datasets require computational resources and expertise. Ensuring the scalability and cost-effectiveness of high-throughput field phenotyping methods remains a challenge that needs to be addressed for wider adoption and implementation.

### 4.6.3 Translation of Research Findings to Practical Applications

While high-throughput field phenotyping generates valuable phenotypic data and research findings, translating these findings into practical applications can be a complex process. The integration of phenotypic data with genetic information, breeding programs, and agronomic practices requires interdisciplinary collaborations and knowledge exchange between scientists, breeders, and farmers (Kusmec et al., 2021). Technical, regulatory and socioeconomic issues must be taken into account in order to close the gap between research and application. Ensuring the accessibility, relevance and applicability of developed technologies and findings in actual agricultural contexts is necessary.

### 4.6.4 Ethical and Social Considerations

The implementation of HTFP technologies raises ethical and social considerations that need to be addressed. Issues related to data privacy, intellectual property rights, and informed consent require careful attention. Researchers must ensure that data collected from farmers' fields or research plots are handled responsibly and respect individual privacy (Costa et al., 2019). Moreover, the potential impact of these technologies on traditional farming practices, labor requirements, and socio-economic dynamics should be considered to ensure equitable and sustainable agricultural development.

### 4.6.5 Phenotypic Complexity and Trait Coverage

The whole spectrum of desirable traits and the complexity of plant phenotypes may still be beyond the reach of high-throughput field phenotyping techniques. While advancements have been made in imaging, sensing, and data analysis techniques, certain traits, such as root architecture, plant-soil interactions, or complex multi-dimensional traits, pose challenges in accurate and high-throughput assessment (Zhao et al., 2019). Developing innovative approaches to capture and quantify these complex traits remains a subject of ongoing research. Engaging stakeholders, collaborating across disciplines, and conducting ongoing research and development are all necessary to address these obstacles and constraints. Overcoming these hurdles will enhance the utility and impact of high-throughput field phenotyping and enable its broader adoption in agricultural research and practice.

## 4.7 FUTURE DIRECTIONS AND OPPORTUNITIES

### 4.7.1 Interdisciplinary Collaboration for Advancements

Strong interdisciplinary partnerships between scientists, engineers, breeders, agronomists, and social scientists are essential for the future of high-throughput field phenotyping (Rahman et al., 2015). Working together can make it easier to share resources, knowledge, and experience, which can result in creative solutions and industry advancements. Integrating biological sciences, data sciences, engineering, and social sciences can address complex challenges, such as data analysis, data integration, technology development, and the translation of research findings into practical applications.

### 4.7.2 Technological Innovations and Emerging Trends

Continued technological advancements will drive the future of high-throughput field phenotyping. Emerging trends include the development of novel sensors, improved imaging systems, miniaturized and portable phenotyping devices, and the integration of multi-omics approaches (Hall et al., 2022). Capabilities for processing and analyzing data will be further improved through the application of big data analytics, machine learning, and artificial intelligence. Additionally, advancements in robotics, automation, and unmanned aerial systems will contribute to increased scalability, efficiency, and cost-effectiveness of high-throughput phenotyping technologies.

### 4.7.3 Policy Implications and Adoption of High-Throughput Field Phenotyping

High-throughput field phenotyping can be adopted and integrated into agricultural research and practice more easily with the help of supportive policy frameworks and regulations. Funding organizations and governments can provide funds to encourage this kind of research and development (Morisse et al., 2022). Incentives can be provided to encourage collaborations between academia, industry, and farmers. Policy considerations may also include data sharing and privacy regulations, intellectual property rights, and guidelines for responsible and ethical data usage. Policy support can help overcome barriers to adoption, ensure technology accessibility, and foster sustainable and inclusive agricultural development.

### 4.7.4 Integration with Digital Agriculture and AgTech

The integration of high-throughput field phenotyping with digital agriculture and agricultural technology (AgTech) holds immense potential (Sarkar et al., 2023). Integration with smart farming systems, precision agriculture technologies, and decision support tools can enable real-time monitoring, adaptive management, and data-driven decision-making. Integration with crop modelling and predictive analytics can enhance crop yield forecasting and optimize resource management. The combination of digital agriculture and high-throughput field phenotyping has the potential to transform farming methods, enhance resource efficiency, and build robust and sustainable agricultural systems.

### 4.7.5 Addressing Global Food Security and Climate Change

Addressing global food security and the problems brought on by climate change requires the use of high-throughput field phenotyping. HTP can aid in the development of climate-resilient crops by facilitating the quick identification and selection of crop varieties with improved traits, such as heat tolerance, disease resistance, and drought tolerance (He and Li, 2020). Additionally, it can help in the development of crop varieties that perform better agronomically and in terms of nutrition. When combined with other environmentally friendly farming methods, high-throughput phenotyping can improve food production, lessen its negative effects on the environment, and lessen the impact of climate change on agriculture. High-throughput field phenotyping has a bright future ahead of it in terms of technological advancements, interdisciplinary teamwork, policy support, and sustainable agriculture. Researchers and other stakeholders can help address global agricultural challenges and pave the way for a more resilient, productive, and sustainable agricultural future by taking advantage of these opportunities.

## 4.8 CONCLUSION

In agricultural research and practice, high-throughput field phenotyping has proven to be a potent and revolutionary tool. Large-scale, accurate, and fast evaluation of plant traits is made possible by it, which supports precision agriculture, genotype-phenotype association research, disease detection, and crop improvement. The advancements in imaging systems, remote sensing technologies, automation, and data analytics have revolutionized the field, providing researchers with unprecedented capabilities to capture and analyze phenotypic data. Despite its numerous benefits, high-throughput field phenotyping also faces challenges and limitations. Issues related to data quality, standardization, scalability, cost-effectiveness, translation of research findings, and ethical considerations need to be addressed for its widespread adoption and practical application. However, these challenges also present opportunities for future advancements and collaborations.

Interdisciplinary collaboration, technological innovations, and policy support are key drivers for the future of high-throughput field phenotyping. By bringing together experts from various disciplines, leveraging emerging technologies, and creating supportive policy frameworks, we can further enhance the capabilities of high-throughput phenotyping. Future research and implementation should focus on integrating digital agriculture and AgTech, addressing global food security and climate change, and promoting sustainable agricultural practices. High-throughput field phenotyping holds promise for revolutionizing crop improvement, optimizing resource management, and contributing to sustainable and resilient agriculture. By harnessing its potential and addressing the challenges, we can unlock new insights, accelerate crop breeding programs, and contribute to global efforts in ensuring food security, mitigating climate change impacts and promoting sustainable agricultural development. Through collaboration, innovation, and a shared commitment to advancing agricultural sciences, HTFP has the potential to shape the future of agriculture.

## REFERENCES


Ahmed, S., & Stepp, J. R. (2016). Beyond yields: Climate change effects on speciality crop quality and agroecological management. *Elementa, 4*, 000092.

Araus, J. L., Kefauver, S. C., Zaman-Allah, M., Olsen, M. S., & Cairns, J. E. (2018). Translating high-throughput phenotyping into genetic gain. *Trends in Plant Science, 23*(5), 451–466.

Arya, S., Sandhu, K. S., Singh, J., & Kumar, S. (2022). Deep learning: As the new frontier in high-throughput plant phenotyping. *Euphytica, 218*(4), 47.

Atefi, A., Ge, Y., Pitla, S., & Schnable, J. (2021). Robotic technologies for high-throughput plant phenotyping: Contemporary reviews and future perspectives. *Frontiers in Plant Science, 12*, 611940.

Bai, G., Ge, Y., Hussain, W., Baenziger, P. S., & Graef, G. (2016). A multi-sensor system for high throughput field phenotyping in soybean and wheat breeding. *Computers and Electronics in Agriculture, 128*, 181–192.

Becker, C. M., Laufer, M. R., Stratton, P., Hummelshoj, L., Missmer, S. A., Zondervan, K. T., … & Zondervan, K. T. (2014). World endometriosis research foundation endometriosis phenome and biobanking harmonisation project: I. Surgical phenotype data collection in endometriosis research. *Fertility and Sterility, 102*(5), 1213–1222.

Boumans, M., & Leonelli, S. (2020). *From Dirty Data to Tidy Facts: Clustering Practices in Plant Phenomics and Business Cycle Analysis* (pp. 79–101). Springer International Publishing.

Chawade, A., van Ham, J., Blomquist, H., Bagge, O., Alexandersson, E., & Ortiz, R. (2019). High-throughput field-phenotyping tools for plant breeding and precision agriculture. *Agronomy, 9*(5), 258.

Cobb, J. N., DeClerck, G., Greenberg, A., Clark, R., & McCouch, S. (2013). Next-generation phenotyping: Requirements and strategies for enhancing our understanding of genotype–phenotype relationships and its relevance to crop improvement. *Theoretical and Applied Genetics, 126*, 867–887.

Costa, J. M., Marques da Silva, J., Pinheiro, C., Barón, M., Mylona, P., Centritto, M., …. & Oliveira, M. M. (2019). Opportunities and limitations of crop phenotyping in southern European countries. *Frontiers in Plant Science, 10*, 1125.

Cummings, I. (2008). *An investigation into the influence of variation in controlled environment plant research facilities on growth responses* (Doctoral dissertation, University of Tasmania).

D'Agostino, N., & Tripodi, P. (2017). NGS-based genotyping, high-throughput phenotyping and genome-wide association studies laid the foundations for next-generation breeding in horticultural crops. *Diversity, 9*(3), 38.

ElMasry, G., Mandour, N., Al-Rejaie, S., Belin, E., & Rousseau, D. (2019). Recent applications of multispectral imaging in seed phenotyping and quality monitoring—An overview. *Sensors, 19*(5), 1090.

Gill, T., Gill, S. K., Saini, D. K., Chopra, Y., de Koff, J. P., & Sandhu, K. S. (2022). A comprehensive review of high throughput phenotyping and machine learning for plant stress phenotyping. *Phenomics, 2*(3), 156–183.

Hall, R. D., D'Auria, J. C., Ferreira, A. C. S., Gibon, Y., Kruszka, D., Mishra, P., & Van de Zedde, R. (2022). High-throughput plant phenotyping: A role for metabolomics? *Trends in Plant Science, 27*(6), 549–563.

He, T., & Li, C. (2020). Harness the power of genomic selection and the potential of germplasm in crop breeding for global food security in the era with rapid climate change. *The Crop Journal, 8*(5), 688–700.

Holman, F. H., Riche, A. B., Michalski, A., Castle, M., Wooster, M. J., & Hawkesford, M. J. (2016). High throughput field phenotyping of wheat plant height and growth rate in field plot trials using UAV based remote sensing. *Remote Sensing, 8*(12), 1031.

Hunter, M. C., Smith, R. G., Schipanski, M. E., Atwood, L. W., & Mortensen, D. A. (2017). Agriculture in 2050: Recalibrating targets for sustainable intensification. *Bioscience, 67*(4), 386–391.

Jangra, S., Chaudhary, V., Yadav, R. C., & Yadav, N. R. (2021). High-throughput phenotyping: A platform to accelerate crop improvement. *Phenomics, 1*(2), 31–53.

Jiang, Y., & Li, C. (2020). Convolutional neural networks for image-based high-throughput plant phenotyping: A review. *Plant Phenomics, 2020.*

Kumar, S., Meena, R. S., Sheoran, S., Jangir, C. K., Jhariya, M. K., Banerjee, A., & Raj, A. (2022). Remote sensing for agriculture and resource management. In M. K. Jhariya, R. S. Meena, A. Banerjee, & S. N. Meena (Eds.), *Natural Resources Conservation and Advances for Sustainability* (pp. 91–135). Elsevier.

Kusmec, A., Zheng, Z., Archontoulis, S., Ganapathysubramanian, B., Hu, G., Wang, L., ... & Schnable, P. S. (2021). Interdisciplinary strategies to enable data-driven plant breeding in a changing climate. *One Earth, 4*(3), 372–383.

Li, D., Quan, C., Song, Z., Li, X., Yu, G., Li, C., & Muhammad, A. (2021). High-throughput plant phenotyping platform (HT3P) as a novel tool for estimating agronomic traits from the lab to the field. *Frontiers in Bioengineering and Biotechnology, 8*, 623705.

Li, L., Zhang, Q., & Huang, D. (2014). A review of imaging techniques for plant phenotyping. *Sensors, 14*(11), 20078–20111.

Liu, H., Bruning, B., Garnett, T., & Berger, B. (2020). Hyperspectral imaging and 3D technologies for plant phenotyping: From satellite to close-range sensing. *Computers and Electronics in Agriculture, 175*, 105621.

Maja, M. M., & Ayano, S. F. (2021). The impact of population growth on natural resources and farmers' capacity to adapt to climate change in low-income countries. *Earth Systems and Environment, 5*, 271–283.

Mir, R. R., Reynolds, M., Pinto, F., Khan, M. A., & Bhat, M. A. (2019). High-throughput phenotyping for crop improvement in the genomics era. *Plant Science, 282*, 60–72.

Morisse, M., Wells, D. M., Millet, E. J., Lillemo, M., Fahrner, S., Cellini, F., ... & Janni, M. (2022). A European perspective on opportunities and demands for field-based crop phenotyping. *Field Crops Research, 276*, 108371.

Negrão, S., & Julkowska, M. M. (2020). Plant phenotyping. *eLS*, 1–14.

Ninomiya, S. (2022). High-throughput field crop phenotyping: Current status and challenges. *Breeding Science, 72*(1), 3–18.

Praveen, B., & Sharma, P. (2019). A review of literature on climate change and its impacts on agriculture productivity. *Journal of Public Affairs, 19*(4), e1960.

Rahman, H., Ramanathan, V., Jagadeeshselvam, N., Ramasamy, S., Rajendran, S., Ramachandran, M., ... & Muthurajan, R. (2015). Phenomics: Technologies and applications in plant and agriculture. *PlantOmics: The Omics of Plant Science*, 385–411.

Rebetzke, G. J., Jimenez-Berni, J., Fischer, R. A., Deery, D. M., & Smith, D. J. (2019). High-throughput phenotyping to enhance the use of crop genetic resources. *Plant Science, 282*, 40–48.

Reynolds, M., Chapman, S., Crespo-Herrera, L., Molero, G., Mondal, S., Pequeno, D. N., ... & Sukumaran, S. (2020). Breeder friendly phenotyping. *Plant Science, 295*, 110396.

Sarić, R., Nguyen, V. D., Burge, T., Berkowitz, O., Trtílek, M., Whelan, J., ... & Čustović, E. (2022). Applications of hyperspectral imaging in plant phenotyping. *Trends in Plant Science, 27*(3), 301–315.

Sarkar, S., Ganapathysubramanian, B., Singh, A., Fotouhi, F., Kar, S., Nagasubramanian, K., ... & Singh, A. K. (2023). Cyber-agricultural systems for crop breeding and sustainable production. *Trends in Plant Science, 29*(2), 130–149.

Shakoor, N., Lee, S., & Mockler, T. C. (2017). High throughput phenotyping to accelerate crop breeding and monitoring of diseases in the field. *Current Opinion in Plant Biology, 38*, 184–192.

Sharma, A., Vans, E., Shigemizu, D., Boroevich, K. A., & Tsunoda, T. (2019). DeepInsight: A methodology to transform a non-image data to an image for convolution neural network architecture. *Scientific Reports, 9*(1), 11399.

Singh, A., Ganapathysubramanian, B., Singh, A. K., & Sarkar, S. (2016). Machine learning for high-throughput stress phenotyping in plants. *Trends in Plant Science, 21*(2), 110–124.

Smith, D. T., Potgieter, A. B., & Chapman, S. C. (2021). Scaling up high-throughput phenotyping for abiotic stress selection in the field. *Theoretical and Applied Genetics, 134*(6), 1845–1866.

Soltani, A. (2012). *Modeling Physiology of Crop Development, Growth and Yield.* CABi.

Van Eeuwijk, F. A., Bustos-Korts, D., Millet, E. J., Boer, M. P., Kruijer, W., Thompson, A., ... & Chapman, S. C. (2019). Modelling strategies for assessing and increasing the effectiveness of new phenotyping techniques in plant breeding. *Plant Science, 282*, 23–39.

Virlet, N., Sabermanesh, K., Sadeghi-Tehran, P., & Hawkesford, M. J. (2016). Field Scanalyzer: An automated robotic field phenotyping platform for detailed crop monitoring. *Functional Plant Biology, 44*(1), 143–153.

Williams, D., Aitkenhead, M., Karley, A. J., Graham, J., & Jones, H. G. (2018). Use of imaging technologies for high throughput phenotyping. *Raspberry: Breeding, Challenges and Advances*, 145–158.

Xiao, Q., Bai, X., Zhang, C., & He, Y. (2022). Advanced high-throughput plant phenotyping techniques for genome-wide association studies: A review. *Journal of Advanced Research, 35*, 215–230.

Xie, C., & Yang, C. (2020). A review on plant high-throughput phenotyping traits using UAV-based sensors. *Computers and Electronics in Agriculture, 178*, 105731.

Xu, Y. (2016). Envirotyping for deciphering environmental impacts on crop plants. *Theoretical and Applied Genetics, 129*, 653–673.

Yang, W., Feng, H., Zhang, X., Zhang, J., Doonan, J. H., Batchelor, W. D., … & Yan, J. (2020). Crop phenomics and high-throughput phenotyping: Past decades, current challenges, and future perspectives. *Molecular Plant, 13*(2), 187–214.

Zhao, C., Zhang, Y., Du, J., Guo, X., Wen, W., Gu, S., … & Fan, J. (2019). Crop phenomics: Current status and perspectives. *Frontiers in Plant Science, 10*, 714.

# 5 Imaging Techniques for High-Throughput Phenotyping

## *Methods and Applications*

*Ezhumalai Sivapragasam, Amooru Harika, Madhurima Das, Seethalaxmi, Viswanathan Chinnusamy, Mahesh Kumar, and Sudhir Kumar*

## LIST OF ABBREVIATIONS

**RGB:** red-green-blue
**UV:** ultraviolet
**VIS:** visible
**IR:** infrared
**NIR:** near-infrared
**SWIR:** short wavelength infrared
**LIDAR:** Light detection and ranging
**MRI:** Magnetic resonance imaging
**X-CT:** X-ray computed tomography
**PET:** Positron emission tomography

## 5.1 INTRODUCTION

Modern plant science heavily relies on high-throughput phenotyping (HTP), as it is crucial for effectively evaluating complex traits across extensive populations. Plant phenotyping is greatly accelerated by these high-throughput techniques, which enable the non-destructive collection of numerous phenotypic data (Li et al., 2021; Sun et al., 2022). High-throughput phenotyping makes it possible to identify unique features that were previously challenging to assess using traditional methods by utilising optical sensors and computer vision. Using different kinds of sensors, high-throughput phenotyping platforms are able to capture images of hundreds of plants every day, enabling the non-destructive measurement of morphological and physio biochemical features (Das Choudhury et al., 2019; Jangra et al., 2021). The core concept of image-based phenotyping will be the relationship between electromagnetic radiation with the plant surface, involving absorption, transmission, fluorescence and reflection. These characteristics differ between control and stressed plants (Jangra et al., 2021). These interactions provide valuable insights for estimating diverse phenotypic traits in plants.

Imaging technology is integral to HTP because the phenotypic characteristics captured are ascertained by the specific imaging techniques and sensors used. Techniques such as visible imaging (RGB) and hyperspectral or multispectral imaging are mostly utilized to record various traits, including those related to morphology, physiology, biochemistry and stress responses. Fluorescence imaging assesses photosynthetic activity, while thermal imaging is employed to monitor surface temperature and other crucial physiological traits. Moreover, technologies like light detection and ranging (LiDAR), CT, MRI and X-ray are predominantly used to obtain


DOI: 10.1201/9781003505631-5

detailed morphological information (Sun et al., 2022). Image-based HTP is a key component of precision agriculture. Each imaging technique has its own strengths and limitations in capturing specific plant traits. The use of these sensors can vary depending on the imaging platform and can be applied either independently or in combination to achieve rapid and precise phenotyping (He et al., 2021). A thorough overview of various imaging methods and the phenotypic characteristics they can assess with possible application in HTP is provided in Table 5.1 along with a visual summary in Figure 5.1. Based on the phenotyping objectives, high-throughput phenotyping can be performed at different phases of crop development, cultivation and post-harvest procedures. This chapter thoroughly reviews the imaging techniques used in high-throughput phenotyping and explores their diverse applications.

**TABLE 5.1**
**Standard Imaging Techniques Used in High-Throughput Phenotyping**

| S. No. | Imaging Technique | Wavelength Range | Description | Traits Measured |
|---|---|---|---|---|
| 1 | RGB color imaging | 400–700 nm | Color imaging with the high-resolution camera from the top and sides of the plants | Biomass, leaf area, greenness (chlorophyll), early vigor, growth rates, phenology |
| 2 | Infrared thermal imaging | 8–4 μm | The amount of thermal energy that an object emits per unit area per second is directly proportional to the object's absolute temperature raised to the fourth power | Tissue temperature, stomatal conductance, plant stress |
| 3 | Fluorescence imaging | – | Light energy is observed by the chlorophyll molecules that drive photosynthesis, and unused energy is re-emitted as heat and fluorescence | Quantum efficiency of PSII photochemistry (photosynthesis and stress phenotyping) |
| 4 | NIR shoot imaging | 900–1,700 nm | Water absorbs NR at three bands between 900 and 1,500 nm; thus, reflectance is inversely related to water content | Tissue water content (drought/salinity phenotyping, grain development) |
| 5 | Root imaging | NIR (900–1,500 nm), VIS (400–700 nm) | Water absorbs NR at three bands between 900 and 1,500 nm. Thus, reflectance is inversely related to water content and combined with color imaging for root distribution | Root distribution and soil moisture depletion |
| 6 | Tomographic imaging | – | Tomographic imaging is increasingly utilized to investigate the three-dimensional structure of root systems | Root growth patterns, root-soil interactions, and spatial distribution |
| 7 | Hyperspectral imaging | VNIR (400–1,000 nm) | The imaging sensor measures the spectral reflectance of an entire plant at 2-nm intervals in the visible near-infrared (VNIR) spectrum, pixel by pixel | Water content, biochemical composition, leaf area index, plant health status, pigment concentration, canopy architecture, nutrient status, and various spectral indices associated with plant performance |
| | | SWIR (1,000–2,500 nm) | The imaging sensor collects spectral reflectance at 2 nm intervals at short wave infrared (SWIR) pixel-wise for the whole plant | |

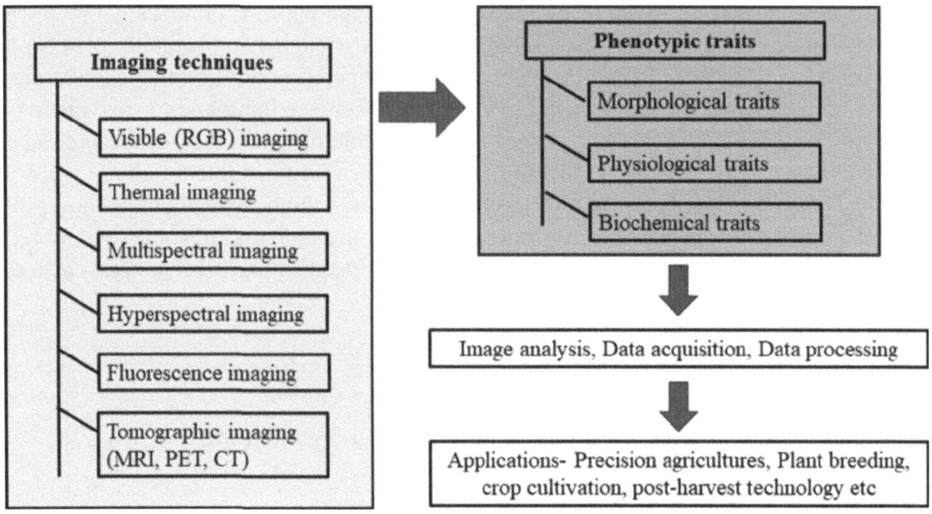


**FIGURE 5.1** High-throughput phenotyping using images: a schematic representation.

## 5.2 VISIBLE IMAGING

Visible imaging sensors operate within the 400–700 nm wavelength range, capturing light reflected in the RGB (red, green and blue) spectrum. Because of its accessibility, cost and convenience of use, this approach is the most commonly used in high-throughput phenotyping (Kim & Chung, 2021). High-resolution RGB images can effectively assess various traits, such as biomass (Zheng et al., 2022), germination rate (Li et al., 2019), plant growth rate (Wasonga et al., 2021; Gang et al., 2022), root system architecture (Brainard et al., 2021), yield (Chen et al., 2022), disease resistance (Calou et al., 2020; Chandel et al., 2021) and reactions to abiotic stresses (Lee et al., 2019). On the other hand, the limited color contrast of leaves against the backdrop and the effect of lighting on automated picture processing represents a basic limitation of visual imaging.

### 5.2.1 Applications

Measurements of structural traits like plant height, canopy architecture, total plant size and leaf area are frequently made using visible imaging. These traits are essential for monitoring growth rates, biomass estimation and plant development over time (Golzarian et al., 2011). The ability to extract these traits from 2D images makes RGB imaging a valuable tool in high-throughput phenotyping pipelines. It also plays a role in detecting visible signs of plant diseases, such as chlorosis, necrosis, or discoloration of leaves. Algorithms can analyze the intensity and distribution of colors within the RGB spectrum to identify disease symptoms. While it is effective at capturing physical damage, RGB imaging is often paired with other techniques like hyperspectral imaging for more detailed disease diagnosis (Mahlein et al., 2012). It is particularly useful in greenhouse and laboratory settings where environmental conditions are controlled where it is used to monitor the development of plant structures and to assess how plants respond to various treatments, such as nutrient changes or stress factors. The images can be captured over time, allowing researchers to track dynamic growth patterns and morphological changes.

### 5.2.2 Advantages

One of the key advantages of RGB imaging is its cost-effectiveness. Cameras used for visible imaging are relatively inexpensive and the technique requires minimal setup. Because of this, a variety of applications from small-scale research to large-scale projects can be used (Walter et al., 2015).

It is well-suited for high-throughput phenotyping, where large numbers of plants need to be analyzed quickly. The simplicity of capturing 2D images means that many plants can be imaged and processed in a short period. This is especially beneficial for breeding programs that require the evaluation of large populations (Tsaftaris & Noutsos, 2016). It allows for non-invasive monitoring of plants over time. Researchers can repeatedly capture images without disturbing the plants, enabling longitudinal studies of growth and development. Additionally, image capture can be automated, providing real-time insights into plant phenotypes (Fiorani & Schurr, 2013).

### 5.2.3 Limitations

While RGB imaging captures information in the visible spectrum, it lacks the capacity to detect non-visible wavelengths, such as near-infrared or ultraviolet light. This limits its ability to provide insights into physiological processes like water content, chlorophyll concentration, or nutrient status, which are better captured using multispectral or hyperspectral imaging (Mahlein, 2016). Environmental factors like shadows and lighting can have an impact on the quality of RGB photos. In field settings, varying sunlight intensity or weather conditions can introduce noise and affect the accuracy of trait measurements. Controlled lighting setups can mitigate this issue, but they are not always feasible in outdoor experiments (Paulus et al., 2014). It is limited to surface-level analysis, meaning it cannot provide information on internal plant features or physiological traits that are not visibly apparent. For example, early-stage disease detection, which often occurs before visual symptoms appear, is not possible with RGB imaging alone. This limitation necessitates the use of additional modalities, such as thermal or hyperspectral imaging, to supplement visible imaging (Römer et al., 2012).

## 5.3 THERMAL IMAGING

The mapping and visualization of infrared radiation across the surfaces of plants or leaves are made effective by thermal imaging. Thermal cameras translate this infrared radiation, which corresponds to the heat emitted by plants into visual data that reveal the spatial distribution of surface temperatures. These cameras capture emitted light in the 3–5 μm or 7–14 μm wavelength ranges, providing temperature information for each pixel. This method is used to assess a plant's physiological condition in response to stress conditions (both biotic and abiotic), including factors like leaf or canopy temperature (Ahmadi et al., 2017), transpiration rate (Prashar et al., 2013), stomatal conductance, and overall water status (Li et al., 2014; Jangra et al., 2021). Plants block their stomata during drought in order to reduce the amount of water lost through transpiration, a process that is highly dependent on soil moisture. The temperature of the leaves increases as a result of this reduction in transpirational cooling. Thermal imaging is highly effective for optimizing irrigation and water management in precision agriculture (Vieira & Ferrarezi, 2021).

### 5.3.1 Applications

Thermal imaging is useful in detecting plant water stress. As plants undergo water deficits, their leaf temperature increases due to reduced transpiration, which can be accurately measured with thermal cameras. This method is an effective tool for water management and drought-resistant breeding programs since researchers have used it to detect early indicators of drought stress. (Fiorani & Schurr, 2013; Zia et al., 2013). It helps in monitoring stomatal behavior. When stomata close in response to environmental stress, the plant's surface temperature rises due to decreased transpiration. By capturing temperature variations across the canopy, researchers can infer stomatal conductance, which is a critical trait for assessing plant water-use efficiency (Jones et al., 2009). This application is particularly useful in studying the effects of environmental stressors, such as high temperatures or low humidity, on plant physiology. In field settings, thermal imaging is often used to measure canopy temperature as a proxy for various physiological traits. For example, canopy temperature depression (CTD) is a widely recognized indicator of crop health under stress conditions. Studies have demonstrated that plants with cooler canopies under hot and dry conditions tend to have higher yields, making CTD a valuable trait for breeding drought-tolerant crops (Prashar et al., 2013). It has shown potential in detecting plant diseases and pest infestations. As diseased plants often exhibit altered physiological processes, including disrupted transpiration, they can show temperature anomalies. While this application is still being explored, there is growing interest in using thermal imaging to detect subtle changes in plant temperature that may signal the onset of disease or pest infestation before visible symptoms occur (Maes & Steppe, 2019).

### 5.3.2 Advantages

Thermal imaging is a non-destructive method that allows for rapid assessment of plant physiological states without physically disturbing the plants. This makes it particularly suitable for HTP in both controlled environments and field trials (Sirault et al., 2020). It has the ability to detect early signs of water stress or stomatal closure, often before visual symptoms appear. This early detection capability makes it a valuable tool for stress phenotyping, where timely interventions can mitigate damage and improve crop performance (Costa et al., 2013). It is versatile and can be used in both controlled greenhouse environments and large-scale field experiments. In the field, it can be mounted on drones or tractors to monitor large areas, providing spatially explicit data on crop water status (Jones et al., 2009). In laboratory settings, it allows for detailed monitoring of individual plants or small groups under precisely controlled conditions (Chaerle et al., 2005).

### 5.3.3 Limitations

A significant limitation of thermal imaging is its sensitivity to external environmental elements, such as wind speed, air temperature and humidity. These factors can influence plant surface temperatures and introduce noise into the data, complicating the interpretation of thermal images. Proper calibration and environmental control are necessary to minimize these effects (Leinonen et al., 2006). It only provides information on the surface temperature of plants, which may not always reflect internal physiological processes. For example, plants with thick or waxy leaves may not exhibit significant surface temperature changes despite undergoing physiological stress. This makes it more difficult for the method to completely reflect the intricacy of plant reactions to stress (Grant et al., 2007). High-resolution thermal imaging systems are still quite costly when compared to other imaging modalities like RGB or multispectral cameras, even if the cost of thermal cameras has dropped recently. This can limit access to thermal imaging technology, particularly in resource-constrained research environments (Sirault et al., 2020). Interpreting thermal data can be complex, as temperature variations can arise from multiple sources, including soil background, shading, or leaf orientation. Careful image analysis and the use of complementary data (such as environmental sensors) are often required to accurately interpret thermal images, which may increase the time and expertise needed for analysis (Prashar et al., 2013).

## 5.4 FLUORESCENCE IMAGING

Light energy is absorbed by chlorophyll and can either be lost as heat, utilized for photosynthesis, or released again as fluorescence. Less than 3% of the radiation emitted by light sources is connected with fluorescence, which is the light that plants emit after absorbing shorter wavelength radiation, mostly through the chlorophyll complex. A critical indication of a plant's general health is the amount of light that is re-emitted (fluorescence), which represents a plant's ability to use light that has been absorbed. (Maxwell & Johnson, 2000). Fluorescence imaging is used to quantify photosynthetic efficiency and other associated metabolic processes influenced by biotic and abiotic stress (Weng et al., 2020; Kumar et al., 2021). Stressed plants show distinct fluorescence patterns compared to healthy ones. These signals are picked up by fluorescence sensors when visible, infrared, or ultraviolet light passes through the plant or tissue. Plant physiological status is assessed under various stress conditions using important chlorophyll fluorescence parameters such as effective quantum yield of PSII (ΦPSII), NPQ (non-photochemical quenching—energy lost as heat from the photosynthetic reaction center), maximum quantum efficiency (Fv/Fm) and relative electron transport rate. A leaf's maximum fluorescence in a dark-adapted condition is represented by Fm and the difference between Fm and a leaf's minimum fluorescence (F0) is represented by Fv (Mutka & Bart, 2015). A primary limitation of fluorescence imaging is its inability to specify the source of signal changes, possibly due to light, temperature, or other environmental influences.

### 5.4.1 Applications

Fluorescence imaging can be useful for evaluation of photosynthetic performance. Chlorophyll fluorescence, particularly from photosystem II, serves as a proxy for the efficiency of photosynthesis. The method provides useful information about the plant's capacity to transform light energy into chemical energy under a variety of circumstances by measuring metrics including the quantum yield of photosystem II (ΦPSII) and NPQ (Baker, 2019). This makes fluorescence imaging an essential tool for studying photosynthetic efficiency and the impacts of stress on plant metabolism. It can detect early signs of stress (both biotic and abiotic) in plants, including drought, salinity and disease. Plants frequently display modified fluorescence patterns during stress because of variations in photosystem activity and chlorophyll concentration. For instance, drought-stressed plants tend to show a decrease in chlorophyll fluorescence, reflecting impaired photosynthesis (Zhu et al., 2021). Similarly, plants infected with pathogens can exhibit changes in fluorescence, making this technique valuable for disease diagnosis and crop monitoring. This imaging is increasingly being incorporated into high-throughput phenotyping platforms for crop breeding. Automated systems equipped with fluorescence sensors allow for the rapid screening of large plant populations to identify individuals with desirable traits, such as stress tolerance or enhanced photosynthetic efficiency (Junker et al., 2018). This has made the technique particularly useful in breeding programs aimed at developing climate-resilient crops. This has also been used to study plant-microbe interactions, particularly symbiotic relationships such as nitrogen fixation in legumes. Fluorescence signals can indicate the health of the photosynthetic apparatus during these interactions, enabling researchers to assess how symbiotic relationships influence plant performance (Perez-Sanz et al., 2022).

### 5.4.2 Advantages

Fluorescence imaging is a non-invasive method that enables the live observation of plant physiological activities without requiring the removal or harm of plants. This makes it ideal for longitudinal studies where plants are observed over time under different environmental conditions (Baker, 2019). Researchers can monitor how photosynthesis and stress responses evolve, making it a powerful tool for dynamic plant studies. Fluorescence imaging has a major benefit in its capacity to identify initial indications of stress, frequently before observable symptoms manifest. Changes in fluorescence patterns can indicate physiological disruptions long before they manifest as visual cues, such as wilting or chlorosis. This allows for timely interventions in agricultural settings and enables breeders to screen for stress-resistant plants more efficiently (Gorbe & Calatayud, 2021). Fluorescence imaging can generate quantitative data on photosynthetic performance and stress indicators, making it highly useful for large-scale phenotyping. Recent advances in automation and image processing have made it possible to apply this technique in high-throughput phenotyping platforms, allowing researchers to evaluate thousands of plants quickly (Junker et al., 2018). This scalability is particularly beneficial for breeding programs and precision agriculture.

### 5.4.3 Limitations

Fluorescence imaging is highly specialized for assessing photosynthetic performance and is not suited for analyzing non-photosynthetic tissues or processes. While it provides detailed information on photosystem efficiency and related stress responses, it cannot capture other important phenotypic traits, such as root development or nutrient uptake, which require complementary imaging technologies (Kalaji et al., 2018). The accuracy of fluorescence imaging can be influenced by environmental factors such as temperature, light intensity and humidity. High ambient light can interfere with the fluorescence signal, leading to inaccurate readings. To mitigate this, fluorescence imaging often requires carefully controlled lighting conditions, which may limit its use in field settings without appropriate adjustments (Rascher et al., 2019). While fluorescence imaging provides valuable insights into plant physiology, it requires sophisticated equipment and technical expertise. Advanced cameras, sensors, and software are needed to capture and process fluorescence signals, making the technique more expensive than other imaging methods, such as RGB or thermal imaging. This can limit its accessibility, particularly for smaller research facilities (Perez-Sanz et al., 2022).

## 5.5 HYPERSPECTRAL IMAGING

Hyperspectral imaging gathers both spectral (λ) and spatial (x, y) information for each pixel, producing a 3D data set known as a hypercube. Over a spectral range of 250–2,500 nm, this hypercube has thousands of images covering wavelengths in the ultraviolet (UV), visible (VIS), near-infrared (NIR), and shortwave infrared (SWIR) regions (Sarić et al., 2022). This technique provides a wealth of data, enabling detailed extraction of various phenotypic characteristics. Despite its potential, managing and analyzing the large datasets generated by hyperspectral imaging can be challenging. It is applied in areas such as nutrient content estimation, disease detection (Abdulridha et al., 2020; Skoneczny et al., 2020), fruit maturity, ripening assessment (Gutiérrez et al., 2019; Shao et al., 2020) and the evaluation of physio biochemical traits critical for understanding plant's growth, development and yield (Sarić et al., 2022).

### 5.5.1 Applications

Hyperspectral imaging (HIS) is highly effective for detecting plant nutrient deficiencies and diseases. By analyzing the spectral signatures of leaves and stems, researchers can identify particular changes in plant physiology caused by nutrient imbalances (such as nitrogen or potassium deficiencies) before visible symptoms appear. Similarly, the technique has an application in identifying fungal, bacterial and viral diseases, as affected plants typically exhibit unique spectral characteristics (Mahlein et al., 2018). This allows for early detection and targeted interventions in agricultural settings. HSI is increasingly employed to assess various abiotic stresses including drought, salinity and heat. By capturing data across different spectral bands, hyperspectral imaging can reveal changes in the plant's physiological status, such as water content, pigment concentration and cell structure. Researchers have successfully used this technology to quantify drought stress in crops by analyzing the water absorption features in the near-infrared (NIR) and shortwave infrared (SWIR) regions of the spectrum (Homolová et al., 2019). Hyperspectral imaging has proven useful for monitoring photosynthesis and estimating chlorophyll content in plants. Chlorophyll strongly absorbs light in certain specific visible spectrum, particularly red and blue wavelengths, while reflecting near-infrared light. By analyzing these absorption and reflectance patterns, HSI can provide detailed information on chlorophyll content, photosynthetic efficiency and overall plant health (Humplík et al., 2015). This technique has been widely adopted in high-throughput phenotyping platforms used in crop breeding programs. Automated systems equipped with hyperspectral sensors can rapidly screen large populations of plants to identify desirable characteristics, such as disease resistance, stress tolerance or enhanced photosynthetic efficiency (Bajwa et al., 2022). This speed up the creation of better varieties of crops and makes it possible for selecting high-performing individuals more effectively.

### 5.5.2 Advantages

Hyperspectral imaging is known for their high spectral resolution, which allows for the detection of physiological and biochemical changes in plants. This capability is crucial for accurate assessment of plant health and stress. Recent studies highlight that hyperspectral imaging can distinguish between different stress conditions and nutrient deficiencies that are not detectable with traditional methods (Lowe et al., 2017). Sensors collect detailed spectral information across a wide wavelength range in a single scan, facilitating the simultaneous assessment of multiple plant traits. This capability significantly enhances the throughput and efficiency of phenotyping processes (Tran et al., 2022). Hyperspectral imaging enables non-invasive monitoring of plant conditions, allowing researchers to observe plant development and stress responses over time. This feature is particularly valuable for longitudinal studies (Li et al., 2021). The ability to detect early signs of stress through hyperspectral imaging is a major advantage. Research has demonstrated that spectral changes associated with stress can be identified before visible symptoms appear, enabling timely management interventions (Lowe et al., 2017).

### 5.5.3 Limitations

The cost and complexity of hyperspectral sensors and their associated data processing systems are significant barriers. This limitation can restrict their use to well-funded research institutions or commercial operations (Pechlivani et al., 2023). The large volumes of data generated by hyperspectral sensors necessitate advanced data processing and storage solutions. The complexity of managing and analyzing these data requires substantial computational resources and expertise (Adão et al., 2017). While hyperspectral sensors provide detailed spectral data, their spatial resolution can be limited compared to other imaging technologies. This can affect the ability to detect fine-scale variations in plant traits. Environmental factors including light and atmospheric interference can affect hyperspectral imaging, creating variability and decreasing data accuracy. (Stuart et al., 2022).

## 5.6 TOMOGRAPHIC IMAGING (MRT, PET, CT)

Modern imaging methods that provide high-resolution, 3D images of a whole plant or just certain portions of it include positron emission tomography (PET), MRI (magnetic resonance imaging) and CT (X-ray computed tomography) (Metzner et al., 2015). With its unique ability to visualize internal plant structures, MRI enables non-invasive evaluation of static as well as dynamic characteristics, including structural, temporal and biochemical changes that take place inside the plant. It can track multiple growth and development phases, including bulb and seed germination, fruit development, root expansion, and seed maturation. It can also track water flow and plant reactions to various stresses, including heat, salinity, drought and light conditions. Moreover, MRI makes it easier to study how plants and pathogens interact (Borisjuk et al., 2012).

X-ray CT, on the other hand, offers detailed three-dimensional imaging of both internal and external plant structures at various scales from micro to macro level. This process involves passing a beam of X-ray through the plant sample, with a detector detecting the portion of the beam that is absorbed by the sample. By collecting multiple two-dimensional images through either sample or sensor movement, these projections are then reconstructed to create a 3D image (Piovesan et al., 2021). The morphological characteristics relevant to the dimensions and form of seeds and fruits have been examined using X-ray CT (Liu et al., 2020; Ahmed et al., 2020). These imaging techniques, however, require a lot of work and are not suitable for larger studies with a large number of samples. Moreover, the considerable size and weight of the equipment make these techniques impractical for use in aerial imaging applications (Sarić et al., 2022).

### 5.6.1 Applications

Tomographic imaging is increasingly utilized to investigate the three-dimensional structure of root systems. To assess plant performance and stress reactions, this application offers comprehensive insights into root growth patterns, root-soil interactions, and spatial distribution (Mairhofer et al., 2017). The ability to visualize and measure root systems non-destructively supports comprehensive root phenotyping and breeding programs. Advances in tomographic imaging allow for the detailed examination of stem and leaf structures. Researchers use this technology to measure parameters such as leaf area, stem diameter and internal tissue organization, which are important for understanding plant architecture and function (Larimer et al., 2020). This application facilitates the evaluation of characteristics linked to the growth and adaptation of plants. Tomographic imaging enables continuous, non-invasive monitoring of plant development over time. This capability is particularly valuable for tracking changes in plant structure and growth dynamics throughout different growth stages, providing a dynamic view of plant development (Herrero-Huerta et al., 2022). This longitudinal approach supports better phenotypic characterization and temporal analysis. The technology is also applied to detect and analyze plant stress and disease conditions. By visualizing internal changes and structural modifications that occur in response to stressors or pathogens, tomographic imaging helps in early diagnosis and understanding of plant health issues (Van Grondelle & Boeker, 2017). This technique is necessary for developing strategies to ameliorate the impacts of stress and improve plant resilience. This technology makes precise biomass estimation possible by offering detailed 3D models of plant structures, which is crucial for assessing plant growth and productivity (Mairhofer et al., 2017). This capability supports research in crop yield prediction and agronomic practices. Recent developments have seen tomographic imaging integrated with other HTP technologies, such as hyperspectral imaging and robotics. This combination improves the capacity to gather multidimensional data and offers a more thorough comprehension of plant characteristics (Nevavuori et al., 2020). Integrative approaches facilitate improved data analysis and more robust phenotypic assessments.

### 5.6.2 Advantages

Tomographic imaging provides comprehensive three-dimensional data of plant structures, allowing researchers to analyze internal features such as root systems and stem anatomy. This capability is crucial for understanding plant growth and development at a detailed level (Moran et al., 2020; (Mathers et al., 2018). This imaging technique is non-invasive, enabling the study of plants over time without causing physical damage. This characteristic is particularly useful for observing developmental processes and monitoring responses to environmental conditions without impacting plant health (Nevavuori et al., 2020). Advances in tomographic imaging technologies have led to improved resolution, allowing for precise measurements of plant components. This combination improves the capacity to gather multidimensional data and offers a more thorough comprehension of plant characteristics (Mairhofer et al., 2017). By providing detailed 3D images, tomographic imaging enhances phenotypic screening capabilities, allowing for more accurate identification and characterization of plant traits. This improvement supports better selection and breeding strategies.

### 5.6.3 Limitations

Tomographic imaging systems are often expensive and require specialized equipment and expertise. The high costs associated with these systems may limit their accessibility, especially for smaller research facilities (Kang et al., 2021; Zhao et al., 2023). The acquisition of high-resolution tomographic images can take a lot of time, particularly for large plant samples or high-throughput scenarios. This can limit the number of samples processed in a given timeframe (Gao et al., 2022; Wang et al., 2024). The large volumes of data generated by tomographic imaging require substantial computational resources for processing and analysis. This challenge necessitates advanced software and significant data management efforts (Chen et al., 2022). Depending on the imaging technique and plant material, there can be limitations in penetration depth and contrast. This issue can affect the visibility of internal structures in some plants (Zhao et al., 2020; Ramasamy et al., 2024).

## 5.7 RASPBERRY PI SERIES AND AFFORDABLE PHENOTYPING

The Raspberry Pi series has emerged as a budget-friendly platform for developing imaging systems in plant phenotyping. Researchers have leveraged its low cost and versatility to create custom imaging setups that provide valuable data for plant analysis without substantial financial investment (Ma et al., 2019; Singh et al., 2021). These systems often incorporate cameras and other sensors to capture high-resolution images, facilitating detailed phenotypic assessments. The modular nature of the Raspberry Pi allows for the easy integration of various sensors and imaging components, making it a flexible tool for phenotyping applications. Specialized setups, ranging from basic image capture to intricate multi-sensor configurations trying to support thorough plant analysis, can be created to satisfy particular research requirements (Zhao et al., 2022). This adaptability enhances the accessibility and scalability of high-throughput phenotyping. The use of Raspberry Pi systems in conjunction with image processing software enables real-time data acquisition and analysis. This capability is particularly beneficial for monitoring plant growth and health in dynamic environments, as it allows for immediate feedback and adjustments (He et al., 2021). Real-time processing supports efficient phenotyping workflows and improves the speed of data collection.

## 5.8 LOW-COST MULTISPECTRAL SENSORS

Recent advancements have introduced affordable multispectral sensors that provide spectral data across several wavelengths. These sensors are used to assess various plant characteristics, including chlorophyll content and water stress, at a lower cost compared to traditional hyperspectral sensors (Kumar et al., 2020; Wang et al., 2022). The availability of these sensors broadens the scope of high-throughput phenotyping by enabling widespread application in different research settings.

## 5.9 LOW-COST ENVIRONMENTAL SENSORS

The integration of affordable environmental sensors with imaging systems provides additional context for plant phenotyping. These sensors take measurements of temperature, humidity, and light intensity, all of which are essential for comprehending how plants react to their surroundings (Lee et al., 2019; Gupta et al., 2023). By combining environmental data with imaging, researchers gain a more comprehensive view of plant performance and stress responses.

## 5.10 AUTOMATED IMAGING SYSTEMS

Advances in automation have led to the emergence of automated imaging systems that integrate with various sensors to streamline the phenotyping process. These systems can capture images and collect data with minimal human intervention, improving efficiency and reproducibility (Zhang et al., 2021; Liu et al., 2024). Automation also supports high-throughput by processing large numbers of samples quickly. Automated imaging systems streamline the process of capturing and analyzing plant images, which boosts the overall efficiency of phenotyping workflows. By minimizing manual intervention, these systems can handle a larger number of plant samples within a shorter time frame, leading to increased throughput and more comprehensive data collection (Liu et al., 2024; Zhang et al., 2021). Automation in imaging systems ensures greater consistency in data acquisition. By standardizing the imaging process and reducing human error, automated systems provide more reliable and reproducible results. This consistency is crucial for accurate phenotypic analysis and comparison across different experiments and conditions (Gao et al., 2023; Singh et al., 2021). Automated imaging systems have recently developed to integrate with other truly innovative technologies like multispectral sensors and machine learning algorithms. This combination enables more in-depth evaluations and insights into plant health and growth patterns by enabling more sophisticated analysis and interpretation of plant traits (Kumar et al., 2020; Niu et al., 2022). Automated systems equipped with real-time imaging capabilities provide immediate feedback on plant conditions. This feature enables researchers to monitor plant development dynamically and make timely adjustments to experimental protocols based on up-to-date information (He et al., 2021; Zhao et al., 2022). These systems are designed to be scalable and adaptable to various research needs. Whether used in small-scale laboratory settings or larger field studies, automated imaging platforms can be customized and scaled to fit different phenotyping requirements, making them versatile tools in plant research (Humplík et al., 2015).

## 5.11 INTEGRATION OF IMAGING MODALITIES

Integrating multiple imaging modalities in high-throughput phenotyping (HTP) allows researchers to capture a comprehensive range of plant traits, enhancing the understanding of plant health, growth, and stress responses. For instance, combining visible (RGB) and hyperspectral imaging provides detailed insights into morphological and biochemical traits. In contrast, integrating thermal and fluorescence imaging offers a dual perspective on plant stress responses and photosynthetic efficiency. However, harmonizing data from these diverse sensors remains challenging, necessitating advanced data fusion techniques. Recent advancements in artificial intelligence (AI) and machine learning make it feasible to analyze and interpret the large datasets generated by these integrated imaging techniques (Sheikh et al., 2024). By recognizing complicated patterns and links between phenotypic traits and genetic components, AI-driven methods are transforming the field of plant breeding and facilitating more rapid and accurate trait development. Despite the challenges, the integration of imaging modalities and AI in HTP holds significant promise for advancing crop improvement and precision agriculture.

## 5.12 FUTURE DIRECTIONS

Advances in next-generation imaging technology promise revolutionary breakthroughs for high-throughput phenotyping. The accuracy and detail of phenotypic data collected should be greatly improved by innovations like real-time multi/hyperspectral imaging, greater resolution in 3D imaging techniques like MRI and X-ray CT, and the integration of AI-driven picture analysis (Metzner et al., 2015). These advancements deepen our understanding of plant biology and stress responses, driving more effective breeding programs. As these technologies evolve, they will enable breeders to screen larger populations more accurately, identifying critical traits related to yield, stress tolerance, and disease resistance more efficiently. In precision agriculture, advanced imaging will facilitate real-time crop monitoring, leading to optimized resource use, improved crop management, and increased productivity. The future of phenotyping will also increasingly depend on interdisciplinary collaborations, uniting experts from plant science, imaging technology, data science, and artificial intelligence. These collaborations are crucial for advancing imaging techniques and managing the vast data generated, ultimately fostering innovation in phenotyping and contributing to more sustainable agricultural practices.

## 5.13 CONCLUSION

Imaging techniques have become indispensable tools in high-throughput phenotyping. With these techniques, a greater range of plant traits can be captured and analyzed at speeds and accuracy never before possible. This chapter has explored various imaging modalities, including visible imaging, thermal imaging, fluorescence imaging, and advanced 3D techniques such as MRI and X-ray CT, all of which contribute to a comprehensive understanding of plant health, growth, development and stress responses.

The integration of these imaging techniques offers a holistic approach to phenotyping, allowing for detailed assessments that are crucial for advancing plant breeding, precision agriculture, and crop management. As imaging technologies evolve, they will further enhance our ability to monitor and optimize plant performance from early development to postharvest. The ongoing advancements in this field and interdisciplinary collaborations promise to drive significant innovations in high-throughput phenotyping, ensuring its continued relevance and impact on sustainable agricultural practices.

## REFERENCES

Abdulridha, J., Ampatzidis, Y., Kakarla, S. C. & Roberts, P. (2020). Detection of target spot and bacterial spot diseases in tomato using UAV-based and benchtop-based hyperspectral imaging techniques. *Precision Agriculture*, *21*, 955–978.

Adão, T., Hruška, J., Pádua, L., Bessa, J., Peres, E., Morais, R. & Sousa, J. (2017). Hyperspectral imaging: A review on uav-based sensors, data processing and applications for agriculture and forestry. *Remote Sensing*, *9*(11), 1110.

Ahmadi, S. H., Agharezaee, M., Kamgar-Haghighi, A. A. & Sepaskhah, A. R. (2017). Comparing canopy temperature and leaf water potential as irrigation scheduling criteria of potato in water-saving irrigation strategies. *International Journal of Plant Production*, *11*(2), 333–348.

Ahmed, M. R., Yasmin, J., Park, E., Kim, G., Kim, M. S., Wakholi, C. & Cho, B. K. (2020). Classification of watermelon seeds using morphological patterns of X-ray imaging: A comparison of conventional machine learning and deep learning. *Sensors*, *20*(23), 6753.

Bajwa, A. A., Mahajan, G., Chauhan, B. S. & Ehsani, R. (2022). Hyperspectral remote sensing: Applications in weed and crop management. *Journal of Plant Diseases and Protection*, *129*(3), 445–461.

Baker, N. R. (2019). Chlorophyll fluorescence: A probe of photosynthesis in vivo. *Annual Review of Plant Biology*, *70*, 89–120.

Borisjuk, L., Rolletschek, H. & Neuberger, T. (2012). Surveying the plant's world by magnetic resonance imaging. *The Plant Journal*, *70*(1), 129–146.

Brainard, S. H., Bustamante, J. A., Dawson, J. C., Spalding, E. P. & Goldman, I. L. (2021). A digital image-based phenotyping platform for analyzing root shape attributes in carrot. *Frontiers in Plant Science*, *12*, 690031.

Calou, V. B. C., dos Santos Teixeira, A., Moreira, L. C. J., Lima, C. S., de Oliveira, J. B. & de Oliveira, M. R. R. (2020). The use of UAVs in monitoring yellow sigatoka in banana. *Biosystems Engineering*, *193*, 115–125.

Chaerle, L., Leinonen, I., Jones, H. G. & Van Der Straeten, D. (2005). Monitoring and screening plant populations with combined thermal and chlorophyll fluorescence imaging. *Journal of Experimental Botany*, 56(411), 1777–1787.

Chandel, A. K., Khot, L. R. & Sallato, B. (2021). Apple powdery mildew infestation detection and mapping using high-resolution visible and multispectral aerial imaging technique. *Scientia Horticulturae*, *287*, 110228.

Chen, R., Zhang, C., Xu, B., Zhu, Y., Zhao, F., Han, S. & Yang, H. (2022). Predicting individual apple tree yield using UAV multi-source remote sensing data and ensemble learning. *Computers and Electronics in Agriculture*, *201*, 107275.

Costa, J. M., Grant, O. M. & Chaves, M. M. (2013). Thermography to explore plant-environment interactions. *Journal of Experimental Botany*, *64*(13), 3937–3949.

Das Choudhury, S., Samal, A. & Awada, T. (2019). Leveraging image analysis for high-throughput plant phenotyping. *Frontiers in Plant Science*, *10*, 508.

Fiorani, F. & Schurr, U. (2013). Future scenarios for plant phenotyping. *Annual Review of Plant Biology*, 64, 267–291.

Gang, M. S., Kim, H. J. & Kim, D. W. (2022). Estimation of greenhouse lettuce growth indices based on a two-stage CNN using RGB-D images. *Sensors*, *22*(15), 5499.

Golzarian, M. R., Frick, R. A., Rajendran, K., Berger, B., Roy, S., Tester, M. & Lun, D. S. (2011). Accurate inference of shoot biomass from high-throughput images of cereal plants. *Plant Methods*, *7*(1), 2.

Gorbe, E. & Calatayud, A. (2021). Applications of chlorophyll fluorescence imaging technique in horticultural research: A review. *Scientia Horticulturae*, *288*, 110352.

Grant, O. M., Tronina, Ł., Jones, H. G. & Chaves, M. M. (2007). Exploring thermal imaging variables for the detection of stress responses in grapevine under different irrigation regimes. *Journal of Experimental Botany*, *58*(4), 815–825.

Gutiérrez, S., Wendel, A. & Underwood, J. (2019). Spectral filter design based on in-field hyperspectral imaging and machine learning for mango ripeness estimation. *Computers and Electronics in Agriculture*, *164*, 104890.

He, F. X., Meng, Q. H., Tang, L., Huang, X., Lu, X. H., Wang, R. Y. & Li, Y. (2021). Research progress in hyperspectral imaging technology for fruit quality detection. *Journal of Fruit Science, 38*(9), 1590–1599.

Herrero-Huerta, M., Raumonen, P. & Gonzalez-Aguilera, D. (2022). 4DRoot: Root phenotyping software for temporal 3D scans by X-ray computed tomography. *Frontiers in Plant Science*, *13*, 986856.

Homolová, L., Malenovský, Z., Clevers, J. G., García-Santos, G. & Schaepman, M. E. (2019). Review of optical-based remote sensing for plant traits estimation in agriculture. *Remote Sensing*, *11*(15), 1803.

Humplík, J. F., Lazár, D., Husičková, A. & Spíchal, L. (2015). Automated phenotyping of plant shoots using imaging methods for analysis of plant stress responses – a review. *Plant Methods*, *11*(1), 1–10.

Jangra, S., Chaudhary, V., Yadav, R. C. & Yadav, N. R. (2021). High-throughput phenotyping: a platform to accelerate crop improvement. *Phenomics*, *1*(2), 31–53.

Jones, H. G., Serraj, R., Loveys, B. R., Xiong, L., Wheaton, A. & Price, A. H. (2009). Thermal infrared imaging of crop canopies for the remote diagnosis and quantification of plant responses to water stress in the field. *Functional Plant Biology*, *36*(11), 978–989.

Junker, A., Muraya, M. M., Weigelt-Fischer, K., Arend, D. & Altmann, T. (2018). High-throughput phenotyping of heterosis in Arabidopsis using machine learning. *Scientific Reports*, *8*(1), 10.

Kalaji, H. M., Schansker, G., Ladle, R. J. & Goltsev, V. (2018). Frequently asked questions about chlorophyll fluorescence, the sequel. *Photosynthesis Research*, *137*(1), 13–66.

Kim, J. & Chung, Y. S. (2021). A short review of RGB sensor applications for accessible high-throughput phenotyping. *Journal of Crop Science and Biotechnology*, 1–5.

Kumar, J., Singh, B., Ghosh, P. & Singh, V. (2020). Hyperspectral imaging for crop stress management: Current status and future perspectives. *Agronomy*, *10*(7), 1042.

Kumar, P., Eriksen, R. L., Simko, I. & Mou, B. (2021). Molecular mapping of water-stress responsive genomic loci in lettuce (Lactuca spp.) using kinetics chlorophyll fluorescence, hyperspectral imaging and machine learning. *Frontiers in G enetics*, *12*, 634554.

Larimer, C. J., Denis, E. H., Suter, J. D. & Moran, J. J. (2020). Optical coherence tomography imaging of plant root growth in soil. *Applied Optics*, *59*(8), 2474.

Lee, U., Silva, R. R., Kim, C., Kim, H., Heo, S., Park, I. S. & Chung, Y. S. (2019). Image analysis for measuring disease symptom to bacterial soft rot in potato. *American Journal of Potato Research*, *96*, 303–313.

Li, B., Xu, X., Han, J., Zhang, L., Bian, C., Jin, L., & Liu, J. (2019). The estimation of crop emergence in potatoes by UAV RGB imagery. *Plant Methods*, *15*, 1–13.

Li, D., Quan, C., Song, Z., Li, X., Yu, G., Li, C. & Muhammad, A. (2021). High-throughput plant phenotyping platform (HT3P) as a novel tool for estimating agronomic traits from the lab to the field. *Frontiers in Bioengineering and Biotechnology*, *8*, 623705.

Li, L., Zhang, Q., & Huang, D. (2014). A review of imaging techniques for plant phenotyping. *Sensors*, *14*(11), 20078–20111.

Liu, W., Liu, C., Jin, J., Li, D., Fu, Y. & Yuan, X. (2020). High-throughput phenotyping of morphological seed and fruit characteristics using X-ray computed tomography. *Frontiers in Plant Science*, *11*, 601475.

Liu, Y., et al. (2024). Automated imaging systems in high throughput plant phenotyping: Recent advances and future directions. *Journal of Agricultural and Food Chemistry*, *72*(10), 2347–2360.

Maes, W. H. & Steppe, K. (2019). Perspectives for remote sensing with unmanned aerial vehicles in precision agriculture. *Trends in Plant Science*, *24*(2), 152–164.

Mahlein, A. K. (2016). Plant disease detection by imaging sensors – Parallels and specific demands for precision agriculture and plant phenotyping. *Plant Disease*, *100*(2), 241–251.

Mahlein, A. K., Oerke, E. C., Steiner, U. & Dehne, H. W. (2012). Recent advances in sensing plant diseases for precision crop protection. *European Journal of Plant Pathology*, *133*(1), 197–209.

Mahlein, A. K., Oerke, E. C., Steiner, U. & Dehne, H. W. (2018). Recent advances in sensing plant diseases for precision crop protection. *European Journal of Plant Pathology*, *152*(4), 745–762.

Mairhofer, S., Pridmore, T., Johnson, J., Wells, D. M., Bennett, M. J., Mooney, S. J., & Sturrock, C. J. (2017). X-ray computed tomography of crop plant root systems grown in soil. *Current Protocols in Plant Biology*, *2*(4), 270–286.

Mathers, A. W., Hepworth, C., Baillie, A. L., Sloan, J., Jones, H., Lundgren, M., Fleming, A. J., Mooney, S. J. & Sturrock, C. J. (2018). Investigating the microstructure of plant leaves in 3D with lab-based X-ray computed tomography. *Plant Methods*, *14*(1), 1–12.

Maxwell, K. & Johnson, G. N. (2000). Chlorophyll fluorescence—a practical guide. *Journal of Experimental Botany*, *51*(345), 659–668.

Metzner, R., Eggert, A., van Dusschoten, D., Pflugfelder, D., Gerth, S., Schurr, U. & Jahnke, S. (2015). Direct comparison of MRI and X-ray CT technologies for 3D imaging of root systems in soil: potential and challenges for root trait quantification. *Plant Methods*, *11*, 1–11.

Mutka, A. M. & Bart, R. S. (2015). Image-based phenotyping of plant disease symptoms. *Frontiers in Plant Science*, *5*, 734.

Nevavuori, P., Narra, N., Linna, P., & Lipping, T. (2020). Crop yield prediction using multitemporal UAV data and spatio-temporal deep learning models. *Remote Sensing*, *12*(23), 4000.

Paulus, S., Schumann, H., Kuhlmann, H. & Léon, J. (2014). High-precision laser scanning system for capturing 3D plant architecture and analysing growth of cereal plants. *Biosystems Engineering*, *121*, 1–11.

Pechlivani, E. M., Papadimitriou, A., Pemas, S., Giakoumoglou, N., & Tzovaras, D. (2023). Low-Cost hyperspectral imaging device for portable remote sensing. *Instruments*, *7*(4), 32.

Perez-Sanz, F., Navarro, P. J. & Egea-Cortines, M. (2022). A review of thermal and chlorophyll fluorescence imaging applications in plant phenotyping. *Sensors*, *22*(6), 2252.

Piovesan, A., Vancauwenberghe, V., Van De Looverbosch, T., Verboven, P. & Nicolaï, B. (2021). X-ray computed tomography for 3D plant imaging. *Trends in Plant Science*, *26*(11), 1171–1185.

Prashar, A., Yildiz, J., McNicol, J. W., Bryan, G. J. & Jones, H. G. (2013). Infra-red thermography for high throughput field phenotyping in Solanum tuberosum. *PLoS One*, *8*(6), e65816.

Rascher, U., Blossfeld, S., Fiorani, F., Jahnke, S., Jansen, M., Kuhn, A. J. & Schurr, U. (2019). Non-invasive approaches for phenotyping of enhanced performance traits in bean. *Plant Methods*, *15*(1), 21.

Römer, C., Wahabzada, M., Ballvora, A., Pinto, F., Rossini, M., Panigada, C. & Mahlein, A. K. (2012). Early drought stress detection in cereals: Simplex volume maximization for hyperspectral image analysis. *Functional Plant Biology*, *39*(11), 878–890.

Sarić, R., Nguyen, V. D., Burge, T., Berkowitz, O., Trtílek, M., Whelan, J. & Čustović, E. (2022). Applications of hyperspectral imaging in plant phenotyping. *Trends in Plant Science*, *27*(3), 301–315.

Shao, Y., Wang, Y., Xuan, G., Gao, Z., Hu, Z., Gao, C. & Wang, K. (2020). Assessment of strawberry ripeness using hyperspectral imaging. *Analytical Letters*, *54*(10), 1547–1560.

Sheikh, M., Iqra, F., Ambreen, H., Pravin, K. A., Ikra, M. & Chung, Y. S. (2024). Integrating artificial intelligence and high-throughput phenotyping for crop improvement. *Journal of Integrative Agriculture*, *23*(6), 1787–1802.

Singh, S., et al. (2021). The role of Raspberry Pi in affordable and scalable plant phenotyping solutions. *International Journal of Applied Earth Observation and Geoinformation*, *96*, 102283.

Sirault, X. R., James, R. A. & Furbank, R. T. (2020). A new screening method for osmotic component of salinity tolerance in cereals using infrared thermography. *Journal of Experimental Botany*, *71*(3), 1092–1102.

Skoneczny, H., Kubiak, K., Spiralski, M., Kotlarz, J., Mikiciński, A. & Puławska, J. (2020). Fire blight disease detection for apple trees: Hyperspectral analysis of healthy, infected and dry leaves. *Remote Sensing*, *12*(13), 2101.

Stuart, M. B., Davies, M., Hobbs, M. J., Pering, T. D., McGonigle, A. J. S., & Willmott, J. R. (2022). High-Resolution hyperspectral imaging using low-cost components: Application within environmental monitoring scenarios. *Sensors*, *22*(12), 4652.

Sun, D., Robbins, K., Morales, N., Shu, Q. & Cen, H. (2022). Advances in optical phenotyping of cereal crops. *Trends in Plant Science*, *27*(2), 191–208.

Tran, T., Keller, R., Trinh, V., Tran, K. & Kaldenhoff, R. (2022). Multi-Channel Spectral Sensors as Plant Reflectance Measuring Devices—Toward the Usability of Spectral Sensors for Phenotyping of Sweet Basil (Ocimum basilicum). *Agronomy*, *12*(5), 1174.

Tsaftaris, S. A. & Noutsos, C. (2016). Plant phenotyping with low-cost digital cameras and image analytics. *Methods in Molecular Biology*, 1405, 209–233.

Van Grondelle, R., & Boeker, E. (2017). Limits on natural photosynthesis. *The Journal of Physical Chemistry*, *121*(30), 7229–7234.

Vieira, G. H. S. & Ferrarezi, R. S. (2021). Use of thermal imaging to assess water status in citrus plants in greenhouses. *Horticulturae*, *7*(8), 249.

Walter, A., Liebisch, F. & Hund, A. (2015). Plant phenotyping: From bean weighing to image analysis. *Plant Methods*, *11*(1), 14.

Wasonga, D. O., Yaw, A., Kleemola, J., Alakukku, L. & Mäkelä, P. S. (2021). Red-green-blue and multispectral imaging as potential tools for estimating growth and nutritional performance of cassava under deficit irrigation and potassium fertigation. *Remote Sensing*, *13*(4), 598.

Weng, H., Zeng, Y., Cen, H., He, M., Meng, Y., Liu, Y., … & He, Y. (2020). Characterization and detection of leaf photosynthetic response to citrus Huanglongbing from cool to hot seasons in two orchards. *Transactions of the ASABE*, *63*(2), 501–512.

Zheng, C., Abd-Elrahman, A., Whitaker, V. & Dalid, C. (2022). Prediction of strawberry dry biomass from UAV multispectral imagery using multiple machine learning methods. *Remote Sensing*, *14*(18), 4511.

Zhu, J., Tremblay, N. & Liang, Y. (2021). Field phenotyping of plant resistance to environmental stresses. *Current Opinion in Plant Biology*, *60*, 101981.

Zia, S., Romano, G., Spreer, W., Sanchez, C., Cairns, J., Araus, J. L. & Müller, J. (2013). Infrared thermal imaging as a rapid tool for identifying water-stress tolerant maize genotypes of different phenology. *Journal of Agronomy and Crop Science*, *199*(2), 75–84.

# 6 High-Throughput Phenotyping for Plant Metabolomics

## *Methods and Applications with a Special Focus on Chemical Ecology*

*Ishita Paul, Swati Hazra, Akbar Hossain, and Pratap Bhanu Singh Bhadoria*

## 6.1 INTRODUCTION

Phenomics, as described by various scholars, pertains to obtaining multi-dimensional data describing the phenotype of an entire organism, such as traits describing growth, composition and structural organization in the case of plants (Yang et al., 2020). The objective for such wide-scale mapping of phenotypes is to bridge the knowledge gap between functional genomics – efficiently standardized due to updated sequencing technology – and the end-points of gene expression, viz. phenotypes. High-throughput phenotyping (HTP) implies the rapid acquisition of multi-dimensional phenotypic datasets at an organismal level from large numbers of specimens using multidisciplinary approaches, including biological variations in response to environmental factors. Phenotyping tools used are preferred to be minimally invasive, and so for crop plants include morphometry, monitoring yield-related traits, optical/fluorescence/hyperspectral imaging, etc. The latest HTP approaches attempt to incorporate the full complement or "fingerprint" of metabolites of the organisms, especially for crop improvement or disease-related studies (Djande et al., 2020; Hall et al., 2022).

Metabolomics, i.e. systematic metabolic profiling of an organism or a cell, is carried out in plants to study the impacts of environmental factors or stress (e.g. disease, drought, nutrient imbalance), genotype, developmental biology (identify and quantify growth stage-dependent biomarkers), and assess yield-related parameters and novel products (Djande et al., 2020; Guo et al., 2021; Manickam et al., 2023). Workflow for a plant metabolomic study typically includes extraction of tissue constituents (especially organic molecules with molecular weights lower than 1,500 Da), analyses using a variety of techniques, followed by interpretation of acquired data (curation, identification of compounds, quantification, and statistical analyses of the obtained profiles). Plant metabolites are generally categorized as primary (directly involved in growth, development, reproduction, photosynthesis and respiratory pathways) and secondary (small molecules produced from primary precursors, and involved in plant-environment interactions). Although secondary metabolites are not absolutely essential for a plant's survival, these are molecules that contribute to stress responses and acclimation or adaptive mechanisms and represent most of the diversity of plant metabolites. Since these molecules often exhibit taxon-specific distribution, these are also referred to as specialized metabolites. Extraction methods depend on whether the study is targeted towards any particular group(s) of metabolites, or untargeted to cover the widest possible array of detectable compounds (including previously unknown components). In general, extraction methods are based on fluid solvents or absorbent matrices in which biological molecules can occur in solution/suspension (Table 6.1).

DOI: 10.1201/9781003505631-6

**TABLE 6.1**
**Examples of Extraction Procedures for Plant Metabolites to be Analyzed by High-Throughput Profiling Methods**

| Target Analytes | Extraction Procedure | Analytical Technique(s) | Reference(s) |
|---|---|---|---|
| Flavonoids (including anthocyanins) | Solvent-mediated extraction from *Rhododendron arboreum* flowers, followed by fractionation by thin-layer chromatography | LC-MS (ESI: positive mode), NMR | Bhatt et al. (2022) |
| Phenolic acids | Solvent-mediated extraction from *Rhododendron arboreum* flowers, followed by fractionation by thin-layer chromatography | LC-MS (ESI: negative mode) | Bhatt et al. (2022) |
| Coumarins (e.g. scopoletin), phenolic acids (e.g. ferulic acid) | Solvent-mediated extraction from fresh *Murraya paniculata* samples, followed by treatment with trimethylsilylating reagents | GC-MS | Paul et al. (2022) |
| Untargeted metabolites in *Curculigo* samples (e.g. phenolic glycosides, triterpenes) | Extraction of *C. orchioides* and *C. latifolia* organs with 70% ethanol | FTIR spectrophotometry, LC-MS (ESI: dual mode, multiple collision energies) | Umar et al. (2021) |
| Fatty acids | Solvent-mediated extraction from algal thalli, followed by treatment with methylating reagents | GC-MS | Gorain et al. (2018) |
| Photopigments specific to algal clades | Unicellular phytoplankton gated towards FACS | FACS (excitation wavelengths: 375, 488, 642 nm) | Roy et al. (2018) |
| Primary metabolites specific to root tip endodermal and cortical cells | Root protoplasts of *Arabidopsis thaliana* transgenic line J0571 sorted by FACS into GFP+ and GFP-populations | GC-TOF-MS | Petersson et al. (2015) |
| Volatile organic compounds (VOCs) emitted by leaves and flowers | Sorption by Tenax TA (60/80 mesh) granules from glass-enclosed headspace of *Brassica rapa* organs, drawn by a push-pull system | GC-MS | Schiestl et al. (2014) |
| Glucosinolates (e.g. glucobrassicanapin) | Solvent-mediated extraction from flash-frozen *Brassica rapa* tissue samples | LC-MS (ESI: negative mode) | Schiestl et al. (2014) |
| Plant hormones (cytokinins, auxins, gibberellins, abscisic acid, salicylic acid, jasmonic acid) | Pulverized fresh tissue extracted sequentially with multiple solvent mixes and eluted through a series of SPE cartridges; derivatization with specific biochemical probe (for gibberellins) and alkaline phosphatase treatment (for cytokinin nucleotides) at required steps | LC-MS | Kojima and Sakakibara (2012) |
| Procyanidins (oligomers of epicatechin and catechin) | Solvent-mediated extraction from *Heritiera fomes* bark, followed by preparative chromatography | $^{13}C$ NMR | Wangensteen et al. (2009) |

## 6.2 HIGH-THROUGHPUT METHODS FOR PLANT METABOLOMICS STUDY

### 6.2.1 Invasive Methods

State-of-the-art metabolome analytical techniques are mostly based on chromatographic separation followed by mass spectrometry, including gas chromatography-mass spectrometry (GC-MS) and liquid chromatography-mass spectrometry (LC-MS), which are most commonly used for large-scale isolation and analyses of plant metabolites (Manickam et al., 2023). While these methods are fast and easy to standardize based on generations of studies, both have specific requirements which delimit the type of samples which can be analyzed (including extraction procedures for the analytes). The main requirement for GC-MS is that the molecules to be isolated should have low molecular weights and very low polarity, so as to be highly volatile. Compounds with polar moieties have to be derivatized, i.e. the polar functional groups rendered non-polar by treatment with alkylating/silylating reagents, and hence separable by gas chromatography (Gorain et al., 2018; Paul et al., 2022).

In the case of LC-MS, ionization techniques need to be different according to the samples to be analyzed. Electrospray ionization (ESI) may be in positive or negative mode, as required for the types of compounds to be identified – and hence used commonly for targeted metabolomic studies, including profiling for phenols and other amphiphilic metabolites (Table 6.1). The use of the atmospheric pressure chemical ionization (APCI) method, rather than ESI, for LC-MS would be more suitable for thermally stable, less polar molecules with heteroatoms, such as plant alkaloids and their derivatives (Crooks and Byrd, 1999). Analytes targeted for separation by LC-MS may also have to be derivatized to increase their specificity, such as the treatment of phytohormones with bromocholine in order to add tetramethylamine moieties to the gibberellin molecules to improve their accuracy of identification and quantification (Kojima and Sakakibara, 2012). Nuclear magnetic resonance (NMR) spectroscopy is applied to elucidate molecular structures of unknown or novel compounds (Wangensteen et al., 2009; Bhatt et al., 2022). However, the lower sensitivity of NMR spectroscopy mandates that the target molecules are isolated priorly in large quantities by chromatographic techniques (Manickam et al., 2023).

Untargeted metabolomics using LC-MS thus requires versatile, flexible strategies that accommodate the chemical affinities and stabilities of various classes of metabolites. Applications of GC-MS are restricted to profiling for VOCs and small amphiphilic molecules (e.g. fatty acids, polyphenols) which might be derivatized with appropriate reagents (Table 6.1). Except for headspace sampling methods, other extraction procedures for such analytes are destructive and may require cold-chain or drying operations. However, these methods are easy to standardize for any particular set of analytes and therefore can be implemented for analyzing large batches of similar samples collected for comparative studies. Phytohormones (e.g. auxins, cytokinins, gibberellins), which play crucial roles in cellular developmental trajectories, needed to be extracted and isolated using sequential steps involving multiple solvents and solid-phase extraction (SPE) matrices before their analysis by the LC-MS method. The entire preparative process was scaled up to take place in large batches of 96-well microplates, with each sample requiring less than 100 mg of fresh tissue (Kojima and Sakakibara, 2012). Although such approaches may not cover the full array of metabolites extractable from any plant specimen, mass spectrometry-based quantitative data provide information about phenotypic variations among plants as functions of developmental and ecological factors.

Another major challenge for metabolomic profiling in HTP is the rapid identification of analytes. Qualitative data derived by the GC-MS and LC-MS are processed *in silico* by matching the mass spectra of analytes with similar data available in the existing metabolomic databases (Schiestl et al., 2014; Vicherová et al., 2020; Paul et al., 2022). Moreover, mass fragmentation patterns of any organic molecule would vary according to the collision energy used for ionization. Therefore, comprehensive databases are required to include such variations. Untargeted metabolomic studies using LC-MS might have to take into account the multiple levels of collision energy to obtain larger datasets (Umar et al., 2021). Databases built by The Scripps Research Institute (METLIN), the National Institute of Standards and Technology (NIST), and the Golm Metabolome Database are continuously being updated to include the mass spectrometry and FTIR data for the small biological molecules, including amino acids, oligopeptides, lipids, steroids, carbohydrates, exogenous drugs and toxicants (Manickam et al., 2023). Identification of VOCs separated by GC-MS is carried out by matching retention indices, calculated from retention times as compared between the analytes and *n*-alkanes separated using the same temperature programs, with values available in the existing scientific literature (Bicchi and Maffei, 2012).

Detection and quantitation of the analytes in organic/aqueous extracts using the metabolite-specific wavelengths are applied in spectrophotometry (Table 6.1). Estimates of certain metabolite contents (e.g. chlorophylls, proteins, phenolics, carbohydrates, carotenoids, anthocyanins) are obtained by reading their wavelength-specific intensities, in some cases following the treatments with suitable reagents, using spectrophotometers in the ultraviolet-visible (UV-Vis) light range (Satpati et al., 2016; Gorain et al., 2018; Umar et al., 2021; Bhatt et al., 2022; Paul et al., 2022). Applications of Fourier-transform infrared (FTIR) spectrophotometry, using wavenumbers specific to the chemical bonds and structures, resulting in the rapid, broad-spectrum profiling of the chemical functional groups in biological samples (Cebi et al., 2023; Satpati et al., 2016; Umar et al., 2021). In general, wavenumbers ranging from 400 to 4,000/cm are scanned to account for the bio-active molecules including proteins, lipids, carbohydrates, organic acids, aldehydes, amines, amides, hydroxylated compounds as well as the common toxins (Cebi et al., 2023). Different bands of wavenumbers within this range can even be applied to identify the distribution patterns of neutral and polar lipids, along with the functional groups of carbohydrates and proteins (Satpati et al., 2016). Wavenumber bands scanned by FTIR were used to distinguish metabolite contents characterized by phenyl, alkane and carbonyl groups (associated with antioxidant and α-glucosidase inhibitory activities) between two medicinal plants of *Curculigo* genus (Umar et al., 2021).

Scanning for molecule-specific fluorescence, under the excitation by specific wavelengths of the incident light and in some cases induced by suitable reagents, is used to image the spatial distribution of metabolites by flow cytometry of cells. Apart from the cell size, internal granularity and refractive index, various optical filters used in flow cytometry have been applied to quantify the distribution of the clade-specific photopigments in phytoplankton assemblages by the fluorescence-assisted cell sorting technique (FACS) technique (Roy et al., 2018). In the case of multicellular plants, living protoplasts derived by the enzymatic breakdown of the cellulose cell walls in organs may be sorted according to their cellular composition or developmental trajectory, as exemplified by the root cells of *Arabidopsis thaliana* transgenic line J0571 (Petersson et al., 2015). Cortical and endodermal cells of *A. thaliana* J0571 root tips are engineered to express a green fluorescent protein (GFP), so that GFP+ and GFP- protoplasts could be sorted by FACS and then analyzed for their primary metabolite contents by GC-TOF-MS (gas chromatography coupled with time-of-flight mass spectrometer). Once validated with wild-type cells (*A. thaliana* line C24) and compared with metabolite profiles of intact J0571 root tissue, the FACS-based preparative method proved to be a robust and sensitive one.

A rapidly isolating method such as FACS is therefore applied for the high-throughput collection of samples for single-cell metabolomics (SCM) studies (Guo et al., 2021). Another advantage of FACS is the isolation of the unicells or protoplasts keeping their morphology intact, so that a large population of cells collected may be studied and, if required, sonicated to extract the endogenous metabolites. Protoplasts isolated from the multicellular organs, however, might have different metabolite profiles than intact organs due to the dissolution of cell walls (Petersson et al., 2015). Cell metabolites can be directly visualized and extracted using a microscopic probe coupled with a nano-spray tip connected to the syringe, which has the advantage of transferring metabolites to the mass spectrometer setup with minimal disruption (Guo et al., 2021). Profiling of plant SCMs following such protocols might contribute to the elucidation of organ-specific cell types and their development, cellular changes responding to stress factors, and unique cell functions that might be exploited for crop breeding.

Microarray technology, in addition to its applications in genomic and proteomic studies, can be used to investigate polysaccharide and oligosaccharide profiles of plant cells (Fangel et al., 2012). The carbohydrate complement of a cell can be immobilized on a microarray slide or membrane to study interactions between the cell wall components, cell surface molecules, or protein-glycan chemistry. Polysaccharides can be directly adsorbed into the porous surfaces (e.g. nitrocellulose membranes), while oligosaccharides may have to be linked covalently with the carbohydrate-binding proteins or carbohydrate-active enzymes; with linking treatments performed before the surface adsorption in some cases. This HTP approach enables comparative studies of cellular carbohydrate chemistry in large batches (such as in 96-well microplates) using very minute quantities of analytes. Since large proportions of phenolic specialized metabolites in plant cells occur as glycosides (Umar et al., 2021; Bhatt et al., 2022; Paul et al., 2022), microarray technology might be similarly applied to study the interactions of glycosylated polyphenols with other biomolecules, especially in the context of their stress-resistant and medicinal properties.

### 6.2.2 Non-Invasive Methods

Research into volatilomics is expected to reveal several aspects of the ecology and stress physiology of plants, such as the interactions of specific organs with pollinators, seed dispersers and opportunistic herbivores, as well as the physiological status of the plant itself under the biotic or abiotic stress factors (Schiestl et al., 2014; Vicherová et al., 2020). Non-invasive methods of sampling and analyzing the VOCs released by crop species may indicate sustainable, environment-friendly strategies for integrated plant protection, or lead to the improvement of crop varieties with enhanced flavors and fragrances (Bicchi and Maffei, 2012). The volatilome emitted naturally by any individual plant (or plant organ) may directly be sampled by capturing it in the headspace, i.e. air around the emitting organ enclosed within a sampling device, the latter most often made of glass or polymers. A solvent-free porous matrix is introduced into this headspace to 'trap' or retain the VOCs. Then, the latter can be sorbed by the solid but porous matrix when the still headspace air gets saturated with the emitted molecules (static headspace), or passed through the matrix by a steady flow of air through the headspace (dynamic headspace). Alternatively, air from the static headspace can directly be sampled by a gas-tight syringe and then injected into a GC-MS setup. The matrices used for static headspace sampling are composed of varying proportions of polydimethylsiloxane, carboxen and other adsorbents coated onto a fine fiber in a device for solid-phase micro-extraction (SPME). The matrices for dynamic headspace sampling are typically packed as a bed of porous material into a cartridge of inert material such as glass. These porous adsorbents used include activated charcoal, silica gel, or synthetic resins such as Porapak Q, which have to be desorbed by eluting with any organic solvents – the resulting liquid mix is then injected for GC-MS analysis. Some adsorbent resins, such as Tenax TA, and SPME fibers can thermally be desorbed and therefore directly introduced into the injection portal for GC-MS analysis. While thermal desorption requires specific and sometimes expensive equipment for implementation, liquid desorption carries the risk of introduction of the contaminants and loss of the analytes. Injections by liquid desorption and gas-tight syringe methods have been automated in the latest GC-MS setups, whereas SPME necessitates manual injection.

Non-invasive imaging of the plant pigments has been attempted through digital photography techniques (Sass et al., 2012; del Valle et al., 2017). Sass et al. (2012) developed a method to quantify the chlorophyll content and photochemical quantum efficiency of leaves, through different stages of maturity till the senescence, as obtained by the processing of red-green-blue (RGB) pixel data in digital photographs using the trademarked image processing software. The RGB data, when converted to the Hue Saturation Values, were found to be the effective predictors of the abovementioned pigment parameters, which were independently assessed using biochemical assays and fluorimetry. This method was proven to be an effective one for a wide range of foliar chlorophyll concentrations and was inferred to be useful for the field studies provided the whiteness of the leaf background would be normalized in the images. Estimation of the anthocyanin contents of plant organs using the RGB data of digital photographs was tried by del Valle et al. (2017). The prerequisites of this method include avoidance of over-exposed or excessively bright images, and standardization of the parameters according to the species/variety-specific anthocyanin composition. After cross-checking the coloration and pigment composition of test samples by the spectral reflectance and biochemical assays respectively, this method was found to effectively assess the relative anthocyanin contents of various plant organs, with correction factors standardized for the chlorophyllous tissues. Spectral reflectance using portable spectrometers was found to be another method to compare effectively the pigmentation patterns of plant organs in a non-invasive way for field studies.

## 6.3 APPLICATIONS OF HTP METABOLOMICS

### 6.3.1 In Systems Biology

High-throughput analyses of specialized metabolites under variable stress factors are expected to reveal complex interactions dictating the biology of crop plants. Often these interactions and metabolomic changes are studied on a whole-plant basis to understand the systemic effects on quality and yield; to predict large-scale distribution of the stress factors, such as drought, disease and pest infestations; and to even identify the sustainable techniques of stress-resistance. The organ-specific analyses of secondary metabolites in white turnip (*Brassica rapa*) under pest infestations have revealed the species-specific defense responses leading to the differential effects on reproductive efficiency (Schiestl et al., 2014). Under the infestation by native (in Europe) cabbage butterflies (*Pieris brassicae*), the flowers of *B. rapa* exhibited an overall lower emission of the VOCs but a higher early-blooming frequency, compensating for any loss in pollinator attraction. However, when infested by the invasive African cotton leafworms (*Spodoptera littoralis*) loss of emitted floral VOCs was accompanied by a higher production of the glucosinolate compound glucobrassicanapin, leading to the loss of pollinators. In both cases, decrements in floral scent emission resulted in the attraction of parasitoid wasps attacking the herbivores.

The evolutionary arms race between the timber-producing conifers and the bark beetles (and associated parasitic organisms) presents a case of complex adaptations on both sides, including the exploitation of constitutive resins and VOCs expressed in Pinaceae – a family that largely co-evolved with the bark beetles (Franceschi et al., 2005). Pinewood nematodes (*Bursaphelenchus xylophilus*), which are the causative agents of the lethal pine wilt disease, are introduced into the Masson pine (*Pinus massoniana*) by the longhorned sawyer beetles (*Monochamus alternatus*). The nematodes at specific juvenile, dispersing stages are attracted to the VOCs (mainly camphor, β-pinene and longifolene) released by *M. alternatus* larvae developing inside the wood, and at their mature, reproductive stages attracted to the scent (mainly α-pinene) emitted by healthy *P. massoniana* resin (Zhao et al., 2007). Reproductively mature *B. xylophilus* was found to be repelled significantly by the odors released from the wood infested with either the blue-stain fungi or the longhorned beetle pupae, in both cases dominated by almost equal proportions of α-pinene and β-pinene. However, *in vitro* experiments suggested that the additional quantities of α-pinene are detrimental to the survival of *B. xylophilus* cultures, and the mortality is enhanced significantly when the nematodes are depleted of symbiotic microbiota by prior antibiotic treatments (Wang et al., 2019). Concentrations of α-pinene in *P. massoniana* shoots were enhanced over two-fold after 24 h of feeding by *M. alternatus* (Niu et al., 2012), which implies a possible antibiotic/anti-feedant role of this monoterpene. Paradoxically the mycelial growth of certain blue-stain fungal strains was inhibited significantly by a terpene mixture (α-pinene:β-pinene::1:0.8) simulating that emitted by fungus-infected wood, indirectly resulting in a decreased propagation of *B. xylophilus* by depleting the nematodes' food resources. Very large concentrations of α-pinene, however, resulted in an increased propagation of *B. xylophilus*, which was thought to be due to its major food source, viz. blue-stain fungal strain *Sporothrix* sp. 1 being resistant to high monoterpene contents. Similarly in Chinese pine

(*Pinus tabulaeformis*), dominance of α-pinene among bark monoterpenes of larger trees led to the increased susceptibility to red turpentine beetles (*Dendroctonus valens*) invasion, whereas lower but equal concentrations of α-pinene and β-pinene in smaller trees conferred the resistance, probably evolved as a protective mechanism for younger, growing saplings (Liu et al., 2011). The major monoterpenes in *P. massoniana* resin and *P. tabulaeformis* bark thus function as the mediators and biocontrol agents in the complex ecological interactions as revealed by the systemic monitoring of metabolites (Zhao et al., 2007; Liu et al., 2011; Niu et al., 2012; Wang et al., 2019).

Conifers other than Pinaceae are less susceptible to bark beetle infestations but express various forms (including inducible but not constitutive resins) of chemical defenses against herbivores and pathogens (Franceschi et al., 2005). The Italian cypress (*Cupressus sempervirens*), for example, is globally vulnerable to the cypress bark canker caused by a fungus (*Seiridium cardinale*), and responds by changing the phloem and foliar terpenoid compositions (Achotegui-Castells et al., 2015). The long-term (30–90 days) responses in infected phloem included a significant increase in the total monoterpenoids (e.g. α-thujene, thymyl methyl ether) and diterpenoids (e.g. manool), at the expense of sesquiterpenoids (e.g. cedrol), whereas the systemic responses (at 10 days after inoculation) in leaves consisted of a decreased sesquiterpenoid (e.g. caryophyllene) content and an increased content and emission of monoterpenoids (e.g. δ-3-carene). The causative agents of cypress bark canker are dependent on the wood-tunneling activity of the cedar bark beetles (*Phloeosinus aubei*), which are oligophagous and lethally attack other Cupressaceae members, such as the Northern white cedar (*Thuja occidentalis*) by taking advantage of the drought conditions (Bozsik et al., 2016). Investigations using electroantennography revealed that *P. aubei* adults are sensitive to several monoterpenoids (e.g. β-thujone, terpinen-4-ol), sesquiterpene (e.g. α-humulene) and oxylipin (e.g. 1-octene-3-ol) VOCs shared by the conifers belonging to *Cupressus*, *Cedrus* and *Juniperus* genera, enabling the cedar bark beetles to colonize new host species using these VOCs as kairomones. The drought-enabled expansion of the cedar bark beetles' geographical range, exploiting the volatilome common among different members of Cupressaceae, might be countered by applying repellent but eco-friendly VOCs, or by creating the 'trap' populations of trees with heightened emission of beetle-attracting kairomones.

### 6.3.2 In Functional Genomics

Rapid advancements in "omics" technology (genomics, transcriptomics, proteomics and metabolomics) have facilitated the comprehensive studies of major biomolecules, establishing the information transfer pathways from genotype to phenotype (Djande et al., 2020; Yang et al., 2020; Manickam et al., 2023). Metabolomics, as one component of these interrelated fields, is applied to systematically study the myriad of small molecules produced due to various enzymatic activities; the enzymes (proteins) in turn were translated from transcript-carried information transcribed from genomes (Figure 6.1). Such functional genomic studies are targeted to understand the biochemical

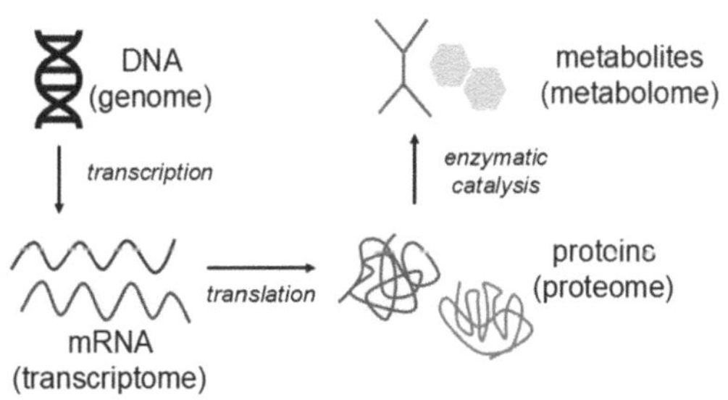


**FIGURE 6.1** The flow of biological information from the genome level through transcriptome and proteome to metabolome level; environmental factors affect genetic information transfer at each step.

pathways, along with their complex regulatory mechanisms, which determine a plant's physiological status, which in turn dictates its ecological and economic values. The state-of-the-art genome editing tools can then be used to perform precise, efficient modifications in gene functions leading to a targeted manipulation of these metabolomic profiles and resultant crop traits. Databases containing the genomic information of crop breeds, such as the "1000 Plant-Genomes Project", enable researchers to improve crop traits selectively.

### 6.3.3 Generation of Hypotheses on Molecular Mechanisms and Metabolic Regulation

The optimal defense strategy is a mechanism theorized to account for the spatiotemporal variations in the defense-specialized metabolites in a species – the phytochemical contents changing according to the incident stress and resources available to the plant as dictated by the environmental and developmental factors. This principle was demonstrated by the inoculation of phloem tissue in Norway spruce (*Picea abies*) with the bark beetle-associated fungus (*Ceratocystis polonica*) resulting in significantly enhanced terpene concentrations under all the tested soil conditions of macronutrient availability (Viiri et al., 2001). The intra-phloem concentrations of monoterpenes were found to be inversely proportional to the distance from the inoculation site, and directly proportional to the lesion size when sampled near the inoculation site, which implied the need-based systemic movement of defense compounds. Total concentrations of terpene VOCs in *P. abies* bark and phloem appeared to rise steadily with time from the day of *C. polonica* inoculation, with the ratios among the individual VOCs changing up to 35 days after inoculation (Zhao et al., 2011). At this time point, the trees with more than 200 milligram terpenes per gram dry phloem, showed only 28.6% of bark beetle colonization, as compared to the trees with 100–200 mg terpenes/g dry phloem. An equivalent to 4.9% of the infestation level was observed in trees with less than 100 mg terpenes/g dry phloem, thus establishing the time-dependent build-up of terpenoid-conferred resistance to the bark beetles. The variations caused by growth stage differences, such as between the bark monoterpene concentrations of large (diameter 30 cm) and small (diameter 10 cm) Chinese pines (*Pinus tabulaeformis*), might also result in a significant difference in the herbivore infestations (Liu et al., 2011). In the case of *P. tabulaeformis*, however, the higher monoterpene concentrations (especially of $\alpha$-pinene) in the larger trees were found to attract the invasive, red turpentine beetles (*Dendroctonus valens*) compared to the smaller trees, indicating an ecological arms' race in which the bark beetles evolved to exploit the conifers' chemical defenses as kairomones signaling for greater nutrient and space availability. This rapid adaptation of *D. valens* for optimal exploitation of its new host species makes a strong case for motivating the HTP metabolomic studies in the context of plant-herbivore interactions.

Genotype- and development-specific responses of parasites to conifer terpenoid contents are also evident in significantly greater biodiversity in uneven-aged mixed boreal forests over even-aged monocultures, as reported by Savilaakso et al. (2021). Monoterpene emission profiles effectively contribute to the interspecific differences in bark beetle resistance in the mixed conifer communities, such as the attraction of the female mountain pine beetles (*Dendroctonus ponderosae*) to the scent of limber pines (*Pinus flexilis*, a typical host species), while their repellence to that of the Great Basin bristlecone pines (*Pinus longaeva*) which successfully avoid *D. ponderosae* infestations (Gray et al., 2015). Attacks by *D. ponderosae* on conifers are often compounded by the infestations of associated blue-stain fungi (e.g. *Grosmannia clavigera*), which have to be countered by the trees' systemic induced resistance (SIR) mechanisms, including the rapidly rising concentrations of cytotoxic monoterpenes (Lusebrink et al., 2011; Goodsman et al., 2013; Keefover-Ring et al., 2015; Arango-Velez et al., 2016; Klutsch et al., 2017). In the Canadian lodgepole pines (*Pinus contorta* var. *latifolia*) and jack pines (*Pinus banksiana*), a deficit of soil water was found to hamper significantly the capacity of emitting monoterpenes, which was still enhanced significantly in the lodgepole pines after one day of the *G. clavigera* inoculation (Lusebrink et al., 2011). Production of these monoterpenes was up-regulated along with the mobilization of soluble sugars (as precursors) after the lodgepole pines were inoculated with *G. clavigera*, so that parts of a tree distant from the crown (where sugars are dominantly manufactured) eventually acquired defensive potential build-up (Goodsman et al., 2013). Being a typical host of the mountain pine beetles, bark monoterpene concentrations in *P. contorta* var. *latifolia* under water deficit conditions were found to be compromised by the *G. clavigera* inoculation within 7 days but significantly enhanced over a longer (28 days) period (Arango-Velez et al., 2016). In contrast, the monoterpene concentrations in the native *P. banksiana* populations showed no significant effects in response to the *G. clavigera* inoculation (Lusebrink et al., 2011; Arango-Velez et al., 2016), and in fact, were inhibited by water deficit under the inoculated condition (Klutsch et al., 2017). However, *G. clavigera* treatment induced a local defensive response rather than the SIR in ponderosa pines (*Pinus ponderosa*), with the multi-fold enhanced level production of monoterpenes (e.g. δ-3-carene), sesquiterpenes (e.g. longifolene), diterpenes (e.g. neoabietic acid), and phenylpropanoids (e.g. estragole) in the phloem within 17 days (peak period of *D. ponderosae* attack). This might explain the lower susceptibility of *P. ponderosa* to the mountain pine beetle (Keefover-Ring et al., 2016). Hence, there seems to be ample evidence that the physiology of many conifers (e.g. *P. abies*, *P. contorta* var. *latifolia*) is adapted to maintain the optimal chemical defense against the herbivores and pathogens under the conditions of abiotic stress, such as drought and nutrient deficit, even if a variable period of lag occurs between invasion and the consequent response (Viiri et al., 2001; Lusebrink et al., 2011; Goodsman et al., 2013; Arango-Velez et al., 2016). Such phenomena and ecological principles are established with the data obtained from the widely ranged, long-term HTP metabolomic studies.

### 6.3.4 Others

Metabolomic fingerprinting, i.e. profiling of taxon- or condition-specific metabolites, is applied as a strategy for the high-throughput metabolomic monitoring of the physiological status of plants subject to genetic, developmental and/or environmental differences (Djande et al., 2020). Both primary and secondary metabolites can be monitored, mapped and statistically analyzed to identify the phenotypic discriminants among crop cultivars, among maturation stages, or between the optimal and stressed conditions. Various accessions of tomato (*Solanum lycopersicum*) and blueberry (*Vaccinium* spp.) crops were assessed for their flavor parameters which could be predicted by the fruit metabolite contents (Colantonio et al., 2022). Fruit VOCs were found to account for respectively 42% and 56% of variances in tomato and blueberry flavors. Sugar/acid ratios are also correlated with the sweetness and sourness parameters. Such data can be used for food quality analysis and crop improvement programs.

Metabolomic footprinting is an HTP approach whereby metabolites (e.g. VOCs, root secretions) released by plants into the environment are monitored to map their biochemical or physiological status in the context of an ecosystem (Djande et al., 2020). Systemic observation of natural odor dynamics revealed ecologically significant plant-plant interactions in the peatland mosses (Vicherová et al., 2020). A rare, protected species *Hamatocaulis vernicosus* constitutively emits apocarotenoid VOCs including β-cyclocitral and methyl 2,6,6-trimethyl-1-cyclohexene-1-carboxylate (MTCC); the proportion of the latter enhanced multi-fold when *H. vernicosus* populations were exposed to VOCs emitted by neighboring *Sphagnum flexuosum* populations. Since *S. flexuosum* is ubiquitous in peatland habitats and grows vigorously, its presence induces MTCC-mediated stress signaling leading to accelerated growth rates in *H. vernicosus*. Enhanced emission of MTCC and β-cyclocitral was also observed under the increased far-red light exposure, indicating the role of the apocarotenoids in activating the shade avoidance mechanisms in this rare bryophyte species. The competition for sun and space between this protected moss with its neighbors thus might be monitored through odor emission profiles.

## 6.4 CONCLUSION

Plant metabolomic studies in the current stage have various HTP tools at their disposal. These include advanced chromatographic and imaging techniques for both primary and secondary metabolites. Major challenges in these operations are still concerned with the separation of structurally similar compounds, and identification of the isolated molecules, though databases are continuously being updated with the molecular characteristics of known compounds. Applications of these studies can reveal ecologically and economically important details of plant-environment and cell-cell biochemical interactions, and developmental botany, and potentially open the paths towards crop improvement and novel biocontrol agents.

## REFERENCES


Achotegui-Castells A, Danti R, Llusià J, Rocca GD, Barberini S, Peñuelas J (2015). Strong induction of minor terpenes in Italian Cypress, *Cupressus sempervirens*, in response to infection by the fungus *Seiridium cardinale*. *Journal of Chemical Ecology* 41, 224–243. https://doi.org/10.1007/s10886-015-0554-1

Arango-Velez A, El Kayal W, Copeland CC, Zaharia LI, Lusebrink I and Cooke JE (2016). Differences in defence responses of *Pinus contorta* and *Pinus banksiana* to the mountain pine beetle fungal associate *Grosmannia clavigera* are affected by water deficit. *Plant, Cell & Environment* 39, 726–744. https://doi.org/10.1111/pce.12615

Bhatt V, Sendri N, Swati K, Shinde BD, Bhandari P (2022). Identification and quantification of anthocyanins, flavonoids, and phenolic acids in flowers of *Rhododendron arboreum* and evaluation of their antioxidant potential. *Journal of Separation Science* 2022, 1–11. https://doi.org/10.1002/jssc.20220.0145

Bicchi C, Maffei M (2012). The plant volatilome: Methods of analysis. In: Normanly J (ed), *High-Throughput Phenotyping in Plants. Methods in Molecular Biology 918*. Humana Press, Totowa, NJ. https://doi.org/10.1007/978-1-61779-995-2_15

Bozsik G, Tröger A, Francke W, Szőcs G (2016). *Thuja occidentalis*: Identification of volatiles and electroantennographic response by the invasive cedar bark beetle, Phloeosinus aubei. *Journal of Applied Entomology* 140, 434–443. https://doi.org/10.1111/jen.12272

Cebi N, Bekiroglu H, Erarslan A (2023). Nondestructive metabolomic fingerprinting: FTIR, NIR and Raman spectroscopy in food screening. *Molecules* 28, 7933. https://doi.org/10.3390/molecules28237933

Colantonio V, Ferrão LFV, Tieman DM, Bliznyuk N, Sims C, Klee HJ, Munoz P, Resende MFR Jr (2022). Metabolomic selection for enhanced fruit flavor. *PNAS* 119:7, e2115865119. https://doi.org/10.1073/pnas.2115865119

Crooks PA, Byrd GD (1999). Chapter 7 - Use of high-performance liquid chromatographic–mass spectrometric (LC–MS) techniques for the determination of nicotine and its metabolites. In: Gorrod JW, Jacob P (eds), *Analytical Determination of Nicotine and Related Compounds and their Metabolites*, Elsevier Science, pp. 225–264, ISBN 9780444500953. https://doi.org/10.1016/B978-044450095-3/50008-5

del Valle JC, Gallardo-López A, Buide ML, Whittall JB, Narbona E (2018). Digital photography provides a fast, reliable, and noninvasive method to estimate anthocyanin pigment concentration in reproductive and vegetative plant tissues. *Ecology and Evolution* 8, 3064–3076. https://doi.org/10.1002/ece3.3804

Djande CYH, Pretorius C, Tugizimana F, Piater LA, Dubery IA (2020). Metabolomics: A tool for cultivar phenotyping and investigation of grain crops. *Agronomy* 10, 831. https://doi.org/10.3390/agronomy10060831

Fangel JU, Pedersen HL, Vidal-Melgosa S, Ahl LI, Salmean AA, Egelund J, Rydahl MG, Clausen MH, Willats WGT (2012). Carbohydrate microarrays in plant science. In: Normanly J (ed), *High-Throughput Phenotyping in Plants. Methods in Molecular Biology 918*. Humana Press, Totowa, NJ. https://doi.org/10.1007/978-1-61779-995-2_19

Franceschi VR, Krokene P, Christiansen E, Krekling T (2005). Anatomical and chemical defenses of conifer bark against bark beetles and other pests. *New Phytologist* 167, 353–376. https://doi.org/10.1111/j.1469-8137.2005.01436.x

Goodsman DW, Lusebrink I, Landhäusser SM, Erbilgin N, Lieffers VJ (2013). Variation in carbon availability, defense chemistry and susceptibility to fungal invasion along the stems of mature trees. *New Phytologist* 197, 586–594. https://doi.org/10.1111/nph.12019

Gorain PC, Sengupta S, Satpati GG, Paul I, Tripathi S, Pal R (2018). Carbon sequestration in macroalgal mats of brackish-water habitats in Indian Sunderbans: Potential as renewable organic resource. *Science of the Total Environment* 626, 689–702. https://doi.org/10.1016/j.scitotenv.2018.01.106

Gray CA, Runyon JB, Jenkins MJ, Giunta AD (2015). Mountain pine beetles use volatile cues to locate host limber pine and avoid non-host great basin bristlecone pine. *PLoS One* 10(9), e0135752. https://doi.org/10.1371/journal.pone.0135752

Guo S, Zhang C, Le A (2021). The limitless applications of single-cell metabolomics. *Current Opinion in Biotechnology* 71, 115–122. https://doi.org/10.1016/j.copbio.2021.07.015

Hall RD, D'Auria JC, Silva Ferreira AC, Gibon Y, Kruszka D, Mishra P, van de Zedde R (2022). High-throughput plant phenotyping: A role for metabolomics? *Trends in Plant Science* 27, 6. https://doi.org/10.1016/j.tplants.2022.02.001

Keefover-Ring K, Trowbridge A, Mason CJ, Raffa KF (2016) Rapid induction of multiple terpenoid groups by ponderosa pine in response to bark beetle-associated fungi. *Journal of Chemical Ecology* 42, 1–12. https://doi.org/10.1007/s10886-015-0659-6

Klutsch JG, Shamoun SF, Erbilgin N (2017). Drought stress leads to systemic induced susceptibility to a necrotrophic fungus associated with mountain pine beetle in *Pinus banksiana* seedlings. *PLoS One* 12:12, e0189203. https://doi.org/10.1371/journal.pone.0189203

Kojima M, Sakakibara H (2012). Highly sensitive high-throughput profiling of six phytohormones using MS-probe modification and liquid chromatography–tandem mass spectrometry. In: Normanly J (ed), *High-Throughput Phenotyping in Plants*. Methods in Molecular Biology 918. Humana Press, Totowa, NJ. https://doi.org/10.1007/978-1-61779-995-2_11

Liu Z, Wang B, Xu B, Sun J (2011). Monoterpene variation mediated attack preference evolution of the bark beetle *Dendroctonus valens*. *PLoS One* 6:7, e22005. https://doi.org/10.1371/journal.pone.0022005

Lusebrink I, Evenden ML, Blanchet FG, Cooke JEK, Erbilgin N (2011). Effect of water stress and fungal inoculation on monoterpene emission from an historical and a new pine host of the mountain pine beetle. *Journal of Chemical Ecology* 37, 1013–1026. https://doi.org/10.1007/s10886-011-0008-3

Manickam S, Rajagopalan VR, Kambale R, Rajasekaran R, Kanagarajan S, Muthurajan R (2023). Plant metabolomics: Current initiatives and future prospects. *Current Issues in Molecular Biology* 45, 8894–8906. https://doi.org/10.3390/cimb45110558

Niu H, Zhao L, Lu M, Zhang S, Sun J (2012). The ratio and concentration of two monoterpenes mediate fecundity of the pinewood nematode and growth of its associated fungi. *PLoS One* 7:2, e31716. https://doi.org/10.1371/journal.pone.0031716

Paul I, Ghosh S, Acharya K, Bhadoria PBS (2022). Bioactivities of phenylpropanoids extracted from ephemeral flowers of medicinal woody species *Murraya paniculata* (Linn.) Jack. *Journal of Biologically Active Products from Nature* 12:4, 353–365. https://doi.org/10.1080/22311866.2022.2125903

Petersson SV, Lindén P, Moritz T, Ljung K (2015). Cell-type specific metabolic profiling of *Arabidopsis thaliana* protoplasts as a tool for plant systems biology. *Metabolomics* 11, 1679–1689. https://doi.org/10.1007/s11306-015-0814-7

Roy AS, Gorain PC, Paul I, Sengupta S, Mondal PK, Pal R (2018). Phytoplankton nutrient dynamics and flow cytometry based population study of a eutrophic wetland habitat in eastern India, a Ramsar site. *RSC Advances* 8:17, 9530–9545. https://doi.org/doi.org/10.1039/C7RA12761H

Sass L, Majer P, Hideg É (2012). Leaf hue measurements: A high-throughput screening of chlorophyll content. In: Normanly J (ed), *High-Throughput Phenotyping in Plants*. Methods in Molecular Biology 918. Humana Press, Totowa, NJ. https://doi.org/10.1007/978-1-61779-995-2_6

Satpati GG, Gorain PC, Paul I, Pal R (2016). An integrated salinity-driven workflow for rapid lipid enhancement in green microalgae for biodiesel application. *RSC Advances* 6:113, 112340–112355. https://doi.org/doi.org/10.1039/C6RA23933A

Savilaakso S, Johansson A, Häkkilä M, Uusitalo A, Sandgren T, Mönkkönen M, Puttonen P (2021). What are the effects of even-aged and uneven-aged forest management on boreal forest biodiversity in Fennoscandia and European Russia? A systematic review. *Environmental Evidence* 10, 1. https://doi.org/10.1186/s13750-020-00215-7

Schiestl FP, Kirk H, Bigler L, Cozzolino S, Desurmont GA (2014). Herbivory and floral signaling: Phenotypic plasticity and tradeoffs between reproduction and indirect defense. *New Phytologist* 203:1, 257–266. https://doi.org/10.1111/ nph.12783

Umar AH, Ratnadewi D, Rafi M, Sulistyaningsih YC (2021). Untargeted metabolomics analysis using FTIR and UHPLC-Q-orbitrap HRMS of two *Curculigo* species and evaluation of their antioxidant and α-glucosidase inhibitory activities. *Metabolites* 11, 42. https://doi.org/10.3390/metabo11010042

Vicherová E, Glinwood R, Hájek T, Šmilauer P, Ninkovic V (2020). Bryophytes can recognize their neighbours through volatile organic compounds. *Scientific Reports* 10, 7405. https://doi.org/10.1038/s41598-020-64108-y

Viiri H, Annila E, Kitunen V, Niemel. P (2001). Induced responses in stilbenes and terpenes in fertilized Norway spruce after inoculation with blue-stain fungus, Ceratocystis polonica. *Trees* 15, 112–122. https://doi.org/10.1007/s004680000082

Wang X, Yu Y, Ge J, Xie B, Zhu S, Cheng X (2019). Effects of α-pinene on the pinewood nematode (*Bursaphelenchus xylophilus*) and its symbiotic bacteria. *PLoS One* 14:8, e0221099. https://doi.org/10.1371/journal.pone.0221099

Wangensteen H, Dang HCT, Jamal Uddin S, Alamgir M, Malterud KE (2009). Antioxidant and antimicrobial effects of the mangrove tree *Heritiera fomes*. *Natural Product Communication* 4, 371–376. https://doi.org/10.1177/19345.78X09.00400.311

Yang W, Feng H, Zhang X, Zhang J, Doonan JH, Batchelor WD, Xiong L, Yan J (2020). Crop phenomics and high-throughput phenotyping: Past decades, current challenges, and future perspectives. *Molecular Plant* 13, 187–214. https://doi.org/10.1016/j.molp.2020.01.008

Zhao LL, Wei W, Kang L, Sun JH (2007). Chemotaxis of the pinewood nematode, *Bursaphelenchus xylophilus*, to volatiles associated with host pine, *Pinus massoniana*, and its vector *Monochamus alternatus*. *Journal of Chemical Ecology* 33, 1207–1216. https://doi.org/10.1007/s10886-007-9289-y

Zhao T, Krokene P, Hu J, Christiansen E, Björklund N, Långström B, Solheim H, Borg-Karlson A-K (2011). Induced terpene accumulation in Norway spruce inhibits bark beetle colonization in a dose-dependent manner. *PLoS One* 6:10, e26649. https://doi.org/10.1371/journal.pone.0026649

# 7 High-Throughput Phenotyping

## *An Effective Tool for Plant Breeders to Dissect Crop Responses to Abiotic Stresses*

*Swati Hazra, Priyanka Joshi, Arkabanee Mukherjee, and Akbar Hossain*

## 7.1 INTRODUCTION

The genetic improvement of crops through extensive plant breeding approaches has contributed significantly to increasing crop yield in recent decades. However, global food security is further threatened by the ever-increasing demand for food caused by population increases, along with other anthropogenic factors, such as climate change, land degradation, and resource limitations. Approximately 3.6 billion people are exposed to the vulnerability associated with accelerated climate change. According to the Intergovernmental Panel on Climate Change (IPCC) report, anthropogenic activities have been major contributors to the increase in the global warming rate since the mid-20th century. Climate change has been identified as one of the most important factors affecting food security. Climate change has altered all climatic factors, including an increase of 1.09°C in global surface temperature from the last decade, as reported in the Sixth Assessment Report, i.e., AR6 (IPCC, 2021). Water is an indispensable natural resource used to sustain life on earth. All kinds of human settlements, including industrial, urban, and rural settlements, require the use of fresh water. Climate change and indiscriminate exploitation have led to the depletion of all freshwater resources, including groundwater reserves. Water scarcity directly affects irrigation practices for agriculture, especially for crops such as rice, which have relatively high water requirements. Drought completely destroys agricultural production systems. In addition to limited water availability, increasing temperature also causes alterations in crop species structure, increased exposure to abiotic stress, and shortening of the cultivation period (Ullah, 2017). Rising temperatures can affect the soil nutrient profile and increase the risk of disease aggravation (Davis and Vincent, 2017). Extreme rainfall events and high temperatures cause soil erosion (Eekhout and de Vente, 2022). Soil salinity is another major abiotic stressor for plants. The increasing salinity of soil compels spatial shifts in crops and alterations in cropping sequences. Salinity reduces soil fertility, transforms cultivated land into barren land and thus gradually leads to the generation of wasteland (Mukhopadhyay et al., 2021).

Ensuring sufficient food with the desired nutritious value with consistent production for every mouth remains the sole purpose of the Food Security mission under the Sustainable Development Goals (SDGs) set by the United Nations (Margulis, 2012). The development of nations with relatively high population growth rates, which are located in southern Asia and sub-Saharan Africa, would be strongly affected by changes in climate. By 2050, further addition of 1.24 billion people in these developing nations will put tremendous pressure on agriculture. To meet food security and safety challenges for large populations, researchers have focused on mitigation and adaptive strategies for all aspects of environmental factors that may affect crop production with respect to climate change.

The introduction of smart technologies in relation to resources such as water, carbon, soil, and energy can reduce their overuse and may stabilize production and increase the resource use

DOI: 10.1201/9781003505631-7

efficiency and sustainability of the system. The basics of these technologies are rooted in the conventional agricultural practices of farmers. Gradually, technologies have evolved from precision farming to smart agriculture. Climate-smart agriculture is a more precise form of smart agriculture because of changing climatic factors (Mizik, 2021). The benefits of smart agriculture include water conservation, lower greenhouse gas (GHG) emissions, reduction in soil erosion, reduction in cultivation expenses, and optimized use of fertilizers and pesticides (Mitra et al., 2022). Sustainable farming practices such as intercropping, crop diversification, organic farming, reduced pesticide application, and integrated farming can benefit both the economy and the environment.

To meet the increased demand for food, crop production requires a boost while optimizing the use of resources efficiently. Implementation of smart agricultural practices and Hi-Tech systems, along with the use of artificial intelligence (AI), machine learning (ML) and Deep Learning (DL) technologies can help to achieve the targeted enhancement of crop yield. For example, smart irrigation systems equipped with soil- and weather-based moisture sensors have been developed (Nagothu, 2016). The sensors help in real-time monitoring of soil moisture, humidity, and water potential parameters that are subsequently utilized to regulate water application through integrated sprinkler systems. Thus, crop growth and water use efficiency (WUE) are optimized. Similarly, the deployment of multiple linear regression algorithms and LoRa point-to-point (P2P) networks to the automated irrigation system of greenhouses has helped the production of organic high-value crops (Chang et al., 2019). On the other hand, automated tractors and unmanned aerial vehicles (UAVs), such as drones, are available in the field for real-time monitoring and management of large agricultural farms. The normalized difference vegetation index (NDVI) is mostly used in these cases to analyze the signals or images to assess crop health and to identify the abiotic and biotic stress factors (Radoglou-Grammatikis et al., 2020). The use of artificial neural networks (ANNs) to identify and manage pests and diseases in agricultural and horticultural farms (Liu et al., 2019) and to detect and eradicate weeds, saves time and labor as compared to the conventional labor-intensive, manual methods. In regards to climate change mitigation, carbon-smart farming lowers greenhouse gas emissions. The former also includes the energy-smart agricultural practices involving the use of power-saving tools for farm management.

In addition to the management of crops in the field, the crop breeding and biotechnology front has also advanced. A detailed study of the whole-genome sequence and its implementation in various crop species have been conducted over the past few years. Currently, high-throughput phenotyping (HTP) is the next important research area. The HTP implies big data acquisition, which is a mammoth task using conventional methods. Traditional methods are laborious and time-consuming and involve destructive/invasive plant sampling (Furbank and Tester, 2011; Chen et al., 2014). However, with advancements in technological applications in agricultural fields, obtaining such data has become easier. Sensors, automated agricultural devices, machine learning, and drones have altered the traditional (manual) data acquisition process and substantially reduced the cost of time and labor. The Internet of Things (IoT) is the next step for wireless sensor networks (WSNs), in which sensor-based datasets are obtained for further decision-making processes in conservation agriculture (Madushanki et al., 2019; Yang et al., 2021). In addition to sensors and automated agricultural machinery, such as power tillers, tractors, and harvesters, fluorescence spectroscopy is used in ground-based HTP data collection. Cameras capturing wavelengths of 3–5 μm and 7–14 μm are the most commonly used ones for thermal imaging. Aerial images obtained by drones are also analyzed via solar radiation reflectance from vegetation to obtain phenomics data. HTP with AI-based data-acquiring techniques is a noninvasive technology for information gathering. In addition to AI/ML, deep learning (DL) models, i.e., generative adversarial networks (GANs), convolutional neural networks (CNNs), recurrent neural networks (RNNs), and multilayer perceptrons (MLPs), have been used (LeCun et al., 2015). The application of HTP includes root phenotyping for better access to nutrients in the crop, detection of pathogens and pests, identification of high-yielding abiotic stress-resistant genotypes, and improvements in crop productivity.

## 7.2 CROP PHENOTYPING AND PHENOMICS FOR IMPROVED AGRICULTURAL PRODUCTIVITY

"Phenotype", the general term used by biologists, originated from two Greek words, "phaínô", i.e., "to show/appear", and "tupos", i.e., "mark/type" (de Vienne, 2022). In 1911, Wilhelm Johannsen characterized the 'phenotype' by stating that "all types of organisms can be distinguished by direct inspection or with finer methods of measurement or description" (Yang et al., 2020a). The term has frequently been linked to another frequently used term, "genotype", which means "genetic makeup" in biology. In the traditional approach, the phenotype represents the "distinguishable characteristics" or macroscopically observable traits of biological organisms that are produced due to interactions between the genotype and the surrounding environment (Jabaudon et al., 2018; Adrish et al., 2022). However, with the advancement of technology, the phenotype concept was later expanded to include various qualitative and quantitative characteristics observed at every dimension of biological organization, starting from the molecular (transcriptomes, proteomes, metabolomes, etc.), chromosomal and anatomical levels to the macroscopic organ level and the whole body of organisms (de Vienne, 2022).

A phenotype is developed through continuous interactions between an organism and its surroundings from the early developmental stage until death. Therefore, quantitative analysis of phenotypes, which demands detailed records of the possible multidimensional interactions among the variables of biological and environmental systems distributed across spaces and times, is very difficult (Houle et al., 2010). The acquisition of phenotypic data at a large scale has been one of the major challenges in studies on crop breeding and functional genomics (Yang et al., 2020a). Moreover, there is "plasticity of phenotypes" in plants, as they possess great potential to adapt to limiting environmental conditions (e.g., drought) by modifying their structure and physiology (Pieruschka and Schurr, 2019). All organs and their molecular assemblies (including those at the cellular and tissue levels) essentially play roles in the regulation of a particular plant phenotype; for example, 'dwarf' or 'tall' phenotypic traits are determined by the patterns of shoot and root growth, which in turn are influenced by the dynamics of carbon allocation to different plant organs.

The term 'phenome' represents the complete set of phenotypes of an organism as a whole and at the cellular, tissue, and organ levels (Dhondt et al., 2013). Limited resource availability and changes in climatic conditions often lead to the increased frequency of occurrence of various abiotic stressors, such as heat, cold, drought, and salinity, which interfere with several physiological and biochemical processes, and consequently hinder plant growth and development. Modern technologies have revolutionized the interdisciplinary field of 'phenomics', which relies on the comprehensive study of phenotypes through the collection and analysis of large-scale, high-throughput data. The use of advanced imaging technologies and sensor networks has made identification and measurement of the traits related to crop growth, development, and stress responses easier. Researchers are now able to find the links between the genetic and environmental variables and the observable traits using data analytics, and thus their level of understanding of plant performance under various conditions is enhanced (Fahlgren et al., 2015). Furthermore, the recent advancements in phenomics have facilitated the discovery of new genes associated with critical traits such as yield, disease resistance, and drought tolerance of crops (Yang et al., 2020b). Accordingly, suitable biotechnological and breeding methods are discovered to manipulate the genomes of crop plants to develop improved, high-yielding, stress-tolerant varieties. The HTP platforms, including drone-based imaging and automated phenotyping systems, have further revolutionized crop breeding by accelerating the evaluation process of large plant populations and by identifying the elite genotypes more efficiently (Gage et al., 2019). The integration of phenomics with AI-ML and big data analytics is playing a big role in the field of precision agriculture, offering new insights into how plants interact with their environment and thus improving crop resilience and productivity (Yang et al., 2020b).

## 7.3 ABIOTIC STRESS-INDUCED CHANGES IN CROP PHENOMICS LEAD TO YIELD LOSS

Various abiotic stressors, for example, extreme temperatures (heat and cold), drought, flood, and salinity, have a significant impact on crop growth, yield, and quality. They cumulatively account for about 51%–82% of agricultural yield loss (Oshunsanya et al., 2019) and, thus pose a serious challenge to global food security. Anthropogenic GHG emission is causing a gradual rise in the global mean surface temperature which is apparently very low, i.e. 0.17°C/decade (IPCC, 2021). However, the extent of temperature rise is not uniform across the globe and there is an increased frequency of occurrence of extreme temperatures, both high and low. The cumulative effect of these stresses often results in decreased crop yield, affecting the food supply and economic stability in agriculture-dependent nations. High temperatures lead to heat stress which denatures proteins and, thus enzymes are inactivated. Low temperatures, on the other hand, can impede cellular processes such as photosynthesis and protein synthesis (Mittler et al., 2022). Plants exposed to high temperatures often exhibit reduced numbers of flowers, fruits, and seeds, whereas low temperatures can lead to frost damage and a reduced rate of germination (Yoshida et al., 2023). Modern phenomics has enabled researchers to monitor the thermal tolerance traits of plants in detail. Recent studies involving infrared sensors and thermal imaging have demonstrated that heat stress can result in yield loss due to the induced changes in crop canopy temperatures and transpiration rates (Chen et al., 2023).

Climate change and global warming also result in uneven patterns and distribution of precipitation, both spatially and temporally. Therefore, moisture stress events in the form of drought and flood have occurred at increased frequency in recent years. Drought is a major abiotic stressor that reduces crop yield by limiting water availability. Moisture stress impairs major metabolic and growth processes such as photosynthesis, nutrient uptake, and cell expansion (Aroca, 2012; Wang et al., 2021). Under restricted water conditions for a short time span, plants demonstrate a range of adaptive responses like stomatal closure and osmotic regulation to conserve water. However, prolonged drought periods can cause a drastic reduction in photosynthetic efficiency and biomass production of plants (Zhao et al., 2017). Thus, drought stress typically results in reduced leaf area, stunted crop growth, and poor seed setting (Farooq et al., 2019). Advances in phenomics have facilitated the identification of drought-tolerance traits and the development of stress-resistant varieties. For example, HTP platforms have enabled the assessment of canopy temperature and leaf water content, which are indicative of drought stress (Sadeghi et al., 2021). By correlating these traits with yield performance, researchers can identify and select varieties with better drought resistance. A notable case is the development of a drought-tolerant wheat variety where the phenomics data have been used for the selection of tolerant physio-morphological traits based on root depth and canopy temperature (Tardieu et al., 2023).

Salinity, another critical abiotic stress, results from high concentrations of soluble salts in the soil, which leads to ion toxicity and osmotic stress. High salinity disrupts a plant's ion balance, affecting enzyme activities and membrane stability, ultimately leading to reduced growth and yield (Roy et al., 2021). By integrating HTP data with the genomic information and focusing on traits such as root morphology, ion accumulation, and leaf chlorophyll content (Khan et al., 2020), researchers can develop salt-tolerant crop varieties that maintain yield under saline conditions.

The recent advances in phenomics along with HTP methods have provided new insights into how crops respond to these stresses. The breeding programs and biotechnological advancements have focused on the development of stress-resistant crop varieties or the enhancement of stress tolerance through genetic modifications to mitigate the effects of abiotic stressors (Bhatnagar-Mathur et al., 2021).

## 7.4 HIGH-THROUGHPUT PHENOTYPING – CURRENT STATUS AND FUTURE PROSPECTS

High-throughput phenotyping (HTP) enables the rapid measurement of a large number of physiological, morphological, structural, biochemical and molecular traits to characterize agronomic biodiversity for breeding stress resilient high yielding cultivars (Gill et al., 2022). To allow the breeder to perform multiple trials under different growth conditions and diverse lines populations mapping, breeding population, mutant population study and germplasm pool assessment. HTP technologies may enhance yield gains for breeding while helping to improve the precision and accuracy of phenotypic data capture. Drawing from progress in the fields of imaging, sensor technology, robotics and data analytics (Thakur et al., 2024), HTP enables an ultra-fast and reliable high-resolution assessment of plant traits. Traditional methods of phenotyping are expensive, time-consuming, slow, invasive and harmful to research subjects, restrictive in scope and thereby not conducive to large-scale studies.

Through the incorporation of phenotyping technologies capable of operating in a non-destructive manner, another dimension is added to data collection thus enhancing the accuracy and speed plus light penetration analysis. The genetic improvements by HTP are foreseen to yield great increases in agricultural productivity according to scientists. In recent years, advances in sensing technology, data processing and analysis have led to a large improvement in field- and crop-management strategies regarding for example the expression of water content, chlorophyll or biomass (Huang et al., 2004), plant growth potential, etc. HTP is an invaluable tool that contributes significantly to both the speed and accuracy of plant breeding, allowing us to produce crops tailored more adeptly toward solving the problems associated with sustainable agriculture and food security under climate change. It is cheap, reliable versatile and scalable.

### 7.4.1 The HTP Concept

The main idea of HTP is to gather high-dimensional, large-scale phenotypic data in an economical and timely way. Conventional phenotyping techniques are labor-intensive, time-consuming, and prone to human error since they frequently need manual measurements. To overcome the challenges of conventional phenotypic techniques, HTP leverages sophisticated sensors and automated systems that simultaneously capture various features, which will also provide researchers access to a huge chunk of the large dataset for analysis. As the HTP system is very versatile and can function in field conditions as well as in controlled environments such as greenhouses and growth chambers therefore it is possible to observe the plant reactions to varying environmental conditions (Li et al., 2014). Moreover, the HTP approach will be very helpful in evaluating numerous phenotypic variables, including morphology, growth, stress tolerance, and physiological responses. The HTP generates a tremendous amount of data that can be used for improving crop yield and resilience to environmental challenges. Additionally, the data generated by HTP can also help in understanding the genetic basis of the traits and will support catering to the needs of the market through proper selection. Integrating multiple HTP platforms such as microscopic level to aerial systems will help in accomplishing comprehensive phenotyping at various levels, i.e. from tissues and organs to individual plants and entire fields. By following this holistic approach, organizations can enhance the pace of identifying desirable traits and developing new crop varieties that will enhance agricultural productivity. There are two major steps in HTP: high-throughput data collection or observations via various platforms and phenotypic data extraction from various platforms and analysis.

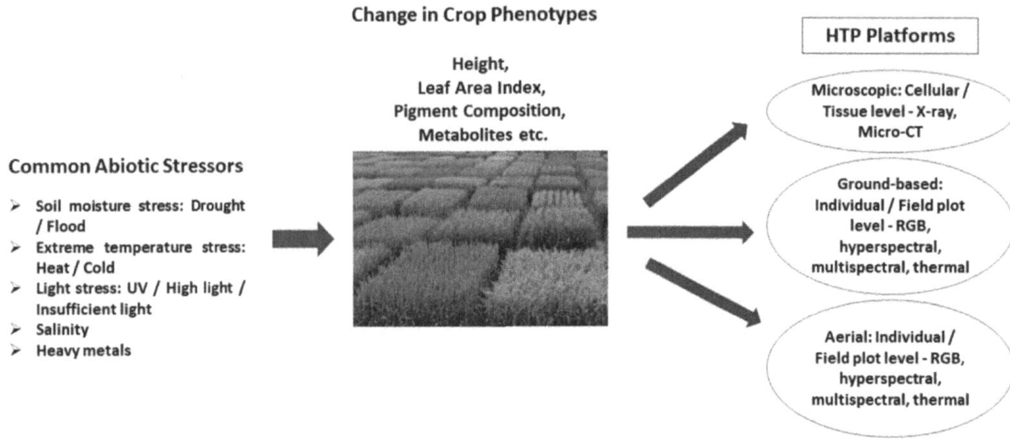


**FIGURE 7.1** Abiotic stress-induced changes in different crop phenotypes at different levels which are studied using various High-Throughput Phenotyping (HTP) platforms.

## 7.4.2 High-Throughput Data Collection or Observations via Various Platforms

This step involves the rapid collection of extensive phenotypic data via advanced imaging technologies and platforms. These platforms enable the characterization of phenotypic features at the tissue, organ, individual plant, plot, and field levels. They can be divided into three categories according to the imaging level: microscopic, ground-based, and aerial phenotyping (Figure 7.1).

### 7.4.2.1 Microscopic Phenotyping

It focuses on the detailed analysis of tissue morphologies and microphenotypes within plants. Advances in imaging technologies, such as X-ray micro-CT and automated image analysis, have enabled the nondestructive and rapid measurement of microphenotypes of kernels, stalks and roots (Table 7.1).

### 7.4.2.2 Ground-Based Phenotyping

There are three categories of ground-based platforms, portable, stationary, and movable, that are well suited for phenotyping in plant as well as field levels. Portable instruments include RGB cameras of high resolution, hyperspectral cameras and thermal infrared imaging cameras, integrated into smartphones and specialized devices such as Leaf-GP and Leaf Spec. High-resolution RGB

**TABLE 7.1**
**Microscopic Traits of Various Plant Organs Studied with the Help of Advanced Technologies and Tools**

| Organ Traits | Technology/Tools | Parameters |
|---|---|---|
| Kernel | Imaging with X-ray micro-CT and 3D morphological processing | Porosity, kernel-to-kernel distance, and kernel aspect ratio in both maize and wheat |
| Stalk | Automated color image analysis and micro-CT technology | Quantification of Lignin, analysis of vascular bundle characteristics in maize |
| Root | Software tools like Root Scan and Root Analyzer<br>3D reconstruction techniques | Semi-automated analysis of anatomical traits in root cross-sections<br>Root tissue structures |

cameras, thermal infrared imaging cameras, and hyperspectral cameras built into smartphones and specialist gadgets like Leaf-GP and Leaf Spec are examples of portable equipment. Stationary platforms can include fixed-site equipment, such as the Crop Design high-throughput phenotyping platform and the Field Scanalyzer, which uses multiple sensors to monitor crop traits in controlled environments and field settings. Movable platforms include agricultural vehicles (including electric carts), high-clearance tractors, and phenotyping robots, which are equipped with sensors such as light detection and ranging (LiDAR) and hyperspectral cameras to measure traits such as canopy height and biomass.

#### 7.4.2.3 Aerial Phenotyping

These include platforms such as satellites and UAVs, which are utilized for large-scale crop phenotyping. Satellites and manned aircraft measure traits such as canopy temperature and chlorophyll content over extensive field areas, although their lower image resolution can be a limitation. The UAV-based platforms offer high flexibility, low cost, and high spatial resolution. They are mostly equipped with sensors such as RGB cameras, infrared thermal imagers, LiDAR, and multispectral cameras to assess traits such as yield prediction, stress quantification, and photosynthetic status.

### 7.4.3 Phenotypic Data Extraction from Various Platforms and Analysis

The vast amounts of phenotypic data generated by high-throughput platforms require robust methods for data extraction, normalization, reconstruction, and analysis. This process involves advanced image understanding algorithms, 3D reconstruction techniques, and the integration of multisensor data.

#### 7.4.3.1 Image Processing and 3D Reconstruction

**2D and 3D Image Analysis**: Algorithms for 2D image and 3D mesh processing, such as the image harvesting framework and Paproki's 3D mesh-based technique, enable the extraction of phenotypic parameters from plant images.

**3D Modelling**: Multiview image-based 3D reconstruction methods, LiDAR, and structured light imaging sensors are used to create detailed 3D models of plant structures, facilitating accurate phenotype measurement.

#### 7.4.3.2 Sensor Integration and Spectral Analysis

Modern sensors (fluorescence, near-infrared, multispectral, and hyperspectral sensors, visible light,) provide dynamic datasets on traits such as chlorophyll content, plant height, and disease susceptibility. Hyperspectral imaging combined with computational algorithms for analysis can reveal nutrient components and detect physiological traits, such as amylose content, in rice and wheat.

#### 7.4.3.3 Crop Modelling and Functional Structural Plant Models (FSPMs)

Mathematical models simulate crop growth, development, and yield formation, providing insights into crop physiology and assisting in decision-making for crop management. FSPMs capture multidimensional phenotypes and predict plant growth and reproduction in various environments, integrating 3D digital technology and remote sensing data to optimize agricultural practices.

### 7.4.4 Role of HTP in Plant Breeding for Crop Improvement under Abiotic Stress

HTP has emerged as a transformative technology in plant breeding, especially for addressing the challenges posed by abiotic stress (Table 7.2). By increasing the efficiency and precision of phenotypic assessments, HTP enables the rapid identification and selection of stress-tolerant traits, paving the way for the development of robust, resilient crop varieties that can sustain productivity in

**TABLE 7.2**
**Studies on the Applications of High-Throughput Phenotyping (HTP) in Addressing Various Plant Abiotic Stresses**

| S. No. | Crop | Platform of HTP | Techniques of HTP | Traits | References |
|---|---|---|---|---|---|
| 1 | *Arabidopsis thaliana* | PHENOPSIS | RGB, Infrared, Fluorescence | Responses to soil water deficit | Granier et al. (2006) |
| 2 | Wheat, Barley | A multisensor field-based high-throughput phenotyping platform Breed Vision | Multispectral, Hyperspectral | Nitrogen status, drought stress responses | Busemeyer et al. (2013) |
| 3 | *Arabidopsis thaliana* | Growscreen FLUO | Fluorescence Imaging | Chlorophyll fluorescence | Hartmann et al. (2011) |
| 4 | Various plants | PSI Open FluorCam | Chlorophyll Fluorescence Imaging | Photosynthesis efficiency | Das Choudhury et al. (2019) |
| 5 | Various plants | WIWAM | RGB, NIR | Growth dynamics, drought stress responses | Li et al. (2021) |
| 6 | Various crops | LemnaTec System | RGB, NIR, Thermal, Fluorescence | Growth rates, stress responses | Li et al. (2021) |
| 7 | Maize | High-throughput root phenotyping platform | Shovelomics | root architecture | Trachsel et al. (2011) |
| 8 | Wheat | [a]Wheat high-throughput phenotyping platform | Phenocart | Canopy temperature, vegetative index, and grain yield | Crain et al. (2016) |
| 9 | Maize, Soybean, and Wheat | Scanning technique | The terrestrial 3D laser | Detect the effect of stress like drought and nutrient deficiency in monocots and dicots | Friedli et al. (2016) |
| 10 | Various crops | Robotic platforms for high-throughput field phenotyping | Vinobot and Vinocule | Large areas under biotic and abiotic stress during day and night | Shafiekhani et al. (2017) |
| 11 | *Arabidopsis thaliana* | An automated large-scale phenotyping platform | Phenoscope | Suitable for large-scale experiments and offers high spatial homogeneity | Tisné et al. (2013) |
| 12 | Wheat and Soybean | Multisensor systems with portable spectrometers and RGB web cameras | | Crop canopy traits from field plots | Bai et al. (2016) |

[a] Integrating several sensors: a GreenSeeker for spectral reflectance, an infrared thermometer (IRT), and a global navigation satellite system (GNSS) receiver into one functional unit.

the face of adverse environmental conditions. The importance of HTP in plant breeding will only increase as the world's agricultural environment changes, guaranteeing that crops produced in the future will be sufficiently prepared to withstand the effects of climate change.

**Accelerating Genetic Gain:** HTP technologies play a crucial role in speeding up genetic gain by increasing the intensity and precision of selection. HTP with its precise measures of a wide range of biochemical, physiological and morphological parameters linked to stress tolerance helps breeders to work more effectively and identify the best-performing genotypes (Thakur et al., 2024). For instance, thermal imaging, near-infrared spectroscopy, and fluorescence imaging help in rapidly assessing the canopy temperature, chlorophyll content, and water use efficiency traits. Additionally, LeasyScan, and Lysimeter facility enable accurate characterization and screening for abiotic stress (drought, heat, and salinity) adaptation (Joshi et al., 2023). Furthermore, to develop stress-susceptible resistant crops in less time, increased precision with HTP plays playing important role by detecting genetic variation more effectively and shortening the breeding cycle (Kumar and Kaushik, 2023).

**Screening of Large Populations:** One of the benefits of HTP is its capacity to effectively and swiftly screen a large number of plant populations. Traditional phenotyping procedures due to the skilled labor and time-constrained evaluate only a limited number of plants, whereas breeders can gather a large amount of information from thousands of plants at the same time by utilizing the HTP technologies such as drones and robotic platforms. The high-throughput capacity is important as there is large genetic variability during the early phases of breeding projects in identifying superior genotypes (Yang et al., 2020a). By efficiently screening large breeding populations, HTP increases the probability of discovering unique traits or rare genes associated with improved stress tolerance.

**Real-Time Monitoring and Non-destructive Measurements:** With the use of HTP technologies breeders can continuously monitor the plants in their different growing stages as it employs non-destructive methods for monitoring in real-time. Berger et al. (2010), concluded that this approach is crucial for researching abiotic stresses i.e. recovery following rehydration and development of drought tolerance. HTP also provides insights from temporal dynamics and stress responses and thereby helping breeders in knowing more about the plant's reaction to stresses over time.

**Identification of Physiological and Morphological Markers:** The precise phenotypic data generated by HTP facilitate the identification of physiological and morphological markers associated with abiotic stress tolerance. For example, thermal imaging can detect variations in leaf temperature, which may indicate differences in transpiration rates and heat tolerance. These markers are invaluable for marker-assisted selection, enabling breeders to select and develop crop varieties that are relatively resilient to stress. For instance, wheat genotypes with greater water-use efficiency and drought tolerance have been found utilizing HTP platforms that use thermal and multispectral imaging (Deery et al., 2014; Tardieu et al., 2023). HTP has also been used in selecting the salt tolerance rice varieties by measuring various traits i.e., leaf senescence and ion accumulation under saline conditions (Negrão et al., 2017).

**Climate-Resilient Crops and Precision Agriculture:** HTP also supports agriculture by providing explicit data about the responses of different crops and varieties toward specific stresses. This data not only helps in making crucial decisions on irrigation and fertilizer scheduling but also optimizes resources and reduces stress. HTP also plays a key role in the production of climate-resilient crops by providing greater information about the various abiotic stresses i.e., heat, drought, salinity, etc., which will ensure that agriculture can adapt to a wider range of weather conditions.

## 7.5 CHALLENGES IN HTP AND PROBABLE SOLUTIONS

Despite all the advantages of HTP, there are also various limitations of HTP that prevent it from achieving its full potential. A few of them are complicated data management, high costs and inconsistent results. As these problems transcend in the fields of technology, biology and operations and there is an urgent need for multidisciplinary collaboration across various disciplines of science. These challenges in usability and accessibility of HTP platforms can be overcome by implementing the following strategies:

**Cutting Expenses with Shared Resources and Innovative Technology**: The HTP systems can become more affordable with new innovations in the field of science and technologies such as low-cost sensors and imaging equipment for HTP platforms. As the new research progresses in the field of material sciences and microelectronics there will be new phenotyping tools with low cost and nearly same efficiency. Pooled HTP facilities in collaboration with multiple organizations will help in cost-cutting and provide small research institutes access to cutting-edge phenotyping facilities (Araus and Cairns, 2014). Through this approach of combining resources, different small breeding groups will be able to obtain the latest phenotyping technologies, which would be difficult to obtain due to financial constraints.

**Increasing the Effectiveness of Data Analysis and Management**: HTP generates a huge amount of data and to handle this large chunk of data effective data management is necessary. To ensure comparability and consistency across research there should be a standardized format and procedure in place for data collection and storage. Data management systems with a combination of cloud computing can be widely used for processing and storing large amounts of datasets. This will also enhance the accessibility of data worldwide among researchers (Fiorani and Schurr, 2013). Additionally, the application of new technologies such as artificial intelligence (AI) and machine learning (ML) in generating new algorithms can analyze complex and large phenotyping data in very less time. This will pace up the process of identifying important traits and enhance the accuracy of trait forecasts by streamlining data processing operations (Singh et al., 2016).

**Enhancing Trait Measuring Methods**: Ongoing innovations in imaging and sensor technology will be able to quantify more phenotypic traits. Tardieu et al. (2017) demonstrated that the use of multisensory platforms containing fluorescence, hyperspectral, and infrared imaging can provide a more comprehensive picture of plant health and stress responses by generating large amounts of data in different wavelength bands. Furthermore, in the near future as science progresses new sensors can be developed that can capture complex processes i.e., biochemical compositions, that are difficult to detect right now.

**Reducing Environmental Variability**: Field-based phenotyping has a significant issue of data inconsistency from uncontrolled contexts. These problems can be reduced under controlled environments, like greenhouses and growth chambers by using precise environmental control techniques and this will also help in reducing variations in measuring phenotypic data. The use of standardized processes and reference plants or plots will reduce environmental variations and thus the data gathered from field phenotyping is more accurate and reliable (Furbank and Tester, 2011).

**Encouraging Integration with Breeding Programs**: Effective Multidisciplinary collaboration is required to integrate HTP data effectively into traditional breeding operations. A team of phenotyping specialists with breeders and data scientists can use the data generated in HTP more effectively. This approach will help in making data-based breeding decisions. HTP insights can be more easily integrated into the breeder's workflow with the help of tools and software that can visualize the HTP data. Moreover, this will also narrow the gap between the collection of data and analysis (Cobb et al., 2013).

**Training and Capacity Building**: The effective and efficient use of HTP technologies requires gaining essential skills by investment in training and capacity building. Research organizations can organize various workshops and courses for breeders and technicians to acquire expertise in HTP systems. For wider applicability of these technologies, capacity building is necessary and important. Furthermore, training and capacity building will increase the use of HTP technologies particularly in locations where access to these technologies is very limited (Fiorani and Schurr, 2013).

**Standardization and Scaling**: The development of various platforms that can accommodate different crops, varieties and environmental scenarios should be developed for wider adoption of HTP systems. Data reproducibility and comparability among different research teams will be achieved by standardizing phenotyping techniques. This will also help in integrating findings from different sources efficiently (Qiu et al., 2019). To implement HTP technologies successfully in various agricultural situations these programs play an important role.

**Taking Care of Ethical and Privacy Issues**: There is a need for transparent laws and regulations about ownership of data, privacy and sharing of data, as a large chunk of data is generated by a wide range of phenotyping tools. In order to preserve sensitive data, a safe framework for managing confidential phenotypic and genetic data is necessary (Tardieu et al., 2017).

**Improving Temporal and Spatial Resolution**: Enhancing Dynamic changes in different plant traits can be monitored with precision by increasing the temporal and geographical resolution of HTP systems. With the advancement in science and technology, new tools i.e., robotics and drones containing high-resolution cameras will help in generating precise and accurate information about the growth and development of plants. Researchers can able to see subtle phenotypic variations with the use of high-resolution instruments which they would not able to capture earlier.

**Improving Accessibility and Adoption**: The flow of knowledge and policy support are very important factors in the adoption of new technology. To encourage wider adoption of HTP technologies, especially in an environment with limited resources, it is important to showcase the advantages of HTP technologies and offer support for their deployment. For instance, low-cost phenotyping technologies cater to the requirements of small breeding programs and projects (Araus and Cairns, 2014). If these technologies are available to a wide range of consumers, it will pace up the development of new crop varieties with superior traits and resilience to abiotic challenges.

## 7.6 CONCLUSION

Crop productivity affected by abiotic stressors will threaten food security worldwide. Recently, huge strides made in phenomics have shed light on how different abiotic stresses can affect crops and how their responses impact production. By using high-throughput technologies and integrating the data with genomics and machine learning, researchers are achieving unprecedented advances in the development of stress-resistant new varieties of crops. However, several challenges stand in the way of using phenomics for the development of crops. One major issue is that the HTP technologies are truly very expensive and complex, hence limiting the researchers and breeders, especially in developing countries, as noted by Kumar et al. (2021). These technologies also generate huge data, meaning there need to be high levels of statistical data analysis methods and appropriate computing powers. Therefore, future research, for increasing the utility of phenomics in breeding applications, has to focus on affordable technologies for phenotyping and refinement in the data integration techniques.

## REFERENCES


Adrish, M., Anand, M.P. and Hanania, N.A., 2022. Phenotypes of asthma–chronic obstructive pulmonary disease overlap. *Immunology and Allergy Clinics*, 42(3): 645–655. https://doi.org/10.1016/j.iac.2022.04.009

Araus, J.L. and Cairns, J.E., 2014. Field high-throughput phenotyping: The new crop breeding frontier. *Trends in Plant Science*, 19(1): 52–61. https://doi.org/10.1016/j.tplants.2013.09.008

Aroca, R., 2012. *Plant Responses to Drought Stress: From Morphological to Molecular Features*. Springer, Berlin. https://doi/10.1007/978-3-642-29021-5

Bai, G., Ge, Y., Hussain, W., Baenziger, P.S. and Graef, G., 2016. A multisensor system for high throughput field phenotyping in soybean and wheat breeding. *Computers and Electronics in Agriculture*, 128: 181–192. https://doi.org/10.1016/j.compag.2016.08.021

Berger, B., Parent, B. and Tester, M., 2010. High-throughput shoot imaging to study drought responses. *Journal of Experimental Botany*, 61(13): 3519–3528. https://doi.org/10.1093/jxb/erq201

Bhatnagar-Mathur, P., Devi, J. and Sharma, K., 2021. Advances in breeding for abiotic stress tolerance in crops. *Current Science*, 120(11): 1733–1745. https://doi.org/10.18520/cs/v120/i11/1733-1745

Busemeyer, L., Mentrup, D., Möller, K., Wunder, E., Alheit, K., Hahn, V., Maurer, H.P., Reif, J.C., Würschum, T., Müller, J. and Rahe, F., 2013. BreedVision—A multisensor platform for nondestructive field-based phenotyping in plant breeding. *Sensors*, 13(3): 2830–2847. https://doi.org/10.3390/s130302830

Chang, Y.C., Huang, T.W. and Huang, N.F., 2019. A machine learning based smart irrigation system with LoRa P2P networks. In: *2019 20th Asia-Pacific Network Operations and Management Symposium (APNOMS)*, September (pp. 1–4). IEEE. https://doi/10.23919/APNOMS.2019.8893034

Chen, D., Neumann, K., Friedel, S., Kilian, B., Chen, M., Altmann, T. and Klukas, C., 2014. Dissecting the phenotypic components of crop plant growth and drought responses based on high-throughput image analysis. *The Plant Cell*, 26(12): 4636–4655. https://doi.org/10.1105/tpc.114.129601

Chen, L., Zhang, Y. and Yang, S., 2023. Thermal imaging and high-throughput phenotyping for assessing heat stress in crops. *Frontiers in Plant Science*, 14: 1045678. https://doi.org/10.3389/fpls.2023.1045678

Cobb, J. N., DeClerck, G., Greenberg, A., Clark, R. and McCouch, S., 2013. Next-generation phenotyping: Requirements and strategies for enhancing our understanding of genotype–phenotype relationships and its relevance to crop improvement. *Theoretical and Applied Genetics*, 126(4): 867–887. https://doi.org/10.1007/s00122-013-2066-0

Crain, J.L., Wei, Y., Barker III, J., Thompson, S.M., Alderman, P.D., Reynolds, M., Zhang, N. and Poland, J., 2016. Development and deployment of a portable field phenotyping platform. *Crop Science*, 56(3): 965–975. https://doi.org/10.2135/cropsci2015.05.0290

Das Choudhury, S., Samal, A. and Awada, T., 2019. Leveraging image analysis for high-throughput plant phenotyping. *Frontiers in Plant Science*, 10: 508. https://doi.org/10.3389/fpls.2019.00508

Davis, C.L. and Vincent, K., 2017. *Climate Risk and Vulnerability: A Handbook for Southern Africa*, 2nd ed. Council for Scientific and Industrial Research, Pretoria, South Africa, p. 202. https://www.csir.co.za/sites/default/files/Documents/SADC%20Handbook_Second%20Edition_full%20report.pdf

de Vienne, D., 2022. What is a phenotype? History and new developments of the concept. *Genetica*, 150(3): 153–158. https://doi.org/10.1007/s10709-021-00134-6

Deery, D., Jimenez-Berni, J., Jones, H., Sirault, X. and Furbank, R., 2014. Proximal remote sensing buggies and potential applications for field-based phenotyping. *Agronomy*, 4(3): 349–379. https://doi.org/10.3390/agronomy4030349

Dhondt, S., Wuyts, N. and Inzé, D., 2013. Cell to whole-plant phenotyping: The best is yet to come. *Trends in Plant Science*, 18(8): 428–439. https://doi.org/10.1016/j.tplants.2013.04.008

Eekhout, J.P. and de Vente, J., 2022. Global impact of climate change on soil erosion and potential for adaptation through soil conservation. *Earth Science Reviews*, 226: 103921. https://doi.org/10.1016/j.earscirev.2022.103921

Fahlgren, N., Gehan, M.A. and Miranda, M.L, 2015. Deep learning methods for phenotypic trait prediction in plants. *Journal of Experimental Botany*, 66(21): 5377–5386. https://doi.org/10.1093/jxb/erv307

Farooq, M., Hussain, M. and Siddique, K.H.M., 2019. Drought stress in plants: Impacts, mechanisms, and management. *Agronomy for Sustainable Development*, 39(1): 26. https://doi.org/10.1007/s13593-018-0548-3

Fiorani, F. and Schurr, U., 2013. Future scenarios for plant phenotyping. *Annual Review of Plant Biology*, 64(1): 267–291. https://doi.org/10.1146/annurev-arplant-050312-120137

Friedli, M., Kirchgessner, N., Grieder, C., Liebisch, F., Mannale, M. and Walter, A., 2016. Terrestrial 3D laser scanning to track the increase in canopy height of both monocot and dicot crop species under field conditions. *Plant Methods*, 12: 1–15. https://doi.org/10.1186/s13007-016-0109-7

Furbank, R.T. and Tester, M., 2011. Phenomics-technologies to relieve the phenotyping bottleneck. *Trends in Plant Science*, 16: 635–644. https://doi.org/10.1016/j.tplants.2011.09.005

Gage, J.L., Anderson, C.J. and Manohar, M., 2019. High-throughput phenotyping and its potential impact on plant breeding, *Frontiers in Plant Science*, 10: 1712. https://doi.org/10.3389/fpls.2019.01712

Gill, T., Gill, S.K., Saini, D.K., Chopra, Y., de Koff, J.P. and Sandhu, K.S., 2022. A comprehensive review of high throughput phenotyping and machine learning for plant stress phenotyping. *Phenomics*, 2(3): 156–183. https://doi.org/10.1007/s43657-022-00048-z

Granier, C., Aguirrezabal, L., Chenu, K., Cookson, S.J., Dauzat, M., Hamard, P., Thioux, J.J., Rolland, G., Bouchier-Combaud, S., Lebaudy, A. and Muller, B., 2006. PHENOPSIS, an automated platform for reproducible phenotyping of plant responses to soil water deficit in *Arabidopsis thaliana* permitted the identification of an accession with low sensitivity to soil water deficit. *New Phytologist*, 169(3): 623–635. https://doi.org/10.1111/j.1469-8137.2005.01609.x

Hartmann, A., Czauderna, T., Hoffmann, R., Stein, N. and Schreiber, F., 2011. HTPheno: An image analysis pipeline for high-throughput plant phenotyping. *BMC Bioinformatics*, 12, 1–9. https://doi.org/10.1186/1471-2105-12-148

Houle, D., Govindaraju, D.R. and Omholt, S., 2010. Phenomics: The next challenge. *Nature Reviews Genetics*, 11(12): 855–866. https://doi.org/10.1038/nrg2897

IPCC, 2021. Summary for policymakers. In: *Climate Change 2021: The Physical Science Basis. Contribution of Working Group I to the Sixth Assessment Report of the Intergovernmental Panel on Climate Change.* https://www.ipcc.ch/report/ar6/wg1/downloads/report/IPCC_AR6_WGI_SPM.pdf

Jabaudon, M., Blondonnet, R., Audard, J., Fournet, M., Godet, T., Sapin, V. and Constantin, J.M., 2018. Recent directions in personalized acute respiratory distress syndrome medicine. *Anaesthesia Critical Care and Pain Medicine*, 37(3): 251–258. https://doi.org/10.1016/j.accpm.2017.08.006

Joshi, P., Gupta, S.K., Ojulong, H., Sharma, R., Vetriventhan, M., Kudapa, H., Choudhary, S., Naresh, D., Kholova, J. and Sajja, S., 2023. Finger millet improvement in post-genomic era: Hundred years of breeding and moving forward. In D. Sharma, S. Singh, S.K. Sharma, R. Singh (Eds.), *Smart Plant Breeding for Field Crops in Post-genomics Era* (pp. 221–253). Singapore: Springer. https://doi.org/10.1007/978-981-19-8218-7_7

Khan, M.A., Khan, M.I. and Khan, M.A., 2020. Advances in phenomics for salinity stress tolerance in crops. *Plant Physiology and Biochemistry*, 150: 158–169. https://doi.org/10.1016/j.plaphy.2020.03.001

Kumar, A. and Kaushik, P., 2023. A review on high throughput phenotyping for vegetable crops. *Journal of Botany Research*, 6(1): 170–175. https://doi.org/10.36959/771/575

Kumar, P., Sharma, N. and Nair, R., 2021. High-throughput phenotyping: A review of technologies and applications. *Journal of Agricultural and Food Chemistry*, 69(50): 15029–15044. https://doi.org/10.1021/acs.jafc.1c04273

LeCun, Y., Bengio, Y. and Hinton, G., 2015. Deep learning. *Nature*, 521(7553): 436–444. https://doi.org/10.1038/nature14539

Li, D., Quan, C., Song, Z., Li, X., Yu, G., Li, C. and Muhammad, A., 2021. High-throughput plant phenotyping platform (HT3P) as a novel tool for estimating agronomic traits from the lab to the field. *Frontiers in Bioengineering and Biotechnology*, 8: 623705. https://doi.org/10.3389/fbioe.2020.623705

Li, L., Zhang, Q. and Huang, D., 2014. A review of imaging techniques for plant phenotyping. *Sensors*, 14(11): 20078–20111. https://doi.org/10.3390/s141120078

Liu, L., Wang, R., Xie, C., Yang, P., Wang, F., Sudirman, S. and Liu, W., 2019. PestNet: An end-to-end deep learning approach for large-scale multiclass pest detection and classification. *IEEE Access*, 7: 45301–45312. https://doi.org/10.1109/access.2019.2909522

Madushanki, A.R., Halgamuge, M.N., Wirasagoda, W.S. and Syed, A., 2019. Adoption of the internet of things (IoT) in agriculture and smart farming towards urban greening: A review. *International Journal of Advanced Computer Science and Applications*, 10(4): 11–28. https://doi.org/10.14569/ijacsa.2019.0100402

Margulis, M., 2012. Global food security governance: The Committee on World Food Security, Comprehensive Framework for Action and the G8/G20. In: R. Rayfuse & N. Weisfelt (Eds.), *The Challenge of Food Security* (pp. 231–254). Edward Elgar Publishing. https://doi.org/10.4337/9780857939388.00025

McGrath, J.M., Siebers, M.H., Fu, P., Long, S.P. and Bernacchi, C.J., 2024. To have value, comparisons of high-throughput phenotyping methods need statistical tests of bias and variance. *Frontiers in Plant Science*, 14: 1325221. https://doi.org/10.3389/fpls.2023.1325221

Mitra, A., Vangipuram, S.L., Bapatla, A.K., Bathalapalli, V.K., Mohanty, S.P., Kougianos, E. and Ray, C., 2022. Everything you wanted to know about smart agriculture. arXiv preprint arXiv:2201.04754. https://doi.org/10.48550/arXiv.2201.04754

Mittler, R., Blumwald, E., 2022. Understanding stress tolerance. *Current Opinion in Plant Biology*, 65: 102–109. https://doi/10.1016/j.pbi.2021.12.008

Mizik, T., 2021. Climate-smart agriculture on small-scale farms: A systematic literature review. *Agronomy*, 11(6): 1096. https://doi.org/10.3390/agronomy11061096

Mukhopadhyay, R., Sarkar, B., Jat, H.S., Sharma, P.C. and Bolan, N.S., 2021. Soil salinity under climate change: Challenges for sustainable agriculture and food security. *Journal of Environmental Management*, 280: 111736. https://doi.org/10.1016/j.jenvman.2020.111736

Nagothu, S.K., 2016. Weather based smart watering system using soil sensor and GSM. In: *2016 World Conference on Futuristic Trends in Research and Innovation for Social Welfare (Startup Conclave)* (pp. 1–3). IEEE. https://doi.org/10.1109/STARTUP.2016.7583991

Negrão, S., Schmöckel, S.M. and Tester, M.J.A.O.B., 2017. Evaluating physiological responses of plants to salinity stress. *Annals of Botany*, 119(1): 1–11. https://doi.org/10.1093/aob/mcw191

Oshunsanya, S.O., Nwosu, N.J. and Li, Y. 2019. Abiotic stress in agricultural crops under climatic conditions. In: Jhariya, M., Banerjee, A., Meena, R., Yadav, D. (Eds.), *Sustainable Agriculture, Forest and Environmental Management*. Springer, Singapore. https://doi.org/10.1007/978-981-13-6830-1_3

Pieruschka, R. and Schurr, U., 2019. Plant phenotyping: Past, present, and future. *Plant Phenomics*. 7507131. https://doi/10.34133/2019/7507131

Qiu, Q., Sun, N., Bai, H., Wang, N., Fan, Z., Wang, Y., Meng, Z., Li, B. and Cong, Y., 2019. Field-based high-throughput phenotyping for maize plant using 3D LiDAR point cloud generated with a "Phenomobile". *Frontiers in Plant Science*, 10: 554. https://doi.org/10.3389/fpls.2019.00554

Radoglou-Grammatikis, P., Sarigiannidis, P., Lagkas, T. and Moscholios, I., 2020. A compilation of UAV applications for precision agriculture. *Computer Networks*, 172: 107148. https://doi.org/10.1016/j.comnet.2020.107148

Roy, S.J., Negrão, S. and Tester, M., 2021. Salinity tolerance in plants: Revisiting the role of potassium and sodium transport. *Plant Physiology*, 186(1): 14–29. https://doi.org/10.1093/plphys/kiab012

Sadeghi, H., Nezami, A. and Javidfar, H., 2021. High-throughput phenotyping for drought stress: Advances and future prospects. *Journal of Agronomy and Crop Science*, 207(3): 275–286. https://doi.org/10.1111/jac.12458

Shafiekhani, A., Kadam, S., Fritschi, F.B. and DeSouza, G.N., 2017. Vinobot and vinoculer: Two robotic platforms for high-throughput field phenotyping. *Sensors*, 17(1): 214. https://doi.org/10.3390/s17010214

Singh, A., Ganapathysubramanian, B., Singh, A.K. and Sarkar, S., 2016. Machine learning for high-throughput stress phenotyping in plants. *Trends in Plant Science*, 21(2): 110–124. https://doi.org/10.1016/j.tplants.2015.10.015

Singhvi, V., Lunga, L., Nidhi, P., Keum, C. and Prakash, V., 2024. High-throughput phenotyping using computer vision and machine learning. arXiv preprint arXiv:2407.06354. https://doi.org/10.48550/arXiv.2407.06354

Tardieu, F., Cabrera-Bosquet, L. and Pridmore, T., 2023. Breeding for drought tolerance using high-throughput phenotyping. *Field Crops Research*, 291: 108783. https://doi.org/10.1016/j.fcr.2023.108783

Tardieu, F., Cabrera-Bosquet, L., Pridmore, T.P. and Bennett, M.J., 2017. Plant phenomics, from sensors to knowledge. *Current Biology*, 27(15): PR770–R783. https://doi.org/10.1016/j.cub.2017.05.055

Thakur, P., Dogra, P., Bairwa, S.K., Meena, S.K. and Chandra, K., 2024. Modern phenotyping: A paradigm of present need in crop improvement. *International Journal of Plant & Soil Science*, 36(1): 24–37. https://doi.org/10.9734/ijpss/2024/v36i14326

Tisné, S., Serrand, Y., Bach, L., Gilbault, E., Ben Ameur, R., Balasse, H., Voisin, R., Bouchez, D., Durand-Tardif, M., Guerche, P. and Chareyron, G., 2013. Phenoscope: An automated large-scale phenotyping platform offering high spatial homogeneity. *The Plant Journal*, 74(3): 534–544. https://doi.org/10.1111/tpj.12131

Trachsel, S., Kaeppler, S.M., Brown, K.M. and Lynch, J.P., 2011. Shovelomics: High throughput phenotyping of maize (*Zea mays* L.) root architecture in the field. *Plant and Soil*, 341: 75–87. https://doi.org/10.1007/s11104-010-0623-8

Ullah, S., 2017. Climate change impact on agriculture of Pakistan-A leading agent to food security. *International Journal of Environmental Sciences & Natural Resources*, 6(3): 76–79. https://doi/10.19080/IJESNR.2017.06.555690

Wang, Y., Wang, X., Zhang, L. and Zhao, X., 2021. Advances in understanding drought stress tolerance mechanisms in plants. *Journal of Plant Growth Regulation*, 40(3): 808–823. https://doi/10.1007/s00344-020-10072-2

Yang, W., Feng, H., Zhang, X., Zhang, J., Doonan, J.H., Batchelor, W.D., Xiong, L. and Yan, J., 2020a. Crop phenomics and high-throughput phenotyping: Past decades, current challenges, and future perspectives. *Molecular Plant*, 13(2): 187–214. https://doi.org/10.1016/j.molp.2020.01.008

Yang, W., Wang, L., Xu, L., Zhang, X. and Wang, G. 2020b. Integrating high-throughput phenotyping and machine learning to boost breeding productivity in crops. *Nature Plants*, 6(4): 435–445. https://doi/10.1038/s41477-020-0646-4

Yang, X., Shu, L., Chen, J., Ferrag, M.A., Wu, J., Nurellari, E. and Huang, K., 2021. A survey on smart agriculture: Development modes, technologies, and security and privacy challenges. *IEEE/CAA Journal of Automatica Sinica*, 8(2): 273–302. https://doi.org/10.1109/JAS.2020.1003536

Yoshida, T., Fujii, H. and Nakashima, K., 2023. Mechanisms of plant responses to cold stress. *Current Opinion in Plant Biology*, 73: 102245. https://doi/10.1016/j.pbi.2023.102245

Zhao, C., Liu, B., Shao, Q., Wang, L., Chen, F. and Dong, Z., 2017. Temperature increase reduces global yields of major crops in four ways. *Nature Climate Change*, 7(4): 221–225. https://doi/10.1038/nclimate2142

# 8 Genetic Dissection of Major Crops by High-Throughput Phenotyping and Phenomics

*Gunasekaran Arthi, Jothi Dheivasikamani Abidharini, Balasubramanian Balamuralikrishnan, Meyyazhagan Arun, Manikantan Pappuswamy, Murugesan Loganathan, Jeyabal Philomenathan Antony Prabhu, and Arumugam Vijaya Anand*

## LIST OF ABBREVIATIONS

**AI**: artificial intelligence
**CNNs**: convolutional neural networks
**GS**: genomic selection
**GWAS**: genome-wide association studies
**G x E**: genotype-environment interactions
**HTP**: high-throughput phenotyping
**IoT**: internet of things
**MAS**: marker-assisted selection
**ML**: machine learning
**NGS**: next-generation sequencing
**NIRS**: near-infrared spectroscopy
**QTL**: genetic mapping and quantitative trait loci
**SNPs**: single nucleotide polymorphisms
**XAI**: explainable AI

## 8.1 INTRODUCTION TO HIGH-THROUGHPUT PHENOTYPING (HTP)

High-Throughput Phenotyping (HTP) represents a transformative approach in plant science and agriculture (Araus & Cairns, 2014), enabling the rapid and accurate measurement of plant traits across large populations. HTP integrates automated data collection methods, advanced imaging technologies (Nield, 2014), and data analysis techniques to provide insights into plant growth, development, and responses to environmental stresses (Furbank & Tester, 2011). This approach is becoming increasingly important as researchers and breeders seek to improve crop productivity, resilience, and quality to meet the global demands of food security and sustainable agriculture (Tardieu et al., 2017a). HTP involves using automated systems to collect vast amounts of phenotypic data-observable characteristics of plants (Li et al., 2014) such as height, leaf area, flowering time, and stress tolerance (Figure 8.1). By capturing these traits in a standardized and efficient manner, HTP provides the necessary data for identifying promising plant varieties and understanding the genetic and environmental factors influencing crop performance (Fiorani & Schurr, 2013).

 DOI: 10.1201/9781003505631-8

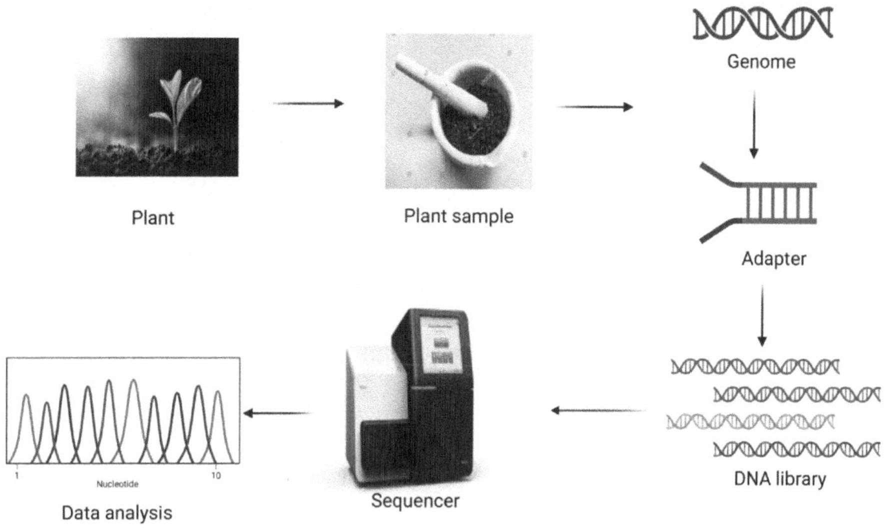


**FIGURE 8.1** High-throughput phenotyping.

### 8.1.1 Key Components of HTP

a. **Imaging Technologies**: HTP utilizes advanced imaging systems (Yang et al., 2013), including visible light (Walter et al., 2015a) multispectral, hyperspectral, and thermal imaging (Mishra et al., 2020) These technologies allow researchers to monitor plant health, structure, and physiological responses in great detail (Mahlein et al., 2012; Jones & Vaughan, 2010).
b. **Sensors and Drones**: Ground-based sensors, drones, and aerial platforms equipped with cameras and spectrometers are commonly used to capture data in both controlled environments (like greenhouses) and in the field (Bucksch et al., 2014). These tools enable large-scale data collection across diverse terrains and environmental conditions.
c. **Automated Data Processing**: Machine learning (ML) and computer vision algorithms play a central role in HTP by processing and analyzing massive datasets quickly (Shakoor et al., 2017). These algorithms can identify patterns and extract meaningful information from complex images and sensor data, helping researchers focus on key phenotypic traits.
d. **Data Integration and Analysis**: To fully understand plant traits, HTP often integrates multimodal data, combining phenotypic information with genetic, environmental, and management data (Araus et al., 2012). This integration enables comprehensive analysis and helps pinpoint the genetic basis of complex traits.

### 8.1.2 Benefits of HTP

- **Increased Efficiency**: HTP allows for the rapid assessment of large plant populations, accelerating the plant breeding process (White et al., 2012a).
- **Precision and Accuracy**: Automated systems minimize human error and provide highly accurate measurements of plant traits (Coppens et al., 2017), improving the reliability of research findings.
- **Enhanced Trait Discovery**: With HTP, researchers can measure a wide range of traits that would be difficult or impossible to assess manually, such as root architecture (Araus & Kefauver, 2018) disease progression, and drought tolerance.

### 8.1.3 Applications of HTP

HTP is widely applied across various fields in plant science and agriculture. Some key applications include:

- **Crop Improvement**: HTP enables breeders to quickly identify plant varieties with desirable traits, such as higher yield, nutrient efficiency (Chen et al., 2014), and stress tolerance (Cai et al., 2018).
- **Stress Tolerance Studies**: Researchers use HTP to monitor plant responses to abiotic stresses (like drought and heat) (Roitsch et al., 2019) and biotic stresses (like pests and diseases), helping to develop resilient crop varieties.
- **Genotype-Phenotype Mapping**: By correlating phenotypic data with genomic information, HTP (Awlia et al., 2016) aids in identifying genetic markers associated with important traits, facilitating precision breeding efforts.

### 8.1.4 Challenges in HTP

Despite its many advantages, HTP also faces several challenges, including the need for high-quality data, standardization across platforms, computational resources, and expertise in data analysis (Montes & Melchinger, 2016c). Addressing these issues is crucial to maximizing the potential of HTP in research and agricultural applications.

## 8.2 PHENOMICS: INTEGRATING PHENOTYPES WITH GENOTYPES

Phenomics, the study of phenotypes on a large scale, focuses on systematically quantifying the observable characteristics of organisms, known as phenotypes (Tackenberg, 2007), and understanding how they relate to the underlying genotypes. In plants, phenomics bridges the gap between genetic (Liu et al., 2017) information and observable traits by integrating phenotypic data with genomic data. This approach provides researchers with a comprehensive view of how genetics, environment, and their interactions drive the traits expressed by plants (Pask et al., 2012). Here's an overview of how phenomics plays a vital role in modern plant science.

Phenomics involves the high-throughput collection, analysis, and interpretation of plant traits (phenotypes) to understand the genetic and environmental influences on these traits. By linking phenotypes (McCouch et al., 2013) with genotypes, researchers can study complex traits like yield, drought tolerance, and disease resistance more effectively. Phenomics employs tools such as imaging technologies, sensors, and advanced data analytics to capture vast amounts of data, enabling researchers to observe trait variations in response to different environmental and genetic factors.

### 8.2.1 Integrating Phenotypes with Genotypes

Integrating phenotypic data with genotypic information is essential to uncovering the genetic basis of complex traits and their environmental interactions. Here's how it works:

a. **Genotype-Phenotype Mapping**: Through quantitative trait locus (QTL) mapping and genome-wide association studies (GWAS), scientists can identify specific genes or genomic regions associated with particular traits. By observing how these genetic markers correlate with phenotypic expression, researchers can select desired traits in breeding programs.
b. **HTP and Genotyping**: Advanced phenotyping (Cichy & Bohn, 2019) platforms collect data on a wide range of traits across thousands of plants, while next-generation sequencing

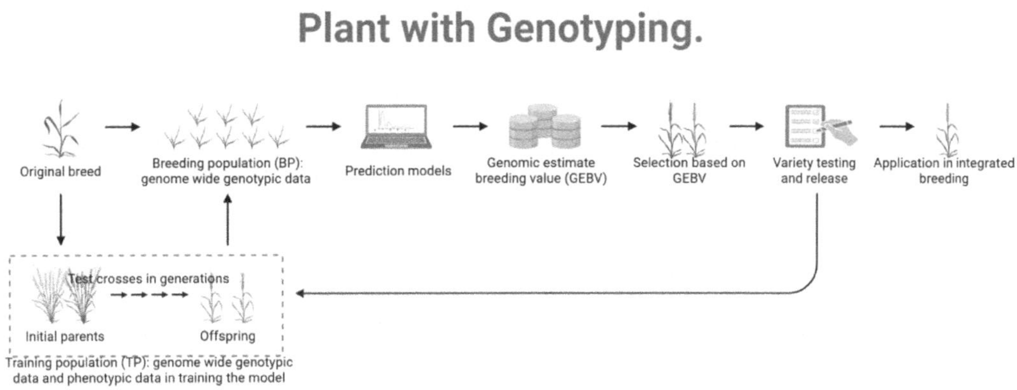


**FIGURE 8.2** Plant with genotyping.

(NGS) technologies allow for detailed genotypic analysis. Together, these technologies enable the identification of genes underlying key traits, such as root structure for water uptake or leaf architecture for photosynthetic efficiency (Figure 8.2).

c. **Environmental and Epigenetic Factors**: Phenomics also considers how environmental conditions and epigenetic changes affect gene expression and phenotypes. Plants adapt to their environments through gene-environment interactions (G x X), and phenomics allows researchers to analyze these interactions and their impact on observable traits.
d. **ML in Data Integration**: ML models are increasingly used to integrate complex multimodal data, including genetic, environmental, and phenotypic data (Das Choudhury et al., 2013). These algorithms can analyze patterns, predict outcomes, and identify novel trait associations that would be difficult to detect with traditional methods.

### 8.2.2 Applications of Phenomics in Plant Science

The integration of phenotypic and genotypic data is transformative for a range of applications, including:

- **Crop Improvement:** By identifying genetic markers associated with desirable traits, researchers can use marker-assisted selection (MAS) or genomic selection (GS) to accelerate the breeding process and produce improved crop varieties with traits such as increased yield, resilience to climate stressors (Wang et al., 2017) and enhanced nutritional content.
- **Studying Abiotic and Biotic Stress Responses:** Phenomics enables scientists to understand how plants respond (Montesinos-López et al., 2019) to stresses like drought, salinity, pests, and diseases. By studying these responses, researchers can identify genes that improve stress tolerance, contributing to the development of more resilient crops.
- **Precision Agriculture:** Integrated (Vergara-Diaz et al., 2016) phenomics and genomic data can guide precision agriculture practices, allowing farmers to tailor cultivation practices to the specific genetic profiles of their crops. This leads to optimized resource use, better yield, and minimal environmental impact.
- **Systems Biology and Functional Genomics:** Phenomics contributes to a systems biology approach, integrating data at multiple biological levels to provide a holistic view of plant biology (Furbank & Sadras, 2017) This helps researchers understand how genes, pathways, and networks collectively influence plant traits.

### 8.2.3 Benefits of Integrating Phenotypes with Genotypes

- **Uncovering Complex Traits**: Many agronomically important traits are complex and controlled by multiple genes. Integrating genotypes with phenotypes helps unravel these traits by identifying genetic loci (Jin et al., 2016) involved in their expression.
- **Enhanced Breeding Efficiency**: By targeting specific traits at the genetic level, researchers can make breeding programs more efficient, reducing the time and resources needed to develop improved crop varieties.
- **Personalized Crop Management**: Genotype-phenotype integration supports more precise crop management strategies, allowing tailored interventions based on specific genetic and phenotypic profiles of plants in different environments (Montes et al., 2011a).

### 8.2.4 Challenges in Phenomics

Despite its promise, phenomics faces several challenges, including:

- **Data Volume and Complexity**: The integration of HTP data with genetic and environmental data generates massive datasets that require robust data storage, processing, and analysis capabilities (Montes et al., 2011b).
- **Standardization of Measurements**: Variability in data collection methods across phenotyping (Cobb et al., 2013) platforms make standardization essential for ensuring consistency and comparability of phenotypic data.
- **Interdisciplinary Expertise**: Phenomics requires expertise in genetics, plant biology, data science, and bioinformatics (Bolger et al., 2014) making interdisciplinary collaboration essential but challenging.

## 8.3 KEY TECHNOLOGIES IN HTP

HTP relies on advanced technologies to efficiently capture and analyze a wide array of plant traits on a large scale (Montesinos-López et al., 2017). These key technologies are essential for enabling rapid, precise, and automated data collection across different environments. Here's an overview of some of the most important technologies driving HTP:

### 8.3.1 Imaging Technologies

Imaging is a cornerstone of HTP, as it allows for non-invasive monitoring of plant traits and health. Key imaging technologies include:

- **Visible Light Imaging**: Used for basic assessments of plant morphology, such as height, leaf area, and color (Ubbens & Stavness, 2017). This technology is straightforward and forms the foundation for many phenotyping platforms.
- **Multispectral Imaging**: Captures images at several wavelengths across the electromagnetic spectrum (e.g., red, green, blue, near-infrared), allowing researchers to assess various (Fiorani et al., 2012) plant health parameters, such as chlorophyll content, which can indicate stress or disease.
- **Hyperspectral Imaging**: Extends multispectral imaging by capturing a continuous spectrum of wavelengths, providing detailed information about plant biochemical composition. Hyperspectral imaging is useful for analyzing (Zhang & Kovacs, 2012) traits related to plant nutrition, stress response, and disease susceptibility.

- **Thermal Imaging**: Detects temperature variations in plants, which can reveal water stress, transpiration rates (Tsaftaris et al., 2016), and other physiological responses. Thermal imaging is particularly valuable in drought studies and in assessing water use efficiency.
- **3D Imaging**: Combines multiple 2D images or uses techniques like structured light to create three-dimensional models (Scharr et al., 2016) of plants, allowing for accurate measurements of plant structure, such as root architecture, leaf angle, and canopy volume.

### 8.3.2 Ground-Based Sensors and Robotic Platforms

Ground-based phenotyping platforms, often equipped with various sensors and robotic arms, allow for the automated collection of detailed data (Sadeghi-Tehran et al., 2017) from plants in controlled and field environments. Key types of ground-based systems include:

- **Automated Gantries**: Structures that move sensors or cameras over a controlled plant population in greenhouses or phenotyping facilities, enabling the collection of multiple plant metrics over time (Rahaman et al., 2015). These systems are ideal for tracking growth and changes in traits in a controlled environment.
- **Robotic Carts**: Mobile robots equipped with cameras and sensors that can navigate through crop rows (Shi et al., 2016) capturing data on individual plants in the field. Robotic carts allow for detailed, close-up phenotyping and are suitable for field environments.
- **Handheld Sensors**: Portable sensors used for smaller-scale studies or where automated systems may not be feasible. These can measure traits like photosynthesis rates, chlorophyll content, and soil moisture, providing flexible data collection (Reynolds et al., 2016)

### 8.3.3 Aerial Platforms (Drones and Satellites)

Aerial platforms play a crucial role in large-scale phenotyping in open field environments, providing a broad, bird's-eye view of plant populations and enabling data collection across large areas (Minervini et al., 2015)

- **Drones (Unmanned Aerial Vehicles)**: Equipped with cameras and multispectral or thermal sensors, drones capture high-resolution images over vast fields, providing data on traits such as canopy cover, plant height, biomass, and health. Drones are highly flexible, can cover large areas quickly, and allow for precise monitoring of individual plots or plants.
- **Satellites**: Used for large-scale crop monitoring, satellite imaging provides lower-resolution but broader data than drones. This technology is suitable for assessing regional or global crop trends (Deery et al., 2014) biomass estimates, and environmental impacts. Satellites provide data on crop health, vegetation indices, and land-use changes, making them useful for precision agriculture and ecosystem management.

### 8.3.4 Spectroscopy and Fluorescence Sensors

Spectroscopy and fluorescence technologies allow for detailed analysis of plant biochemical and physiological traits, often at the cellular or molecular level.

- **Chlorophyll Fluorescence Sensors**: Used to assess photosynthetic efficiency, these sensors provide insights into plant health and stress tolerance. Chlorophyll fluorescence is a key indicator of photosynthetic performance and can be used to detect responses to drought or disease.

- **Near-Infrared Spectroscopy (NIRS)**: Measures the chemical composition of plant tissue, such as moisture and protein content, without damaging the plant. NIRS is widely used in assessing nutrient (Sankaran et al., 2015) content and stress markers.
- **Laser-Induced Breakdown Spectroscopy**: Analyzes elemental composition in plant tissues (Schirrmann et al., 2016) useful for studying nutrient uptake and mineral content. This technology is valuable for understanding soil-plant interactions and crop nutrition.

### 8.3.5 Environmental Sensors

Environmental data is essential for understanding the complex interactions between plants and their surroundings (Choudhury et al., 2017). Environmental sensors provide context to phenotypic data by monitoring conditions such as:

- **Soil Sensors**: Measure soil moisture, temperature, pH, and nutrient levels. These parameters are critical for assessing water use efficiency, root-zone conditions, and nutrient availability.
- **Weather Stations**: Provide data on ambient temperature, humidity, wind speed, and solar radiation (Choudhury et al., 2017) Integrating this information helps in analyzing how environmental factors affect plant growth, stress responses, and overall health.
- **$CO_2$ Sensors**: Track atmospheric $CO_2$ levels around plants, important for studies on photosynthetic efficiency and climate change effects on plant physiology.

### 8.3.6 Data Processing and ML Tools

Collecting high-quality data is only the first step in HTP; the vast datasets generated require advanced processing and analysis.

- **Computer Vision**: Algorithms analyze images to quantify traits like leaf area, root structure, and canopy cover. Computer vision is essential for converting raw images into measurable data (Pieruschka & Schurr, 2012) and identifying patterns within large datasets.
- **ML and Deep Learning**: ML algorithms classify, segment, and interpret complex phenotypic data (Singh et al., 2015), allowing for the automated identification of traits and patterns. Deep learning models can analyze hyperspectral data, predict stress responses, and identify trait-genotype relationships.
- **Data Integration Platforms**: Tools and software that allow for the integration of phenotypic data with genotypic and environmental data, creating comprehensive models that account for genetic, phenotypic, and environmental interactions.

### 8.3.7 Internet of Things (IoT) and Cloud Computing

IoT and cloud-based systems enable real-time monitoring and data storage for large-scale phenotyping.

- **IoT Networks**: Connect various sensors and devices, enabling continuous data collection and communication between phenotyping (Denny et al., 2016) equipment. This network allows for real-time data transfer and monitoring (Bucksch et al., 2014), which is essential for time-sensitive studies on plant responses to environmental stressors.
- **Cloud Computing and Storage**: Facilitates the storage and processing of vast datasets generated from HTP (Montes et al., 2011a) allowing researchers to access and analyze data remotely. Cloud computing also supports collaborative research, as data can be easily shared and accessed by different teams.

## 8.4 APPLICATION OF HTP IN MAJOR CROPS

HTP has proven valuable for enhancing the productivity, resilience, and quality of major crops through large-scale, precise measurements of critical traits. By enabling faster and more efficient data collection on traits (Huang et al., 2020, Kang et al., 2019, Shah et al., 2018) like growth rates, yield potential, disease resistance, and environmental tolerance, HTP applications are revolutionizing breeding programs and crop management strategies. Here are some notable examples of how HTP is applied to key staple and cash crops:

1. **Rice (*Oryza sativa*)**
   - **Objective**: Improve rice yield, resilience to drought, salinity, and nutrient efficiency.
   - **HTP Techniques Used**: Drones, multispectral imaging, and root phenotyping platforms help assess traits like canopy cover, root architecture, and chlorophyll content.
   - **Applications**:
     - **Drought and Salinity Tolerance**: By measuring root depth and water-use efficiency using multispectral and thermal imaging, HTP aids in identifying drought- and salt-tolerant rice varieties suited to challenging environments.
     - **Nutrient Use Efficiency**: Using hyperspectral imaging, researchers analyze nutrient content in leaves, improving nutrient uptake traits and reducing fertilizer reliance.
     - **Disease Resistance**: Visible and multispectral imaging allows for early detection of diseases such as rice blast, enabling the selection of resistant varieties.
2. **Wheat (*Triticum aestivum*)**
   - **Objective**: Enhance wheat's adaptability to climate variability, improve yield, and develop disease resistance.
   - **HTP Techniques Used**: Aerial drones, thermal imaging, and chlorophyll fluorescence sensors help track growth and health metrics.
   - **Applications**:
     - **Heat and Drought Resilience**: By using thermal imaging to detect temperature regulation and chlorophyll fluorescence for photosynthetic efficiency, HTP supports the development of wheat varieties that maintain productivity under high temperatures and drought.
     - **Disease Resistance**: HTP enables rapid screening for resistance to fungal diseases like rusts and Fusarium head blight. High-resolution imagery helps detect early-stage infections, promoting timely interventions and aiding in selecting resistant strains.
     - **Yield Prediction**: Canopy height and biomass are monitored with drones and 3D imaging, helping breeders predict yield potential early in the growth cycle.
3. **Maize (*Zea mays*)**
   - **Objective**: Develop high-yielding, drought-resistant, and pest-resistant maize varieties.
   - **HTP Techniques Used**: Drone-based multispectral imaging, soil sensors, and thermal cameras.
   - **Applications**:
     - **Drought Tolerance**: Root phenotyping and thermal imaging help identify varieties with deep rooting systems and efficient water use, improving drought resilience.
     - **Nutritional Quality**: NIRS assesses protein and starch content, enabling the selection of varieties with enhanced nutritional profiles for human consumption and livestock feed.
     - **Pest and Disease Resistance**: High-resolution imaging detects pest damage and disease symptoms early, helping breeders develop resistant varieties.
4. **Soybean (*Glycine max*)**
   - **Objective**: Enhance soybean yield, improve resistance to pests and diseases, and increase oil and protein content.

- **HTP Techniques Used**: Robotic platforms, multispectral imaging, and fluorescence sensors.
- **Applications**:
  - **Yield Optimization**: HTP tracks canopy cover, leaf area, and biomass to identify high-yield soybean lines in breeding trials.
  - **Disease Resistance**: Imaging technologies detect diseases like Asian soybean rust early, allowing for the breeding of resistant lines and reducing pesticide use.
  - **Nutritional Enhancement**: NIRS and hyperspectral imaging quantify oil and protein content in soybeans, supporting the development of high-quality, nutritionally dense varieties.

5. **Sorghum (*Sorghum bicolor*)**
   - **Objective**: Develop drought-tolerant, high-yield, and nutritionally enriched varieties for arid and semi-arid regions.
   - **HTP Techniques Used**: Drone-based multispectral imaging, root scanners, and chlorophyll fluorescence sensors.
   - **Applications**:
     - **Drought Tolerance**: Root architecture analysis using ground-based sensors and imaging identifies varieties with deep and expansive root systems that can access water more efficiently in dry conditions.
     - **Yield and Biomass Estimation**: Drones with multispectral sensors capture canopy cover and height, enabling accurate biomass estimation and selection of high-yielding lines.
     - **Nutritional Value**: Spectroscopic analysis helps select varieties with enhanced protein and mineral content, contributing to food security in nutrient-poor regions.
6. **Barley (*Hordeum vulgare*)**
   - **Objective**: Improve barley's resilience to abiotic stresses and quality traits for food and beverage production.
   - **HTP Techniques Used**: 3D imaging, chlorophyll fluorescence, and multispectral imaging.
   - **Applications**:
     - **Heat and Drought Tolerance**: Thermal imaging and soil moisture sensors are used to monitor water stress responses, supporting the breeding of varieties that thrive in hot, dry climates.
     - **Disease Resistance**: High-resolution imaging helps breeders detect fungal diseases like powdery mildew early, aiding the selection of resistant strains.
     - **Quality Traits**: Hyperspectral imaging assesses traits critical for malting and brewing, such as grain protein content and uniformity, to select high-quality barley varieties.
7. **Tomato (*Solanum lycopersicum*)**
   - **Objective**: Enhance tomato yield, improve flavor and nutrient content, and develop disease-resistant varieties.
   - **HTP Techniques Used**: Visible and multispectral imaging, fluorescence sensors, and robotic platforms.
   - **Applications**:
     - **Disease Resistance**: Imaging techniques enable the early detection of fungal and bacterial diseases like late blight, helping select resistant varieties and reducing crop losses.
     - **Yield Improvement**: HTP assesses flowering and fruit sets by monitoring plant structure and leaf health, leading to the selection of high-yielding varieties.
     - **Nutritional Quality**: Hyperspectral imaging and NIRS are used to analyze lycopene, vitamin, and mineral content, supporting the breeding of nutritionally enriched tomato varieties.

8. **Potato (*Solanum tuberosum*)**
   - **Objective**: Improve yield, disease resistance, and tolerance to climate variability.
   - **HTP Techniques Used**: Drone imaging, soil sensors, and thermal imaging.
   - **Applications**:
     - **Disease Resistance**: Thermal and multispectral imaging detect signs of diseases such as late blight, allowing for the early selection of resistant varieties.
     - **Drought and Heat Tolerance**: Soil moisture sensors and root phenotyping help identify potato varieties with efficient water use and heat tolerance.
     - **Yield Enhancement**: 3D imaging monitors tuber formation and growth, enabling breeders to predict yield and identify high-producing varieties.
9. **Cotton (*Gossypium* spp.)**
   - **Objective**: Enhance cotton fiber quality, pest resistance, and yield under water-limited conditions.
   - **HTP Techniques Used**: Drones, thermal cameras, and fluorescence sensors.
   - **Applications**:
     - **Fiber Quality**: Hyperspectral imaging assesses fiber length, strength, and uniformity, supporting the selection of high-quality cotton varieties.
     - **Pest and Disease Resistance**: Multispectral imaging helps detect early signs of pests and diseases, aiding in the development of resistant varieties.
     - **Water Use Efficiency**: Thermal imaging and soil sensors measure water stress responses, helping develop varieties that are productive in arid conditions.

## 8.5 GENETIC MAPPING AND QUANTITATIVE TRAIT LOCI (QTL)

Genetic mapping and quantitative trait loci (QTL) (Abdurakhmonov et al., 2008) analysis are critical tools in modern plant breeding and genetics, helping researchers understand the genetic basis of complex traits (Figure 8.3). These methods enable breeders to identify specific genes or genomic regions associated with desirable characteristics, such as yield, stress tolerance, and disease resistance. Here's a breakdown of how genetic mapping and QTL analysis work and their applications in crop improvement (Abdurakhmonov et al., 2009).

### 8.5.1 Overview of Genetic Mapping

Genetic mapping is the process of determining the relative positions of genes or genetic markers on a chromosome (Abdurakhmonov et al., 2009). By understanding the location of specific genes and their linkage with observable traits, genetic mapping helps researchers identify regions of the genome (Meuwissen et al., 2001) responsible for specific traits. There are two main types of genetic mapping:

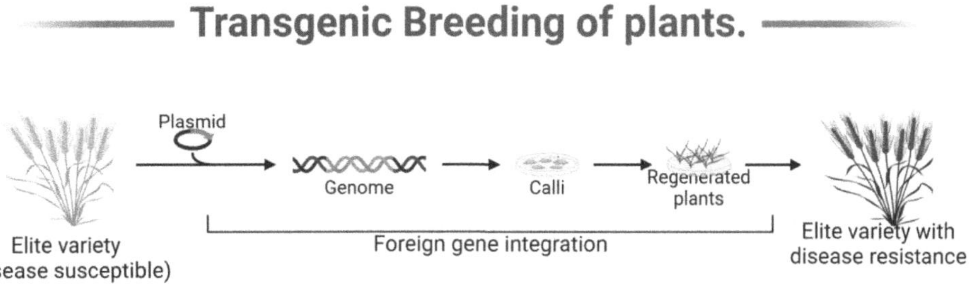


**FIGURE 8.3** Transgenic breeding of plant.

- **Linkage Mapping**: Based on the principle that genes or markers located close to each other on a chromosome are more likely to be inherited together. Linkage mapping uses data from crosses between genetically distinct lines to establish the order and distance between genes or markers (Griffith & Crowfoot, 1934).
- **Association Mapping (GWAS)**: This method uses natural populations and identifies correlations between genetic markers and traits of interest across a wide range of genotypes (Nicholson, 1960). GWAS is effective for studying complex traits influenced by multiple genes.

### 8.5.2 QTL

Quantitative traits are traits that vary continuously, such as plant height, yield, and drought tolerance, and are typically influenced by multiple genes and environmental factors (Muthukumaran, 2016). QTL are specific regions of the genome associated with these quantitative traits. QTL analysis identifies these regions, helping to locate and study the genetic factors that contribute to trait variation.

### 8.5.3 Steps in QTL Mapping

QTL mapping (Preetha & Raveendren, 2008) involves several key steps, often using a population derived from a cross between two genetically distinct parents:

1. **Develop a Mapping Population**: The mapping population is usually a set of offspring derived from two contrasting parental lines, such as recombinant inbred lines (RILs) or backcross populations.
2. **Phenotyping**: Measure the trait(s) of interest across the mapping population under controlled conditions to obtain accurate data on trait variation.
3. **Genotyping**: Use molecular markers [such as single nucleotide polymorphisms (SNPs)] to genotype the mapping population, providing genetic information on each individual in the population.
4. **Linkage Analysis**: Determine the relationship between markers and phenotypic traits using statistical methods, which helps identify (Yu et al., 2014) QTL regions associated with the trait.
5. **Mapping QTL**: Use statistical models to map the QTL (Xu et al., 2017), identifying specific genomic regions associated with the quantitative trait and calculating their effect on the trait.

### 8.5.4 Applications of QTL Mapping in Plant Breeding

QTL mapping has broad applications in plant breeding (Riaz et al., 2017), helping to identify genetic markers that can be used in breeding programs to develop improved crop varieties. Key applications include:

- **Disease Resistance**: QTL mapping (Kang et al., 2019) helps identify genomic regions associated with resistance to diseases like rust in wheat, late blight in potatoes, and blast in rice. Breeders can use these markers to develop disease-resistant varieties, reducing the need for chemical control methods.
- **Stress Tolerance**: QTLs associated with drought, salinity, and heat tolerance help breeders select varieties that perform well under environmental stresses. This is crucial for adapting crops to climate change.

- **Yield Improvement**: Yield is a complex trait controlled by many genes. QTL mapping (Ashraf & Harris, 2004) allows breeders to identify and select yield-related traits like grain size, plant height, and flowering time, helping to increase overall productivity.
- **Quality Traits**: Traits like oil content in soybean, protein content in wheat, and fiber quality in cotton are important for the market value of crops. QTLs associated (Tuberosa et al., 2011) with these traits help breeders develop varieties with enhanced quality attributes.
- **Nutritional Enhancement**: QTL mapping helps locate genes involved in traits like vitamin, mineral, and antioxidant content, supporting the development of nutrient-rich crops for better human health.

### 8.5.5 Challenges in QTL Mapping

Despite its advantages, QTL mapping has some challenges:

- **Complexity of Polygenic Traits**: Many traits are controlled by numerous small-effect QTLs (Zhang et al., 2020), making it challenging to pinpoint specific genes or markers.
- **Environmental Influence**: Environmental factors can significantly impact the expression of quantitative traits, complicating the identification of stable QTLs across different environments.
- **Genotype x Environment (G x E) Interactions**: QTLs for certain traits may only be expressed in specific environmental conditions, requiring multi-environment trials to identify stable QTLs (Liu et al., 2019, Silva et al., 2018).

### 8.5.6 Recent Advances in QTL Mapping and Applications in HTP

Recent advancements, particularly the integration of QTL mapping with high-throughput HTP and genomics (Hussain et al., 2020), have improved the resolution and application of QTL analysis:

- **High-Resolution Mapping**: With the advent of high-throughput sequencing, researchers can now map QTL (Li et al., 2017) at a higher resolution, identifying smaller genetic intervals and even individual genes associated with traits.
- **Speed Breeding and GS**: MAS and GS use QTL (Takeda & Matsuoka, 2008) markers to speed up breeding cycles, accelerating the development of high-performing crop varieties.
- **Integration with ML**: ML models can analyze complex datasets from QTL (Saxena & Verma, 2017, Abdurakhmonov & Reddy, 2010) mapping and HTP, enabling researchers to identify patterns and predict trait outcomes with greater accuracy.

## 8.6 GWAS and HTP

GWAS and HTP together represent a powerful approach to uncovering the genetic basis of complex traits in plants. By integrating GWAS (Gao et al., 2020) which identifies genetic associations across the genome, with HTP, which enables rapid and precise phenotyping on a large scale, researchers can accelerate the identification of genes that contribute to valuable agronomic traits. This integration is particularly useful for improving crop resilience, yield, and quality under diverse environmental conditions.

### 8.6.1 Overview of GWAS

GWAS is a method used to identify associations between genetic markers, typically SNPs, and phenotypic traits. Unlike traditional QTL mapping, which uses controlled crosses and specific populations, GWAS is performed on diverse, natural populations and considers the entire genome. This provides a higher resolution and can identify associations (Voss-Fels & Snowdon, 2019) between traits and small-effect genes across a wide range of genetic backgrounds (Korte & Farlow, 2013).

- **Population Structure**: GWAS often requires a genetically diverse population that reflects natural genetic variation, which can include landraces, breeding lines, or wild relatives.
- **Marker Density**: GWAS relies on high-density SNP arrays or whole-genome sequencing to cover the entire genome, allowing for precise association mapping.
- **Statistical Models**: Statistical models, such as linear mixed models, help control for population structure and relatedness, reducing false positives and improving association accuracy.

### 8.6.2 Role of HTP in GWAS

HTP enhances the power of GWAS by providing accurate, detailed, and large-scale phenotypic data across multiple traits and environments (Atkinson et al., 2019). Traditional phenotyping methods are time-consuming and can limit the scope of GWAS studies, but HTP enables:

- **High Precision and Accuracy**: HTP technologies, such as imaging (e.g., multispectral, hyperspectral, and 3D imaging), environmental sensors, and robotic platforms, capture complex traits like drought tolerance, biomass, and canopy structure with greater precision.
- **Temporal Phenotyping**: HTP allows for repeated measurements over time, creating dynamic datasets that show how traits develop and respond to environmental changes. This is especially useful in GWAS studies that focus on traits affected by growth stages or environmental interactions.
- **Environmental Control**: By conducting HTP under controlled conditions (greenhouses or growth chambers) and field conditions, researchers can assess G x E and pinpoint loci that contribute to trait stability across environments (Tardieu et al., 2017b).

### 8.6.3 Key Steps in GWAS-HTP Integration

Combining GWAS with HTP involves several key steps to maximize the discovery of genetic associations and trait-related loci:

a. **Population Selection and Genotyping**: Selecting a genetically diverse population is crucial. Researchers use SNP arrays or whole-genome sequencing to genotype each individual, ensuring comprehensive genome coverage (Zhang et al., 2021)
b. **HTP**: Use HTP platforms to collect trait data at scale. Multiple traits, including yield, plant structure, stress tolerance, and photosynthetic efficiency, can be measured simultaneously.
c. **Data Processing and Analysis**: Image processing and ML models analyze phenotypic data from HTP to quantify traits. This data is then integrated with genotype data for GWAS.
d. **Association Analysis**: Statistical models analyze the association between SNP markers and phenotypic traits, identifying significant markers and candidate genes (Agrama et al., 2012).
e. **Validation and Functional Analysis**: Once potential loci are identified, further studies validate the function of candidate genes, confirming their role in trait development (Singh et al., 201).

### 8.6.4 Applications of GWAS-HTP in Crop Improvement

The integration of GWAS and HTP has multiple applications across various crops, helping to identify genes and loci responsible for key traits:

- **Stress Tolerance**: GWAS-HTP has been widely used to identify genes associated with abiotic stresses (Li et al., 2019) like drought, heat, and salinity. HTP platforms can measure traits like canopy temperature, transpiration, and root depth, helping identify stress-resilient genes that contribute to water-use efficiency and heat tolerance (Kebede et al., 2020).
- **Yield and Biomass**: High-yielding varieties are critical for food security. By associating yield-related traits (Granier & Vile, 2014) like plant height, biomass, and leaf area with genetic markers, GWAS-HTP helps discover loci linked to high productivity (Korte & von Haeseler, 2012).
- **Nutritional Quality**: GWAS-HTP can reveal genes linked to nutritional quality traits, such as protein, oil, and vitamin content. Using HTP to measure these traits enables breeders to develop bio-fortified crops that improve human nutrition (Banziger et al., 2019).
- **Disease Resistance**: By measuring disease symptoms and plant responses over time, HTP captures the dynamic nature of disease resistance. GWAS-HTP can identify genetic loci associated with resistance to diseases like rice blasts, rusts in wheat, and blights in potatoes, aiding the development of more resilient varieties (Cui et al., 2018).
- **G x E and Trait Stability**: GWAS-HTP helps researchers understand G x E interactions, identifying QTLs that contribute to stable trait expression across environments. This is particularly valuable for breeding (Haug et al., 2019) climate-resilient varieties that can perform consistently in different conditions.

### 8.6.5 Challenges in GWAS-HTP Integration

While GWAS and HTP complement each other, their integration presents certain challenges:

- **Data Complexity and Management**: HTP generates vast amounts of data, which requires efficient processing, storage, and analysis infrastructure. Integrating these datasets with genotypic data for GWAS can be complex (Korte & Farlow, 2013).
- **Environmental Variability**: Environmental conditions can significantly affect phenotypic expression. Ensuring consistency in HTP measurements across different locations and environmental conditions can be challenging, particularly for field studies (Tardieu et al., 2017a).
- **Cost and Resource Requirements**: HTP platforms and genome-wide genotyping can be costly. Implementing both at scale requires significant investment in technology and expertise (Friedrichs et al., 2017).
- **Statistical and Computational Demands**: The sheer volume of data from GWAS and HTP requires advanced statistical models and computational power to accurately analyze and interpret associations, especially when accounting for G x E interactions (Voss-Fels & Snowdon, 2019).

### 8.6.6 Advances and Future Directions in GWAS-HTP

Recent advances and future directions in the field aim to overcome challenges and expand the capabilities of GWAS-HTP:

- **ML and AI**: Advanced ML algorithms help process and analyze large phenotypic datasets, allowing for automated trait extraction, feature selection, and pattern recognition (Voss-Fels & Snowdon, 2019). AI also enhances the prediction of genetic associations based on phenotypic and genotypic data.
- **Precision Field Phenotyping**: Improved field-based HTP platforms, such as drone-based multispectral imaging and mobile sensor systems, allow researchers to phenotype crops in natural settings, increasing the relevance of GWAS (Al-Tamimi et al., 2016) results for field conditions.

- **Multi-Omics Integration**: Combining GWAS-HTP (Anderson et al., 2019) with transcriptomics, metabolomics, and proteomics provides deeper insights into the molecular mechanisms underlying traits, enabling researchers to link genetic loci with specific pathways and biochemical processes.
- **Longitudinal Studies**: Repeated HTP measurements over time allow researchers to perform temporal GWAS, identifying loci associated with developmental stages or seasonal changes, which is particularly valuable for understanding complex traits like growth rates and stress responses (Araus et al., 2020).

## 8.7 ML AND DATA ANALYTICS IN PHENOMICS

ML and data analytics are transforming the field of plant phenomics by enabling the extraction of meaningful insights from massive and complex datasets generated by HTP (Bachem et al., 2019). These advanced tools allow researchers to analyze phenotypic, environmental, and genetic data, ultimately aiding in the identification of traits and genes linked to yield, resilience, and quality in crops. Here's a closer look at how ML (Cabrera-Bosquet et al., 2012) and data analytics are revolutionizing plant phenomics.

### 8.7.1 Role of ML in Phenomics

ML algorithms excel at identifying patterns, making predictions, and processing large datasets that would be challenging to analyze with traditional methods. In phenomics, ML can enhance various stages of data collection (Chapman et al., 2014), processing, and interpretation by automating complex processes and uncovering hidden relationships.

- **Pattern Recognition**: ML algorithms can identify patterns in phenotypic data related to traits like plant growth, disease symptoms, and stress responses.
- **Prediction**: Using past data (Das et al., 2018) ML models can predict phenotypic outcomes based on environmental conditions, genotype, or both.
- **Data Processing and Feature Extraction**: ML simplifies image and sensor data processing, extracting key traits like leaf area, biomass, and plant architecture from HTP images or sensor outputs.

### 8.7.2 Common ML Techniques in Phenomics

Several ML techniques are commonly used in phenomics, each suited for different types of data and research objectives:

- **Supervised Learning**: These models, such as support vector machines (SVM), random forests, and neural networks, are trained with labeled data. Supervised learning is useful for classifying traits (Fahlgren et al., 2015) (e.g., disease symptoms) and predicting quantitative traits like yield based on phenotypic and environmental inputs.
- **Unsupervised Learning**: Techniques like clustering (k-means) and principal component analysis do not require labeled data (Flood et al., 2016). They help group plants based on phenotypic similarity, identify outliers, and reduce dimensionality in large datasets.
- **Deep Learning**: Convolutional neural networks (CNNs), a type of deep learning model, are particularly effective for image-based phenomics (Fuentes et al., 2018), as they can identify features like leaf area, canopy cover, and disease symptoms with high accuracy.
- **Reinforcement Learning**: Though still emerging in phenomics, reinforcement learning can be used to optimize experimental conditions or resource allocation by continuously learning from feedback and adjusting accordingly.

### 8.7.3 Data Analytics Techniques in Phenomics

Data analytics in phenomics involves using statistical and computational methods to analyze data, extract trends, and make data-driven decisions (Knecht & Campbell, 2020a). These techniques help manage and interpret large phenotypic datasets generated by HTP systems.

- **Descriptive Analytics**: Summarizes phenotypic data, providing insights into mean, variance, and distribution. This is often the first step in analyzing HTP data to understand basic trends.
- **Predictive Analytics**: Combines phenotypic data with environmental and genetic information to predict outcomes like growth rates, yield (Lobos & Hancock, 2015) potential, or stress response.
- **Prescriptive Analytics**: Suggests actionable insights for breeders based on predictive models, recommending the best genotypes or phenotypes to pursue in breeding programs.

### 8.7.4 Applications of ML and Data Analytics in Phenomics

ML and data analytics have broad applications across various phenotyping tasks, including trait measurement, stress detection, and G x E:

- **Automated Trait Measurement**: ML algorithms analyze images from HTP platforms to measure traits like leaf size, plant height, root architecture, and biomass. CNNs, for example, process images to estimate leaf area (Montes & Melchinger, 2016b), while other algorithms use 3D modeling to calculate plant volume and canopy structure.
- **Disease Detection and Classification**: Using image recognition and classification algorithms, ML can detect early signs of diseases like rust, blight, or mildew from leaf or canopy images. By automating disease identification, ML (Moore et al., 2021a) allows for rapid screening of large populations, which is especially valuable in field conditions.
- **Drought and Stress Tolerance Analysis**: ML models analyze traits like leaf temperature, transpiration, and soil moisture to identify plants with high drought tolerance. Thermal imaging data, for instance, can be processed by ML (Salas Fernandez & Rossini, 2021) algorithms to detect early signs of heat or water stress, guiding selection in breeding programs.
- **G x E Interaction Analysis**: By analyzing phenotypic responses across multiple environments, ML models can help identify stable genotypes and understand how different genetic lines respond to varying conditions. This is useful for breeding climate-resilient varieties.

### 8.7.5 Challenges and Limitations of ML and Data Analytics in Phenomics

Although ML and data analytics offer powerful tools for phenomics, there are several challenges:

- **Data Quality and Variability**: HTP data can be noisy, especially when collected in the field where environmental factors (Spindel et al., 2016a) vary. Ensuring data quality and consistency is essential to produce reliable ML models.
- **Computational Requirements**: Processing high-resolution images and large datasets requires significant computational power, which can be a constraint for some institutions.
- **Interpretability**: Complex ML models, like deep learning, can function as "black boxes" with limited interpretability. Breeders may need to understand the basis of trait predictions to apply findings effectively.
- **Training Data**: ML models, particularly supervised models, require large amounts of labeled data for training (Tardieu et al., 2017b). However, labeling data can be time-consuming and resource-intensive in phenomics.

### 8.7.6 Advances and Future Directions

The field of ML and data analytics in phenomics is evolving rapidly, with several exciting advances and future directions on the horizon:

- **Explainable AI (XAI)**: XAI methods make ML models more interpretable by showing how they arrive at their predictions. This transparency helps breeders trust and apply ML insights in their programs.
- **Integration with Multi-Omics Data**: Combining phenomic data with other omics data (genomics, transcriptomics, metabolomics) (Becker et al., 2021) enables researchers to explore the full complexity of trait (Tester & Langridge, 2010) development and response to environmental conditions, providing a comprehensive understanding of traits.
- **Edge Computing and IoT**: Mobile phenotyping and edge computing reduce data transmission and processing needs (Voss-Fels & Snowdon, 2019), enabling real-time analysis in the field. IoT devices can capture environmental data alongside phenotypic data, enhancing G x E studies.
- **Automated Breeding Pipelines**: ML can support fully automated breeding pipelines by combining HTP (White et al., 2012b), genotyping, and predictive analytics to continuously refine breeding selections. This could accelerate breeding cycles, allowing for the rapid development of improved crop varieties.

## 8.8 CHALLENGES AND LIMITATIONS OF HTP AND PHENOMICS

HTP and phenomics have the potential to revolutionize plant science by enabling rapid and detailed analysis of plant traits at scale. However, the integration of these advanced technologies into plant breeding and research is met with various challenges and limitations. Here's an exploration of the key obstacles facing HTP and phenomics.

### 8.8.1 High Cost and Resource Demands

Setting up HTP platforms and supporting technologies requires substantial financial investment in equipment, infrastructure, and skilled personnel. HTP systems often include specialized sensors, imaging devices, drones, and automated data-processing systems, all of which can be expensive to install and maintain (Montes & Melchinger, 2016b).

- **Cost of Advanced Equipment**: Sophisticated devices like multispectral or hyperspectral cameras, thermal sensors, and 3D imaging equipment are costly.
- **Infrastructure Needs**: HTP requires physical space, such as greenhouses or dedicated fields, and must be adapted for different environmental conditions and crops.
- **Skilled Workforce**: Operating HTP platforms and analyzing complex phenotypic data requires trained personnel with expertise in imaging, ML, and data analysis.

### 8.8.2 Data Volume and Management

HTP generates vast amounts of data across multiple modalities (e.g., images, sensor readings, environmental data) and time points. Managing, storing, and processing this data is a challenge, especially in large-scale field studies (Fahlgren et al., 2015).

- **Data Storage**: Storing high-resolution images and sensor data from large plant populations can be overwhelming, demanding robust data storage solutions.

- **Data Processing**: Processing raw data into useful metrics, such as leaf area or canopy structure, requires significant computational power and specialized software.
- **Data Integration**: HTP data needs to be integrated with genetic, environmental, and management data to be useful in phenomics, which often requires complex databases and pipelines.

### 8.8.3 Standardization and Data Quality

Phenotyping protocols, data acquisition methods, and environmental conditions can vary significantly across studies, leading to inconsistencies in data quality and reproducibility. Lack of standardization in HTP (Al-Tamimi et al., 2016) processes creates challenges for comparing and integrating data across different projects.

- **Measurement Variability**: Differences in equipment, lighting, and environmental factors can introduce variability in measurements, particularly in field phenotyping.
- **Protocol Standardization**: Establishing standard protocols for data acquisition and processing is essential to ensure that HTP data is comparable and reliable.
- **Noise and Artifacts**: Field phenotyping is particularly prone to noise from environmental conditions (e.g., shadows, soil moisture), which can distort measurements.

### 8.8.4 Environmental Variability and Field Challenges

While controlled environments, such as greenhouses or growth chambers, offer standardized conditions, field phenotyping is often affected by environmental variability, making it difficult to ensure consistent and accurate measurements.

- **Unpredictable Weather**: Variations in sunlight, temperature, humidity, and wind can influence phenotypic measurements, particularly for traits like canopy structure or leaf temperature (Araus et al., 2020).
- **Soil Heterogeneity**: Soil properties can vary across a field, affecting plant growth and complicating the interpretation of HTP data.
- **Dynamic Plant Interactions**: Plants in field conditions interact dynamically with each other and their environment, creating complex datasets that are challenging to analyze accurately.

### 8.8.5 Limitations in Trait Measurement

HTP platforms excel at measuring above-ground traits (e.g., canopy cover, height, and leaf color) but often have limited capacity to capture below-ground traits, physiological responses, and biochemical compositions (Spindel et al., 2016b).

- **Root Phenotyping**: Current HTP systems struggle to capture root traits in soil, which are crucial for understanding water and nutrient uptake.
- **Complex Traits**: Traits like drought tolerance or disease resistance are complex and involve multiple physiological and biochemical processes that are challenging to measure using standard HTP sensors.
- **Time-Dependent Traits**: Many traits, such as flowering time, photosynthesis rates, and stress responses, vary over time. HTP platforms require sophisticated temporal tracking capabilities to capture these changes accurately.

### 8.8.6 Data Analysis and Interpretation Challenges

HTP data is high-dimensional, often involving numerous interrelated traits that vary across time and environments. Analyzing and interpreting these complex datasets requires advanced statistical and computational (Knecht & Campbell, 2020c). Tools, which are not always accessible or easily applied.

- **Data Complexity**: Analyzing multi-modal data (e.g., combining image data with environmental and genotypic data) is challenging, requiring specialized data integration and analysis techniques.
- **Limited Interpretability of ML Models**: ML models, especially deep learning, can serve as "black boxes," providing predictions without clear explanations of underlying patterns. This lack of transparency can limit the interpretability and applicability of results for breeding.
- **Computational Resources**: Processing and analyzing large-scale HTP data can demand high-performance computing resources that may not be available to all research institutions.

### 8.8.7 Scaling and Field Deployment

While controlled environments allow for reliable and consistent HTP data collection, scaling these platforms for large-scale field (Moore et al., 2021b) use presents unique challenges. Field HTP systems need to be mobile, robust, and able to withstand harsh weather conditions.

- **Mobility and Adaptability**: Field-based HTP systems, such as drones and tractor-mounted sensors, need to adapt to various field layouts and crop types, adding complexity to the design and operation.
- **Operational Limitations**: Factors like battery life, equipment durability, and sensor calibration can limit the ability of field HTP platforms to collect consistent data over large areas.
- **Speed and Coverage**: Covering large agricultural areas with detailed phenotyping data is time-intensive, requiring optimization of data acquisition speed without compromising accuracy.

### 8.8.8 G x E Complexities

The interaction between G x E is a major factor influencing phenotypic traits. HTP can measure these interactions to some extent, but it is difficult to capture all environmental (Tardieu et al., 2017a,b) variables accurately, especially in open-field conditions.

- **Environmental Confounding**: Without precise control over environmental conditions, it can be challenging to disentangle genetic effects from environmental effects in field phenotyping data.
- **Trait Stability Across Environments**: Determining the stability of trait expression across diverse environmental conditions requires extensive multi-location testing, which is costly and labor-intensive.

### 8.8.9 Ethical and Regulatory Concerns

The use of advanced technologies in phenomics raises certain ethical and regulatory considerations, especially concerning data privacy, intellectual property, and environmental impacts.

- **Data Privacy**: In cases where HTP data involves GPS-tagged or location-specific information, there may be privacy and security considerations.

- **Intellectual Property**: As phenotypic data and derived models gain commercial value, intellectual property rights and data ownership issues may arise, particularly between research institutions and commercial companies.
- **Environmental Impact**: Field-based HTP platforms, particularly drones and sensors, must be used responsibly to minimize potential environmental impacts, such as noise pollution or habitat disturbance.

### 8.8.10 Future Directions to Overcome HTP Challenges

Despite these challenges, ongoing advancements in technology and methodology offer promising solutions:

- **Improved Data Management Systems**: Cloud-based data storage, big data analytics, and block chain could enhance data management, security, and accessibility.
- **Standardized Protocols and Open Databases**: Efforts to develop standardized protocols for data collection and processing, along with open-access databases, will facilitate data sharing and collaboration across research communities.
- **Advanced Sensors and Robotics**: New sensor technologies and robotic systems are under development to improve the accuracy, mobility, and range of HTP platforms, including those capable of measuring root and below-ground traits.
- **AI and Model Interpretability**: Advances in XAI could improve the interpretability of ML models, making results more actionable and transparent for plant breeders and researchers.
- **Multi-Environment and Temporal Studies**: Increasing the capacity for multi-environment testing and longitudinal studies will help address G x E complexities, providing a deeper understanding of trait stability across conditions.

## 8.9 FUTURE PERSPECTIVES AND INNOVATIONS

Future perspectives and innovations in HTP and plant phenomics hold the promise to greatly enhance plant breeding, crop improvement, and sustainable agriculture. As new technologies, methodologies, and data science techniques continue to evolve, the future of HTP and phenomics will likely center on increasing accuracy, efficiency, and accessibility for researchers worldwide. Here are some anticipated advancements and directions for innovation:

a. **Integration of Multi-Omics Data:** The convergence of phenomics with other “omics” data such as genomics, transcriptomics, metabolomics, and proteomics will enable a more holistic understanding of how genetic and molecular mechanisms shape plant traits (Zheng & Yang, 2015). Integrating these layers of information can uncover insights into how different genes and pathways contribute to observable phenotypes under various environmental conditions.
    - **Multi-Omics Platforms**: Combining phenotypic data with genomics and transcriptomics will help researchers dissect complex traits more precisely.
    - **Systems Biology Approach**: By creating models that link multiple biological systems, researchers can gain a deeper understanding of G x E interactions, accelerating trait discovery and selection in breeding programs (Fiorani & Schurr, 2013).
b. **Precision Agriculture and IoT:** HTP will benefit significantly from precision agriculture tools and IoT devices that monitor and manage crops in real time. These technologies can provide continuous, in-situ phenotyping data, such as soil moisture, plant temperature, and nutrient levels, improving the context in which phenotypic traits are analyzed (White et al., 2012a,b).

- **Smart Sensors**: Advances in sensor technology (e.g., moisture, pH, and nutrient sensors) will enable real-time phenotyping and environment monitoring, feeding directly into HTP data systems (Smart & Ferris, 2020).
- **Automated Data Collection**: Drones, robots, and automated ground vehicles equipped with IoT sensors can operate continuously, gathering data on plant traits and environmental conditions, and reducing labor costs.

c. **Artificial Intelligence (AI) and XAI:** Artificial intelligence (AI), especially ML and deep learning, will continue to play a crucial role in HTP, with a growing emphasis on XAI. XAI allows researchers and breeders to understand (Liakos et al., 2018, Minervini et al., 2015). The decision-making process within models, which is critical in scientific research and applied breeding programs.
   - **Predictive Modeling**: AI-driven predictive models will help breeders forecast how plants will perform in specific environments, streamlining selection for desirable traits.
   - **Explain Ability for Breeding Decisions**: By revealing the features and data patterns that influence model predictions, XAI can make AI applications in HTP more interpretable, helping breeders make better-informed decisions.

d. **Advanced Imaging Techniques:** New imaging modalities and techniques will provide more precise measurements and capture complex traits that were previously difficult to quantify. Advances in imaging will allow the phenotyping of traits with a level of detail that was previously unattainable.
   - **Hyperspectral and 3D Imaging**: Hyperspectral imaging captures detailed information across multiple wavelengths, making it possible to detect biochemical and physiological traits. 3D imaging (Araus & Cairns, 2014) and reconstruction techniques allow for accurate measurement of canopy structure, root architecture, and biomass.
   - **Fluorescence Imaging**: New imaging methods, such as fluorescence imaging, can capture photosynthetic efficiency (Mahlein, 2016; Walter et al., 2015b) and other biochemical responses, helping to identify plants with optimal photosynthetic performance and stress resilience.

e. **Edge Computing and Real-Time Data Processing:** Edge computing, which processes data near the data source rather than transmitting it to a central server, will allow for faster analysis and real-time decision-making. This innovation will be particularly impactful in field phenotyping where connectivity and computational power may be limited.
   - **On-Device Analysis**: Edge devices can process images, sensor readings, and other phenotypic data in real time, which is especially beneficial for mobile phenotyping platforms like drones and robotic phenotyping systems.
   - **Reduced Data Latency**: By processing data locally, edge computing reduces the time between data collection and actionable insights, enabling immediate responses to environmental changes or plant stress.

f. **Robotics and Automation in Field Phenotyping:** Automation will continue to improve the scalability and consistency of HTP by minimizing the need for human labor and enhancing data acquisition efficiency. Robots, drones, and autonomous vehicles are becoming more sophisticated, and able to handle diverse field conditions while collecting consistent, high-quality data (Furbank & Tester, 2011; Knecht & Campbell, 2020b).
   - **Autonomous Robots**: Ground-based robots can traverse fields, capturing high-resolution images and sensor data from the ground level for precise trait measurements.
   - **Swarm Robotics**: Coordinated teams of drones or robots can cover large fields simultaneously, allowing for faster and more comprehensive phenotyping across large plots or multi-location trials.
   - **Hybrid Approaches**: Combining aerial (drone-based) and ground-based robotic phenotyping will improve trait measurement accuracy, providing a multi-dimensional view of plants from different angles.

g. **Improved Data Standardization and Sharing Protocols:** To facilitate collaboration and data integration, there will be an increased focus on standardizing HTP protocols, data formats, and ontologies across research institutions. Standardized data-sharing practices will allow researchers to share and analyze data across studies and platforms more easily.
   - **Unified Data Platforms**: Centralized repositories with standardized formats will allow researchers worldwide to contribute and access phenotypic datasets, promoting cross-project analysis.
   - **Open-Source Software and Protocols**: Open-source tools for data processing, analysis, and sharing can help standardize practices in phenomics, making it easier to compare results across studies.

h. **Advances in Remote Sensing and Satellite Phenotyping:** Remote sensing technologies, including satellite imagery, are making it possible to monitor crop phenotypes over large geographic areas. These advancements will improve phenotyping for large-scale crop monitoring and breeding efforts, especially for tracking growth stages and stress responses over time.
   - **High-Resolution Satellite Imagery**: Modern satellites equipped with high-resolution sensors can capture detailed phenotypic data on crop health, growth stages, and stress patterns across vast areas.
   - **Integration with Field-Level Data**: By combining satellite data with field-based phenotyping data, researchers can achieve a comprehensive view of plant health and development on multiple scales, from single plants to large agricultural fields.

i. **Real-Time Feedback for Precision Breeding:** In the future, breeders could receive real-time feedback from HTP systems, enabling them to make immediate adjustments in breeding trials. This adaptive feedback loop will optimize breeding programs, making them faster and more responsive to observed phenotypic data.
   - **Dynamic Selection Adjustments**: Real-time feedback on trait expression and environmental performance allows breeders to modify selection criteria mid-trial, focusing on plants that show the most promise.
   - **Enhanced Decision Support**: AI-based decision-support systems, integrated with HTP data, will provide breeders with actionable insights for accelerating breeding cycles.

j. **Sustainable Agriculture and Environmental Impact Monitoring:** HTP will increasingly be used to promote sustainable agriculture by identifying phenotypes that contribute to resource efficiency, climate resilience, and soil health. This focus aligns HTP research with global (Arasu et al., 2020) goals for sustainability and climate action.
   - **Water and Nutrient Efficiency**: By identifying plants that are naturally efficient in water and nutrient use, HTP can help breeders develop crops that require fewer resources.
   - **Resilience to Climate Change**: HTP enables the selection of plants with resilience to extreme conditions such as drought, heat, and flooding, helping to future-proof agriculture against climate variability (Tradieu et al., 2017).
   - **Carbon Sequestration**: Advanced HTP methods can measure plant biomass and soil health, enabling breeders to select crops that contribute to carbon sequestration and soil quality improvements.

## 8.10 CONCLUSION

The rapid advancements in HTP and plant phenomics are reshaping the landscape of plant science, breeding, and agriculture. By enabling the large-scale, precise measurement of complex traits, HTP provides insights that bridge the gap between genotype and phenotype, making it possible to accelerate breeding programs, optimize crop yield, and improve resilience to climate challenges. The integration of HTP with other technologies like ML, robotics, IoT, and multi-omics has opened new horizons in precision agriculture, allowing researchers and breeders to study plants more comprehensively and efficiently than ever before. Despite its transformative potential, HTP faces challenges in cost, data management, standardization, and environmental variability, especially in large-scale field applications. Yet, emerging solutions including edge computing, advanced imaging, and automated field phenotyping are continuously enhancing the reliability and accessibility of these tools. Looking to the future, innovations such as XAI, real-time monitoring, and standardized data-sharing platforms will address many of these barriers, further democratizing HTP and enabling broader adoption.

HTP is revolutionizing the way researchers and breeders approach plant science, enabling a deeper understanding of plant traits and the development of high-performing crops. As HTP technologies and methodologies continue to evolve, they hold great promise for addressing critical challenges in food security, climate resilience, and sustainable agriculture. Phenomics is revolutionizing plant science by providing a deeper understanding of how genotypes and environments shape phenotypes. By integrating phenotypic and genotypic data, phenomics enables researchers to explore the genetic basis of complex traits, enhance crop improvement efforts, and develop crops tailored for specific environments and climates. As the field advances, phenomics will play a central role in addressing global agricultural challenges, from improving food security to adapting to climate change.

These key technologies enable HTP to provide valuable insights into plant traits, physiology, and responses to environmental stressors on a large scale. By combining advanced imaging, sensing, data processing, and IoT connectivity, HTP platforms can accelerate the development of improved crop varieties, optimize agricultural practices, and contribute to sustainable agriculture. As these technologies continue to advance, HTP will play an increasingly vital role in meeting the global challenges of food security, climate change, and sustainable resource management. HTP applications in major crops enhance breeding efficiency, crop resilience, and nutritional quality, contributing to food security and sustainable agricultural practices. By leveraging advanced imaging, sensor technology, and data analysis, HTP allows breeders and researchers to identify traits that are vital for high-performing crop varieties. As HTP technologies continue to evolve, their applications will play a crucial role in addressing global agricultural challenges, ensuring the future resilience of the world's food supply.

Genetic mapping and QTL analysis are essential tools for understanding the genetic basis of quantitative traits in crops. By identifying QTLs associated with desirable characteristics, breeders can develop crop varieties with improved yield, stress tolerance, and nutritional quality. With advancements in genomics and HTP, QTL mapping is becoming increasingly precise and effective, empowering breeders to meet the growing demands for sustainable agriculture and food security in a changing climate. Integrating GWAS with HTP creates a comprehensive approach for dissecting the genetic basis of complex traits, providing a pathway to developing crops that are high-yielding, resilient, and nutritionally enhanced. By leveraging advanced genotyping and phenotyping technologies, researchers can accelerate the breeding process and develop crop varieties that meet the demands of a growing global population. As HTP and data analysis techniques continue to improve, GWAS-HTP integration will play a central role in the future of crop genomics and sustainable agriculture.

ML and data analytics are pivotal to advancing phenomics, providing the tools necessary to analyze the complex, high-dimensional data generated by HTP systems. By automating trait measurement, detecting stress responses, and analyzing G x E interactions, ML and data analytics contribute to the development of crops that are high-yielding, stress-tolerant, and suitable for various climates. As computational and ML techniques continue to evolve, they promise to play an even more significant role in shaping the future of crop improvement and sustainable agriculture. While HTP and phenomics face several technical, logistical, and ethical challenges, they offer invaluable potential to transform plant breeding and crop research. By addressing these limitations through technological innovation, improved data management, and interdisciplinary collaboration, HTP and phenomics can continue to advance, driving breakthroughs in crop improvement and agricultural sustainability.

The future of HTP and plant phenomics is poised for transformative advancements that will reshape agriculture. With the integration of multi-omics, AI, IoT, robotics, and other cutting-edge technologies, HTP will become increasingly efficient, accessible, and impactful. These innovations will not only enhance crop breeding and trait discovery but also contribute to global efforts for sustainable and climate-resilient agriculture. As HTP continues to evolve, it will play a pivotal role in meeting the challenges of food security, environmental sustainability, and climate resilience in the years ahead. As global agriculture confronts issues like climate change, resource scarcity, and food security, HTP and phenomics stand as crucial allies in developing sustainable, high-yielding, and climate-resilient crops. By fostering interdisciplinary collaboration, embracing new technologies, and focusing on sustainable practices, HTP will play an instrumental role in advancing plant science, meeting the needs of an ever-growing population, and paving the way for a resilient agricultural future.

## ACKNOWLEDGEMENTS

All the authors are thankful to their respective Universities and Institutions for their valuable supports.

## DATA AVAILABILITY

The authors confirm that the data supporting the findings of this study are available within the book/chapter.

## REFERENCES

Abdurakhmonov, I. Y., & Reddy, M. S. (2010). Marker-assisted selection in cotton: Prospects and challenges. *Cotton Science, 21*(2), 118–126.

Abdurakhmonov, I. Y., Yu, J., & Zhang, Q. (2008). Cotton genetic resources and molecular breeding: Potential for genetic improvement of fiber yield and quality. *Field Crops Research, 107*(3), 108–115.

Abdurakhmonov, I. Y., et al (2009). Molecular marker-assisted breeding for high-yielding cotton cultivars. *Euphytica, 170*(2), 155–166.

Agrama, H. A., Zhang, H., & Choi, I. R. (2012). Genomic prediction of yield and quality traits in rice. *Field Crops Research, 127*, 112–120.

Al-Tamimi, N., Brien, C., Oakey, H., Berger, B., Saade, S., Ho, Y. S., Schmöckel, S. M., Tester, M., & Negrão, S. (2016). Salinity tolerance loci revealed in rice using high-throughput non-invasive phenotyping. *Nature Communications, 7*, 13342.

Anderson, S. L., Shelden, M. C., & Tran, L. P. (2019). Genotypic diversity and phenotypic plasticity: Opportunities and challenges for crop improvement. *Trends in Plant Science, 24*(6), 558–564.

Araus, J. L., & Cairns, J. E. (2014). Field high-throughput phenotyping: The new crop breeding frontier. *Trends in Plant Science, 19*(1), 52–61.

Araus, J. L., Cairns, J. E., & Das, B. (2020). Phenotyping for adaptation in climate-resilient crop breeding. *Current Opinion in Plant Biology, 56*, 204–210.

Araus, J. L., & Kefauver, S. C. (2018). Breeding to adapt agriculture to climate change: Affordable phenotyping solutions. *Current Opinion in Plant Biology, 45*, 237–247.

Araus, J. L., Serret, M. D., & Edmeades, G. O. (2012). Phenotyping maize for adaptation to drought. *Frontiers in Physiology, 3*, 305.

Araus, J. L., Serret, M. D., & Edmeades, G. O. (2020). Phenotyping maize for adaptation to drought. *Frontiers in Plant Science, 11*, 583835.

Ashraf, M., & Harris, P. J. C. (2004). Potential for the improvement of salt tolerance in crop plants: A review. *Biotechnology Advances, 22*(1–2), 51–53.

Atkinson, B., Baret, F., & Lahr, D. (2019). Advances in high-throughput phenotyping of plant traits for breeding and genomics. *Frontiers in Plant Science, 10*, 1–8.

Awlia, M., Nigro, A., Fajkus, J., Schmidhalter, U., Neumann, K., & Granier, C. (2016). High-throughput non-destructive phenotyping of traits associated with plant response to drought. *Scientific Reports, 6*, 24745.

Bachem, C. W. B., van der Hoeven, R. S., & de Bruijn, S. M. (2019). Plant phenotyping in the genomics era. *Trends in Plant Science, 24*(3), 267–278.

Banziger, M., Betrán, F. J., & Beck, D. (2019). Genetic improvement of maize for nutritional quality: A review. *Field Crops Research, 239*, 1–12.

Becker, C., Weigel, D., & Albrecht, T. (2021). High-throughput phenotyping: Emerging opportunities for the genomic selection of complex traits. *Trends in Plant Science, 26*(4), 392–405.

Bolger, A. M., Lohse, M., & Usadel, B. (2014). Trimmomatic: A flexible trimmer for Illumina sequence data. *Bioinformatics, 30*(15), 2114–2120.

Bucksch, A., Atta-Boateng, A., Zhang, J., & Terrestrial Systems Modeling (2014). Data integration in phenomics. *Frontiers in Plant Science, 5*, 24

Cabrera-Bosquet, L., Molero, G., & Bort, J. (2012). High-throughput phenotyping and genome-wide association studies: A review of the state of the art. *Theoretical and Applied Genetics, 125*(4), 603–618.

Cai, X., Xu, H., & Li, X. (2018). Integrating high-throughput phenotyping and genomics for crop improvement. *Plant Biotechnology Journal, 16*(1), 20–30.

Chapman, S. C., Merz, T., Chan, A., Jackway, P., Hrabar, S., Dreccer, M. F., Holland, E., Zheng, B., & Ling, T. J. (2014). Phenotyping of crops in field conditions: Applications of UAV imagery. *Frontiers in Plant Science, 5*, 759.

Chen, D., Neumann, K., Friedel, S., Kilian, B., Chen, M., Altmann, T., & Klukas, C. (2014). Dissecting maize phenotypic variation using a small Unmanned Aerial Vehicle (UAV)-based phenotyping platform. *Plant Methods, 10*, 44.

Choudhury, B. U., Li, J., & Jones, H. G. (2017). High-throughput phenotyping: Pipelines and platforms. *Plant Methods, 13*, 37.

Cichy, K. A., & Bohn, M. (2019). High-throughput phenotyping in plant breeding and genomics. *Frontiers in Plant Science, 10*, 1–11.

Cobb, J. N., DeClerck, G., Greenberg, A., Clark, R., & McCouch, S. (2013). Next-generation phenotyping: Requirements and strategies for enhancing breeding. *The Plant Cell, 25*(2), 378–391.

Coppens, J., Wuyts, N., Inzé, D., Dhondt, S., & Van Loo, D. (2017). A novel image-based high-throughput methodology for precise measuring of plant growth parameters. *Plant Methods, 13*, 59.

Cui, H., Zhang, Y., & Wang, Q. (2018). High-throughput phenotyping for disease resistance in plants. *Journal of Experimental Botany, 69*(19), 4895–4909.

Das, A., Abdel-Ghani, A. H., & Brocklehurst, S. (2018). High-throughput phenotyping for mapping quantitative trait loci in crop plants. *Trends in Genetics, 34*(7), 543–555.

Das Choudhury, S., Samal, A., & Awada, T. (2013). Leveraging image analysis for high-throughput plant phenotyping. *Frontiers in Plant Science, 4*, 218.

Deery, D. M., Rebetzke, G. J., Jimenez-Berni, J. A., James, R. A., Condon, A. G., Bovill, W. D., & Furbank, R. T. (2014). Methodology for high-throughput field phenotyping of canopy temperature using airborne thermography. *Frontiers in Plant Science, 5*, 123.

Denny, R., Shepherd, S. P., & Leung, H. (2016). Integrating genomics and high-throughput phenotyping in plant breeding. *Trends in Genetics, 32*(12), 827–834.

Fahlgren, N., Gehan, M. A., & Baxter, I. (2015). Lights, camera, action: High-throughput plant phenotyping is ready for a close-up. *Current Opinion in Plant Biology, 24*, 93–99.

Fiorani, F., Rascher, U., Jahnke, S., & Schurr, U. (2012). Imaging plants dynamics in the 21st century. *Plant Biology, 14*(6), 789–791.

Fiorani, F., & Schurr, U. (2013). Future scenarios for plant phenotyping. *Annual Review of Plant Biology, 64*, 267–291.

Flood, P. J., Kruijer, W., Schnabel, S. K., van der Wijst, M. G., Keurentjes, J. J., van Eeuwijk, F. A., & Aarts, M. G. (2016). Combining GWAS and expression QTL analyses to identify candidate genes in Arabidopsis thaliana. *Plant Journal, 86*(4), 481–492.

Friedrichs, S., Kusterer, B., & Rötter, R. P. (2017). The integration of genomic selection and high-throughput phenotyping in crop breeding. *Crop Science, 57*(6), 2496–2506.

Fuentes, S., De Bei, R., Pech, J., & Tyerman, S. (2018). Unmanned aerial systems in phenotyping: An Australian viticulture perspective. *Frontiers in Plant Science, 9*, 1544.

Furbank, R. T., & Sadras, V. O. (2017). Breeding for efficiency in the use of carbon and nitrogen resources. *Functional Plant Biology, 36*(11), 885–892.

Furbank, R. T., & Tester, M. (2011). Phenomics–Technologies to relieve the phenotyping bottleneck. *Trends in Plant Science, 16*(12), 635–644.

Gao, L., Chen, Y., & Wang, T. (2020). Integration of high-throughput phenotyping with GWAS for plant breeding. *Journal of Integrative Plant Biology, 62*(1), 11–24.

Granier, C., & Vile, D. (2014). Phenotyping and beyond: Modelling the relationships between traits. *Current Opinion in Plant Biology, 18*, 96–102.

Griffith, C. P., & Crowfoot, D. (1934). Studies on cotton breeding and genetics. *Journal of the American Society of Agronomy, 26*, 406–415.

Haug, S. R., Arora, S., & Meier, M. (2019). Application of high-throughput phenotyping in plant breeding. *Agronomy, 9*(8), 1–14.

Huang, W., Tian, X., & Jiang, X. (2020). Advances in hyperspectral imaging for monitoring crop nutrient use efficiency. *Precision Agriculture Journal*, 21(3), 365–378.

Hussain, M., et al (2020). Identification of QTLs for heat tolerance in maize using genomic and phenotypic data. *Scientific Reports, 10*(1), 5010.

Jin, X., Wang, L., Gao, Y., Huang, W., Liu, L., Liu, X., & Zhang, H. (2016). Hyperspectral indices for monitoring wheat growth status and grain yield prediction. *International Journal of Applied Earth Observation and Geoinformation, 36*, 154–162.

Jones, H. G., & Vaughan, R. A. (2010). *Remote Sensing of Vegetation: Principles, Techniques, and Applications*. Oxford University Press.

Kang, J., et al (2019). Identification of QTLs associated with salt tolerance in rice. *Rice*, 12(1), 1–13.

Kang, Y., Kim, J., & Oh, J. (2019). Early detection of rice blast using multispectral imaging. *Crop Protection Journal*, 112, 55–64.

Kebede, A. S., Crossa, J., & Tarekegne, A. (2020). Genomic prediction of drought resistance in maize. *Theoretical and Applied Genetics, 133*(3), 789–801.

Knecht, K., & Campbell, M. T. (2020a). Developing robust machine learning pipelines for plant phenomics. *Plant Methods, 16*, 3–12.

Knecht, K., & Campbell, M. T. (2020b). Explainable artificial intelligence in plant phenotyping. *Plant Science, 300*, 110629.

Knecht, K., & Campbell, M. T. (2020c). Robotics in field phenotyping for precision agriculture. *Agricultural Systems, 186*, 102944.

Korte, A., & Farlow, A. (2013). The advantages and limitations of genome-wide association studies. *Nature Reviews Genetics, 14*(1), 81–94.

Korte, A., & von Haeseler, A. (2012). A study of the genetic basis of complex traits using statistical models. *Nature Reviews Genetics*, 13(7), 505–517.

Li, J., Liang, X., & Sun, L. (2019). High-throughput phenotyping for rice stress tolerance. *Rice, 12*(1), 1–12.

Li, L., Zhang, Q., & Huang, D. (2014). A review of imaging techniques for plant phenotyping. *Sensors, 14*(11), 20078–20111.

Li, Y., et al. (2017). Genomic insights into cotton breeding for high fiber quality. *Frontiers in Plant Science, 8*, 594.

Liakos, K. G., Busato, P., Moshou, D., Pearson, S., & Bochtis, D. (2018). Machine learning in agriculture: A review. *Sensors, 18*(8), 2674.

Liu, S., Zhang, M., & Xu, X. (2017). Exploring the genetic architecture of complex traits using high-throughput phenotyping. *Nature Genetics, 49*(6), 784–793.

Liu, X., et al. (2019). The application of QTL mapping in rice breeding: A review. *Rice Science, 26*(1), 1–10.

Lobos, G. A., & Hancock, J. F. (2015). Breeding applications of high-throughput phenotyping. *Trends in Plant Science, 20*(9), 495–504.

Mahlein, A. K. (2016). Plant disease detection by imaging sensors – Parallels and specific demands for precision agriculture and plant phenotyping. *Plant Disease, 100*(2), 241–251.

Mahlein, A. K., Oerke, E.-C., Steiner, U., & Dehne, H.-W. (2012). Hyperspectral imaging for small-scale analysis of symptoms caused by different sugar beet diseases. *Plant Methods, 8*(1), 3.

McCouch, S. R., Zhao, K., Wright, M., Tung, C.-W., Ebana, K., Thomson, M., Reynolds, A., Wang, D., DeClerck, G., Ali, M. L., & Bustamante, C. (2013). Development of genome-wide SNP markers for rice genotyping. *Nature Communications, 4*, 2676.

Meuwissen, T. H. E., Hayes, B. J., & Goddard, M. E. (2001). Prediction of total genetic value using genome-wide dense marker maps. *Genetics, 157*(4), 1819–1829.

Minervini, M., Scharr, H., & Tsaftaris, S. A. (2015). Image analysis: The new bottleneck in plant phenotyping. *IEEE Signal Processing Magazine, 32*(4), 126–131.

Mishra, P., Asaari, M. S. M., Herrero-Langreo, A., Boutrot, F. F. G., Moreau, D. L., & Oberti, R. (2020). High-throughput phenotyping of plant traits using hyperspectral imaging and artificial intelligence: A review. *Current Plant Biology, 23*, 100210.

Montes, J. M., Dhillon, B. S., Singh, J., & Melchinger, A. E. (2011a). High-throughput field phenotyping for genetic studies. *Field Crops Research, 124*(2), 246–253.

Montes, J. M., & Melchinger, A. E. (2016b). Genome-wide prediction models and high-throughput phenotyping. *Heredity, 116*(4), 395–408.

Montes, J. M., & Melchinger, A. E. (2016c). High-throughput phenotyping platforms for plant breeding: Review of main components. *Plant Methods, 12*(5), 1–16.

Montes, J. M., Technow, F., Dhillon, B. S., Mauch, F., & Melchinger, A. E. (2011b). High-throughput non-destructive biomass determination during early plant development in maize under field conditions. *Field Crops Research, 121*(2), 268–273.

Montesinos-López, A., Montesinos-López, O. A., Crossa, J., Gianola, D., Hernández-Suárez, C. M., & Pérez-Rodríguez, P. (2019). Genomic-enabled prediction of plant phenotypes. *Heredity, 123*(3), 208–221.

Montesinos-López, O. A., Montesinos-López, A., Crossa, J., Gianola, D., Hernández-Suárez, C. M., & Manes, Y. (2017). A review of deep learning applications for genomic selection and crop phenotyping. *Frontiers in Genetics, 8*, 171.

Moore, C. R., Basu, S., Johnson, L. S., Zhang, D., & Bryant, D. W. (2021a). Advancing crop improvement through high-resolution phenotyping. *Plant Physiology, 186*(3), 1247–1260.

Moore, R., Spielmeyer, W., & Taylor, J. (2021b). Drones, deep learning, and phenomics: Advanced tools for plant breeding. *Theoretical and Applied Genetics, 134*(3), 775–789.

Muthukumaran, A. (2016). Cotton genetic mapping and marker-assisted selection. *Biotechnology Advances, 34*(5), 666–680.

Nicholson, W. (1960). Advances in cotton breeding and genetics. *Crop Science, 6*(5), 314–319.

Nield, J. (2014). Advances in root imaging techniques: A critical review. *Soil Biology and Biochemistry, 78*, 273–284.

Pask, A. J. D., Pietragalla, J., Mullan, D. M., & Reynolds, M. P. (Eds.) (2012). *Physiological Breeding II: A Field Guide to Wheat Phenotyping*. CIMMYT.

Pieruschka, R., & Schurr, U. (2012). High-resolution phenotyping for agriculture. *Functional Plant Biology, 39*(11), 882–892.

Preetha, P., & Raveendren, M. (2008). Molecular markers in crop improvement: Current trends and future prospects. *Journal of Plant Breeding and Genetics, 15*(1), 31–46.

Rahaman, M. M., Chen, D., Gillani, Z., Klukas, C., & Chen, M. (2015). Advanced phenotyping and phenomics analysis to improve crop breeding. *Heredity, 115*(4), 289–295.

Reynolds, M., Langridge, P., & Grando, S. (2016). Phenotyping for drought tolerance in crop plants. *Current Opinion in Plant Biology, 31*, 94–100.

Riaz, A., et al. (2017). QTL mapping and genomic selection in wheat: A review. *Frontiers in Plant Science*, 8, 463.

Roitsch, T., Cabrera-Bosquet, L., Fournier, C., Ghamkhar, K., Jimenez-Berni, J. A., Pinto, F., & Ober, E. S. (2019). Review: Phenotyping plant–environment interactions under controlled conditions and in the field. *Journal of Experimental Botany, 70*(9), 2439–2453.

Sadeghi-Tehran, P., Virlet, N., Sabermanesh, K., & Hawkesford, M. J. (2017). Multi-feature machine learning model for automatic segmentation of green fractional vegetation cover in wheat. *Frontiers in Plant Science, 8*, 1479.

Salas Fernandez, M. G., & Rossini, L. (2021). Multispectral imaging: Transforming phenotyping for precision agriculture. *Current Opinion in Biotechnology, 70*, 27–34.

Sankaran, S., Khot, L. R., Mishra, A., & Maja, J. M. (2015). A review of advanced techniques for detecting plant diseases. *Computers and Electronics in Agriculture, 125*, 12–20.

Saxena, M. S., & Verma, M. (2017). QTL mapping and genomic selection for improving nutritional content in staple crops. *Plant Biotechnology Journal, 15*(5), 743–754.

Scharr, H., Minervini, M., & French, A. P. (2016). Image analysis for plant phenotyping: Opportunities and challenges. *Plant Methods, 12*, 44.

Schirrmann, M., Hamdorf, A., Garz, A., Giebel, A., & Uto, K. (2016). Quantification of stress responses in sugar beet using hyperspectral imaging. *Computers and Electronics in Agriculture, 124*, 100–110.

Shah, N., Singh, R., & Gupta, K. (2018). Applications of thermal imaging in identifying drought tolerance in rice. *Plant Science Today*, 7(4), 201–213.

Shakoor, N., Lee, S., & Mockler, T. C. (2017). High-throughput phenotyping to accelerate crop breeding and monitoring of diseases in the field. *Current Opinion in Plant Biology, 38*, 184–192.

Shi, Y., Thomasson, J. A., Murray, S. C., Pugh, N. A., Rooney, W. L., Shafian, S., Rajan, N., Maiti, D., Rouze, G., & Neely, H. (2016). Unmanned aerial vehicles for high-throughput phenotyping and agronomic research. *PLoS One, 11*(7), e0159781.

Silva, G. F., et al. (2018). Marker-assisted selection in the development of disease-resistant soybean cultivars. *Journal of Crop Improvement, 32*(3), 320–330.

Singh, A., Jones, H. G., Ganapathysubramanian, B., & Sarkar, S. (2015). Integration of multi-sensor data in phenotyping platforms. *Journal of Experimental Botany, 66*(18), 5429–5440.

Smart, J., & Ferris, M. (2020). Internet of Things (IoT) in precision agriculture. *Computers and Electronics in Agriculture, 175*, 105528.

Spindel, J. E., Begum, H., Akdemir, D., Collard, B., Redoña, E., Jannink, J. L., & McCouch, S. R. (2016a). Genome-wide prediction models that incorporate de novo GWAS are a powerful new tool for tropical rice improvement. *Heredity, 116*(4), 395–408.

Spindel, J. E., McCouch, S. R., & Swamy, B. P. M. (2016b). Combining high-throughput phenotyping and GWAS for precision breeding. *Current Opinion in Plant Biology, 30*, 39–48.

Tackenberg, O. (2007). A new method for non-destructive measurement of biomass, growth rates, and phenology. *Ecological Modelling, 204*(3–4), 579–589.

Takeda, S., & Matsuoka, M. (2008). Genetic approaches to crop improvement. *Nature Reviews Genetics, 9*(6), 379–389.

Tardieu, F., Cabrera-Bosquet, L., Pridmore, T., & Bennett, M. (2017a). Plant phenomics, from sensors to knowledge. *Current Biology, 27*(15), R770–R783.

Tardieu, F., Cabrera-Bosquet, L., Pridmore, T., & Bennett, M. (2017b). Plant phenotyping as a bridge between genomic and crop models. *Nature Plants, 3*(8), 697–706.

Tester, M., & Langridge, P. (2010). Breeding technologies to increase crop production in a changing world. *Science, 327*(5967), 818–822.

Tsaftaris, S. A., Minervini, M., & Scharr, H. (2016). Machine learning for plant phenotyping needs image processing. *Trends in Plant Science, 21*(12), 989–991.

Tuberosa, R., et al. (2011). Genetic and genomic approaches to improve drought tolerance in wheat and maize. *Journal of Experimental Botany, 62*(1), 1325–1337.

Ubbens, J. R., & Stavness, I. (2017). Deep plant phenomics: A deep learning platform for complex plant phenotyping tasks. *Frontiers in Plant Science, 8*, 1190.

Vergara-Diaz, O., Kefauver, S. C., Nieto-Taladriz, M. T., Araus, J. L., & Bort, J. (2016). Remote sensing of water and nitrogen management in Mediterranean wheat breeding. *Plant Science, 217–218*, 80–90.

Voss-Fels, K. P., & Snowdon, R. J. (2019). Prospects for genomic selection in crop breeding. *Trends in Plant Science, 24*(2), 236–250.

Walter, A., Liebisch, F., & Hund, A. (2015a). Plant phenotyping: From bean weighing to image analysis. *Plant Methods, 11*(1), 1–11.

Walter, A., Scharr, H., Gilmer, F., Zierer, R., Nagel, K. A., Ernst, M., & Wojciechowski, T. (2015b). Visualizing root system architecture: Quantifying phenotypic plasticity at the cellular, organ, and whole plant levels. *Plant Physiology, 169*(3), 1601–1613.

Wang, Y., Thorup-Kristensen, K., & Jensen, L. S. (2017). Mapping root architecture in the field using 3D imaging techniques. *Frontiers in Plant Science, 8*, 1034.

White, J. W., Andrade-Sanchez, P., Gore, M. A., Bronson, K. F., Coffelt, T. A., Conley, M. M., Feldmann, K. A., French, A. N., Heun, J. T., Hunsaker, D. J., Jenks, M. A., Kimball, B. A., Roth, R. L., Strand, R. J., & Thorp, K. R. (2012b). Field-based phenomics for plant genetics research. *Field Crops Research, 133*, 101–112.

White, J. W., Andrade-Sanchez, P., Gore, M. A., & French, A. N. (2012a). Field-based phenomics: Enabling breeding for complex adaptive traits. *Crop Science, 52*(3), 591–611.

Xu, Y., et al. (2017). QTL mapping and marker-assisted selection for improving drought tolerance in maize. *Theoretical and Applied Genetics, 130*(4), 777–792.

Yang, W., Duan, L., Chen, G., Xiong, L., & Liu, Q. (2013). Plant phenomics and high-throughput phenotyping: Accelerating rice functional genomics using multidisciplinary technologies. *Current Opinion in Plant Biology, 16*(2), 180–187.

Yu, J., Abdurakhmonov, I. Y., & Zhang, Q. (2014). Advances in cotton genetics and genomics: Applications for the improvement of fiber yield and quality. *Crop Science, 54*(3), 906–915.

Zhang, C., & Kovacs, J. M. (2012). The application of small unmanned aerial systems for precision agriculture: A review. *Precision Agriculture, 13*(6), 693–712.

Zhang, L., Liu, D., & Yang, M. (2021). High-throughput phenotyping for genomics and plant breeding. *Plant Breeding, 140*(4), 559–568.

Zhang, W., et al (2020). High-throughput phenotyping for QTL mapping of plant resistance to biotic stresses. *Frontiers in Plant Science, 11*, 512.

Zhang, X., Sun, J., & Zhou, Y. (2021). Advances in high-throughput phenotyping for crop improvement. *Frontiers in Plant Science, 12*, 1–15.

Zheng, H., & Yang, L. (2015). Integrating multi-omics data in plant phenomics. *Trends in Plant Science, 20*(10), 693–702.

# 9 High-Throughput Phenotyping and Phenomics for Exploration of Plant Growth and Development

*Shivani Chauhan and Arshdeep Singh*

## 9.1 INTRODUCTION

Projections indicate that the global human population will reach 9 billion by the year 2050. Resolving the issues arising from limited resources, climate change, and the growing world population, it is imperative to enhance crop productivity and quality in a sustainable manner in the next decades (United Nations report, 2018). To enhance crop productivity, the most effective approach is genetic enhancement through breeding (Alston et al., 2009). The rapid advancement of functional genomics has led to the sequencing of a growing number of crop genomes and the identification of several genes that determine important agronomic characteristics (Garett et al., 2006). Genetic technology has made significant progress, enabling crop breeders to produce high-yielding, stress-tolerant plants by targeted genetic selection (Phillips and Godshalk, 2010). Nevertheless, the absence of availability of phenotypic data impedes this endeavor. Historically, plant breeders and farmers have been using phenotypes to make cultivar selections well in advance of the identification of DNA and molecular genetic markers (Araus and Cairns, 2014). Sophisticated high-throughput phenotyping (HTP) techniques provide expedited and cost-effective genomic data, enabling the creation of populations for mapping consisting of hundreds of recombinant inbred lines (McMullen et al., 2009). Due to their inability to deliver sufficiently rapid and precise measurements of the traits of significant lines, current phenotyping methods have become the major limitation for crop breeding. The existing phenotyping methods are characterized by their slowness, high cost, labor-intensive nature, and frequent destructiveness (Furbank and Tester, 2011). The development, quality, and production of plants are influenced by the features of the surrounding environment, including various climatic factors, plant disease, weeds, and insect pests, as well as the nature of the genome. Numerous current studies have demonstrated that these parameters can be effectively evaluated using a variety of imaging techniques (Sankaran et al., 2015). The quantification of the photosynthetic process, shoot biomass estimation using visible imaging, identification of physiological changes caused by nutrient deficits using visible-near infrared spectroscopy, and detection of water stress using thermal imaging (Munns et al., 2010).

## 9.2 PLANT PHENOTYPING

Plant phenotyping involves the thorough evaluation of various intricate characteristics of plants, including their growth, development, tolerance, resistance, architecture, physiology, ecology, and yield. It also includes the fundamental measurement of specific quantitative factors that serve as the foundation for intricate features (Li et al., 2014). The plant phenotype encompasses intricate features, such as root shape (Walter et al., 2009; Kumar et al., 2014; Flavel et al., 2012; Clark et al., 2011), biomass

DOI: 10.1201/9781003505631-9

(Menzel et al., 2009; Golzarian et al., 2011), leaf characteristics (Jansen et al., 2009; Arvidsson et al., 2011), fruit characteristics (Monforte et al., 2013), yield-related traits (Duan et al., 2011), photosynthetic efficiency (Bauriegel et al., 2011) and biotic & abiotic stress response, which can be directly measured (Rao and Laxman, 2013).

Plant phenotyping can be done using two methods: conventional phenotyping and modern phenotyping. Traditional phenotyping methods involved the use of destructive sampling and human visual or instrument-assisted measurements. This technique requires a significant amount of time and physical effort. Modern phenotyping techniques involve the use of non-destructive samples and automated devices. There is a simultaneous visualization of multi-parameter data. Rahman et al. (2015) provide an instance from modern phenotyping. Plants cultivated in greenhouses are subsequently transported by robotic systems via a conveyor belt to the inspection unit for examination. Various imaging systems encompass a wide range of techniques, such as visual, thermal, fluorescence, X-ray, MRI, and others. After image processing, data will be assessed and interpreted. In the field, information is gathered using stationary or mobile sensors like aerostats and phenocopters. Finally, the data is analyzed and interpreted.

## 9.3 PLANT PHENOTYPICAL TRAITS AND THEIR SIGNIFICANCE

The traits of the crop can be phenotyped into groups according to their morphology, physiological processes and various pathogens. Morphologic qualities refer to straightforward measurements of the physical characteristics of plant organs, including their dimensions, hue, sensory quality, arrangement and size. Physiological attributes refer to the specific assessments of the activities performed by plant organs, including photosynthetic rate, respiratory process, nutrition, hormonal changes, stress resilience and plant water relations. Pathological characteristics refer to unambiguous indicators of plant illnesses resulting from various pathogens encompassing fungi, bacteria, viruses, nematodes, insects and parasitic plants. To overcome challenges in direct measurement, physiological and pathological characteristics are commonly evaluated indirectly via morphological measurement. For instance, the level of stress experienced by a plant can be indirectly assessed by measuring the fluorescence emission of chlorophyll, as the light energy captured by a leaf is harnessed to support photosynthesis. Higher amount of fluorescent chlorophyll is an indicator of reduced photosynthesis.

### 9.3.1 Morphological Traits

Leaf area and fruit appearance are characteristic examples of morphological features. The green part of crops, including spikes, stems, sheaths, and leaf lamina are associated with the plant biomass, transpiration/water loss surfaces, photosynthetic potential and light interception mechanisms (Pask et al., 2012), measurements of the leaf are crucial for plant growth studies. The Leaf Area Index (LAI) is a characteristic that quantifies the extent to which the canopy intercepts radiation and represents the photosynthetic processes occurring in the crop canopy (Gao et al., 2012). A valuable characteristic that reflects the process of transpiration and carbon dioxide exchange, which offers quantitative data on the energy transfer between plants and their external surroundings (Wu et al., 2008). Variations in stem and leaf pigmentation at various growth phases might serve as important indicators of the growing condition of a crop. Nutritional state, population structure and illnesses affect the growth and development of crops, causing color variations in different regions of the plant. Hence, by integrating color identification with several classification techniques, it is possible to establish numerous phenotypic factors related to colors. Plant roots often exhibit intricate, three-dimensional structures that encompass several conventional characteristics (e.g., root length, no of roots, root size, root surface, root growth response, average root radius, average root length, and number of lateral roots) (Zhu et al., 2011).

### 9.3.2 Physiological Traits

The physiological characteristics of plants include metrics such as canopy temperature, leaf water potential, stomatal conductance, and chlorophyll content. Cellular water metabolism is an essential physiological process in agricultural plants. It exerts a significant impact on the growth, development, productivity, and quality of crops. Tissue transpiration is the process by which crops release water, therefore removing heat and thereby controlling the temperature of the plant body (canopy temperature).

Hence, the temperature of the canopy is strongly correlated with the water condition of crops and has the potential to impact water usage efficiency, stomatal conductance, photosynthetic activity, and transpiration rate (Kumar et al., 2015). The measurement of canopy temperature is a crucial technique employed to monitor crop water conditions. Leaf water potential is a characteristic associated with the water-holding capacity of plants. Water is transported through the xylem system in plants under negative pressure, exactly proportionate to the degree of water stress (Villa et al., 2011). Stomatal conductance is a characteristic associated with the flow of gas between leaves and the tree canopy. It is rapidly responsive to soil water conditions, it regulates photosynthetic activity and growth (Munns et al., 2010). More density, larger size, and a greater level of stomatal opening provide enhanced conductance and therefore potentially elevated rates of photosynthesis and transpiration. The importance of stomatal conductance in breeding resides in its fundamental association with heritability and grain production. Enhanced leaf conductivity during high temperatures has been linked to lower canopy temperatures (Kumar et al., 2015). Chlorophyll content in leaves is positively correlated with dry root biomass, seed output, and resistance to iron deficit chlorosis. Leaf chlorophyll monitoring can also be used to determine nodulation and nitrogen-fixing status (Kumar et al., 2015).

### 9.3.3 Pathological Traits

Plant disease causes significant menace to agriculture, which resulted in at least a 10% reduction in worldwide yield (Strange and Scott, 2005). The disease symptoms have devastating effects on the development of plants, thereby reducing their productivity and making them unfit for consumption (Mutka and Bart, 2015). Plant diseases affect the growth and development of crops limiting crop yields and making agricultural products unfit for consumption. We lack knowledge of the physiological mechanisms that link pathogen infection and the disease symptoms in the host (Mutka and Bart, 2015). Plant roots play a crucial role in the growth and development of plants. The roots of a plant, being essential components of its life, absorb vital elements such as water and nutrients. Concurrently, plant roots serve as repositories for carbohydrates and other compounds, while also facilitating the production of hormones necessary for growth (Stoll et al., 2008). Hence, the potential consequences of illnesses on the structure of roots might have an impact on the productivity of crops (Iyer-Pascuzzi et al., 2010). Presently, there is ongoing development of several quantitative high-throughput image-based techniques for phenotyping plant growth and development.

## 9.4 PHENOMICS APPROACHES

High-throughput phenotyping is a crucial and innovative approach employed for the advancement of new and improved plants, the examination of specific genes' functionality, and the analysis of genetic traits in both forward and reverse directions. Furthermore, high-throughput phenotyping is essential for the comprehensive analysis of many lines, including genotype collection, breeding populations, mapping populations and mutant populations under diverse growth conditions. The plant phenotype is an intricate interplay between the genotype and the current environmental circumstances. The plant phenotype can be understood through its structural qualities, whereas plant

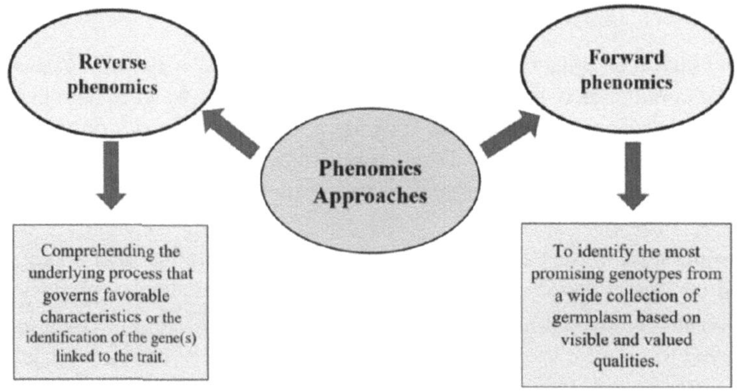


**FIGURE 9.1** Phenomics approaches: forward and reverse phenomics.

functioning can be explained by its biological traits (Tardieu and Tuberosa, 2010). Comprehending the relationship between genotype and phenotype is crucial for predicting plant performance (Houle et al., 2010). High-throughput phenotyping of plants employs two distinct approaches: forward phenomics and reverse phenomics. The forward phenomics strategy involves the use of several phenotyping tools and procedures to identify the most promising genotypes from a wide collection of germplasm based on visible and valued qualities (Figure 9.1). It eventually enables the selection of the most superior line/genotype. It accelerates the phenotyping of a large number of plants or germplasm lines using automated imaging technologies, as well as the rapid identification of desirable traits, making it less need to bring plants to field maturity (Kumar et al., 2015). It permits the evaluation of hundreds of plants (grown in pots) prelabeled with barcodes that run on a conveyor belt and pass through a chamber containing automated imaging devices for evaluating distinctive characteristics (Furbank and Tester, 2011).

Then the identified plants with desirable target traits can be grown for future investigation and breeding. In contrast, reverse phenomics focuses on comprehending the underlying process that governs favorable characteristics. In addition, the identified mechanisms are employed to take advantage of new methods, hence aiding in the development of superior genotypes with enhanced traits/genes. This is accomplished through extensive physiological and biochemical investigation, followed by the correlation of the data with genes that regulate the corresponding biochemical or physiological pathway. After identifying the potential genes, we will compare their expression pattern with those of other genotypes. Therefore, through a meticulous examination of the mechanisms that underlie specific traits, reverse phenomics allows for the utilization of these mechanisms or the identification of the gene(s) linked to the trait. These findings can then be integrated into new plant varieties or transferred to different plant species through genetic engineering (McCoy et al., 2021; Kumar et al., 2015; Furbank and Tester, 2011).

## 9.5 HIGH-THROUGHPUT PHENOTYPING IMAGING TECHNIQUES

High-throughput plant phenomics encompasses a variety of imaging techniques that enable the rapid phenotyping of extensive populations of plants at an individual plant level. High-throughput characterization devices employ several optical sensors to capture images of numerous plants daily. This enables the non-destructive assessment of morphological, physiological, biochemical, and performance parameters (Choudhury et al., 2019; Chawade et al., 2019; Jangra et al., 2021). The primary objective is to statistically assess a phenotypic by examining the interaction between light and plants, including the reflection, absorption and transmission of photons. These interactions will enable the measurement and quantification of several phenotypic features of the plant using optical sensors.

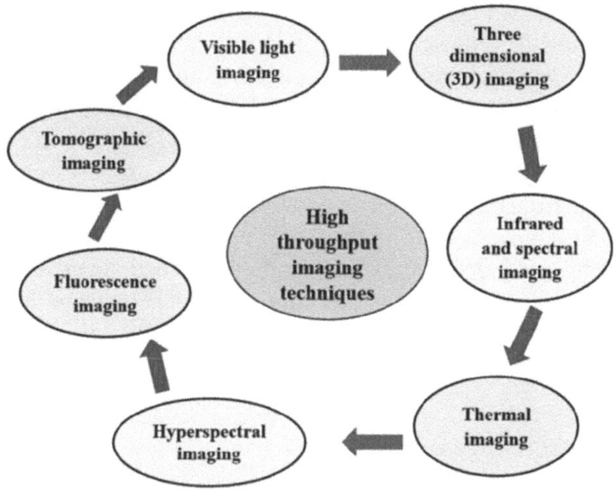


**FIGURE 9.2** Different high-throughput imaging techniques.

Some of the currently available sensors for plant phenotyping include visible light (red–green–blue), thermal, fluorescence, hyperspectral, multispectral, light detection and ranging (LiDAR), magnetic resonance (MRI) imaging, X-ray computed tomography (X-ray CT) and positron emission tomography (PET) (Sun et al., 2022; He et al., 2021) (Figure 9.2).

## 9.6 VISIBLE LIGHT IMAGING

Visible light sensors are capable of detecting light within the wavelength range of 400–700 nm and they offer measurements of the reflected values of red, green, and blue (RGB).

Visible light imaging is most widely used for high-throughput phenotyping because it is easily accessible, simple and affordable. Plant biomass may be precisely measured using high-resolution RGB photographs (Zheng et al., 2022; Sinde-Gonzalez et al., 2022), root architecture (Alaguero-Cordovilla et al., 2018; Brainard et al., 2022), plant growth rate (Wasonga et al. 2021; Gang et al., 2022), germination rate (Li et al., 2021), yield (Elsayed et al., 2021), disease detection and quantification (Osuna-Caballero et al., 2023) and abiotic stresses (Lee et al., 2019). The practical use of these techniques may be influenced by subtle variations in hue between the leaf and the backdrop, as well as the influence of light on the automated processing of images (Kim and Chung, 2021).

## 9.7 THREE-DIMENSIONAL (3D) IMAGING

A 3D image is formed by combining digital photos of the upper and lateral views of plants.

Measurable attributes that can be derived from a three-dimensional image include shoot mass, leaf count, leaf morphology and orientation, leaf pigmentation, and leaf vigor.

In a technical sense, containers with plants are transported on a moving belt through a chamber where images are taken. A computer program then automatically creates three-dimensional models based on these images. The acquired images are sent to the program where necessary modifications, such as color enhancement and optimization, are performed.

Digital photos offer benefits such as easy capture, transmission, and storage in a database. Nevertheless, algorithms are important for the purpose of collecting and examining vast quantities of data (Tsaftaris and Noutsos, 2009).

The Tray Scan system uses barcoded trays to enable effortless recognition of plants and capture precise 3D photographs. Two techniques, specifically LiDAR (Light Detection and Ranging) and stereo photography, are used for capturing three-dimensional images (Li et al., 2014). LiDAR is a sensor technology that utilizes laser light to ascertain the distance to a target by illuminating it and subsequently detecting the reflected pulses. LiDAR technology generates a three-dimensional image during the entire crop cycle and aids in collecting diverse phenotypic data from multiple sources. Stereo vision employs multiple cameras to analyze three-dimensional structures and movements. It is crucial for reconstructing three-dimensional scene geometry in computer science. The utilization of advanced vision technology can greatly decrease expenses by promptly identifying unhealthy plants. An experiment was conducted in 2016, which involved 49 potato plants infected with bacteria and 20 control plants. During the whole period of plant growth, a regular procedure of hyperspectral imaging and a detailed 3D scan were performed, together with a top-down scan that captured both color and depth information. Initial findings indicate that plants impacted by bacterial illnesses can be differentiated from healthy plants by utilizing a combination of three data modalities (Polder et al., 2017).

### 9.7.1 Infrared and Spectral Imaging

This device employs light within the far-infrared (FIR) range of the electromagnetic spectrum, specifically between 15 µm and 1 mm, to investigate temperature. Temperature fluctuations can be employed to examine the tolerance of organisms to saltiness, their water consumption patterns, and the effectiveness of their photosynthesis, among other factors. Plants with lower temperatures possess more efficient root systems and have a higher capacity for water absorption. Far infrared light can be used to generate `heat maps' of individual leaves by illuminating plants cultivated in advanced growth cabinets. These heat maps facilitate the analysis of temperature variations. Essentially, everything emits infrared radiation as a result of the chemical motion occurring therein. Infrared imaging is conducted using two distinct wavelength ranges: one between 0.9 and 1.5 µm, known as near-infrared (NIR), and the other between 7.5 and 13.5 µm, known as far-infrared (FIR).

Near-infrared cameras are used to investigate the moisture content and mobility in soil and foliage, whereas far-infrared cameras are utilized for evaluating temperature. The plants that undergo NIR imaging are cultivated in transparent containers to facilitate root measuring. Near-infrared (NIR) images are employed to quantify the pace at which water is absorbed and to assess their application in plants. Far-infrared imaging was utilized to quantify and compute the thermal discrepancy inside or between the canopy (Sirault et al., 2009). Additionally, they are utilized for quantifying and computing stomatal conductance, providing us with an understanding of the rate of photosynthesis.

### 9.7.2 Thermal Imaging

Thermal infrared imaging enables the identification and monitoring of infrared rays over the outer surface of the plant. A thermal camera converts infrared radiation, which is the thermal energy emitted by an object, into visual images that illustrate the dispersion of surface temperature throughout the object. The thermal sensor detects and measures the infrared radiation emitted by the object within the thermal spectrum of 3–5 µm or 7–14 µm. It produces a picture that displays temperature values for each pixel. Thermal imaging is a valuable technique for assessing the physiological condition of a plant in reaction to organic and non-organic stresses, such as variations in the canopy or leaf temperature (Ahmadi et al., 2017), transpiration and stomatal conductance (Prasher et al., 2013) and plant water status (Jangra et al., 2021). During water shortage, plants respond by closing their stomata, which helps to minimize water loss through transpiration. This response is closely associated with the level of soil moisture. A decrease in transpirational cooling leads to an elevation in leaf temperature. In precision agriculture, thermal imaging can be effectively employed to optimize irrigation and water supply management (Vieira and Ferrarezi, 2021).

### 9.7.3 Hyperspectral Imaging

Hyperspectral imaging is a method that captures both the electromagnetic spectra ($\lambda$) and spatial ($x$, $y$) data at every pixel in a picture. This enables the creation of a three-dimensional data matrix called a hypercube, with a large assortment of photographs covering the spectral range of 250–2,500 nm. This range encompasses ultraviolet (UV), visible (VIS), near-infrared (NIR), and shortwave infrared (SWIR) wavelengths (Sarić et al., 2022). The system provides a substantial quantity of data, enabling the retrieval of a diverse set of observable characteristics, however managing and analyzing the extensive hyperspectral data poses difficulties. Some of its uses encompass the determination of nutrient content, detection of diseases (Abdulridha et al., 2020), assessment of fruit maturity and ripening (Shao et al., 2020), and analysis of various physiological and biochemical variables that are utilized for predicting plant growth and production.

### 9.7.4 Fluorescence Imaging

Fluorescence refers to the emission of light by a plant when it absorbs radiation with shorter wavelengths, mostly through the chlorophyll complex. This emitted light is quite modest, accounting for less than 3% of the total radiation emitted by the light sources.

Fluorescence is the emission of light by a plant when it absorbs radiation of a shorter wavelength. This phenomenon occurs when a plant or chemical absorbs light of a specific wavelength and subsequently emits light of a different wavelength. The chlorophyll complex is typically the primary factor accountable for fluorescence in plants (Lee et al., 2010).

Fluorescence occurs when chlorophyll absorbs actinic or blue light in the chloroplast and re-emits a portion of the light. There is a direct relationship between the amount of light that is released after being received as radiation and a plant's capacity to metabolize the captured light.

The fluorescence imaging system comprises three main components: The system consists of an optical transmission component and an excitation light source to stimulate the fluorescent signal and an amplification and detection of the signal (Gorbe and Calatayud, 2012). The fluorescence imaging system employs customized pulse light and spectral cut-off filters in its cameras to precisely capture and study the kinetics of fluorescence emission. This allows the system to have a high level of sensitivity to particular spectral bands where fluorescent signals are emitted. Fluorescence imaging illuminates plants with a blue light (wavelength less than 500 nm), causing them to release fluorescence light in the range of 600–750 nm. The fluorescence is subsequently captured in photographs and then transformed into spurious signals using a specialized computer program, enabling scientists to observe the fluctuations in fluorescence. Fluorescence imaging is employed to assess photosynthetic activity and plant health in relation to pathological and physiological characteristics.

They are utilized to examine the plant's respiration process in order to identify the impact of plant diseases and insect resistance genes, as well as to monitor the presence of plant pathogens.

Furthermore, they are employed to assess the plant's reaction to both abiotic and biotic factors such as salt and water stress. Currently, modern imaging technologies such as LIFT (Laser-induced Fluorescence Transients) and sun-induced fluorescence are employed for the analysis of the entire canopy or plant (Meroni et al., 2009).

## 9.8 TOMOGRAPHIC IMAGING

Magnetic resonance imaging (MRI), X-ray computed tomography (X-ray CT), and positron emission tomography (PET) are further imaging modalities that offer detailed 3D images of individual plants or plant components (Metzner et al., 2015). MRI utilizes nuclear magnetic resonance to produce images and can detect nuclear resonance signals originating from 13C, 1H, 15N and 14N. MRI utilizes a combination of radio waves and a magnetic field to capture images and is frequently used for imaging plant roots. MRI has offered a non-destructive approach for analyzing the entire plant, including the distribution and measurement of water in different organs. MRI can examine the structural characteristics of plant roots that are grown in pots with a soil combination, whereas previously the plants were cultivated in transparent agar. In addition, MRI visualizes the symptoms caused by cyst nematodes in sugar beetroot and bean root nodulation. MRI acquires three-dimensional images of the internal structures of the sample, allowing for non-invasive measurement of both stationary and changing characteristics, such as the structure, biochemistry, and timing of changes within the plant. MRI is capable of monitoring alterations in growth and development, such as the germination of seeds and bulbs, seed maturation, fruit enlargement, and root expansion. It can also assess the movement of water within plants, as well as their reactions to various forms of stress, including drought, salt, cold, and heat stress. Additionally, MRI can be used to study the interaction between plants and pathogens (Borisjuk et al., 2012).

## 9.9 POSITRON EMISSION TOMOGRAPHY (PET)

PET is a nuclear imaging technology capable of generating a three-dimensional picture or image of a functional process. PET utilizes the detection of a pair of gamma rays that are indirectly released from radio-nuclei that emit positrons. Non-invasive imaging is used to distribute labeled chemicals such as 52Fe, 13N, or 11C. Photosynthesis involves the consumption of $CO_2$, resulting in the generation of a three-dimensional image of 11C-labeled photo-assimilates movement via PET. Additionally, it aids in the examination of plant metabolism. The combination of PET and MRI allows for the analysis of both functional and structural characteristics. These techniques are employed independently to study the movement of water and labeled substances (Jahnke et al., 2009).

## 9.10 X-RAY COMPUTED TOMOGRAPHY (X-RAY CT)

X-ray computed tomography (CT) is employed to visually represent the three-dimensional (3D) formations of both the internal and external morphology of the plant, whether observed at the microscopic or macroscopic level. During the passage of the X-ray beam through the material, a fraction of the beam is absorbed, while the detector records the radiograph. This procedure entails acquiring several two-dimensional projections by either manipulating the sample or the sensor. The projected shapes are then used to rebuild the 3D images (Piovesan et al., 2021). For instance, it has been employed to analyze the size and shape-related morphological features of seeds and fruits (Liu et al., 2020; Ahmed et al., 2020). These imaging approaches are laborious and impractical for analyzing a huge quantity of samples. Furthermore, the equipment's substantial dimensions and significant mass render it unsuitable for deployment on aerial imaging platforms.

### 9.10.1 Phenomics in Field

Presently, the plant phenotyping community appears to be somewhat split between two approaches: more throughput, less resolution phenotyping and detailed phenotyping with less throughput but higher resolution (Dhondt et al., 2013). Phenotyping systems and technologies employed across different sizes give priority to distinct important qualities. Automated phenotyping systems in controlled environments and high-throughput techniques in the field environments prioritize efficiency while phenotyping that includes organ, tissue, and cellular analysis focuses on detailed analysis and higher resolution. In this section, we provided a comprehensive overview of the crop phenotyping methods, ranging from the cellular and tissue level to the field level. We also examined the applicability and practical challenges associated with these technologies in crop research. Modern computer-based imaging systems can be used extensively in numerous disciplines, not just limited to controlled conditions or small areas. Remote sensing phenotyping technology enables researchers to directly study plants in their natural habitat. Multiple plants can be simultaneously measured throughout an entire growing season. The Phenonet sensor network, phenomobile, and phenotower are crucial equipment for doing plant phenotyping in the field, enabling the simultaneous observation of a large number of plants.

### 9.10.2 Phenomobile

The robot operates autonomously. The movement of the Phenomobile is governed by a computer system that utilizes data from the GPS position and accelerometer to control the intended speed and direction. This allows the vehicle to follow a specific course set by an operator. The pathway is automatically formed by intersecting points. The robot is outfitted with a telescoping arm that holds the measurement head. The latter includes a mechanical platform that is furnished with sensors that are connected to the measurement system. The sensors are Sick LMS400 Lidars, RGB Cameras, Flashs LUMIX FR60, RTK GPS, IMU (SBG Ellipse) and Windsonic Anemometer.

## 9.11 PHENONET SENSOR NETWORK

A Phenonet is an advanced field laboratory for phenotyping that utilizes cutting-edge IoT technologies to undertake various crop investigations.

This encompasses sensor networks, IP cameras, mobile devices, and associated data analytics. These technologies provide the collection of crop data in almost real-time to evaluate and forecast crop performance under specific climatic circumstances, the factors to consider are meteorological conditions, canopy temperature, the moisture content in the soil, soil quality and saline soils, application of fertilizers and irrigation. Measurements are taken at 10-second intervals and then averaged over 5 minutes (Jayaraman et al., 2015).

### 9.11.1 Phenotower

This is an advanced field phenotyping equipment with super high-throughput capabilities. It is capable of analyzing data at a height of approximately 16 meters above ground level. This platform consists of a PlantEye camera, Field scales and Weather sensors and can analyze 5,000 plants/hr. The PlantEye camera provides data on plant growth and weight, while the Field scales provide data on irrigation. The Weather sensors collect environmental data. The platform allows for the comparison of temperature of the plant canopy, leaf greenness and crop cover among different plants in a row.

## 9.12 UNMANNED AERIAL VEHICLES

A drone, or unmanned aerial vehicle (UAV), is an aircraft that functions autonomously, without the need for a human pilot, crew, or passengers within its cabin. Unmanned Aerial Vehicles (UAVs) are revolutionizing the agricultural industry by offering intelligent farming solutions that utilize real-time data collection and analysis to enhance overall decision-making and productivity on farms. This technology saves valuable resources, such as time and money, which can be allocated to other areas of farm operations. These devices are outfitted with various sensors and have lately emerged as a crucial method for an efficient and non-intrusive method of high-throughput phenotyping. Unmanned Aerial Vehicle Remote Sensing Platforms (UAV-RSPs) are highly effective instruments for phenomics and genomics (Yang et al., 2017).

### 9.12.1 Phenoscope

The Phenoscope is an advanced instrument used for high-throughput phenotyping. It is designed to efficiently handle and monitor a large number of individual pots.

It is commonly employed to observe and track the development of plants under various environmental conditions across a wide range of genetic variations.

The pots undergo constant rotation (with an average cycle of 4 hours) on a closed-circuit track table, ensuring that the 700+ plants are uniformly exposed to consistent circumstances. The Phenoscope is specifically engineered to manipulate tiny plants, specifically Arabidopsis, during their vegetative stage.

### 9.12.2 Rhizotron

Rhizotrons are instruments used to conduct non-invasive and repetitive observations and measurements of root systems. A rhizotron, obtained from the Greek words "rhizos" meaning root and "tron" meaning instrument, refers to a facility or structure that allows for the observation and measurement of underneath plant components using transparent surfaces.

These facilities provide researchers with simultaneous access to both the underground part and aerial shoots of plants growing in a field-like soil environment. A rhizotron is a permanent structure that can be constructed using transparent walls that can be positioned either directly against a continuous native soil profile or as separate compartments that are isolated from both the native profile and each other.

Usually, these facilities are equipped with a drainage system that eliminates surplus soil water from the vicinity of the facility and/or from inside the compartments. A computer program utilizes algorithms to derive knowledge from photographs by quantifying root length and diameter, as well as analyzing the patterns of root branching. GROWSCREEN-Rhizo has the ability to automatically capture images of the roots and shoots of plants that are grown in soil-filled rhizotrons. It can handle rhizotrons with a volume of approximately 18 L and has the capacity to process 60 rhizotrons per hour. The examination of plants grown in this configuration is restricted to a particular plant size, with a maximum shoot height of 80 cm and a root system that extends up to a depth of 90 cm (Nagel et al., 2014).

The present work utilized a scanner-based rhizotron system to investigate the effects of exogenous tryptophan on root development and morphological responses of soybean `Golden Harvest') seedlings.

The application of tryptophan was carried out at doses of 1.9 and 3.8 mg/plant either by soil drenching or surface spray. The canopy and root projected area was assessed by examining photograms of the canopy and rhizotron using ImageJ software. The samples were taken at the first trifoliate stage, 18 days after sowing (DAS) and root morphology was assessed by analyzing washed root images using WinRHIZO software. The results indicate that the addition of tryptophan externally stimulates auxin-like actions in the growth of roots. Implementing a soil drenching technique with tryptophan seems to be a successful approach to enhance the soybean establishment. Critically, this approach is readily executable by commercial growers without any adverse consequences (Sanada and Agehara, 2023).

## 9.13 APPLICATION OF VARIOUS HIGH-THROUGHPUT PHENOTYPING IMAGING TECHNIQUES IN VEGETABLE CROPS (TABLE 9.1)

**TABLE 9.1**
**Application of Various High-Throughput Phenotyping Imaging Techniques in Vegetable Crops**

| Crop | Improved Traits | Technique | Environment | Place | References |
|---|---|---|---|---|---|
| **Tomato** | Root architecture | RGB | Controlled | Spain | Cordovilla et al. (2018) |
| | Shoot biomass | RGB | Controlled | IIHR, Bangalore | Laxman et al. (2018) |
| | Fruit morphology | RGB | Laboratory | China | Zhu et al. (2022) |

(*Continued*)

**TABLE 9.1 (*Continued*)**
**Application of Various High-Throughput Phenotyping Imaging Techniques in Vegetable Crops**

| Crop | Improved Traits | Technique | Environment | Place | References |
|---|---|---|---|---|---|
| | Nitrogen, Phosphorus and potassium content | MS | Controlled | China | Sun et al. (2019) |
| | Chlorophyll content | RGB, MS | Controlled | China | Sun et al. (2019) |
| | Yield prediction | RGB, MS | Field | Texas (USA) | Chang et al. (2021) and Tatsumi et al. (2021) |
| **Pepper** | Seed quality | X-ray CT | Laboratory | Korea | Ahmed et al. (2020) |
| **Cabbage** | Seed morphology | RGB | Laboratory | Taiwan | Huang and Cheng (2017) |
| | Plant growth | RGB | Field | Korea | Kim et al. (2018) |
| | Shoot biomass | RGB | Field | Ecuador | Sinde-Gonzalez et al. (2022) |
| **Potato** | Crop emergence | RGB | Field | China | Li et al. (2019) |
| | Plant growth | RGB | Controlled | France | Musse et al. (2022) |
| | Tuber growth and development | X-ray CT | Controlled | Germany | Van-Harsselaar et al. (2021) |
| | Tuber size | RGB | Laboratory | China | Si et al. (2017) |
| | Tuber skin color | RGB | Laboratory | Madison, USA | Craza et al. (2020) |
| | Shoot biomass | RGB, HS | Field | China | Liu et al. (2022) |
| | Nitrogen content | RGB | Field | China | Fan et al. (2022) |
| **Lettuce** | Seed morphology | RGB | Laboratory | Japan | Seki and Toda (2022) |
| | Plant growth | RGB | Controlled | China | Du et al. (2022) |
| | Shoot biomass | ChfL | Controlled | USA | Kim and van Iersel (2022) |
| | Salt stress | ChlF | Controlled | USA | Adhikari et al. (2019) |
| **Watermelon** | Seed quality | X-ray | Controlled | Korea | Ahmed et al. (2020) |
| **Carrot** | Root shape attributes | Digital image based | Controlled | USA | Brainard et al. (2021) |

### 9.13.1 Visible Imaging

The visual inspection of horticultural crops relies on the assessment of morphological qualities such as color, size, shape, and surface texture, which influence the visual characteristics and overall quality of the crop. While eye evaluation is commonly employed as a non-destructive way for the purpose of analyzing and categorizing horticultural crops, the use of high-throughput phenotyping tools is crucial for obtaining reliable, expedited, and unbiased outcomes. Currently, image analysis is being progressively employed to quantitatively quantify features like as color, size, shape, and surface texture in various horticultural crops (Haque et al., 2021). Tomato fruit phenotypes play a crucial role in tomato breeding as a standard for agronomic features. Conventional measurement techniques that rely on manual observation, unfortunately, restrict the efficient collection of large-scale data on the physical characteristics of tomato fruits. For this investigation, samples were chosen from ten distinct tomato cultivars that exhibited significant variations in fruit color, size, and other morphological characteristics. The tomato fruit samples were photographed under constant illumination conditions. This study's findings offer technological assistance for the quantitative analysis and assessment of tomato fruit phenotyping, a crucial aspect of tomato breeding. Moreover, the technology has the potential to be applied to the phenotyping of additional fruit crops (Zhu et al 2022). Potato image analysis was employed to measure the skin color and skin set, both of which are crucial factors determining the marketability of novel potato types. An assessment was conducted over two years at a single site to evaluate a collection of 15 red potato types and advanced breeding lines. Following the mechanical harvest and grading process, a total of seven to eight representative tubers per plot were captured in photographs, which were subsequently subjected to analysis. Utilizing image-based phenotyping provides novel prospects for enhancing genetic improvement and comprehension of tuber appearance aspects that were previously challenging to accurately quantify (Caraza Harter and Endelman, 2020).

The Plant Nitrogen Content (PNC) is an essential parameter used to evaluate the nitrogen nutritional condition of crops. Timely and quick acquisition of PNC information is crucial for effectively controlling fertilization and making well-informed decisions in the realm of contemporary precision agriculture. The objective of the work was to establish the feasibility of improving the accuracy of quantifying potato plant nitrogen content (PNC) during critical growth phases by integrating visible light vegetation indices (VIs) and morphological parameters (MPs) derived from a cost-effective unmanned aerial vehicle (UAV) digital camera. Integrating visible light and morphological data obtained by a UAV digital camera was shown to be a feasible method for monitoring crop growth and evaluating plant nitrogen levels (Fan et al., 2022).

### 9.13.2 Hyperspectral Imaging

Accurate measurement of the growth period and prediction of color coordinates are crucial for comparing the quality of postharvest fruit. The present study introduces a novel model for predicting tomato development time and color coordinates using hyperspectral imaging technology. At several stages of tomato development (green-ripe, color-changing, half-ripe, and full-ripe), hyperspectral pictures were captured and color coordinates were acquired using a colorimeter with a prediction accuracy of 93.1%. Hyperspectral imaging has been shown to non-destructively determine the growth period and color parameters of tomatoes, which is highly relevant for developing a tomato quality rating system (Shao et al., 2024). The comprehensive quality index (CQI) value was proposed utilizing hyperspectral imaging and chemometrics to forecast the alterations in tomato quality. The outcome showed that the MLR model had a superior ability to predict the overall quality of tomatoes and was capable of conducting non-destructive testing using hyperspectral imaging. Moreover, it exerts a substantial guiding influence on fruit grading, transportation, refrigeration, and processing (Shao et al., 2022). Processing tomatoes are rich in vitamins and lycopene, which is much favored by consumers. This investigation used near-infrared hyperspectral imaging (HSI) technology to determine the hardness, soluble solids, lycopene, and titratable acid content of processed tomatoes (980–1660 nm) and the fruits was classified at three maturity stages. The findings indicated that the integration of deep learning coupled with near-infrared hyperspectral imaging technology accurately forecasted the quality and maturity of processed tomatoes. This study offers insights into the use of High-Speed Imaging (HSI) as a protocol for the non-destructive examination of diverse agricultural goods (Zhao et al., 2023). Effective irrigation management remains a crucial concern for tomato plantations. Precise and prompt evaluation of the water content of tomato leaves is crucial for facilitating suitable irrigation, therefore preventing the need for unnecessary use of nutrition solutions and labor. The aim of this study was to develop a method for real-time measurement of the water status of tomato leaves, namely the relative water content (WC) and equivalent water thickness (MC), for five distinct tomato cultivars using hyperspectral imaging. This study aims to enhance the understanding of the correlation between hyperspectral reflectance and leaf hydration state (Zhao et al., 2020).

### 9.13.3 Thermal Imaging

Thermography enables the estimate of the canopy temperature (Tcanopy), which serves as an indicator of water stress (Costa et al., 2020; Gomes et al., 2021). When integrated with automated picture analysis, it achieves a high level of precision and is capable of quantifying the relative temperature rather than the absolute temperature. The availability of affordable, high-precision instruments has expanded the potential for utilizing this technology (Garcia-Tejero et al., 2018). In agriculture, infrared thermography is a commonly employed imaging technology that can accurately identify even the slightest temperature variations.

In tropical semi-arid environments, the water status of melon plants (*Cucumis melo* L.) was assessed using thermal imaging. The evaluated parameters comprised canopy temperature (Tcanopy), water potential, air temperature (Tair), soil moisture content, crop yield and thermal index ($\Delta$T), which is determined by subtracting Tair from Tcanopy. The high correlations observed between the thermal index and both crop production and water consumption suggest that thermography is a dependable method for assessing the water quality of melon plants. It can be utilized to enhance the efficiency of irrigation timing in tropical semi-arid regions. Furthermore, thermal pictures revealed the advantageous impacts of soil cover on the hydration level of leaves and the productivity of crops, particularly when subjected to moderate deficiency irrigation. The results suggest that the application of mulching is crucial for enhancing melon production and water conservation in tropical climates (Facundo et al., 2023). The viability of using mobile phones for thermal imaging and unmanned aerial vehicles (UAVs) for multispectral imaging was assessed to evaluate the irrigation efficiency of African eggplant. The leaf moisture content was assessed as a function of soil moisture levels during the early, vegetative, and full vegetative stages. The findings indicate that the crop water stress index (CWSI) obtained from thermal pictures captured by mobile phones is highly responsive to the leaf moisture content (LMC) in I80 and I60 across all phases of growth (Mwinuka et al., 2021). To investigate the intricate interactions between soil and roots, it is necessary to have non-invasive techniques that allow for the Simultaneous observation of geographical and temporal variations in physical and chemical characteristics in the roots and rhizosphere. A study of the laser-induced fluorescence imaging system reveals that the combinations of fluorescence intensity and temporal evolution profiles provide valuable information on root activity and the interactions between roots and soil (Alony and Linker, 2013).

### 9.13.4 X-Ray CT

The internal anatomy of a seed is crucial for the process of germination. A CT scanner with X-ray technology was employed to produce CT pictures of pepper seeds that had naturally matured over a period of five years. The germination test findings were utilized to categorize the seeds into two distinct groups: (1) viable seeds that are considered normal and (2) seeds that are either nonviable or viable but exhibit aberrant characteristics. Out of the entire number of seed samples, 37 were determined to be nonviable and aberrant, while the remaining 63 seeds were confirmed to be normal and viable. The results establish that X-ray CT imaging is an effective approach for classifying seeds based on their internal quality attributes (Ahmed et al., 2020).

## 9.14 UNMANNED AERIAL VEHICLE

Biomass and output are crucial factors for evaluating the productivity and effectiveness of agricultural systems. At the farm level, accurately estimating and forecasting the amount of plant matter and crop production for each individual plant is a significant difficulty in precision agriculture, especially when considering the potential impact of salinity and other non-living factors on plant health. Both field and unmanned aerial vehicle (UAV)-based phenotyping approaches were used to systematically analyze a diversity panel of the *Solanum pimpinellifolium* wild tomato species.

An analysis was conducted on a total of 600 control plants and 600 salt-treated plants. The goal was to forecast the weight of new shoots, the number of tomato fruits, and the total weight of the crop by utilizing several characteristics obtained from the UAV imagery. RGB images captured by unmanned aerial vehicles (UAVs) at different time intervals (1, 2, 4, 6, 7, and 8 weeks prior to harvest) were utilized to assess if the accuracy of predictions differed between plants treated with salt and those in the control group. The results suggested that the predictive accuracy of salt-stressed plant output was greater when the models did not use control plants. Nevertheless, the presence of salt-affected plants did not affect the ability to forecast the yield of control plants. These findings are important for predicting crop production in both healthy and salt-stressed tomato plants. This information can then be used to drive agricultural methods, logistical preparation, and marketing strategies (Johanssen et al., 2020).

### 9.14.1 Rhizotron

The rhizotron technique was employed in tomatoes to examine the salt tolerance at various phases of development, aiming to identify any particular responses to salt tolerance in the chosen varieties. A total of seven *Solanum lycopersicum* genotypes were evaluated i.e., Walter, Moneymaker and LA0147-cultivars and H8504, H9661, H1015 and H5003-hybrids (Spain). The researchers determined that rhizotrons offer a cost-effective and efficient method for studying the root characteristics of tomato seedlings. H1015 had the highest tolerance to salinity in both of the developmental phases that were examined. The technology allows for the automated simulation of the plant's natural habitat on a small scale (Gandullo et al., 2021).

### 9.14.2 High-Throughput Plant Phenotyping Facilities in India

In emerging nations such as India, phenotyping facilities are becoming progressively crucial. Several phenotypic platforms have been lately developed by our nation.

The Plant Phenomics National Facility was initiated in November 2015 at ICAR-IIHR (Indian Institute of Horticulture Research). This state-of-the-art facility aids in the research of identifying resilient genotypes through non-destructive means. Its primary focus is on developing strategies to mitigate the impact of environmental change on horticultural crops.

The Phenospex high-throughput field phenotyping facility, named Field Scan, was inaugurated at ICRISAT, India, in April 2014. The capacity of this platform is limited to 10,000 plants, and it can process 5,000 plants in 1 hour. The Leasy Scan phenotyping platform, which is located at ICRISAT, utilizes the Plant Eye camera to capture three-dimensional images. This machine has the ability to scan 3,200–4,800 plots per 2 hours. The Central Research Institute for Dryland Agriculture (CRIDA) inaugurated high-throughput phenotyping facilities in July 2014. The installation of this facility was carried out as part of the National Initiative on Climate Resilient Agriculture (NICRA) project initiated by ICAR. This method of phenotyping enables precise and non-invasive analysis of cultivars and genetic material in specified environments.

The Indian Council of Agricultural Research (ICAR) created a cutting-edge plant phenomics laboratory at the Indian Agricultural Research Institute in New Delhi, with the support of the National Agricultural Science Fund (NASF). This testing facility is the biggest in India and is considered one of the most technologically sophisticated among publicly funded organizations globally, particularly in terms of its analytical capabilities. The center also includes a "Climate Controlled Facility" consisting of eight distinct greenhouse chambers that allow for precise regulation of temperature and CO2 levels. This research will be immensely valuable for investigating the synergistic impacts of increased carbon dioxide levels in combination with increasing temperatures and other climatic factors. The facility is equipped with four advanced climate-controlled greenhouses, specifically designed for cultivating plants under precise environmental conditions. The plant culture facility is outfitted with 1,200 plant carriers that have RFID chip tags. The plant carrier, integrated with the moving field conveyor system, introduces randomness to the arrangement of plants within the greenhouse. It is responsible for transporting plants to perform automated tasks such as weighing, watering, and imaging at different imaging platforms. The facility is equipped with five automated weighing and watering stations, allowing for accurate application of drought stress to plants and evaluation of transpirational water loss and plant water consumption efficiency.

## 9.15 CONCLUSION

Phenotyping is a highly effective method for rapidly screening and selecting resilient germplasm under various abiotic challenges, significantly faster than traditional phenotyping techniques. The utilization of noninvasive high-throughput sensors has significantly improved the capacity of the phenomics platform to get numerous phenotypic characteristics, in contrast to the limited attributes assessed by traditional approaches. The utilization of phenomics in analytical breeding is anticipated to overcome the limitations in agricultural productivity and tolerance to stressful environments. The majority of plant phenomics technologies currently accessible are limited to controlled conditions, which presents difficulty in gathering phenotypic data in a dynamic microclimate. The next focus for field phenomics imaging technologies may be agricultural germplasm, breeding lines, and mutant populations in field banks, in order to obtain accurate, precise, multi-dimensional, and trustworthy phenotypic data.

Recently, a range of advanced field imaging technology sensors, including the phenonet sensor network, phenomobile, phenotower, and phenoscope, have been employed to generate very accurate data. In order to effectively utilize plant phenomics technology for the identification of climate-resilient varieties, it is essential to possess equipment that is both user-friendly and cost-effective, together with fundamental image processing software and straightforward statistical analysis schemes.

## REFERENCES

Abdulridha J, Ampatzidis Y, Kakarla SC, Roberts P. Detection of target spot and bacterial spot diseases in tomato using UAV-based and benchtop-based hyperspectral imaging techniques. *Precis. Agric.* 2020;21:955–978.

Adhikari ND, Simko I, Mou B. Phenomic and physiological analysis of salinity effects on lettuce. *Sensors.* 2019;19:4814.

Ahmadi SH, Agharezaee M, Kamgar-Haghighi AA, Sepaskhah AR. Comparing canopy temperature and leaf water potential as irrigation scheduling criteria of potato in water-saving irrigation strategies. *Int. J. Plant Prod.* 2017;11:333–348.

Ahmed MR, Yasmin J, Park E, Kim G, Kim MS, Wakholi C, Mo C, Cho B-K. Classification of watermelon seeds using morphological patterns of X-ray imaging: a comparison of conventional machine learning and deep learning. *Sensors.* 2020;20(23):6753. https://doi.org/10.3390/s20236753

Alony A, Linker R. Development of a laser-induced fluorescence imaging system for root activity and rhizosphere visualisation. *Biosyst. Eng.* 2013;114(4):466–473. https://doi.org/10.1016/j.biosystemseng.2012.08.006

Alston JM, Beddow JM, Pardey PG. Agricultural research, productivity, and food prices in the long run. *Science.* 2009;325(5945):1209–1210.

Araus JL, Cairns JE. Field high-throughput phenotyping: the new crop breeding frontier. *Trends Plant Sci.* 2014;19(1):52–61.

Arvidsson S, Pérez-Rodríguez P, Mueller-Roeber B. A growth phenotyping pipeline for Arabidopsis thaliana integrating image analysis and rosette area modeling for robust quantification of genotype effects. *New Phytol.* 2011;191(3):895–907.

Bauriegel E, Giebel A, Herppich WB. Hyperspectral and chlorophyll fluorescence imaging to analyse the impact of Fusarium culmorum on the photosynthetic integrity of infected wheat ears. *Sensors.* 2011;11(4):3765–3779.

Borisjuk L, Rolletschek H, Neuberger T. Surveying the plant's world by magnetic resonance imaging. *Plant J.* 2012;70:129–146.

Brainard SH, Bustamante JA, Dawson JC, Spalding EP, Goldman IL. A digital image-based phenotyping platform for analyzing root shape attributes in carrot. *Front. Plant Sci.* 2021;12:690031. https://doi.org/10.3389/fpls.2021.690031

Brainard SH, Ellison SL, Simon PW, et al. Genetic characterization of carrot root shape and size using genome-wide association analysis and genomic-estimated breeding values. *Theor. Appl. Genet.* 2022;135:605–622. https://doi.org/10.1007/s00122-021-03988-8

Caraza Harter MV, Endelman JB. Image-based phenotyping and genetic analysis of potato skin set and color. *Crop Sci.* 2020;60:202–210. https://doi.org/10.1002/csc2.20093

Chang A, Jung J, Yeom J, Maeda MM, Landivar JA, Enciso JM, Avila CA, Anciso JR. Unmanned aircraft system-(UAS-) based high-throughput phenotyping (HTP) for tomato yield estimation. *J. Sens.* 2021;2021:8875606.

Chawade A, van Ham J, Blomquist H, Bagge O, Alexandersson E, Ortiz R. High-throughput field-phenotyping tools for plant breeding and precision agriculture. *Agronomy.* 2019;9:258.

Choudhury S, Samal A, Awada T. Leveraging image analysis for high-throughput plant phenotyping. *Front. Plant Sci.* 2019:10. https://doi.org/10.3389/fpls.2019.00508

Clark RT, MacCurdy RB, Jung JK, Shaff JE, McCouch SR, Aneshansley DJ, et al. Three-dimensional root phenotyping with a novel imaging and software platform. *Plant Physiol.* 2011;156(2):455–465.

Cordovilla A, Gran-Gómez FJ, Tormos-Moltó S, Pérez-Pérez JM. Morphological characterization of root system architecture in diverse tomato genotypes during early growth. *Int. J. Mol. Sci.* 2018;19:3888. https://doi.org/10.3390/ijms19123888

Costa JdO, Coelho RD, Barros THdS, Fraga Júnior EF, Fernandes ALT. Canopy thermal response to water deficit of coffee plants under drip irrigation. *Irrig. Drain.* 2020;69:472–482. https://doi.org/10.1002/ird.2429

Dhondt S, Wuyts N, Inze D. Cell to whole-plant phenotyping: the best is yet to come. *Trends Plant Sci.* 2013;18:428–439. https://doi.org/10.1016/j.tplants.2013.04.008

Du J, Li B, Lu X, Yang X, Guo X, Zhao C. Quantitative phenotyping and evaluation for lettuce leaves of multiple semantic components. *Plant Methods.* 2022;18:54.

Duan L, Yang W, Huang C, Liu Q. A novel machine vision-based facility for the automatic evaluation of yield-related traits in rice. *Plant Methods* 2011b;7(1):44.

Elsayed S, El-Hendawy S, Khadr M, Elsherbiny O, Al-Suhaibani N, Alotaibi M, Tahir MU, Darwish W. Combining thermal and RGB imaging indices with multivariate and data-driven modeling to estimate the growth, water status, and yield of potato under different drip irrigation regimes. *Remote Sens.* 2021;13:1679.

Facundo AM, Pinheiro Neto L, Viana TVA, Manzano J, Lacerda C, Nascimento J, Lima J, Azevedo B. Evaluation of crop water status of melon plants in tropical semi-arid climate using thermal imaging. *Rev. Bras. Eng. Agríc. Ambient.* 2023;27:447–456. https://doi.org/10.1590/1807-1929/agriambi.v27n6p447-456

Fan Y, Feng H, Jin X, Yue J, Liu Y, Li Z, Feng Z, Song X, Yang G. Estimation of the nitrogen content of potato plants based on morphological parameters and visible light vegetation indices. *Front. Plant Sci.* 2022;13:1012070. https://doi.org/10.3389/fpls.2022.1012070

Flavel RJ, Guppy CN, Tighe M, Watt M, McNeill A, Young IM. Non-destructive quantification of cereal roots in soil using high-resolution X-ray tomography. *J. Exp. Bot.* 2012;63(7):2503–2511.

Furbank RT, Tester M. Phenomics—technologies to relieve the phenotyping bottleneck. *Trends Plant Sci.* 2011;16(12):635–644.

Gandullo J, Ahmad S, Darwish E, Karlova R, Testerink C. Phenotyping tomato root developmental plasticity in response to salinity in soil rhizotrons. *Plant Phenom.* 2021;2021:2760532. https://doi.org/10.34133/2021/2760532

Gang M-S, Kim H-J, Kim D-W. Estimation of greenhouse lettuce growth indices based on a two-stage CNN using RGB-D images. *Sensors.* 2022;22:5499. https://doi.org/10.3390/s22155499

Gao M, Heijden GVD, Vos J, Eveleens BA, Marcelis LFM. Estimation of leaf area for large scale phenotyping and modeling of rose genotypes. *Sci. Hortic.* 2012;138:227–234.

Garcia-Tejero IF, Rubio AE, Viñuela I, Hernández A, GutiérrezGordillo S, Rodríguez-Pleguezuelo CR, Durán-Zuazo VH. Thermal imaging at plant level to assess the crop-water status in almond trees (cv. Guara) under deficit irrigation strategies. *Agric. Water Manage.* 2018;208:176–186. https://doi.org/10.1016/j.agwat.2018.06.002

Garrett KA, Dendy SP, Frank EE, Rouse MN, Travers SE. Climate change effects on plant disease: genomes to ecosystems. *Annu. Rev. Phytopathol.* 2006;44(1):489–509.

Golzarian MR, Frick RA, Rajendran K, Berger B, Roy S, Tester M, et al. Accurate inference of shoot biomass from high-throughput images of cereal plants. *Plant Methods.* 2011;7:2.

Gomes FFL, Lima JS, Pinheiro Neto LG. Aplicações da termografia na agricultura. *Agrar. Acad.* 2021;8:91–105. https://doi.org/10.18677/Agrarian_Academy_2021B9

Gorbe E, Calatayud A. Applications of chlorophyll fluorescence imaging technique in horticultural research: a review. *Sci. Hortic.* 2012;138:24–35.

Haque S, Lobaton E, Nelson N, Yencho GC, Pecota KV, Mierop R, Kudenov MW, Boyette M, Williams CM. Computer vision approach to characterize size and shape phenotypes of horticultural crops using high-throughput imagery. *Comput. Electron. Agric.* 2021;182:106011.

He F, Meng Q, Tang L, Huang X, Lu X, Wang R, Zhang K, Li Y. Research progress in hyperspectral imaging technology for fruit quality detection. *J. Fruit Sci.* 2021;38:1590–1599.

Houle D, Govindaraju DR, Omholt S. Phenomics: the next challenge. *Nat. Rev. Genet.* 2010;11:855–866.

Huang KY, Cheng JF. A novel auto-sorting system for Chinese cabbage seeds. *Sensors.* 2017;17:886.

Iyer-Pascuzzi AS, Symonova O, Mileyko Y, Hao Y, Belcher H, Harer J, Weitz JS, Benfey PN. Imaging and analysis platform for automatic phenotyping and trait ranking of plant root systems. *Plant Physiol.* 2010;152(3):1148–1157.

Jahnke S, Menzel M, Dusschoten D, Roeb G, Bühler J, Minwuyelet S, Blümler P, Temperton V, Hombach T, Streun M, Beer S, Khodaverdi M, Ziemons K, Coenen H, Schurr U. Combined MRI-PET dissects dynamic changes in plant structures and functions. *Plant J. Cell Mol. Biol.* 2009;59:634–644. https://doi.org/10.1111/j.1365-313X.2009.03888.x

Jangra S, Chaudhary V, Yadav RC, Yadav NR. High-throughput phenotyping: a platform to accelerate crop improvement. *Phenomics.* 2021;1:31–53.

Jansen M, Gilmer F, Biskup B, Nagel KA, Rascher U, Fischbach A, et al. Simultaneous phenotyping of leaf growth and chlorophyll fluorescence via grow screen fluoro allows detection of stress tolerance in Arabidopsis thaliana and other rosette plants. *Funct. Plant Biol.* 2009;36:902–914.

Jayaraman PP, Palmer D, Zaslavsky A, Georgakopoulos D. Do-it-yourself digital agriculture applications with semantically enhanced IoT platform. In *Proceedings of the 2015 IEEE Tenth International Conference on Intelligent Sensors, Sensor Networks and Information Processing (ISSNIP)*, Singapore, 7–9 April 2015; pp. 1–6.

Johansen K, Morton MJL, Malbeteau Y, Aragon B, Al-Mashharawi S, Ziliani MG, Angel Y, Fiene G, Negrão S, Mousa MAA, Tester MA, McCabe MF. Predicting biomass and yield in a tomato phenotyping experiment using UAV imagery and random forest. *Front. Artif. Intell.* 2020;3:28. https://doi.org/10.3389/frai.2020.00028

Kim C, van Iersel MW. Morphological and physiological screening to predict lettuce biomass production in controlled environment agriculture. *Remote Sens.* 2022;14:316.

Kim DW, Yun HS, Jeong SJ, Kwon YS, Kim SG, Lee WS, Kim HJ. Modeling and testing of growth status for Chinese cabbage and white radish with UAV-based RGB imagery. *Remote Sens.* 2018; 10(4):563.

Kim J, Chung YS. A short review of RGB sensor applications for accessible high-throughput phenotyping. *J. Crop Sci. Biotechnol.* 2021;24:495–499.

Kumar J, Pratap A, Kumar S. *Phenomics in Crop Plants: Trends, Options and Limitations.* New Delhi: Springer International. 2015.

Kumar J, Pratap A, Kumar S. Plant phenomics: an overview. In J. Kumar, P. Aditya, &S. Kumar (Eds.), *Phenomics in crop plants: Trends, Options and Limitations.* 2015. New Delhi: Springer. https://doi.org/10.1007/978-81-322-2226-2_1.

Kumar P, Huang C, Cai J, Miklavcic SJ. Root phenotyping by root tip detection and classification through statistical learning. *Plant Soil.* 2014;380(1–2):193–209.

Laxman RH, Hemamalini P, Bhatt RM, et al. Non-invasive quantification of tomato (*Solanum lycopersicum* L.) plant biomass through digital imaging using phenomics platform. *Ind. J. Plant Physiol.* 2018;23:369–375. https://doi.org/10.1007/s40502-018-0374-8

Lee U, Silva RR, Kim C, Kim H, Heo S, Park IS, Kim W, Jansky S, Chung YS. Image analysis for measuring disease symptom to bacterial soft rot in potato. *Am. J. Potato Res.* 2019;90:303–313.

Lee WS, Alchanatis V, Yang C, Hirafuji M, Moshou D, Li C. Sensing technologies for precision specialty crop production. *Comput. Electron. Agric.* 2010;74(1):2–33.

Li C, Adhikari R, Yao Y, Miller AG, Kalbaugh K, Li D, Nemali K. Measuring plant growth characteristics using smartphone based image analysis technique in controlled environment agriculture. *Comput. Electron. Agric.* 2020;168:105123.

Li L, Zhang Q, Huang D. A review of imaging techniques for plant phenotyping. *Sensors.* 2014;14(11):20078–20111.

Liu W, Liu C, Jin J, Li D, Fu Y, Yuan X. High-throughput phenotyping of morphological seed and fruit characteristics using X-ray computed tomography. *Front. Plant Sci.* 2020;11:601475.

Liu Y, Feng H, Yue J, Li Z, Jin X, Fan Y, Feng Z, Yang G. Estimation of aboveground biomass of potatoes based on characteristic variables extracted from UAV hyperspectral imagery. *Remote Sens.* 2022;14:5121. https://doi.org/10.3390/rs14205121

McCoy J, Spicer J, Tills O. Phenomics as an approach to comparative developmental physiology. *Front. Physiol.* 2023;14. https://doi.org/10.3389/fphys.2023.1229500

McMullen MD, Kresovich S, Villeda HS, Bradbury P, Li H, Sun Q, Flint-Garcia S, Thornsberry J, Acharya C, Bottoms C, Brown P, Browne C, Eller M, Guill K, Harjes C, Kroon D, Lepak N, Mitchell SE, Peterson B, Pressoir G, Romero S, Rosas M O, Salvo S, Yates H, Hanson M, Jones E, Smith S, Glaubitz JC, Goodman M, Ware D, Holland JB, Buckler ES. Genetic properties of the maize nested association mapping population. *Science.* 2009;325(5941):737–740.

Menzel MI, Tittmann S, Buehler J, Preis S, Wolters N, Jahnke S, et al. Non-invasive determination of plant biomass with microwave resonators. *Plant Cell Environ.* 2009;32(4):368–379.

Meroni M, Rossini M, Guanter L, Alonso L, Rascher U, Colombo R, et al. Remote sensing of solar-induced chlorophyll fluorescence: review of methods and applications. *Remote Sens. Environ.* 2009;113(10):2037–2051.

Metzner R, Eggert A, van Dusschoten D, Pflugfelder D, Gerth S, Schurr U, Uhlmann N, Jahnke S. Direct comparison of MRI and X-ray CT technologies for 3D imaging of root systems in soil: potential and challenges for root trait quantification. *Plant Methods.* 2015;11:17.

Monforte AJ, Diaz A, Caño-Delgado A, Van Der Knaap E. The genetic basis of fruit morphology in horticultural crops: lessons from tomato and melon. *J. Exp. Bot.* 2013;65(16):4625–4637.

Munns R, James RA, Sirault XR, Furbank RT, Jones HG. New phenotyping methods for screening wheat and barley for beneficial responses to water deficit. *J. Exp. Bot.* 2010;61(13):3499–3507.

Musse M, Hajjar G, Ali N, Billiot B, Joly G, Pépin J, Quellec S, Challois S, Mariette F, Cambert M, et al. A global non-invasive methodology for the phenotyping of potato under water deficit conditions using imaging, physiological and molecular tools. *Plant Methods.* 2021;17:81.

Mutka AM, Bart RS. Image-based phenotyping of plant disease symptoms. *Front Plant Sci.* 2015;5:734. https://doi.org/10.3389/fpls.2014.00734.

Mwinuka PR, Mbilinyi BP, Mbungu WB, Mourice SK, Mahoo HF, Schmitter P. The feasibility of hand-held thermal and UAV-based multispectral imaging for canopy water status assessment and yield prediction of irrigated African eggplant (*Solanum aethopicum* L). *Agricult. Water Manage.* 2021;245:06584. https://doi.org/10.1016/j.agwat.2020.106584

Nagel KA, Putz A, Gilmer F, Heinz K, Fischbach A, Pfeifer J, Faget M, Blossfeld S, Ernst M, Dimaki C, Kastenholz B, Kleinert AK, Galinski A, Scharr H, Fiorani F, Schurr U. GROWSCREEN-Rhizo is a novel phenotyping robot enabling simultaneous measurements of root and shoot growth for plants grown in soil-filled rhizotrons. *Funct Plant Biol.* 2012;39(11):891–904. https://doi.org/10.1071/FP12023

Osuna-Caballero S, Olivoto T, Jiménez-Vaquero MA, Rubiales D, Rispail N. RGB image-based method for phenotyping rust disease progress in pea leaves using R. *Plant Methods.* 2023;19(1):86. https://doi.org/10.1186/s13007-023-01069-z

Pask A, Pietragalla J, Mullan D, Reynolds M. *Physiological Breeding II: A Field Guide to Wheat Phenotyping.* Mexico: CIMMYT. 2012.

Phillips RL, Godshalk B. Mobilizing science to break yield barriers. *Crop Sci.* 2010;50(S1):S99–S10.

Piovesan A, Vancauwenberghe V, Van De Looverbosch T, Verboven P, Nicolaï B. X-ray computed tomography for 3D plant imaging. *Trends Plant Sci.* 2021;26:1171–1185.

Polder G, Blok P, De Villiers H, Wolf J, Kamp J. Potato virus Y detection in seed potatoes using deep learning on hyperspectral images. *Front. Plant Sci.* 2019;10. https://doi.org/10.3389/fpls.2019.00209

Rahman H, Ramanathan V, Jagadeeshselvam N, Ramasamy S, Rajendran S, Ramachandran M, et al. Phenomics: technologies and applications in plant and agriculture. *Plant Omics: Omics Plant Sci.*; c2015. p. 385–411.

Rao NKS, Laxman RH. Phenotyping horticultural crops for abiotic stress tolerance. In *Climate-Resilient Horticulture: Adaptation and Mitigation Strategies.* 2013; pp. 147–157.

Sanada A, Agehara S. Characterizing root morphological responses to exogenous tryptophan in soybean (*Glycine max*) seedlings using a scanner-based rhizotron system. *Plants.* 2023;12:186. https://doi.org/10.3390/plants12010186

Sankaran S, Khot LR, Espinoza CZ, Jarolmasjed S, Sathuvalli VR, Vandemark GJ, Miklas PN, Carter AH, Pumphrey MO, Knowles NR, Pavek MJ. Low-altitude, high-resolution aerial imaging systems for row and field crop phenotyping: a review. *Eur. J. Agron.* 2015;70:112–123.

Sarić R, Nguyen VD, Burge T, Berkowitz O, Trtílek M, Whelan J, Lewsey MG, Čustović E. Applications of hyperspectral imaging in plant phenotyping. *Trends Plant Sci.* 2022;27(3):301–315. https://doi.org/10.1016/j.tplants.2021.12.003

Seki K, Toda Y. QTL mapping for seed morphology using the instance segmentation neural network in Lactuca spp. *Front. Plant Sci.* 2022;13:949470.

Shao Y, Ji S, Shi Y, Xuan G, Jia H, Guan X, Chen L. Growth period determination and color coordinates visual analysis of tomato using hyperspectral imaging technology. *Spectrochim. Acta A: Mol. Biomol. Spectrosc.* 2024;319:124538. https://doi.org/10.1016/j.saa.2024.124538

Shao Y, Shi Y, Qin Y, Xuan G, Li J, Li Q, Yang F, Hu Z. A new quantitative index for the assessment of tomato quality using Vis-NIR hyperspectral imaging. *Food Chem.* 2022;386:132864. https://doi.org/10.1016/j.foodchem.2022.132864

Shao Y, Wang Y, Xuan G, Gao Z, Hu Z, Gao C, Wang K. Assessment of strawberry ripeness using hyperspectral imaging. *Anal. Lett.* 2020;54:1547–1560.

Si Y, Sankaran S, Knowles NR, Pavek MJ. Potato tuber length-width ratio assessment using image analysis. *Am. J. Potato Res.* 2017;94:88–93.

Sinde-González I, Gómez-López JP, Tapia-Navarro SA, Murgueitio E, Falconí C, Benítez FL, Toulkeridis T. Determining the effects of nanonutrient application in cabbage (Brassica oleracea var. capitate L.) using spectrometry and biomass estimation with UAV. *Agronomy.* 2022;12:81.

Sirault XR, James RA, Furbank RT. A new screening method for osmotic component of salinity tolerance in cereals using infrared thermography. *Funct. Plant Biol.* 2009;36(11):970–977.

Stoll M, Schultz HR, Berkelmann-Loehnertz B. Early pathogen detection under different water status and the assessment of spray application in vineyards through the use of thermal imagery. *Precis. Agric.* 2008;9(6):407–417.

Strange RN, Scott PR. Plant disease: a threat to global food security. *Annu. Rev. Phytopathol.* 2005;43:83–116. https://doi.org/10.1146/annurev.phyto.43.113004.133839.

Sun D, Robbins K, Morales N, Shu Q, Cen H. Advances in optical phenotyping of cereal crops. *Trends Plant Sci.* 2022;27:191–208.

Sun G, Ding Y, Wang X, Lu W, Sun Y, Yu H. Nondestructive determination of nitrogen, phosphorus and potassium contents in greenhouse tomato plants based on multispectral three-dimensional imaging. *Sensors.* 2019;19(23):5295. https://doi.org/10.3390/s19235295

Tardieu F, Tuberosa R. Dissection and modelling of abiotic stress tolerance in plants. *Curr. Opin. Plant Biol.* 2010;13:206–212. [PubMed] [Google Scholar] [Ref list]

Tatsumi K, Igarashi N, Mengxue X. Prediction of plant-level tomato biomass and yield using machine learning with unmanned aerial vehicle imagery. *Plant Methods.* 2021;17:77.

Tsaftaris S, Noutsos C. Plant phenotyping with low-cost digital cameras and image analytics. In N.A. Ioannis, E.R. Andrea, A.M. Pericles, M.G. Jorge (Eds.), *Information Technologies in Environmental Engineering* (pp. 238–251). Berlin/Heidelberg: Springer. 2009.

United Nations Department of Economic and Social Affairs Population Division. *World Population Projected to Reach 9.7 Billion.* 2050. Available at the United Nations website on Oct. 2, 2018.

Van Harsselaar JK, Claußen J, Lübeck J, Wörlein N, Uhlmann N, Sonnewald U, Gerth S. X-ray CT phenotyping reveals bi-phasic growth phases of potato tubers exposed to combined abiotic stress. *Front. Plant Sci.* 2021;12:613108.

Vieira GHS, Ferrarezi RS. Use of thermal imaging to assess water status in citrus plants in greenhouses. *Horticulturae.* 2021;7:249.

Vila H, Hugalde I, Filippo MD. Estimation of leaf water potential by thermographic and spectral measurements in grapevine. *Rev. Investig. Agropec.* 2011;37(1):46–53.

Walter A, Silk WK, Schurr U. Environmental effects on spatial and temporal patterns of leaf and root growth. *Annu. Rev. Plant Biol.* 2009;60:279–304.

Wasonga DO, Yaw A, Kleemola J, Alakukku L, Mäkelä PSA. Red-green-blue and multispectral imaging as potential tools for estimating growth and nutritional performance of cassava under deficit irrigation and potassium fertigation. *Remote Sens.* 2021;13:598. https://doi.org/10.3390/rs13040598

Wu CY, Niu Z, Tang Q, Huang W. Estimating chlorophyll content from hyperspectral vegetation indices: modeling and validation. *Agric. Forest Meteorol.* 2008;148(8–9):1230–1241.

Yang G, Liu J, Zhao C, Li Z, Huang Y, Yu H, Xu B, Yang X, Zhu D, Zhang X, Zhang R, Feng H, Zhao X, Li Z, Li H, Yang H. Unmanned aerial vehicle remote sensing for field-based crop phenotyping: current status and perspectives. *Front. Plant Sci.* 2017;8:1111. https://doi.org/10.3389/fpls.2017.01111

Zhao M, Cang H, Chen H, Zhang C, Yan T, Zhang Y, Gao P, Xu W. Determination of quality and maturity of processing tomatoes using near-infrared hyperspectral imaging with interpretable machine learning methods, *LWT*. 2023;183:114861. https://doi.org/10.1016/j.lwt.2023.114861

Zhao T, Nakano A, Iwaski Y, Umeda H. Application of hyperspectral imaging for assessment of tomato leaf water status in plant factories. *Appl. Sci.* 2020;10:4665. https://doi.org/10.3390/app10134665

Zheng C, Abd-Elrahman A, Whitaker V, Dalid C. Prediction of strawberry dry biomass from UAV multispectral imagery using multiple machine learning methods. *Remote Sens.* 2022;14:4511.

Zhu J, Ingram PA, Benfey PN, Elich T. From lab to field, new approaches to phenotyping root system architecture. *Curr. Opin. Plant Biol.* 2011;14(3):310–317.

Zhu Y, Gu Q, Zhao Y, Wan H, Wang R, Zhang X, Cheng Y. Quantitative extraction and evaluation of tomato fruit phenotypes based on image recognition. *Front. Plant Sci.* 2022;13:859290. https://doi.org/10.3389/fpls.2022.859290

# 10 Root Phenotyping for Breeding of Climate-Resilient Crops

*Sreehari V. Santhosh, Sudhir Kumar, Madhurima Das, Seethalaxmi, Ezhumalai Sivapragasam, Arumugam Tamilselvan, Sachin Bellubbi, Raman Rajasekar, and Viswanathan Chinnusamy*

## 10.1 INTRODUCTION

Plant morphology demonstrates considerable variability under diverse environmental conditions, even among plants with identical genetic backgrounds. Environmental factors like temperature and light intensity have a profound influence on plant architecture (Tardieu et al., 2017), which makes phenomics studies in plants more complex than in animals. Historically, breeding programs have concentrated on shoot traits such as plant height, time of anthesis, and the number of flowers and tillers to enhance crop yields (Mathan et al., 2016). However, despite their vital role in water and nutrient uptake, plant roots have been relatively ignored in research (Mathan et al., 2016). The first green revolution primarily involved soil-based crop cultivation supported by the use of high levels of chemical fertilizers (Tilman, 1998). The prospect of a second green revolution lies in enhancing crop tolerance to poor soil conditions, as both moisture-deficit and low soil fertility are major constraints on crop production in developing countries (Lynch, 2007). Given the importance of plant roots in nutrient absorption and their extensive soil penetration, improving root architecture is a potential strategy for advancing this revolution. Root system architecture or RSA pertains to the configuration and spatial distribution of root systems in the soil, determined by genetic factors and environmental aspects such as water and mineral nutrient availability, as well as the rhizosphere dimension (Rogers & Benfey, 2015).

Grasping root system architecture (RSA) is crucial for comprehending how plants adjust to varying conditions and for improving agricultural output. However, assessing root systems presents several challenges. Firstly, root evaluation often results in destruction or partial loss. Secondly, identifying the optimal root systems for enhancing plant growth is challenging because they vary significantly across different environmental conditions. Researchers have utilized various phenotyping approaches, including both nondestructive and efficient methods (Tardieu et al., 2017).

Advancements in root phenotyping techniques and the development of analysis software have made it easier to gather RSA data. RSA includes various root traits that are susceptible to environmental influences. Therefore, we examine the key root traits necessary for a thorough understanding of RSA. To achieve this, we employ root modeling to reconstruct RSA. Notably, recent advances include the development of root phenotyping pipelines (Atkinson et al., 2019) and the creation of multiple root models (Barczi et al., 2018). Researchers might still hold old root data in outdated or unpublished formats. This chapter aims to make such data accessible and improve the interpretation of individual researcher data.

## 10.2 THE HIDDEN BLUEPRINT: ROOT SYSTEM ARCHITECTURE (RSA)

The overarching notion of root system architecture (RSA) encompasses both the physical structure and spatial arrangement of roots. Root structure, which pertains to the organization and characteristics of root segments, is shaped by developmental processes governing root elongation, orientation,

DOI: 10.1201/9781003505631-10

aging, and the initiation of new roots (Rumyana et al., 2021). Conversely, root shape plays a pivotal role in determining RSA, affecting the capacity for resource uptake through the strategic positioning of roots (Orman-Ligeza, Beata, 2014). Moreover, water transportation mechanisms, modulated by local factors such as aquaporins and the properties of cell walls, contribute to the functionality of the root system, operating independently of RSA (Javot & Maurel, 2002).

The intricate nature of root system architecture (RSA) has led to a variety of interpretations across different research domains, encompassing diverse interests, scales, and methodologies (Li et al., 2021). Despite this variability, a cohesive framework for describing RSA remains elusive. Recent endeavors aimed at achieving this objective prioritize dynamic phenotyping approaches, which underscore developmental processes influenced by factors such as root age, plant phenology, and environmental conditions (Mostafa et al., 2023). Leveraging noninvasive observation techniques, these approaches seek to construct response curves that elucidate the relationship between development and environmental variables, thereby facilitating comprehension of genotype-environment interactions in RSA.

Adama et al. (2017) assert that the utilization of mathematical models to integrate developmental processes and environmental factors represents a significant advancement in the understanding of RSA. These models have been effectively employed in studying water and nutrient dynamics, carbon allocation, and anchorage, offering valuable tools for simulating RSA under various conditions and investigating genotype reactions (Lobet, 2014).

The spatial arrangements of roots (RSA) have been significantly influenced by domestication and breeding efforts, as noted by Maccaferri et al. (2016). Li et al. (2020) indicate that there is an intersection of quantitative trait loci (QTLs) associated with root traits and productivity, implying that RSA plays a crucial role in improving crop performance. Nevertheless, the utilization of RSA's genetic factors encounters hurdles such as polygenic control characterized by low heritability, pronounced gene-environment interactions, and challenges in the validation and cloning of QTL (Ren et al., 2017).

Identifying breeding objectives for root system architecture (RSA) presents a challenge due to variations in optimal architectures influenced by environmental factors and crop needs (Verhoeven, 2006). Achieving successful RSA manipulation entails tasks such as isolating functional quantitative trait loci (QTL) sequences, establishing uniform phenotyping protocols, and merging omics methodologies with plant physiology and breeding approaches (Moreira et al., 2020). This amalgamation will authenticate the functions of QTL and guide the formulation of breeding tactics tailored to distinct environmental circumstances.

### 10.2.1 Dynamic Adaptations of Root Systems under Stress

The behavior of plant roots under various environmental stresses is essential for their survival and adaptation. Different environmental stresses, such as water-deficit stress, salinity, and heat, significantly influence root system architecture (RSA) and functionality. Understanding these responses can help us understand the underlying mechanisms bringing stress tolerance and adaptations. The effects of different environmental conditions on primary and lateral roots are presented in Table 10.1. Under moisture-deficit stress conditions, plants often develop deeper root systems to access water from lower soil layers, a key adaptation that enhances drought resistance (Uga et al., 2013). The suppression of

**TABLE 10.1**
**Effect of Different Environmental Conditions on Primary and Lateral Roots**

| Stress | Effect on Primary Root | Effect on Lateral Root |
|---|---|---|
| **Drought** | Root length increases | Root length, number and angle decrease |
| **Salinity** | Root length decreases | Root length and number decreases |
| **Nitrogen deficiency** | Root length increases | Root length increases, angle and root number decrease |
| **Phosphorous deficiency** | Root length increases | Root length, number and angle increase |

lateral root growth under drought stress results in a streamlined root system that minimizes water loss and focuses growth toward deeper soil layers (Lynch, 2013). Root angle adjustments, leading to steeper, more vertical growth, promote deeper soil penetration (Yoshida et al., 2016). Additionally, drought stimulates the development of root hairs, increasing surface area for better water and nutrient uptake (Comas et al., 2013). Plants may also shift resource allocation towards root growth, increasing the root-shoot ratio to enhance water exploration (Blum, 2011). These root system adaptations are regulated by hormonal signaling pathways, including abscisic acid (ABA), auxins, and cytokinins (Smith & De Smet, 2012).

Root system architecture (RSA) adapts to salinity stress by altering root growth patterns, structure, and distribution to manage water uptake and ion toxicity. Salinity stress commonly results in reduced overall root growth, including shorter roots due to osmotic and ionic stress (Munns & Tester, 2008). Lateral root growth is often reduced as a protective measure to limit exposure to high salinity (Geng et al., 2013). Similar to drought conditions, salinity can stimulate the development of an increased number of longer root hairs, which enhance water and nutrient absorption in saline soils (Zolla et al., 2010). An increased root-shoot ratio under salinity stress helps maintain water uptake and reduce the translocation of toxic ions to shoots (Shabala & Munns, 2012). Additionally, roots may grow more horizontally to avoid highly saline deeper soil layers or to exploit less saline surface layers (Galvan-Ampudia & Testerink, 2011).

Waterlogging, or soil saturation with water, induces anaerobic conditions that restrict oxygen availability to roots. A key response to waterlogging is the development of adventitious roots, which usually emerge from the stem above the waterlogged soil and are more suited to low-oxygen environments than the original roots (Sauter, 2013). These adventitious roots often contain aerenchyma tissues, which facilitate oxygen transport from the shoot to the roots, aiding plant survival in flooded conditions (Visser et al., 2003). To manage oxygen deficiency, plants frequently increase aerenchyma formation in roots. Aerenchyma are specialized tissues with large air spaces that enable oxygen diffusion from the shoot to the root, supporting respiration in hypoxic conditions (Yamauchi et al., 2018).

Heat stress, marked by elevated temperatures, impacts root growth and function by affecting water and nutrient uptake and disrupting root cell metabolism. High temperatures can hinder root elongation and decrease overall root biomass due to the disruption of cellular processes, leading to reduced cell division and expansion in root meristems (Bokhari & Trent, 2022).

Under nitrogen deficiency, plants adjust their root system architecture (RSA) to optimize nitrogen uptake. In low nitrogen conditions, plants often increase lateral root growth and increase root branching, which enlarges the root surface area in the upper soil layers where nitrogen is more accessible (Forde & Lorenzo, 2001). Plants facing nitrogen stress tend to allocate more resources to root growth, leading to a higher root-to-shoot ratio. This strategy allows the plant to explore more soil for nitrogen, even at the expense of above-ground growth (Hermans et al., 2006).

Phosphorus, another essential nutrient often limited in soils, also triggers RSA adaptations. When phosphorus availability is low, plants may adjust their root growth direction to favor horizontal growth in the topsoil, where phosphorus is more accessible. This shallow rooting pattern enhances the plant's ability to capture phosphorus from the soil surface (Richardson et al., 2011). Low phosphorus conditions often promote the initiation and growth of lateral roots, particularly in the topsoil, enabling plants to exploit localized phosphorus-rich layers in the soil (Lynch, 2011).

### 10.2.2 Root Phenotyping: Unveiling the Hidden Half

Root phenotyping refers to the quantitative evaluation of root characteristics aimed at comprehending the relationships between genotype and phenotype and pinpointing traits linked to plant performance. It holds significant importance in plant studies by offering insights into root growth, structure, and functionality. Through the utilization of diverse measurement methodologies and technological tools, root phenotyping facilitates the efficient and precise assessment of root attributes, thus aiding progress in agricultural practices, crop enhancement, and environmental stewardship (Tracy et al., 2020).

Root phenotyping pipelines include root imaging and digitization of root traits, with a specific focus on root system visualization in this analysis. Various methods for root imaging have been documented, yet no standardized approach exists due to the inherent advantages and limitations of each method (McGrail et al., 2020). Before embarking on studies related to Root System Architecture (RSA), it is essential to ascertain whether the crop is cultivated under field conditions or controlled environments. Advances in remote sensing tools, including drones and sensors, have significantly boosted shoot phenotyping in both high-throughput plant phenotyping platforms and field-based studies (Chen et al., 2019b; Liu et al., 2019).

Despite the progress in shoot assessments, root evaluations in high-throughput phenotyping lag behind due to the challenges associated with visualizing underground areas in the field. The method of using trenches to uncover RSA in the field, producing detailed root images, was pioneered by Weaver et al. (1922). However, some trenches exceeded a depth of 2 m, necessitating manual excavation of roots. Despite improvements in trenching methods, acquiring root images remains time-intensive, making trenches less suitable for large-scale studies. To address the time challenge, Teramoto and Uga (2020) applied a convolutional neural network for root segmentation in trench profile images.

Root phenotyping can be classified into 2-Dimension, 3-Dimension and 4-Dimension root phenotyping. When the root parameters are analyzed by restricting the growth in a 2-D growth system or are analyzed with 2-D images, it is called 2-D root phenotyping and when root traits are measured by considering the root growth in all directions in space, it is called 3-D root phenotyping. 2-D root phenotyping can be done using analyzing software such as WinRHIZO (Arsenault et al., 1995) and RhizoVision Explorer (Seethepalli et al., 2021a) and provides reliable measurements such as root length, root diameter, root surface area, and root volume. TopoRoot (Zeng et al., 2021) is a software application designed for 3-D root phenotyping using XCT data by processing 2D image slices. It utilizes computer graphics techniques like topological simplification and 3D skeletonization to identify traits such as root length, number, angle, thickness, tortuosity (growth pattern waviness), and hierarchy. By integrating advanced imaging technologies with root phenotyping, 4-D root phenotyping aims to enhance our comprehension of plant growth and adaptation. The 4Droot software can process time-series 3D CT scans to investigate the spatial and temporal dynamics of roots (Herrero-Huerta et al., 2022). The approach is fine-tuned to precisely, automatically, and reliably measure traits, enabling high-throughput (HTP) root phenotyping with X-ray CT scans. Various imaging techniques for root phenomics are given below in Table 10.2.

**TABLE 10.2**
**Imaging Techniques for Root Phenomics**

| Imaging Technique | Description | References |
|---|---|---|
| **X-ray Computed Tomography (CT)** | This non-invasive technique provides high-resolution, three-dimensional images of root systems in their natural environment. It allows the examination of root architecture without disturbing the root-soil interface. | Mooney et al. (2012) |
| **Magnetic Resonance Imaging (MRI)** | MRI is used to image root structure and function in vivo, providing detailed insights into root size, shape, and water content in the soil. | Pflugfelder et al. (2022) |
| **Visible Light and Fluorescence Imaging** | These methods involve using light in the visible region of the spectrum or utilizing the property of fluorescence by the roots. They are useful for analyzing root growth patterns and dynamics. | Wasson et al. (2016) |

## 10.3 GROWTH SYSTEMS AND PLATFORMS FOR ROOT PHENOTYPING

To phenotype the root system, the plants should be grown in a suitable system. They can be in field conditions, laboratory conditions, or in controlled environment chambers (Wasaya et al., 2018). Field condition provides physiological and practical relevance but is labor-time-intensive and destructive. Laboratory methods include the easy handling of the roots and the phenotyping methods. In this, the conditions and the experiment can be repeated but may not give the response that we get from the actual field conditions. As an intermediate between the two, there is the greenhouse method. Various growth systems for establishing Root System for Phenotyping are given below.

1. **Soil-Free Growth Systems:** Soil-free methods for root phenotyping are used for laboratory-based works mainly for seedlings. This includes aeroponics, hydroponics, germination paper, and agar gel-based systems.
   - **Aeroponics-based Methods:** The aeroponic system, first proposed by Carter (1942), consists of a water pump, an air compressor, and an incubator with adjustable parameters such as air composition, nutrient solution, and ejection pressure (Soto, 1982). It is primarily employed for investigating vegetable root structures (Tiwari et al., 2020). A modified aeroponic system has been developed to facilitate the study of root systems in small-seeded legumes and other plants. This system, described by Cai et al. (2023) utilizes a simple design involving a modified trash can and a waterproof industrial humidifier, allowing for precise control over the growth medium and easy observation of root development. The system is particularly advantageous for phenotyping roots and monitoring nodule development in legumes, although it may alter root physiology compared to traditional soil growth.
   - **Hydroponics-based Systems:** The hydroponic system consists of growing the plants in water-based nutrient solutions. Seeds after germination in filter papers can be transferred into trays or transparent containers containing nutrient solutions for further growth and phenotyping (Ma et al., 2017). Semihydroponics (Chen et al., 2011; Kawai et al., 2022) and Rhizoponics (Mathieu et al., 2015) are modified hydroponic-based systems for root phenotyping. The Semihydroponic system consists of growing plant roots in growth units arranged in a large bin and irrigation is given periodically by using a fountain pump (Chen et al., 2011; Kawai et al., 2022). Combining hydroponics with rhizotrons, rhizoponics features a system made of nylon held by an aluminum frame, which is submerged in a tank with water-based nutrient media (Mathieu et al., 2015). Glass bead cylinder method is another modification of the hydroponic system in which acrylic tubes are filled with glass beads in nutrient solution. Germinated seeds are transplanted to this and are covered with an additional layer of glass beads (Slota et al., 2016).
   - **Germination Paper-based Methods:** Germination paper-based methods mainly include the germination paper pouch method in which the germination paper is inserted into polypropylene sheets and clamped to acrylic sheets with rubber bands and clips. The seeds are then placed in the germination sheets in a way that they germinate and roots grow properly (Richard et al., 2015; Adeleke et al., 2020). The pouch-and-wick system, another 2D method, offers a cost-effective way to observe roots in situ using germination paper. Rhizoslides (Marié et al., 2014) and RhizoChamber-Monitor (Wu et al., 2018a,b) are non-invasive, high-throughput root phenotyping platforms based on the pouch-and-wick system. Rhizoslides consist of large plates layered with germination paper, enabling the growth of various root types while maintaining visibility for imaging (Marié et al., 2014). The RhizoChamber Monitor automates the monitoring of root growth, utilizing a custom imaging system that captures sequential images of roots growing on a substrate within specialized rhizoboxes (Wu et al., 2018 a,b).

Early-stage root development studies can also be conducted in petri dishes by placing seeds on filter paper or germination paper moistened with distilled water and incubating (Singh et al., 2010).

- **Agar Gel-based Systems:** Bengough et al. (2004) created an agar chamber-based method to assess seedling root traits, where seedlings grow between transparent gel layers, allowing non-destructive measurement of traits like root length and elongation rate via flatbed scanning. Agar or gellan gum-based chambers can be prepared in transparent containers which provides a transparent solid media for proper growth as well as phenotyping. Provision to avoid light and temperature fluctuations should be provided during growth and lighting should be given for image acquisition (Iyer-Pascuzzi et al., 2010; Uga et al., 2018). Although seedling root traits may not always correspond to those of mature plants, they can be reliable predictors. A significant drawback of soil-less systems is their potential to misrepresent the root characteristics of plants grown in the natural environment (Tuberosa et al., 2002; McPhee, 2005; Cai et al., 2015a,b; Kuijken et al., 2015).

2. **Soil-based Growth Systems:** Soil-based growth systems can be used for root phenotyping for different phenological stages and can be utilized for fields, laboratories as well as green houses. The various soil-based growth systems are explained below.
   - **Field-based Methods:** Soil excavation can be done by carefully excavating the plant to expose the root system, which is then washed and analyzed. This technique provides a detailed view of root architecture but can be labor-intensive and potentially damaging to the roots (Fang et al., 2012; Maeght et al., 2013). Soil trenching can be done by taking large trenches alongside the root and mapping them (Azevedo et al., 2011; Maeght et al., 2013). Shovelomics is the method of manually excavating plants using a shovel and separating the root system from the shoots. Only the upper portion of the roots can be used but will provide traits like root angle (Cope et al., 2024; Bucksch et al., 2014; Siddiqui et al., 2023; Boudiar et al., 2021; Colombi et al., 2015a). Soil coring is done by pushing or hammering the soil core into the soil where the roots are present. This technique fails to provide a complete view of the root system since it does not allow for the assessment of the root spread (Maeght et al., 2013; Wasson et al., 2014). Plastic pasta strainers were filled with suitable soil and buried in the field in which seeds are sown and are destructively phenotyped by digging out the strainers along with the roots (El Hassouni et al., 2018).
   - **Pot Culture:** Plants are grown in soil-filled pots with pre-weighed suitable soil. This can be mainly utilized for destructive phenotyping. In this, the plants are uprooted and washed and data is collected either manually or by affordable root phenomics using various imaging platforms and analyzing software (Ghimire et al., 2024; Batool et al., 2019).
   - **Modified Pot Culture:** In this technique, pots are modified in their shape or made transparent so as to allow the capture of root images. Clear pots involve using transparent containers filled with soil or other media, where seeds are sown near the pot's wall to facilitate root image acquisition. These pots can be kept inside black pots while roots are developing to avoid interference by light (Khalid et al., 2024; Richard et al., 2015). Root boxes or Rhizoboxes are another modification in which the transparent pots are rectangular in outline to enable viewing and imaging of roots (Joshi et al., 2017; Singh et al., 2010). The Growth and Luminescence Observatory is a technique that integrates custom growth vessels with advanced image analysis algorithms to monitor root system architecture (RSA) development non-destructively over both space and time. It utilizes luminescence imaging of roots that express the luciferase enzyme (Rellán-Álvarez et al., 2015).
   - **PVC Pipe Culture:** Plants are grown in PVC pipes of suitable length and diameter. The base of the PVC pipe can be fitted with a steel mesh and hung in a purpose-built rack such that the top portions are at the same level of height. Alternatively, pipes can

be closed at the bottom end and can be arranged in custom-made stands and loaded in the ground (Behrend et al., 2024; Kim et al., 2022; Singh et al., 2010).

- **Rhizo-lysimetry:** Rhizo-lysimetry is a sophisticated setup featuring an underground corridor along with concrete silos and pipes designed to contain plant roots, equipped with windows or walls for direct observation of the roots. Sensors can be fixed to phenotype the roots efficiently (Eberbach et al., 2013).
- **Rhizotrons:** Rhizotrons are a system of transparent tubes that allows non-destructive observation and measurements of root systems under controlled environmental changes. They can be vertical or horizontal and are modified like minirhizotrons as per the requirement of the research work (Iversen et al., 2011; Chen et al., 2019a).

### 10.3.1 Data Collection Approaches for Root Phenotyping

After the establishment of Root System, the root data has to be collected and the different methods for root data acquisition is given in the Table 10.3.

### 10.3.2 Data Acquisition for Image Traits

Segmentation and analysis of root images play a crucial role in phenotyping roots for traits linked to moisture deficit stress tolerance in wheat. This procedure involves capturing detailed root images and applying various image processing and analysis techniques to isolate the roots from the background and evaluate these images to quantify root traits (Li et al., 2022). These traits include root length, diameter, architecture, and volume, all of which offer valuable insights into a plant's capacity to access water and nutrients under drought conditions. Root imaging and analysis are essential for understanding plant responses to environmental stresses like drought and are vital in both agricultural and ecological research (Liao, 2023). The process begins with capturing root images, which can be achieved using 2D methods such as scanning or photography against a contrasting background, or more sophisticated 3D techniques like X-ray Computed Tomography (CT) or Magnetic Resonance Imaging (MRI) to uncover the intricate three-dimensional structure of root systems within the soil (Ushizima et al., 2023).

After capturing the images, the subsequent step is image segmentation, which is crucial for separating the roots from their background or surrounding soil. This process employs various techniques

**TABLE 10.3**
**Data Acquisition Methods for Root Phenotyping**

| Data Acquisition Method | Description |
|---|---|
| Manual Data Collection | Data can be collected manually by either mapping the roots, as employed in the trench method (Maeght et al., 2013), or by washing the roots after excavation using techniques such as shovelomics, soil coring, or other methods. Various root traits can then be manually measured using tools such as measuring scales, weighing balances, and similar instruments. |
| Affordable Root Phenomics | These are semi-automated root imaging systems where roots are carefully taken and washed manually and images are captured using different cameras or scanners and are processed using software to obtain different traits. |
| Root Phenomics | These systems utilize conveyor belts to move potted plants through imaging stations, where the roots are automatically scanned using different imaging techniques. This approach is well-suited for rapidly screening large numbers of plants for in situ root observation. |
| Drones and Aerial Imaging | For large field studies, drones armed with advanced sensors can be used to infer root traits indirectly by assessing plant health and soil properties (Zhang et al., 2024) |

such as thresholding to enhance image contrast for better differentiation, edge detection methods like Sobel or Canny to outline the roots, and region-based segmentation techniques like watershed segmentation that identify root areas based on variations in color or intensity (Jing et al., 2011).

Following segmentation, the feature extraction stage involves quantifying specific root characteristics. This includes measuring root length, root surface area, root volume, root diameter, and analyzing branching patterns and root tips. These metrics are vital for evaluating the root system's mass, water uptake capacity, and overall growth potential (Wu et al., 2023).

The final phase is data analysis, where the extracted data is examined to determine the connection between root traits and moisture deficit stress tolerance. Statistical tools and machine learning models are employed to predict drought tolerance based on these root characteristics (Pappula et al., 2024).

Several software tools facilitate these processes, including ImageJ/FIJI, a widely used platform for root analysis, WinRHIZO for comprehensive morphological analysis, and RootNav for the digital tracing of complex root systems. The insights derived from these analyses assist breeders in selecting crop varieties with favorable traits for drought resilience, while also enriching ecological studies by modeling plant adaptation to environmental changes (Seethepalli et al., 2021a).

### 10.3.3 Softwares for Root Image Analysis

Several open-source tools for segmenting and analyzing 2D root system architecture (RSA) include semi-automatic software like RootTrace (French et al., 2009), SmartRoot (Lobet et al., 2011a), and RootNav (Pound et al., 2013), as well as fully automatic options like EZ-Root-VIS (Shahzad et al., 2018). For 3D RSA imaging, pipelines such as RootReader3D (Clark et al., 2011) and iRoCS Toolbox are available. A summary of these 2D and 3D root phenotyping software options is provided in the Table 10.4.

**TABLE 10.4**
**Software for Root Image Analysis**

| Software | Open Source | Automation | Operating System | Measured Variables | References |
|---|---|---|---|---|---|
| **archiDART** | Yes | Automated | Windows, Mac, Linux | Data analysis, Data visualization | Delory et al. (2018) |
| **ARIA** | Yes | Automated | Windows, Mac | Length, convex-hull, shape, count | Pace et al. (2014) |
| **ARTT (Analyser for Root Tip Tracks)** | No | Automated | Windows | Velocity-profile, localization, orientation | Russino et al. (2013) |
| **BRAT** | Yes | Automated | Windows, Mac, Linux | Length, diameter, orientation | Slovak et al. (2014) |
| **DART (Data Analysis of Root Tracings)** | Yes | Manual | Windows, Mac, Linux | Length, topology, insertion | Le Bot et al. (2010) |
| **DIRT (Digital imaging of root traits)** | Yes | Automated | Windows, Mac, Linux | Length, shape, diameter | Bucksch et al. (2014) |
| **DynamicRoots** | Yes | Automated | Windows | Length, topology, growth, count, gravitropism | Symonova et al. (2015) |
| **EZ-Rhizo** | Yes | Semi-automated | Windows | Length, insertion-angle, n-branches, insertion, topology | Armengaud et al. (2009) |

(*Continued*)

**TABLE 10.4 (*Continued*)**
**Software for Root Image Analysis**

| Software | Open Source | Automation | Operating System | Measured Variables | References |
|---|---|---|---|---|---|
| **GiA Roots (General Image Analysis of Roots)** | Yes | Automated | Windows, Mac, Linux | Convex-hull, length, perimeter, surface, volume, n-branches, depth | Galkovskyi et al. (2012) |
| **GLO-RIA (GLO-Roots Image Analysis)** | Yes | Automated, Semi-automated | Windows, Mac, Linux | Length, surface, convex-hull, color, angle, shape | Rellán-Álvarez et al. (2015) |
| **GROWSCREEN-Root** | No | Semi-automated | Windows | Length, insertion-angle, n-branches | Nagel et al. (2012) |
| **Growth Explorer** | No | Automated | Windows | Velocity-profile | Basu et al. (2012) |
| **GT-RootS (Global Traits of Root System)** | Yes | Automated | Windows, Mac, Linux | Shape, length, size | Borianne et al. (2018) |
| **ImageJ** | Yes | Automated, Semi-automated, Manual | Windows, Mac, Linux | Length, shape, color, convex hull, surface | Tajima et al. (2013) |
| **KineRoot** | Yes | Automated | Windows, Mac, Linux | Growth, gravitropism | Basu et al. (2007) |
| **NMRooting** | Yes | Automated | Windows, Mac, Linux | Length, angle, topology | van Dusschoten et al. (2016) |
| **MyRoot** | Yes | Semi-automated | Windows | Length | Betegón-Putze et al. (2018) |
| **MorphoSnake** | Yes | Automated | Windows, Mac, Linux | Length, n-branches, topology | Reeb et al. (2018) |
| **PlantVis** | No | Automated | Windows, Mac, Linux | Velocity-profile | Roberts et al. (2010) |
| **PlantRoot** | Yes | Semi-automated | Windows, Mac | Length, shape, topology | Wu et al. (2018 a,b) |
| **REST (Root Estimator for Shovelomics Traits)** | Yes | Automated, Semi-automated | Windows, Mac | Width, depth, shape, diameter | Colombi et al. (2015b) |
| **rhizoTrak** | Yes | Automated | Windows, Mac, Linux | Length, diameter, topology | Möller et al. (2019) |
| **RootAnalyzer** | Yes | Automated | Windows | Surface, count, recognition | Chopin et al. (2015) |
| **RhizoVision Explorer** | Yes | Automated | Windows | Length, surface, shape, count, convex-hull | Seethepalli et al. (2021a) |
| **Rootfly** | Yes | Manual | Windows | Length, diameter, color | Zeng et al. (2008) |
| **RootGraph** | Yes | Automated | Windows | Length, diameter, count, topology | Cai et al. (2015a,b) |
| **RootNAV** | Yes | Semi-automated | Windows, Mac | Length, count, convex-hull, insertion, insertion-angle | Pound et al. (2013) |
| **RootPainter** | Yes | Automated, Semi-automated | Windows, Mac, Linux | Length, surface, diameter, count, segmentation | Smith et al. (2022) |
| **RooTrak** | Yes | Semi-automated | Windows | 3D-reconstruction | Mairhofer et al. (2013) |

*(Continued)*

**TABLE 10.4 (*Continued*)**
**Software for Root Image Analysis**

| Software | Open Source | Automation | Operating System | Measured Variables | References |
|---|---|---|---|---|---|
| **RootReader2D** | Yes | Semi-automated | Windows, Mac, Linux | n-branches, length, depth, topology | Clark et al. (2013) |
| **RootReader3D** | No | Automated | Windows, Mac, Linux | Length, width, depth, volume, surface, convex-hull, n-branches, orientation, insertion-angle | Clark et al. (2011) |
| **RootScape** | Yes | Semi-automated | Windows, Mac, Linux | Shape | Ristova et al. (2013) |
| **Root System Analyser** | Yes | Automated | Windows, Mac | Length, insertion-angle, insertion, diameter, count | Leitner et al. (2014) |
| **RootTrace** | Yes | Automated | Windows | Length, curvature, n-branches | Naeem et al. (2011) |
| **saRIA (Semi-automated Root Image Analysis)** | Yes | Semi-automated | Windows, Mac, Linux | Length | Narisetti et al. (2019) |
| **SegRoot** | Yes | Automated | Windows, Mac, Linux | length | Wang et al. (2019) |
| **SmartRoot** | Yes | Semi-automated | Windows, Mac, Linux | Diameter, insertion-angle, orientation, length, n-branches, insertion, topology | Lobet et al. (2011a) |
| **WinRHIZO** | No | Automated | Windows | Color, diameter, length, topology, volume, surface | Martin et al. (2022) |

### 10.3.4 Description of the Softwares

1. **RooTrak**: X-ray Micro Computed Tomography (μCT) is crucial for visualizing plant root systems in soil, though challenges include variations in X-ray attenuation and overlap between roots and soil. RooTrak processes μCT data as a series of images, tracking root structures through the z-axis and using local models to identify new root material. With minimal user input, it adapts to changes in root density and employs a visual tracking framework to improve accuracy and streamline root phenotyping (Mairhofer et al., 2013).
2. **RootSnap**: CID Bio-Science's RootSnap offers a faster, more reliable solution for root image analysis. The CI-690 system eliminates tedious tracing with its user-friendly multi-touch LCD screen, allowing easy root tracing with a "Snap-to-Root" function that auto-aligns tracing points. Integrated image enhancement optimizes scanned images without needing additional software. RootSnap simplifies monitoring root growth and disease dynamics, with an intuitive interface for efficient image and file management, supporting common formats like XML and data export to Excel.
3. **RhizoVision Explorer**: This open-source software estimates root traits from high-contrast images using a flatbed scanner or camera. Built in C++ with a QT interface and OpenCV

for image processing, it offers accuracy validated through comparisons and simulations. It features modes for disconnected ("Broken Roots") and intact roots ("Whole Root"), with capabilities for batch processing and customizable export options.

4. **RootTrace**: Developed by CPIB, this tool automates high-throughput measurement of root length and curvature. Users provide starting points and background patches; the software then traces roots in time series images, measuring lengths and counting lateral roots. Enhanced to track curved and agravitropic roots, it has also been adapted for additional functionalities.
5. **GROWSCREEN-Root**: This software examines root architecture by processing images of root systems cultivated in agar-filled Petri dishes. It constructs a tree model of the root system, calculating parameters such as primary root length, total lateral root length, and branching angles, which are shown in a false-color image.
6. **rhizoTrak**: A free Fiji plugin for manually annotating time-series minirhizotron images across various operating systems. It allows for importing, processing, and annotating image sets, and includes tools for consistency checks, treeline depiction, and data export/import.
7. **Rootfly**: This open-source software simplifies minirhizotron image analysis, allowing measurement of root length, diameter, and color. It integrates automated processing to reduce manual intervention and stores data in a single RFY file. Users can track root development, monitor growth trends, and export data for further analysis.
8. **RootScape**: This innovative allometric method, using landmarks and the AAMToolbox software, rapidly phenotypes Root System Architecture (RSA) by modeling RSA with 20 key landmarks. RootScape outperforms traditional methods by capturing root architecture variations five to ten times faster and revealing new gene-by-environment interactions in Arabidopsis mutants under hormone treatments.
9. **RootGraph**: A fully automated software tool for detailed root trait analysis, RootGraph uses graph optimization to analyze scanned root images. It efficiently separates and quantifies primary and lateral roots, establishing accurate connections between them. Comparative tests show RootGraph surpasses manual measurement methods in precision and efficiency.
10. **RootReader2D**: RootReader2D is a Java-based application that measures root length from digital images, enabling high-throughput analysis of overall root length and specific root categories. It also offers lateral root counting features and is compatible with various growth mediums. The software is user-friendly and can be adapted to different root systems and analytical requirements.
11. **ARIA**: Designed to overcome challenges in root trait measurement, ARIA is a software framework that transforms seedling root images into graphs to extract multiple traits. It simplifies trait extraction and can extend to 3D tomography data, making it versatile for both seedling and mature root analysis.
12. **BRAT**: Optimized for early root growth stages, BRAT's segmentation plugin efficiently measures primary root traits, including length, width, and growth direction from single or time series images. It operates with minimal user input and provides summary statistics and quality control, compatible with Fiji image processing software on multiple platforms.
13. **WinRHIZO Tron**: This software requires manual tracing to analyze root images, automatically measuring and displaying root traits. It allows adjustments and updates morphological data in real time. WinRHIZO Tron provides detailed diameter distribution, unlike programs that give average measurements.
14. **Root-VIS**: Root-VIS is designed to analyze root system architecture (RSA) features from selected root images, offering means, variances, and visual reconstructions. It integrates with EZ-Rhizo for genotype-environment interaction studies and gene identification in Arabidopsis.

15. **RootflowRT**: Developed for high-precision analysis of Arabidopsis root growth, RootflowRT calculates spatial velocity profiles from well-focused root images, but is unsuitable for highly curved or out-of-focus roots.
16. **SmartRoot**: This semi-automated software quantifies root growth and architecture using a vectorial representation and robust tracing algorithm. It provides detailed information on selected roots and is compatible with multiple operating systems through ImageJ.
17. **PHIV-RootCell**: Designed for analyzing transverse section images of rice roots, PHIV-RootCell measures anatomical traits such as root section areas and cell counts. It operates within ImageJ and is platform-independent.
18. **Root System Analyser**: This software quantifies various 2D root traits, including length, radius, connectivity, and branching angles from root images or sequences.
19. **RootPainter**: A graphical user interface (GUI) tool designed for training deep neural networks to interpret intricate root images. It supports both automatic and semi-automatic segmentation, making it user-friendly for those without programming expertise.
20. **SegRoot**: Utilizing convolutional neural networks, SegRoot segments roots from soil backgrounds without complex feature design. It adapts to diverse root images by learning morphological features directly from images.
21. **DART**: DART identifies and analyzes roots in time-series data, generating datasets of root system architecture (RSA) for visualization and study. Developed in Java, it mitigates errors and biases through manual procedures.
22. **NMRooting**: Extracts root structural parameters from 3D MRI images of plant roots. This Python 2.7-based tool requires specific Python packages, which can be downloaded separately.
23. **RootScan**: Automated program for analyzing root cross-sections. It distinguishes cross-sections from backgrounds, measures areas, and counts objects. Featuring an easy-to-use interface, it offers comprehensive data and operates more efficiently than Photoshop.
24. **WinRHIZO**: Specialized for root measurement, including morphological, topological, and color analyses. It offers automated analysis with color-coded root diameters and is compatible with standard computers.
25. **DART**: Analyzes root system architectures (RSAs) from images using various software tools. Batch analysis with environments like R can enhance data throughput.
26. **ARTT**: Analyzes root tip kinematics using segmentation and imaging filters. It autonomously tracks root tips and is applicable to various plant species with minimal user interaction.
27. **DIRT**: Automated software for quantifying root traits from digital photos, using grid computing for high-throughput analysis. It calculates over 30 traits and is available online.
28. **GROW Map-Root**: Analyzes root growth dynamics using optical flow analysis on image sequences to generate growth maps.
29. **RootNav**: Simplifies the reconstruction and measurement of complex root architectures with semi-automated features, and supports bulk data analysis.
30. **RootReader3D**: Analyzes 3D root system architecture from 2D rotational images, currently nearing completion.
31. **Deep Plant Phenotyping**: Offers pre-trained neural networks for plant phenotyping tasks and enables users to customize and train models for particular requirements.
32. **iRoCS Toolbox**: An interactive tool for aligning Root Coordinate Systems with 3D confocal images of Arabidopsis roots. Features include 3D visualization, annotation, and automatic cell detection.
33. **Phenotiki**: Offers a cost-effective, easy-to-deploy phenotyping solution with a sensor and software suite for automatic plant imaging and analysis.
34. **HTPheno**: An open-source image analysis workflow for high-throughput plant phenotyping, utilizing ImageJ. Analyzes color images from multiple perspectives to compute various phenotypic parameters.

35. **Hypophen**: Semi-automatic software for analyzing seedling growth from time-lapse images, calculating elongation and bending, with some manual adjustments.
36. **GiA Roots**: Automates the comprehensive analysis of root networks, supporting large-scale assessments without requiring advanced mathematical or computer science expertise.
37. **PlantCV**: A Python package for high-throughput phenotyping image analysis, applicable to various imaging platforms and open to community contributions.
38. **RAPA**: Automated phenotyping toolkit for assessing plant growth and fitness traits across large populations, using micro-computers for high-throughput imaging.
39. **RhizoScan**: Automated tool for extracting complete root architecture from large datasets, integrated with OpenAlea for further analysis.
40. **LemnaGrid**: LemnaTec's image analysis component for phenotyping, allowing users to create a graphical process chain for automatic image analysis without programming knowledge.

## 10.4 CURRENT CHALLENGES AND FUTURE PROSPECTS

Recently, there has been a considerable surge in the adoption of Root System Architecture (RSA) models to explore various plant species, root attributes, and physiological functions. These models now include advanced simulations of soil environments, focusing on the interaction between environmental factors and plant phenotypes. Initially used in breeding, agronomy, and ecology, RSA studies have provided valuable insights into how phenotypes interact with their environment (Postma & Lynch, 2008). Unlike root length density models that assume uniform root distribution in soil layers, RSA models simulate individual root placement, revealing nutrient competition among species. For example, simulations suggest that a uniform root distribution leads to higher nitrate uptake compared to distributions influenced by root architecture, indicating that root architecture limits foraging efficiency (Postma & Lynch, 2012). This prompts investigation into the ideal root architecture for nutrient acquisition and how nutrient uptake varies with different root architectural parameters (Rubio et al., 2001). One key trait under scrutiny is the root growth angle, which influences the depth of the root system and consequently nutrient uptake. Simulations have demonstrated that optimal nitrate uptake depends on the rate of nitrate leaching, with varying root growth angles achieving favorable performance across different environments (Dathe et al., 2016). Similarly, studies have shown that shallow root growth angles may be advantageous for phosphorus (P) uptake in certain species (Miguel et al., 2015), highlighting the trade-offs between shallow and deep root placements for nutrient acquisition.

Studies on the number of root axes and root fork density have provided valuable insights into their impact on nutrient uptake efficiency. Simulations have shown trade-offs among phenotypes with various lateral root branching densities, indicating that some configurations are better for phosphorus uptake while others are more suitable for nitrate uptake (Postma & Lynch, 2014). Additionally, root anatomical characteristics such as cortical aerenchyma and root hairs have been explored for their effects on nutrient uptake and overall plant physiology. Simulation results suggest that root systems with high metabolic efficiency, particularly in maize with widespread aerenchyma formation, offer considerable benefits (Fan et al., 2003).

RSA models have thus deepened our understanding of root trait functionality, their interactions with the environment, and their significance for plant nutrition and growth. Although challenges like root growth plasticity remain, these models have the potential to advance plant breeding and crop improvement (Zhao et al., 2017).

## 10.5 CONCLUSION

The exploration of root system architecture (RSA) and root phenotyping represents a crucial endeavor in advancing our understanding of plant biology and agricultural sustainability. Throughout this review, we have delved into the intricate world of root systems, examining their structural organization, functional traits, and significance in plant-environment interactions. From the spatial arrangement of roots within the soil matrix to the quantitative assessment of root traits, our exploration has highlighted the multifaceted nature of root biology and its profound implications for crop productivity, soil health, and ecosystem functioning (Lynch & Brown, 2021)

Through the lens of root structure architecture, we have uncovered the complex interplay between genetic factors, environmental conditions, and soil properties in shaping root development and distribution. We have explored the diverse array of root components, from the primary root to lateral roots and root hairs, and examined their roles in nutrient uptake, water acquisition, and plant anchorage. By understanding the mechanisms underlying root architecture, we gain valuable insights into plant adaptation strategies, crop resilience, and soil-plant-microbe interactions (Munir & Farooq, 2022).

Simultaneously, our examination of root phenotyping has underscored the importance of quantitative assessment in unraveling genotype-phenotype relationships and identifying traits associated with plant performance. Root phenotyping employs a range of methods, from classic root excavation and staining techniques to advanced imaging technologies like X-ray computed tomography (CT) and magnetic resonance imaging (MRI), providing a comprehensive toolkit for analyzing root traits and their functional roles. By quantifying root length, branching patterns, and biomass, root phenotyping enables researchers to decipher the complexities of root development and responses to environmental stimuli, paving the way for targeted breeding strategies and agronomic innovations (Pflugfelder et al., 2017).

Looking ahead, the future of root research holds promise for continued advancements in technology, methodologies, and interdisciplinary collaborations. Emerging imaging techniques, such as high-resolution 3D imaging and automated root tracking systems, will offer unprecedented insights into root architecture and growth dynamics, while the integration of multi-omics approaches will illuminate the molecular mechanisms underlying root development and function (Piñero & Figueroa, 2016). However, these advancements must be accompanied by efforts to address key challenges, such as standardizing protocols for root phenotyping, overcoming technical barriers in field research, and scaling up phenotyping efforts to study root systems at the landscape scale (Garcia & Verhees, 2015).

In conclusion, the exploration of root structure architecture and root phenotyping represents a dynamic and interdisciplinary field of research with far-reaching implications for agriculture, environmental management, and global food security. By harnessing the power of roots as a cornerstone of sustainable agriculture and ecosystem resilience, we can unlock new opportunities for enhancing crop productivity, mitigating environmental stresses, and promoting soil health for future generations.

## REFERENCES

Adama, B., Jones, L. & McKenzie, R. (2017). Mathematical models for integrating root developmental processes and environmental influences. *Plant and Soil*, *412*(1–2), 1–16.

Adeleke, E., McClendon, S. & Singh, B. (2020). Pouch-and-wick system for high-throughput phenotyping of root traits in cowpea. *Plant Science Today*, *7*(2), 120–125.

Alay, F. D., İlhan, N. & Güllüoğlu, M. T. (2024). A Comparative study of data mining methods for solar radiation and temperature forecasting models. *Journal of Universal Computer Science*, *30*(6), 847.

Armengaud, P., Zambaux, A., Hills, A., Sulpice, R., Pattison, R. J., Blatt, M. R. & Amtmann, A. (2009). EZ-Rhizo: An image analysis pipeline for measuring root traits in 2D. *Plant Physiology*, *150*(4), 1784–1795.

Arsenault, J. L., Pouleur, S., Messier, C. & Guay, R. (1995). WinRHIZO, a root-measuring system with a unique overlap correction method. *HortScience*, *30*(4), 906.

Atkinson, J. A., Pound, M. P., Bennett, M. J. & Wells, D. M. (2019). Uncovering the hidden half of plants using new advances in root phenotyping. *Current Opinion in Biotechnology*, *55*, 1–8.

Azevedo, R. A., Lancien, M. & Lea, P. J. (2011). The aspartic acid metabolic pathway, an exciting and essential pathway in plants. *Frontiers in Plant Science*, *2*, 136.

Barczi, J. F., Rey, H., Griffon, S. & Jourdan, C. (2018). DigR: A generic model and its open source simulation software to mimic three-dimensional root-system architecture diversity. *Annals of Botany*, *121*(5), 1089–1104.

Basu, P., Pal, A. & Chakrabarti, S. (2012). Growth Explorer: A new image analysis tool for quantifying seedling growth. *Journal of Plant Physiology*, *169*(2), 151–161.

Basu, P., Pal, A. & Guha, S. (2007). KineRoot: A software tool for analysing root tip growth kinetics. *Journal of Plant Physiology*, *164*(4), 529–539.

Batool, A., Jackson, A. & Wang, Y. (2019). High-throughput phenotyping of roots in soil: An overview of methods and their applications. *Agronomy*, *9*(11), 681.

Behrend, A., Häder, J. & Meyer, C. (2024). PVC pipe culture for root phenotyping in controlled environments. *Plant Methods*, *20*, 303.

Bengough, A. G., McKenzie, B. M., Hallett, P. D. & Valentine, T. A. (2004). Root elongation, water stress, and mechanical impedance in sandy soil. *Journal of Experimental Botany*, *55*(414), 157–162.

Betegón-Putze, I., Pascual, L. & González-Melendi, P. (2018). MyRoot: A semi-automated tool for root length measurement in 2D images. *Plant Methods*, *14*(1), 90.

Blum, A. (2011). *Plant Breeding for Water-Limited Environments*. Springer.

Bokhari, S. A. & Trent, D. W. (2022). Impacts of heat stress on root growth and function in plants. *Plant Stress*, *4*, 100077.

Borianne, P., Lobet, G. & Roose, T. (2018). GT-RootS: A new software to model the growth of root systems. *Annals of Botany*, *121*(2), 299–312.

Boudiar, R., Lobet, G. & Draye, X. (2021). Shovelomics: A tool for high-throughput phenotyping of root traits under field conditions. *Plant and Soil*, *464*(1), 171–185.

Bucksch, A., Burridge, J., York, L. M., Das, A., Nord, E., Weitz, J. S. & Lynch, J. P. (2014). Image-based high-throughput field phenotyping of crop roots. *Plant Physiology*, *166*(2), 470–486.

Cai, J., Gao, D. & Chen, X. (2015a). RootGraph: A software platform for root image processing and analysis. *Plant Methods*, *11*(1), 27.

Cai, W., Zhang, H., Chen, L. & Zhao, Y. (2023). A modified aeroponic system for root phenotyping in legumes. *Journal of Plant Growth Regulation*, *42*(3), 456–465.

Cai, W., Zhang, H. & Zhao, Y. (2015b). A novel agar-gel-based growth system for root phenotyping. *Plant Methods*, *11*(1), 1–9.

Carter, G. A. (1942). Aeroponics: The effect of an air mist environment on root structures. *Plant Physiology*, *17*(2), 209–221.

Chen, D., Neumann, K., Friedel, S. & Chen, M. (2019a). Rhizotron phenotyping: Advances and challenges in underground root phenotyping. *Plant Methods*, *15*, 28.

Chen, T., Bao, A., Zhao, J., Hu, C., Zhou, W., Wang, J. & Xu, C. (2019b). High-throughput phenotyping of cotton plant height using a drone in breeding trials. *Remote Sensing*, *11*(3), 302.

Chen, X., Wang, D., Shi, X., Liu, X. & Lin, X. (2011). A semi-hydroponic system for the study of root phenotypes in maize. *Journal of Plant Growth Regulation*, *65*(1), 45–52.

Chopin, J., Le Gall, H., Esteban, E., Jannin, L., Garnica, M., Cuevas, J. & Ducousso, M. (2015). RootAnalyzer: A tool for analyzing the morphology and architecture of root systems. *Frontiers in Plant Science*, *6*, 183.

Clark, J., Jones, C. & Van Zanten, M. (2013). RootReader2D: A software tool for semi-automated analysis of 2D root images. *Journal of Experimental Botany*, *64*(7), 1951–1961.

Clark, J., Wells, D. & Johnson, J. (2011). RootReader3D: Automated software for the analysis of 3D root system images. *BMC Plant Biology*, *11*(1), 1–14.

Clark, R. T., MacCurdy, R. B., Jung, J. K., Shaff, J. E., McCouch, S. R., Aneshansley, D. J. & Kochian, L. V. (2011). Three-dimensional root phenotyping with a novel imaging and software platform. *Plant Physiology*, *156*(2), 455–465.

Colombi, T., Kirchgessner, N., Le Marié, C. A., York, L. M. & Hund, A. (2015a). Next generation shovelomics: Set up a low-cost phenotyping tool for maize root architecture. *Plant and Soil*, *388*(1–2), 1–20.

Colombi, T., Walter, A. & Keller, T. (2015b). REST: A new tool for shovelomics-based root trait estimation. *Plant Methods*, *11*(1), 14.

Comas, L. H., Becker, S. R., Cruz, V. M. V., Byrne, P. F. & Dierig, D. A. (2013). Root traits contributing to plant productivity under drought. *Frontiers in Plant Science*, *4*, 442.

Cope, J., Blaylock, M. & Gibson, K. (2024). Shovelomics and the use of phenotypic traits in field conditions for enhanced crop yield. *Field Crop Research*, *265*, 108190.

Dathe, A., Hargreaves, M. & Mulligan, D. (2016). Optimal root growth angles for nitrate uptake under various leaching scenarios. *Plant Physiology*, *170*(1), 128–137.

Delory, B. M., Baudson, C., Brostaux, Y., Lobet, G. & du Jardin, P. (2018). archiDART: An R package for the automated computation of plant root length and surface area from images. *Plant Methods*, *14*(1), 1–12.

Eberbach, P., Ward, P. R. & Dunin, F. X. (2013). Lysimeters in the environmental sciences. *Soil Science Society of America Journal*, *77*(5), 1433–1442.

El Hassouni, K., Alahmad, S., Belkadi, B., Filali-Maltouf, A., Hickey, L., Bassi, F. M. & Tadesse, W. (2018). Root system architecture and its association with yield under different water regimes in durum wheat. *Crop Science*, *58*(6), 2331–2346.

Fan, X., Wu, L. & Sun, Z. (2003). The role of root cortical aerenchyma in nutrient uptake efficiency in maize. *Plant Cell Reports*, *22*(3), 218–226.

Fang, S., Clark, R. T., Zheng, Y. & Iyer-Pascuzzi, A. S. (2012). Genotypic recognition and root growth inhibition in response to neighbors in Arabidopsis thaliana. *Plant Physiology*, *158*(1), 280–292.

Forde, B. G. & Lorenzo, H. (2001). The nutritional control of root development. *Plant and Soil*, *232*(1–2), 51–68.

French, A., Ubeda-Tomás, S., Holman, T. J., Bennett, M. J. & Pridmore, T. P. (2009). High-throughput quantification of root growth using a novel image-analysis tool. *Plant Physiology*, *150*(4), 1784–1795.

Galkovskyi, T., Mileyko, Y., Bucksch, A., Moore, B., Symonova, O., Price, C. A. & Weitz, J. S. (2012). GiA Roots: Software for the high throughput analysis of plant root system architecture. *BMC Plant Biology*, *12*(1), 116.

Galvan-Ampudia, C. S. & Testerink, C. (2011). Salt stress signals shape the plant root. *Current Opinion in Plant Biology*, *14*(3), 296–302.

Garcia, S. & Verhees, F. (2015). Challenges in root phenotyping and scaling up efforts for field studies. *Plant Science*, *237*, 146–156.

Geng, Y., Wu, R., Wei, X., Du, Y., Wang, Y. & Cui, R. (2013). Salinity stress inhibits lateral root development in Arabidopsis by modulating auxin homeostasis. *Plant Physiology*, *162*(3), 1356–1376.

Ghimire, S. R., Sharma, P. & Upadhyay, R. (2024). Pot culture-based root phenotyping: Techniques and applications. *Plant Methods*, *20*, 302.

Hermans, C., Hammond, J. P., White, P. J. & Verbruggen, N. (2006). How do plants respond to nutrient shortage by biomass allocation? *Trends in Plant Science*, *11*(12), 610–617.

Herrero-Huerta, M., Gutiérrez, S., Damby, D. E., Parsons, A. J. & Mark, D. (2022). 4Droot: A software for high-throughput analysis of the temporal dynamics of root growth using X-ray computed tomography. *Plant Methods*, *18*(1), 1–12.

Iversen, C. M., Hooker, T. D., Classen, A. T., Norby, R. J. & Zak, D. R. (2011). Substantial allocation to surface roots in an upland oak forest. *Ecology*, *92*(8), 1858–1867.

Iyer-Pascuzzi, A. S., Symonova, O., Mileyko, Y., Hao, Y., Belcher, H., Harer, J. & Benfey, P. N. (2010). Imaging and analysis platform for automatic phenotyping and trait ranking of plant root systems. *Plant Physiology*, *152*(3), 1148–1157.

Javot, H. & Maurel, C. (2002). The role of aquaporins in root water uptake. *Annals of Botany*, *90*(3), 301–313.

Jing, G., Li, S. & Wu, P. (2011). Image segmentation methods for plant root system architecture analysis. *Computers and Electronics in Agriculture*, *75*(2), 219–232.

Joshi, D., Billen, L. P. & Singh, V. (2017). Rootbox: A low-cost, modular rhizobox for in situ imaging of soil-grown roots. *Plant Methods*, *13*, 47.

Kawai, S., Shibata, K. & Tanaka, S. (2022). Development of a semi-hydroponic system for high-throughput phenotyping of rice root architecture. *Journal of Plant Nutrition and Soil Science*, *185*(2), 248–258.

Khalid, S., Chen, Y. & Zhang, Y. (2024). Transparent pot culture for root phenotyping in soil: A guide to protocols and applications. *Plant Methods*, *20*, 304.

Kim, N., Kim, M. & Lee, Y. (2022). Modified PVC pipe system for root imaging in phenotyping studies. *Journal of Plant Growth Regulation*, *41*(1), 100–112.

Kuijken, R. C., van Eeuwijk, F. A., Marcelis, L. F. & Bouwmeester, H. J. (2015). Root phenotyping: From component trait in the lab to breeding. *Journal of Experimental Botany*, *66*(17), 5389–5401.

Le Bot, J., Serra, V., Fabre, J., Draye, X. & Pagès, L. (2010). DART: A software to analyze root system architecture in plants. *Computers and Electronics in Agriculture*, *74*(1), 99–103.

Leitner, D., Schmid, M. & Müller, L. (2014). Root System Analyser: A tool for automated root trait measurements. *Plant Methods*, *10*(1), 40.

Li, A., Wang, Z. & Zhang, H. (2022). Advances in root image analysis for assessing drought tolerance in wheat. *Plant Methods*, *18*(1), 123.

Li, Y., Wang, X. & Zhang, S. (2020). Quantitative trait loci for root system architecture and their implications in crop performance. *Theoretical and Applied Genetics*, *133*(2), 503–518.

Liao, W. (2023). Root imaging techniques in agricultural research: Applications and innovations. *Journal of Agronomy and Crop Science*, *209*(4), 567–580.

Liu, S., Azzarello, E., Bacelar, L. & Caruso, G. (2019). Use of UAVs in the measurement of plant height for agriculture. *International Journal of Remote Sensing*, *40*(22), 8455–8475.

Lobet, G. (2014). Image analysis and mathematical modeling of root system architecture: Applications and perspectives. *Plant Physiology*, *166*(4), 1472–1483.

Lobet, G., Couvreur, V. & Meunier, F. (2011a). SmartRoot: A semi-automated software for the analysis of root system architecture. *Plant Methods*, *7*(1), 5.

Lynch, J. P. (2007). Roots of the second green revolution. *Australian Journal of Botany*, *55*(5), 493–512.

Lynch, J. P. (2011). Root phenes for enhanced soil exploration and phosphorus acquisition: Tools for future crops. *Plant Physiology*, *156*(3), 1041–1049.

Lynch, J. P. (2013). Steep, cheap and deep: An ideotype to optimize water and N acquisition by maize root systems. *Annals of Botany*, *112*(2), 347–357.

Lynch, J. P. & Brown, K. M. (2021). Root system architecture and plant productivity: A review. *Plant and Soil*, *466*(1–2), 1–20.

Ma, Z., Bielenberg, D. G., Brown, K. M. & Lynch, J. P. (2017). Regulation of root elongation under phosphorus stress involves changes in ethylene responsiveness. *Plant Physiology*, *144*(3), 1566–1583.

Maccaferri, M., Sanguineti, M. C. & Tuberosa, R. (2016). Breeding roots: RSA modifications during domestication and its impact on crop yield. *Field Crops Research*, *197*, 130–140.

Maeght, J.-L., Rewald, B. & Pierret, A. (2013). How to study deep roots—And why it matters. *Frontiers in Plant Science*, *4*, 299.

Mairhofer, S., Kutzbach, L. & Zink, M. (2013). RooTrak: A software for semi-automated 3D root reconstruction. *Plant Methods*, *9*(1), 5.

Marié, B., Lobet, G., Pages, L. & Draye, X. (2014). Rhizoslides: An improved tool for root phenotyping under controlled conditions. *Plant Methods*, *10*(1), 1–13.

Martin, T., Weber, J. & Chen, J. (2022). WinRHIZO: Automated software for root measurement and analysis. *Plant Methods*, *18*(1), 3.

Mathan, J., Bhattacharya, J. & Ranjan, A. (2016). Enhancing crop yield by optimizing plant developmental features. *Development*, *143*(18), 3283–3294.

Mathieu, L., Lobet, G., Tocquin, P. & Périlleux, C. (2015). Rhizoponics: A novel hydroponic system to study root system architecture and microdevelopment. *Plant Methods*, *11*(1), 1–13.

McGrail, D. J., Downey, C., Parasher, A., Marks, D. L. & Seetharam, A. S. (2020). Automated 3D root phenotyping using light sheet microscopy and machine learning. *Plant Methods*, *16*(1), 1–10.

McPhee, K. E. (2005). Variation in seedling root architecture of pea germplasm. *Crop Science*, *45*(5), 1758–1763.

Miguel, M., Garcia, M. & Velasco, R. (2015). The impact of root growth angle on phosphorus uptake efficiency. *Field Crops Research*, *172*, 89–98.

Möller, B., Pradal, C. & Godin, C. (2019). rhizoTrak: A versatile software platform for high-throughput analysis of root traits in 2D images. *Computers and Electronics in Agriculture*, *162*, 55–66.

Mooney, S. J., Pridmore, T. P., Helliwell, J. & Bennett, M. J. (2012). Developing X-ray computed tomography to non-invasively image 3-D root systems architecture in soil. *Plant and Soil*, *352*(1–2), 1–22.

Moreira, F. M., dos Santos, J. & Lima, A. (2020). Omics integration in root system architecture breeding: Strategies and challenges. *Plant Breeding Reviews*, *44*, 189–215.

Mostafa, A., Zhang, Y. & Du, Q. (2023). Dynamic phenotyping of root system architecture: Analyzing genotype-environment interactions. *Journal of Integrative Plant Biology*, *65*(4), 829–841.

Munir, N. & Farooq, M. (2022). Root system architecture: Implications for plant adaptation and soil interactions. *Journal of Plant Growth Regulation*, *41*(4), 1235–1250.

Munns, R. & Tester, M. (2008). Mechanisms of salinity tolerance. *Annual Review of Plant Biology*, *59*, 651–681.

Naeem, M., Kim, H. & Fiedler, M. (2011). RootTrace: Automated software for root curvature and branching analysis. *Plant Physiology*, *157*(2), 968–978.

Nagel, K. A., Schurr, U. & Walter, A. (2012). GROWSCREEN-Root: A novel phenotyping tool for analysing root growth and architecture in soil-filled rhizotrons. *Functional Plant Biology*, *39*(10–11), 891–904.

Narisetti, N., Subramanian, S. & Karthik, R. (2019). saRIA: Semi-automated root image analysis software for length measurement. *Journal of Plant Physiology*, *235*, 132–142.

Orman-Ligeza, B. & Beata, W. (2014). Root shape and its importance in RSA: A structural analysis. *Plant and Soil*, *378*(1–2), 1–18.

Pace, S., Carmo-Silva, A. E., Bernard, C. & Evers, J. B. (2014). ARIA: An open-source software for root image analysis. *Computers and Electronics in Agriculture*, *104*, 105–111.

Pappula, V. R., Kumar, S. & Singh, R. (2024). Machine learning approaches for predicting drought tolerance in plants based on root traits. *Agricultural Systems*, *212*, 103557.

Pflugfelder, D., Holz, M. & Vereecken, H. (2022). Advances in magnetic resonance imaging of roots: From structure to function. *Trends in Plant Science*, *27*(2), 148–161.

Pflugfelder, D., Zhang, H. & Min, R. (2017). Advances in root phenotyping technologies: From traditional methods to modern imaging techniques. *Plant Methods*, *13*(1), 8.

Piñero, M., & Figueroa, J. (2016). Emerging technologies in root phenotyping and their applications. *Frontiers in Plant Science*, *7*, 528.

Postma, J. A. & Lynch, J. P. (2008). Root system architecture: Opportunities and constraints for genetic improvement of root traits in crops. *Journal of Experimental Botany*, *59*(3), 831–841.

Postma, J. A. & Lynch, J. P. (2012). Theoretical evidence for the importance of root architecture in nutrient acquisition. *Journal of Experimental Botany*, *63*(2), 595–608.

Postma, J. A. & Lynch, J. P. (2014). Root branching density and its impact on nutrient acquisition and plant fitness. *Journal of Plant Nutrition and Soil Science*, *177*(5), 825–836

Pound, M. P., French, A. P., Atkinson, J. A., Wells, D. M., Bennett, M. J. & Pridmore, T. P. (2013). RootNav: Navigating images of complex root architectures. *Plant Physiology*, *162*(4), 1802–1814.

Reeb, P. D., Zobel, R. W. & Archontoulis, S. V. (2018). MorphoSnake: A versatile tool for the analysis of root system architecture using image-based approaches. *Computers and Electronics in Agriculture*, *153*, 95–106.

Rellán-Álvarez, R., Lobet, G., Lindner, H., Pradier, P. L., Sebastian, J., Yee, M. C., Geng, Y., Trontin, C., LaRue, T., Schrager-Lavelle, A., Haney, C. H., Nieu, R., Maloof, J. N. & Vogel, J. P. (2015). GLO-Roots: An imaging platform enabling multidimensional characterization of soil-grown root systems. *eLife*, *4*, eo7597.

Ren, Y., Li, X. & Yang, M. (2017). Genetic challenges in root system architecture: QTL mapping and implications for breeding. *Plant Biotechnology Journal*, *15*(6), 763–774.

Richard, C. A., Hickey, L. T., Fletcher, S., Jennings, R., Chenu, K. & Christopher, J. (2015). High-throughput phenotyping of seminal root traits in wheat. *Plant Methods*, *11*(1), 1–11.

Richardson, A. E., Lynch, J. P., Ryan, P. R., Delhaize, E., Smith, F. A., Smith, S. E. & Lambers, H. (2011). Plant and microbial strategies to improve the phosphorus efficiency of agriculture. *Plant and Soil*, *349*(1–2), 121–156.

Ristova, D., Kusnierczyk, A. & Clarke, D. (2013). RootScape: A semi-automated tool for root shape analysis. *Computers and Electronics in Agriculture*, *93*, 45–55.

Roberts, D., Pal, A. & Guha, S. (2010). PlantVis: A software for visualizing and analyzing growth in plants. *Journal of Plant Physiology*, *167*(5), 408–416.

Rogers, E. D. & Benfey, P. N. (2015). Regulation of plant root system architecture: Implications for crop advancement. *Current Opinion in Biotechnology*, *32*, 93–98.

Rubio, G., Rengel, Z. & Monroy, J. (2001). The influence of root architecture on nutrient uptake and competition. *Plant and Soil*, *230*(1), 71–80.

Rumyana, S., Smith, G. & White, P. J. (2021). Root system architecture: From concept to application in crop improvement. *Journal of Experimental Botany*, *72*(6), 1931–1946.

Russino, D., Boccaccino, F., Busetto, E., Guerra, L. & Zorzan, S. (2013). ARTT: A tool for analyzing root tip tracks in image sequences. *Plant Methods*, *9*(1), 20.

Sauter, M. (2013). Root responses to flooding. *Current Opinion in Plant Biology*, *16*(3), 282–286.

Seethepalli, A., Dhakal, K., Griffiths, M., Guo, H., Freschet, G. T. & York, L. M. (2021a). RhizoVision Explorer: Open-source software for root image analysis and phenotyping. *Plant Physiology*, *187*(1), 10–24.

Shabala, S. & Munns, R. (2012). Salinity stress: Physiological constraints and adaptive mechanisms. *Plant Stress Physiology* (pp. 59–93), CABI, Wallingford, UK.

Shahzad, Z., Amtmann, A. & Bentley, A. R. (2018). EZ-Root-VIS: An integrated platform for quantifying root system traits in 2D images. *Plant Methods*, *14*(1), 18.

Siddiqui, I., Rasheed, A. & Mubeen, M. (2023). Advances in shovelomics for root phenotyping under field conditions. *Journal of Experimental Botany*, *74*(2), 523–534.

Singh, V., Tiwari, R. S. & Singh, B. (2010). Early root growth analysis of seedlings grown in petri dishes using germination paper. *Indian Journal of Plant Physiology*, *15*(3), 201–205.

Slota, M., Cho, Y. J. & Giese, G. (2016). A glass bead hydroponic system for rapid phenotyping of root system architecture. *Plant Methods*, *12*(1), 1–11.

Slovak, R., Göschl, C., Seren, Ü. & Busch, W. (2014). The BRAT platform: A high-throughput automated phenotyping platform for Arabidopsis roots. *Plant Journal*, *77*(4), 879–890.

Smith, A. G., McCallum, A. & Smith, J. (2022). RootPainter: A software tool for automated and semi-automated root image analysis. *Plant Methods*, *18*(1), 25.

Smith, S. & De Smet, I. (2012). Root system architecture: Insights from Arabidopsis and cereal crops. *Philosophical Transactions of the Royal Society B: Biological Sciences*, *367*(1595), 1441–1452.

Soto, M. A. (1982). Aeroponics: A new tool for root structure research. *Plant and Soil*, *67*(2–3), 289–294.

Symonova, O., Topp, C. N. & Roose, T. (2015). DynamicRoots: A software for root image analysis in time-lapse microscopy. *Plant Methods*, *11*(1), 29.

Tajima, R., Kato, H. & Shiwachi, H. (2013). ImageJ: A free software for root image analysis. *Computers and Electronics in Agriculture*, *97*, 48–53.

Tardieu, F., Cabrera-Bosquet, L., Pridmore, T. & Bennett, M. (2017). Plant phenomics, from sensors to knowledge. *Current Biology*, *27*(15), R770–R783.

Teramoto, S. & Uga, Y. (2020). Efficient extraction of root systems from trench profile images using convolutional neural networks. *Plant Root*, *14*, 35–45.

Tilman, D. (1998). The greening of the green revolution. *Nature*, *396*(6708), 211–212.

Tiwari, K., Pandey, B., Singh, B. & Tiwari, A. (2020). Aeroponics: A novel approach for vegetable root research. *Plant Science Today*, *7*(2), 150–158.

Tracy, S. R., Black, C. R., Roberts, J. A., Sturrock, C. J., Mairhofer, S., Craigon, J. & Mooney, S. J. (2020). Quantifying the genetic and environmental determinants of root system architecture using novel imaging and analysis techniques. *Journal of Experimental Botany*, *71*(8), 2159–2170.

Tuberosa, R., Salvi, S., Sanguineti, M. C., Maccaferri, M., Giuliani, S. & Landi, P. (2002). Mapping QTLs regulating morpho-physiological traits and yield: Case studies, shortcomings and perspectives in drought-stressed maize. *Annals of Botany*, *89*(1), 941–963.

Uga, Y., Sugimoto, K. & Ogawa, S. (2018). The control of root system architecture by DEEPER ROOTING 1 increases rice yield under drought conditions. *Plant and Cell Physiology*, *54*(1), 34–38.

Ushizima, D., Azevedo, A. R. & Xavier, A. (2023). Advanced imaging for 3D root system architecture studies: Current trends and future prospects. *Plant Physiology*, *193*(2), 890–899.

Van Dusschoten, D., Metzner, R., Kochs, J., Postma, J. A., Pflugfelder, D., Bühler, J. & Schurr, U. (2016). NMRooting: A software tool for root length and angle measurement in images. *Computers and Electronics in Agriculture*, *125*, 237–246.

Verhoeven, K. (2006). Optimal root system architecture: Breeding objectives and environmental adaptation. *Plant and Soil*, *282*(1–2), 27–38.

Visser, E. J. W., Bögemann, G. M., van der Sman, A. J. M. & Voesenek, L. A. C. J. (2003). Ethylene accumulation in waterlogged plants: Regulation of adventitious root formation and cortical aerenchyma development. *Plant Physiology*, *132*(1), 185–196.

Wang, Q., Li, Z. & Zhang, Z. (2019). SegRoot: Automated software for root length measurement in images. *Plant Methods*, *15*(1), 8.

Wasaya, A., Zhang, X., Fang, Q. & Yan, Z. (2018). Root phenotyping for drought tolerance: A review. *Agronomy*, *8*(11), 241.

Wasson, A. P., Rebetzke, G. J., Kirkegaard, J. A., Christopher, J. & Richards, R. A. (2016). A portable fluorescence imaging system for rapid screening of chlorophyll content in wheat seedlings. *Journal of Experimental Botany*, *67*(2), 4203–4213.

Wasson, A. P., Richards, R. A., Chatrath, R., Misra, S. C., Prasad, S. V. S., Rebetzke, G. J., Kirkegaard, J. A., Christopher, J. & Watt, M. (2014). Traits and selection strategies to improve root systems and water uptake in water-limited wheat crops. *Journal of Experimental Botany*, *63*(9), 3485–3498.

Weaver, J. E. & Bruner, W. E. (1922). *Root Development of Vegetable Crops*. McGraw-Hill Book Company.

Wu, Q., Pagès, L., Wu, J. & Draye, X. (2018a). RhizoChamber-Monitor: A tool for automatic analysis of root growth dynamics from high-resolution images. *Plant Methods*, *14*(1), 1–11.

Wu, Q., Zhang, L. & Chen, X. (2018b). PlantRoot: An image processing tool for analyzing root system architecture. *Computers and Electronics in Agriculture*, *150*, 400–410.

Wu, Q., Zhang, L. & Chen, X. (2023). Quantitative analysis of root architecture traits under drought stress using imaging techniques. *Plant Science*, *326*, 111526.

Yamauchi, T., Shimamura, S., Nakazono, M. & Mochizuki, T. (2018). Aerenchyma formation in crop species: A review. *Field Crops Research*, *222*, 54–70.

Yoshida, S., Eguchi, G., Iwata, Y. & Takahashi, Y. (2016). The regulation of root angle and its implications for root system architecture under drought conditions. *Plant and Cell Physiology*, *57*(8), 1771–1780.

Zeng, Y., Wu, J., Kennedy, M. & Jacobs, D. (2008). Rootfly: A manual software tool for root system measurement. *Plant Physiology*, *147*(1), 1014–1019.

Zeng, Z., Ashraf, U., Tang, X. & Zheng, X. (2021). TopoRoot: A topology-based root phenotyping tool using X-ray computed tomography for maize root systems. *Plant Methods*, *17*(1), 1–15.

Zhang, X., Huang, Y., Li, Y. & Wang, J. (2024). Advances in drone-based remote sensing for monitoring crop health and soil properties: A review. *Precision Agriculture*, *25*(1), 45–68.

Zhao, H., Wu, H. & Yang, X. (2017). Advances in root system architecture modeling and its applications in crop improvement. *Computers and Electronics in Agriculture*, *137*, 192–204.

Zolla, G., Heimer, Y. M. & Barak, S. (2010). Mild salinity stimulates a stress-induced morphogenic response in Arabidopsis thaliana roots. *Journal of Experimental Botany*, *61*(7), 2111–2119.

# 11 High-Throughput Plant Phenotyping Platforms for Cereal Crop Breeding to Improve Abiotic Stress Tolerance

*Raman Rajasekar, Arumugam Tamilselvan, Seethalaxmi, Sreehari V. Santhosh, Allimuthu Elangovan, Madhurima Das, Sudhir Kumar, Mahesh Kumar, and Viswanathan Chinnusamy*

## 11.1 INTRODUCTION

Abiotic stresses such as drought, flood, heat, cold, and excess salt limit geographical distribution, productivity, and food security. Exacerbated climate change and a rising population in the future further aggravate the need for stress-tolerant crops. Especially cereal crops are highly susceptible to abiotic stress and account for two-thirds of caloric consumption worldwide (Caparas et al., 2021).

Plants adjust the stress condition by either escape, avoidance, or tolerance mechanisms. Stress tolerance is an active, reversible response that is commonly referred to as acclimatization, whereas stress avoidance is a physiologically non-active phase similar to mature seeds. Tolerance is caused by adjustments occurring at the molecular level (epigenetic, transcriptional, translational, and post-translational levels), cell, organ, plant, and canopy levels (Zhang et al., 2022). So, in a crop breeding program, identifying tolerance germplasm as well as genes underlying complex tolerance mechanisms is tedious and time-consuming.

The integration of automated, high-throughput plant phenotyping (HTPP) systems with artificial intelligence has greatly improved the ability to measure complex traits related to crop tolerance under stress. At the same time, advances in genomics have lowered the cost of genotyping and molecular breeding, providing more opportunities to understand the genetic basis of stress tolerance (Xiao et al., 2022). This knowledge, gained from studies like GWAS and network analysis using data from HTP, can help develop crop varieties that are more tolerant to stress.

This chapter explores the world of high-throughput phenotyping platforms, their applications in cereal crop breeding and cultivation, and their potential to address the pressing need for abiotic stress tolerance in a changing climate. We will delve into the technologies that make HTPP systems possible, examine their implementation across various scales, and discuss the data analysis techniques that turn raw phenotypic data into valuable insights for breeders.

## 11.2 HIGH-THROUGHPUT PLANT PHENOTYPING PLATFORM (HTPPP)

Conventional low-throughput destructive phenotyping platforms were replaced by high-throughput, non-destructive methods for abiotic stress tolerance phenotyping. An HTPPP contains devices such as automatic robotics with sensors, imaging sensors, imaging processing, and trait extraction

DOI: 10.1201/9781003505631-11 

techniques for precise phenotyping. In climatic control, an HTPPP was established to phenotype abiotic stress tolerance traits in cereal crops.

For instance, the "LemnaTec 3D Scanalyzer System" contains a climatic control greenhouse facility with a conveyor belt system for automatic watering and weighing and multiple imaging sensors (RGB, thermal, chlorophyll fluorescence, and VNIR) and automatic image data processing and extraction of image-based traits and storage of both images and extracted traits. It is used to phenotype large breeding populations under drought, heat, and salinity stress. Like this, many HTPP platforms, such as "PHENODYN", "HyperART", "PhenoBox", "PHENOVISION", "PlantScreenTM Robotic XYZ System", "PlantScreenTM Modular System", "PhenoImage," and "Rice Automatic Phenotyping Platform," were utilized for precise phenotyping of cereal crops under different abiotic stresses.

Recent advancements in the development of field-based HTPPP enabled the precise phenotyping of abiotic stress responses by either ground- or aerial-based phenotyping. Ground-based HTPPPs contain automatic robots that operate in trails fixed in the field, mounted with imaging sensors such as RGB, thermal, chlorophyll fluorescence, VNIR (hyperspectral), etc., and an automatic image processing system. For example, the "LemnaTec Field Scanalyzer" was utilized for drought stress phenotyping in wheat (Arya et al., 2024). Nevertheless, in a "PhénoField" structure like a rain-out shelter with different imaging sensors to phenotype drought stress response (Beauchêne et al., 2019). In aerial-based phenotyping and satellite imaging, drones mounted with imaging sensors and deep learning and machine learning algorithms extract image data for precise phenotyping of cereals under abiotic stress in field conditions. In Table 11.1 the key features of HTPPPs used for abiotic stress tolerance phenotyping in cereals are given.

## 11.3 HIGH-THROUGHPUT IMAGING SYSTEM TO PHENOTYPE ABIOTIC STRESS RESPONSES

A high-throughput imaging system is the main component in HTPPPs, both in control and field settings, to acquire precise phenotypic under various abiotic stress conditions and non-destructively capture stress-induced acclimation in crop plants at multiple time points. Capturing spectral information from plant systems when illuminated with the electromagnetic spectrum at a fixed distance is the basis for non-destructive phenotyping. The spectral difference is due to the interaction between the EMS and plant system, which differs according to the physiological state of the plant. The spectral signature varies according to the interaction between illuminated light and the plant system at each level of stress, type of tissue or organ, and growth stage. For example, under drought stress conditions, plant water content reduces due to stomatal closure, which reduces the absorption of EMS at 1,450 nm, but higher water content in the control condition increases the absorption of EMS at 1,450 nm. So NIR sensors capture this difference in absorption by the amount of 1,450 nm reflected (Danzi et al., 2022). Imaging sensors are classified according to the light wavelength illuminated, namely, visible light, near-infrared, long-wave infrared, gamma, X-ray, and UV (Al-Tamimi et al., 2022).

In visible spectrum imaging, two types of sensors, such as RGB and LIDAR, capture multiple angles and sides of the plant to get morphological and geometrical traits. For example, in drought stress conditions, traits like biomass-related traits such as total projected area give shoot weight or biomass. Green-related traits Green projected area, light and dark green projected area reflect stay green traits, Morphological traits such as height and width and their ratio reflect leaf-rolling traits. Nevertheless, in this field, "LIDAR" is used to capture traits like chlorophyll fluorescence, photochemical reflectance index, and leaf temperature.

In imaging spectroscopy, sensors obtain a spectral signature between 400 and 2,500 nm in each pixel. Fewer band numbers are called multispectral, usually 10–15 bands and more band numbers are called hyperspectral, usually 100–200 band numbers. These sensors capture alteration in plant internal as well as external properties under abiotic stress conditions by changes in spectral signature due

**TABLE 11.1**
**High-Throughput Plant Phenotypic Platforms (HTPPPs) for Abiotic Stress Tolerance Studies in Cereals**

| S. No. | HTPPP | Stress | Crop | Key Features | References |
|---|---|---|---|---|---|
| | | | | **Controlled Condition (Green House)** | |
| 1. | PHENODYN | Heat | Wheat | This platform contains sensors to measure plant transpiration, leaf expansion rate, climate condition, and an automated drip system for precise watering | Leveau et al. (2021) |
| 2. | LemnaTec 3D Scanalyzer | Salinity<br>Drought | Rice Wheat | Platforms contain a climate control green house facility with a conveyor belt system for automatic watering and weighing and multiple imaging sensors (RGB, thermal, chlorophyll florescence, and VNIR) in series to acquire phenotypic data | Hairmansis et al. (2014)<br>Arya et al. (2024) |
| 3. | HyperART | Drought | Wheat | This plate form is equipped with hyper-spectral sensors. It has the capability to non-destructively examine a whole leaf that is still attached to a plant, providing information on biochemical characteristics such as the chlorophyll content of the leaf | Bergsträsser et al. (2015)<br>Cai et al. (2016) |
| 4. | PhenoBox | Salinity<br>Light intensity | Maize<br>Rice | This plate form includes a chamber that automatically captures and processes plant images from various angles, as well as the extraction of images using 'PhenoPipe' | Czedik-Eysenberg et al. (2018)<br>Tang et al. (2024) |
| 5. | PHENOVISION | Drought | Maize | This plate form contains a conveyor belt system that transports plants from the growth area to the automated weighing and irrigation stations, as well as three imaging camera systems: an RGB camera, a thermal infrared camera, and proximal top-view hyperspectral imaging | Asaari et al. (2019)<br>Mertens et al. (2021) |
| 6. | PlantScreen™ Robotic XYZ System | Drought | Rice | This plate form is a sensor-to-plant type, and it can measure RGB, fluorescence, and 3D images | Kim et al. (2020) |
| 7. | PlantScreen™ Modular system | D+H | Wheat | This plate form has automatically moving on a conveyor belt between the acclimation unit, robotic-assisted imaging units, and the weighing and watering unit | Abdelhakim et al. (2021) |
| 7. | PhenoImage | Drought | Wheat, Sorghum | This platform has an open-source, graphical user interface (GUI)-based cross-platform solution for large-scale data processing which is not only convenient to use but highly precise and effective at the same time | Zhu et al. (2021) |
| 8. | Automated video tracking platform | Heat | Wheat | This plate form can capture precise diurnal change under heat stress | Rascio et al. (2020) |

*(Continued)*

**TABLE 11.1 (*Continued*)**
**High-Throughput Plant Phenotypic Platforms (HTPPPs) for Abiotic Stress Tolerance Studies in Cereals**

| S. No. | HTPPP | Stress | Crop | Key Features | References |
|---|---|---|---|---|---|
| 9. | rice automatic phenotyping platform (RAP) | drought | Rice<br>Maize | This platform included greenhouse, transportation, inspection unit two devices: a color-imaging device and a linear X-ray computed tomography (CT) | Yang et al. (2014) |
| | | | | **Field Condition** | |
| 9. | BreedVision | Drought | Triticale | This platform is equipped with a 3D camera, a light curtain, a laser distance sensor, a hyperspectral camera, and an and an RGB camera to monitor lodging, plant moisture content, biomass yield, or tiller density | Busemeyer et al. (2013) |
| 10. | LemnaTec Field Scanalyzer | drought | Wheat | It can only travel within the rail area and operates on rail lines. It can acquire phenotypic data using a thermal camera, chlorophyll fluorescence, imager, 3D laser scanner, RGB camera, and hyperspectral camera | Virlet et al. (2017)<br>Jansen et al. (2018) |
| 11. | PhénoField | Drought, | Wheat | The infrastructure for field-based phenotyping uses rainout shelters, soil water sensors, and fine soil characterization. LiDAR sensor, spectroradiometer, and RGB camera to monitor green cover fractions, green area index, average leaf angle, Meris terrestrial chlorophyll index, and plant height | Beauchêne et al. (2019) |
| 12. | Unmanned Aerial Vehicle Drones | Drought | Rice, Wheat, Sugarcane | This platform uses drones mounted with various sensors (RGB, multispectral, and hyperspectral) to monitor stress responses in field conditions | Jiang et al. (2021)<br>Qin et al. (2022)<br>Hoffma et al. (2024) |

to changes in EMS reflection. For example, leaf photosynthetic pigments, water status, biochemical composition, nutrient content, morphology of the leaf, and internal structure change under salt stress, can be captured using a spectroscopy imaging system (Jaysinghe et al., 2021). In some sensors, only the spectral signature between 1,000 and 2,500 nm is called the NIR or SWIR. We have already seen that maximum water absorption occurs at 1,450 nm. Which is specifically utilized for plant water status and stomatal conductance traits under drought stress conditions (Holzman et al., 2021).

Thermal infrared imaging uses more than 2,500 nm, which is higher than NIR. This spectral signature also gives traits from NIR such as crop water status, stomatal conductance, canopy temperature under drought, salinity, and heat stress. However, more precise phenotypic data were acquired compared to NIR because of the minimum error in absorption of IR due to the higher wavelength than thermal IR, usually more than 2,500 nm.

In UV or blue light spectrum imaging, chlorophyll fluorescence sensors are used to capture the reflectance of fluorescent. From the imaging traits, we get minimum fluorescence ($f_0$), maximal fluorescence ($F_m$), and derived indices such as variable fluorescence ($F_m - F_0$), $F_v/F_m$, or maximum quantum efficiency of photosystem II (PSII), etc. This reflects photosynthesis and is utilized in many abiotic stress studies to examine phenotypes under abiotic stress conditions in cereals.

X-ray imaging works by producing X-rays, sending them through an object, and then capturing the energy of the X-rays that come out the other side. As the X-rays pass through the object, they are absorbed or scattered based on how each molecule in the object interacts with the X-rays at an atomic level. This imaging gives a 2D shadow of a 3D object. This data, combined with computed tomography (CT) and ML algorithms, was reconstructed in 3D. So, X-ray CT captures root architectural traits found underground, aerenchyma formation in root cells under waterlogging stress, and the morphology of whole plants, leaves, flowers, and seeds precisely under drought or heat stress.

Magnetic resonance imaging uses radio waves to identify alterations in the energy state of a proton at the atomic level, which is then applied to detect plant metabolism in response to abiotic stress. This sensor utilizes nuclear magnetic resonance signals emitted by many atomic nuclei ($^{1}H$, $^{13}C$, $^{14}N$, and $^{15}N$) in order to provide three-dimensional phenotyping pictures. High sensitivity to detect the presence of water, metabolic processes, and the transportation of nutrients (Borisjuk et al., 2012).

Positron Emission Tomography (PET) detects gamma photons from β+-emitting nuclei like $^{11}C$, $^{18}F$, and $^{22}Na$, etc. These photons form lines between detectors, which are used to create a 3D image of how a radiotracer is distributed in a plant system. PET allows us to track the uptake and excretion of radiolabeled molecules in real time and study molecular interactions. It is particularly useful for monitoring nutrient movement, like sodium and potassium ions, under salinity stress conditions such as salinity (Arino-Estrada et al., 2019). The key traits under abiotic stress in cereals using different imaging systems are summarized in Table 11.2.

**TABLE 11.2**
**Overview of Imaging System with Key Traits for Abiotic Stress Resilience/Tolerance in Cereals**

| S. No. | Cereal Crop | Spectral Range | Abiotic Stress | Crop | Key Traits | References |
|---|---|---|---|---|---|---|
| 1. | PET | >10pm | Salinity | Rice | $Na^+$ and $K^+$ transport rate | Arino-Estrada et al. (2019) |
| 2. | X-ray CT | 10 nm–10 pm | Drought | Wheat | Root volume, architectural traits over time | Khalil et al. (2020) |
| | | | D+H | Wheat | Seed weight per ear distinguish poorly and well filled seeds | Schmidt et al. (2020) |
| | | | Submergence | Barley | Aeranchyma formation | Kehoe et al. (2022) |

*(Continued)*

**TABLE 11.2 (*Continued*)**
**Overview of Imaging System with Key Traits for Abiotic Stress Resilience/Tolerance in Cereals**

| S. No. | Cereal Crop | Spectral Range | Abiotic Stress | Crop | Key Traits | References |
|---|---|---|---|---|---|---|
| 3. | Chlorophyll Fluorescence imaging | 10–400 nm | Drought, | Wheat | variable to maximum fluorescence ratio ($F_v/F_m$), projected $F_v/F_m$ ratio area | Arya et al. (2024) |
| | | | Heat, | Wheat | $F_v/F_m$ ratio, half time to reach $F_v/F_m$ ratio ($T_{1/2}$) and Sun induced chlorophyll fluorescence, | Song et al. (2018) Sherstneva et al. (2021) |
| | | | Cold | Rice | NPQ and $F_v/F_m$ ratio | Zhao et al. (2020) |
| | | | Salinity | Rice | Maximal fluorescence $F_m$ and $F_v/F_m$ ratio | Pradhan et al. (2019) |
| 4. | Visible imaging | 400–700 nm | Drought | Rice | Leaf rolling, projected leaf area, projected biomass, geometrical i-traits, Projected green area etc. | Guo et al. (2018) |
| | | | Salinity | Maize | Digital biomass, height | Guo et al. (2023) |
| | | | Heat | Wheat | Leaf wilting, rolling, senescence rate, leaf angle | Rascio et al. (2020) |
| | | | Cold | Sorghum | Plant architecture, projected leaf area and biomass | Chiluwal et al. (2018) |
| 5. | LIDAR | 400–700 nm | Drought | Maize | 3D morphological phenotypes (e.g., plant height, stem diameter, crown diameter, leaf area, leaf inclination angle, leaf length, and leaf width) | Su et al. (2019) |
| 6. | NIR/SWIR | 1,000–2,500 nm | Drought, | Wheat | Plant water content and its distribution within plants | Danzi et al. (2022) |
| 7. | Hyper spectral | 350–2,500 nm | Drought, | Rice, Maize, Wheat | Spectral indices, stomatal conductance, water composition and distribution, biochemical composition such as chlorophyll content, | Guo et al. (2018) Mertens et al. (2021) Wu et al. (2021) |
| | | | D+H | Wheat | Spectral indices | Correia et al. (2022) |
| | | | Cold | Maize | Spectral indices, | Xie et al. (2017) |
| | | | Salinity | Rice | Spectral indices, $Na^+$ and $K^+$ ion distribution, | Pabuayon et al. (2022) |
| 8. | Thermal infrared | >2,500 nm | Drought | Wheat | Crop water stress index, stomatal conductance index and stomatal resistance index | Banerjee and Roychoudhury (2022) |
| | | | Heat | Rice | The Spikelet and flag leaf tissue temperature depression | Karwa et al. (2020) |
| | | | Salinity | Wheat Rice | CWST, SCI | Bayoumi et al. (2014) Siddiqui et al. (2014) |

Various imaging methods capture multidimensional data sets at diverse scales, including metabolic, cellular, organ, plant, and canopy levels. These data sets are collected under varying environmental circumstances. The picture data from the imaging equipment was automatically analyzed and interpreted using machine learning and deep learning techniques.

## 11.4 MACHINE LEARNING AND DEEP LEARNING FOR DATA ANALYSIS AND COMPREHENSION

Machine learning combines fields like probability, decision theories, visualization, and optimization. It's especially effective in handling large data sets, enabling plant researchers to identify patterns by analyzing multiple traits together instead of one at a time. This ability to detect feature hierarchies and draw general trends from data makes ML tools highly successful. Supervised and unsupervised are the main learning techniques used, particularly in studying how HTTP-derived image data behaves under different environmental stresses (Zhang et al., 2019).

In supervised ML models, a label for the data is provided during the training procedure. In an automated fashion, the algorithms can comprehend the relationship between a various trait and plant systems. It does not assume any fundamental assumptions regarding the distribution of data. Various algorithms, including "Random Forest", "Support Vector Machines", "K-Nearest Neighbor", "Simplex Volume Maximization", Partial Least Squares Regression", "Support Vector Classifier" and "Artificial Neural Networks", and, were employed to analyze image data under abiotic stress conditions. Conversely, the unsupervised model does not use labeling over the learning process; rather, it learns similarities and dissimilarities to independently discriminate datasets. As an illustration, in order to identify genes, significant traits were selected for GWAS analysis under drought stress conditions using clustering (hierarchical clustering) and dimensionality reduction (PCA) of i-traits (Guo et al., 2018).

Deep learning (DL) is a machine learning technique that employs advanced computation and extensive datasets to accumulate data in layers. Manual feature design is not necessary with DL, as it autonomously learns features from the data, in contrast to traditional methods. "Multilayer perceptions", "Generative Adversarial Networks", "Convolutional Neural Networks" (CNN), and "Recurrent Neural Networks" are key deep learning models . Especially, Deep CNNs are principally employed in the classification of images, object recognition, and image segmentation, with a particular emphasis on the examination of how plants respond to environmental stress.

## 11.5 ADVANCEMENTS IN HIGH-THROUGHPUT PHENOTYPING IN CEREAL FOR CROP IMPROVEMENT

Recent advancements in HTPP techniques have revolutionized the screening to utilize natural evolutionary variation in crop plants enabling precise, rapid, and reliable identification of breeding donor lines for abiotic stress tolerance. Integrating HTPP with genome-wide association studies (GWAS) and QTL mapping, researchers can better understand the genetic architecture of cereals under various abiotic stress conditions. Nevertheless, the ability to precisely quantify complex quantitative traits at different time intervals under stress enhances the robustness of these genetic insights (Wang et al., 2023). Further, these genetic insights can be utilized in molecular breeding and genetic engineering approaches to improve abiotic stress tolerance in cereals. This section will explore the latest advancements in HTPP techniques and their applications in improving crop resilience and productivity. In the following section, we discuss recent advancements in high-throughput plant phenotypic techniques and their application in crop improvement for different abiotic stresses in cereals. In Figures 11.1 and 11.2 given overview of HTPP for precise phenotyping of diverse germplasm under different abiotic stress in controlled and field conditions respectively are given.

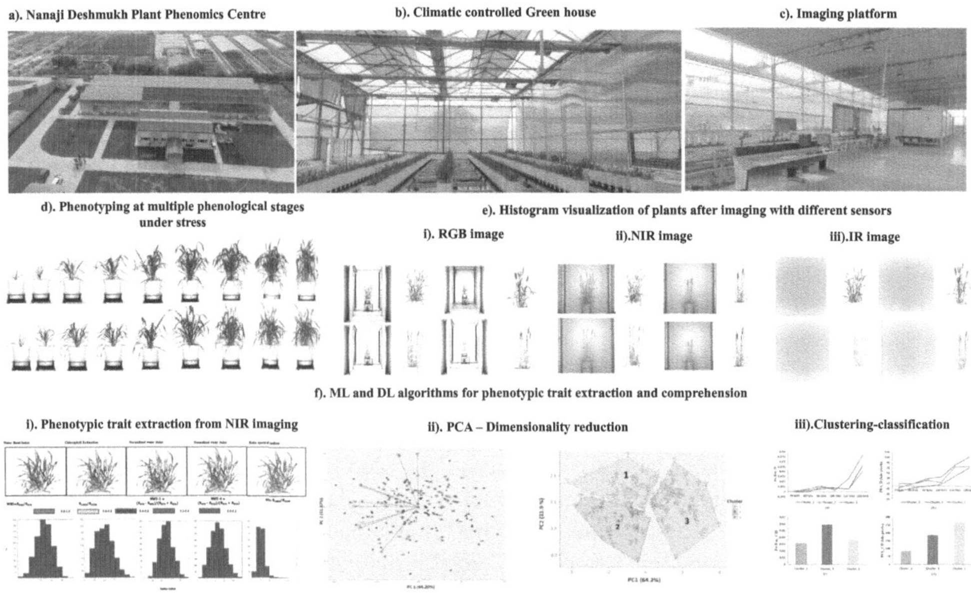


**FIGURE 11.1** Overview of HTPP for precise phenotyping of diverse germplasm under different abiotic stress in controlled conditions. (a) Bird-eye view of an HTPPP—Nanaji Deshmukh Plant Phenomics Centre (NDPPC), ICAR-IARI, New Delhi, (b) climatic controlled greenhouse facility in NDPPC, (c). Imaging chamber with different imaging sensors, (d) Phenotyping of different phenological stages in wheat under control (top) and drought (bottom), (e) Histogram Visualization of wheat after imaging with different sensors (i) RGB, (ii) NIR, (iii) IR imaging, (f) ML and DL algorithms for phenotypic trait extraction and comprehension (i) phenotypic trait extraction from NIR imaging, (ii) principal Components analysis for dimensionality reduction (iii) clustering algorithms for stress tolerance classification. (Sources: Images a to f (i) were taken from Nanaji Deshmukh Plant Phenomics Centre (NDPPC), ICAR-IARI, New Delhi, Data f(ii), and (iii) shown were taken from Arya et al., 2024.)

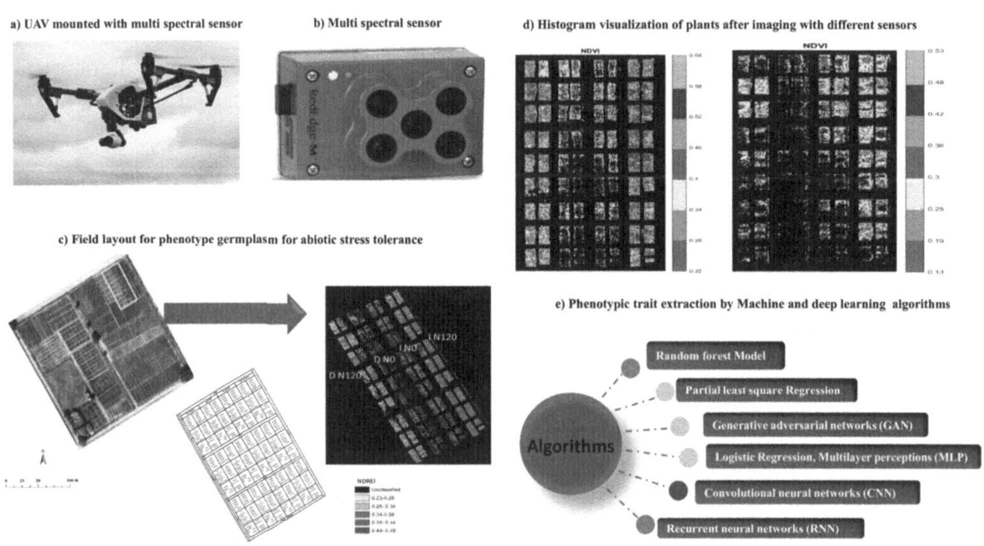


**FIGURE 11.2** Overview of aerial-based HTPP for precise phenotyping of diverse germplasm under different abiotic stress in field conditions. (a). Unmanned aerial vehicle mounted with multispectral sensors for field phenotyping, (b) Multispectral sensor, (c). Field layout for phenotype germplasm for abiotic stress tolerance, (d) Histogram visualization of plants after imaging with different sensors, and (e) Phenotypic trait extraction by machine and deep learning algorithms. (Sources: Images a to e were taken from field plot of Nanaji Deshmukh Plant Phenomics Centre (NDPPC), ICAR-IARI, New Delhi.)

### 11.5.1 Drought

Cereals tolerate drought stress through numerous molecular, physiological, and biochemical morphological changes at different scales. This mechanism evolved over a period of time. Therefore, plant researchers can utilize natural variation to identify various tolerance mechanisms that have evolved over time through a forward genetic approach. They can also identify genes that underlie the physiological and biochemical mechanisms of stress tolerance, which can then be used to develop improved tolerance varieties through genome editing or a transgenic approach.

Recent studies have demonstrated the use of high-throughput plant phenotyping platforms to capture phenotypic data, environmental condition data, and various stress indices or machine learning algorithms to infer or get valuable information from big phenotypic data, such as imaging systems under drought stress conditions, at a particular time interval (internal and intraday) to screen genotypes and identify candidate genes. It gives traits at the morphological, physiological, and biochemical levels, such as leaf morphology, root architectural traits, and spike. Panicle architectural trait, grain filling rate, fertility percentage, relative water content, stomatal conductance, transpiration rate, photosynthetic efficiency, osmolyte accumulation, and nutrient content.

In rice, using RGB image data to acquire drought responses at morphological traits such as biomass, greenness level of leaf, leaf rolling, and yield traits of diverse germplasm set at different time intervals, we identified the drought resistance gene OsPP15 using GWAS analysis (Guo et al., 2018). Although wheat combined RGB and NIR images to capture morphological and physiological traits to identify better germplasm to tolerate drought (Danzy et al., 2021) and also in wheat, Arya et al. (2024) investigated drought-induced chlorophyll fluorescence trait changes using chlorophyll fluorescence imaging sensor in an automated HTPPP ML model to identify tolerant RIL under drought conditions. Furthermore, in maize, combining RGB, HIS, and CT captures the drought responses at the physiological and biochemical levels. Thereby, they identified novel *ZmcPGM2* and *ZmFAB1A* genes, which play a role in osmotic adjustment under drought stress (Wu et al., 2021).

Despite the challenges, combining RGB and hyperspectral imaging for close-range phenotyping on high-throughput plant phenotyping platforms (HTPPPs) allows for precise monitoring of drought stress responses over time and across different plant traits. However, the method faces issues with non-biological variations in reflectance caused by changes in lighting due to the plant's shape, which can obscure subtle biological differences, such as daily patterns and drought reactions. To address this, hyperspectral data can be enhanced by integrating it with detailed 3D plant models. This improvement can be achieved by averaging the plant data, correcting it with linear reflection adjustments, applying the Standard Normal Variate (SNV) transformation, and using variable sorting normalization to reduce noise.

For instance, in maize, an automated high-throughput phenotyping system using hyperspectral imaging combined with 3D plant architecture, reflection adjustments, and averaging was used to capture physiological traits like photosynthesis, transpiration, stomatal conductance, PSII activity, and leaf water content under drought stress with an accuracy of R2=0.92, 0.88, and 0.86 for water potential, $F_v/F_m$, and PSII efficiency, respectively, using the PLSR model (Mertens et al., 2021. Nevertheless, in wheat, using RGB and hyperspectral imaging, the DCNN model and random forest model captured the physiological and biochemical responses under drought. Thereby, they identified two significant SNP regions, AX-109084840 and AX-110502456, and novel *TaPP2C6* genes, which are protein phosphatase 2C homologs of *HAI3* and play a role in 1,000 grain weight under drought stress (Zhang et al., 2024).

Advancements in HTPP, such as robotics, especially unmanned aerial vehicles mounted with all types of imaging and global positioning sensors and ML and DL algorithms for image segmentation and trait extraction under field conditions, enabled the identification of tolerance genotypes for drought stress tolerance as well as genetic architecture.

For instance, in rice, UAV-RGB and models such as DCNN, canopy height model, and empirical linear regression models were used to capture physiological traits like leaf rolling score and plant water content, and their ratios were calculated for the drought resistance index at a particular time interval under drought stress. Furthermore, using GWAS, haplotype, and gene expression analysis, the OsSROlc allele was found to play a critical role in drought tolerance (Jiang et al., 2021). Nevertheless, in sugarcane, using UAV with RGB, NIR, and thermal images, physiological traits like vegetation index, stomatal conductance, water relations, and canopy temperature were captured non-destructively under drought stress in field conditions to screen large genotypes (Hoffman et al., 2024).

### 11.5.2 Water Logging

Under waterlogged conditions, only the root zone is completely filled with water, while in flooded conditions, some portion of the plant's shoot is also submerged. This creates a low-oxygen (hypoxic) environment, limiting the oxygen available for root respiration, which is essential for their functioning. This oxygen shortage affects the plant's ability to absorb water and nutrients, leading to reduced yields in cereal crops like rice (Colmer and Voesenek, 2009).

However, some plants can cope with waterlogging by forming aerenchyma cells—specialized cells that store oxygen and supply it to the roots. This adaptation helps the roots continue to respire even in low-oxygen conditions. In cereals, aerenchyma can form through two processes: schizogenous (where two cells separate) or lysigenous (where cells undergo programmed cell death) (Nishiuchi et al., 2012).

Identifying and studying aerenchyma in the roots is important for finding germplasm that is tolerant to waterlogged conditions (Abiko et al., 2012; Zhang et al., 2015). Recent advances in X-ray computed tomography (X-ray CT) and automated image analysis provide powerful tools to study aerenchyma formation in detail, both in 2D and 3D. By capturing images at different time points, researchers can analyze various structural traits of aerenchyma, such as their volume, surface area, and shape. This detailed information can help identify candidate genes and tolerant cereal varieties that can thrive under waterlogged conditions (Kehoe et al., 2022).

### 11.5.3 Salinity

The sodium component of sodium chloride (NaCl), the predominant salt in saline soils, primarily causes saline stress. We hypothesize that elevated $Na^+$ ion concentrations induce osmotic and ionic stresses, leading to saline toxicity, along with downstream oxidative stresses that disrupt cellular function and signaling.

The ratio of $Na^+$ and $K^+$ significantly influences the magnitude of $Na^+$ toxicity and osmotic stress in plant cells. Traditional analytical methods involve taking destructive samples of tissue and analyzing them chemically, so it is not possible to watch real-time experiments that span genetic or mating populations. This method is also prone to erroneous biological interpretation and can be highly inaccurate.

HTPP analysis, like PET imaging and hyperspectral imaging, is a new, non-destructive way to track the nutrient status of plants over time. It is more accurate and costs less than chemical analysis. Despite PET's increased accuracy, its limitations, such as its higher sensitivity and smaller sample size, limit its use in large-scale germplasm screening for $Na^+$ and $K^+$ dynamics in salt tolerance breeding (Ario-Estrada et al., 2018).

Nevertheless, we used HTPP with RGB and 3D lasers to capture physiological traits such as digital biomass, plant height, and normalized vegetation index at multiple salinity time intervals during the seedling stage. Furthermore, hypothesized that Zm00001d009708, Zm00001d018204, and Zm00001d042608 play a critical role in salt tolerance using GWAS analysis (Jiang et al., 2021).

However, recent studies have focused more on hyperspectral imaging to predict $Na^+$ and $K^+$ content in tissues, as demonstrated in rice by using HSI and a combined PLSR and PC model trained with flame photometer data (Das et al., 2020). Also, it was found that using HIS to predict the levels of $Na^+$ and $K^+$ in different tissues over time and space by training a PLSR model with flame photometry in two different salt tolerance RILs in rice worked well (Pabuayon et al., 2022).

### 11.5.4 Heat Stress

Wheat growth is influenced by heat in a multitude of ways. It increases the rate of Heat Degree Days accumulation, which results in a more rapid completion of phenophases. This leads to a significant reduction in the crop's duration. It diminishes the metabolism of both the source (photosynthesis) and the receiver (starch synthesis). It causes male sterility (Freeha et al., 2008; Kaur and Behl, 2010) and increases the number of structurally abnormal and nonfunctional florets (Hedhly et al., 2009), reduces green (photosynthetic) area, accelerates senescence, and shortens grain filling duration, reduces grain number and grain weight, and increases photorespiration at the whole plant level (Kuroyanagi and Paulsen, 1985; Viswanathan and Khanna-Chopra, 2001; Chinnusamy and Khanna-Chopra, 2003; Cossani and Reynolds, 2012; Jagadish et al., 2016). In the end, these modifications have a negative impact on yield components, including the reduction of grain weight, seed number, and grain filling duration.

Capturing morphological, physiological and biochemical changes under heat stress nondestructively in plants at multiple time intervals is essential, especially in temperature-controlled environments. This is because, in field conditions, it's difficult to impose the same level of heat stress on all genotypes at the same growth stages, even with delayed sowing. As a result, only a few genes related to heat tolerance can be identified in the field. Additionally, the interaction between day length (photoperiod) and temperature, along with differences in the growth stages of the genotypes used in genetic studies, makes it challenging to accurately assess heat tolerance in the field. This lack of precision limits the identification of genes associated with heat tolerance.

Phenotyping of heat stress tolerance in wheat by diurnal change in leaf morphology due to wilting process under high temperature by tracking quantities description of wilting phenomena like angle changes over time, computed various kinematic data such as angular velocity, centripetal acceleration, and total degrees of displacement using a mobile camera in a semiautomated manner (Agata et al., 2020).

However, for instance, in wheat, an automated high-throughput phenotyping system using RGB, fluorescence, and thermal imaging was used to capture growth dynamics, physiological traits like PSII activity, transpiration efficiency under heat and drought, as well as the combined D+H stress condition. Thereby, they identified the robust NS3-tolerant genotype under all stress conditions, especially heat stress, from the diverse set of germplasm studied (Abdelhakim et al., 2021). Furthermore, in wheat, HTPP uses RGB, thermal and multispectral images, and gravimetric data to capture morphological and physiological traits like plant biomass, transpiration rate, and spectral reflectance indices for water content, carbohydrate and anti-oxidant metabolism, as well as water deficit and heat stress. Further, CMH82 is highly tolerant to both heat and drought, as well as fast recovery. Which can further be utilized as a breeding line for drought and heat stress tolerance (Correia et al., 2022).

### 11.5.5 Cold Stress

Cold stress happens when plants are exposed to low temperatures. Chilling stress happens at temperatures between 0°C and 15°C, which are above freezing but below what plants usually need to grow. When temperatures drop below freezing (0°C), plants experience freezing stress, which is more harmful than chilling stress. During freezing, ice crystals can form inside or outside the cells, damaging the cell structure. When these ice crystals grow large, they can kill the cell (Bracale et al., 2003). Cold stress can also cause the production of reactive oxygen species (ROS) like superoxide radicals, hydrogen peroxide, hydroxyl radicals, and singlet oxygen in chloroplasts, mitochondria, and peroxisomes. These ROS can damage proteins and DNA, lead to lipid peroxidation in membranes, inhibit enzymes, or even cause cell death (Apel et al., 2004).

Cold tolerance is a complex trait that involves morphological, physiological, and biochemical changes such as a reduction in biomass, the accumulation of osmolytes, etc. Precise phenotyping of these changes at the cell, organ, plant, and canopy levels at a given time interval is crucial to identifying tolerant and genetic architecture using GWAS/QTL analysis.

For example, in Sorghum, HTPP uses RGB and gravimetric data to capture morphological and physiological traits like plant biomass, height, transpiration rate, RGR, and WUE under cold stress. Furthermore, using GWAS analysis, 2,305 highly significant SNPs were identified; nevertheless, the Sobic.007G033300 gene, which encodes *Plastocyanin biosynthesis*, was found to have a critical role in cold tolerance (Agnew et al., 2024).

Nevertheless, in wheat UAVs use multispectral imaging with spectral bands at 475, 560, 668, 717, and 840 nm to capture physiological responses like biomass, chlorophyll, and water status under cold stress. Furthermore, using GWAS analysis, significant frost resistance regions such as the Fr-A1/VRN1 locus and QTL hotspots on 5A, 5B, and 7B 2 were identified; nevertheless, the new frost-resistant TraesCS5B03G100850 was *functionally validated, which encodes a cold acclimation protein that plays a role in maintaining cell structure under* cold stress (Wu et al., 2024). In Table 11.3 summary of key advancement and application of HTPP for abiotic stress resilience/tolerance in cereals was given.

## 11.6 CURRENT CHALLENGES AND FUTURE PROSPECTS

Precise phenotyping using HTTP under different abiotic stress conditions in cereals and their current advancement clearly elucidated the use of hyperspectral, RGB, computed tomography, and various ML and Dl algorithms for automatic image extraction and comprehension of the results. Only a few studies have validated biochemical and physiological traits at the cellular, organ, and plant levels, including water content, PSII activity, stomatal conductance, transpiration rate, carbohydrate metabolism, and $Na^+$ and $K^+$ content. Simultaneously, these studies aim to forecast non-destructively numerous metabolic and physiological changes resulting from various abiotic stresses. However, these predictions remain imprecise due to errors caused by geometrical, diurnal, and growth and development changes in plants under abiotic stress conditions, highlighting the critical need for robust biochemical and nutritional data validation. Therefore, we need to develop robust ML and DL algorithms that can accurately segment images, analyze them, and extract phenotypic data.

Image acquisition at similar stress levels, similar phenological stages, and the same time of day at multiple time points is important for telling the difference between changes that happen during the day, changes that happen during growth and development, and changes that are caused by stress. This helps scientists figure out how tolerable crop plants are and find important new genes. However, despite the HTPP system maintaining similar phenological and stress levels, capturing diurnal variation remains challenging for large germplasms.

In the field, using HTPP systems like UAVs with ML and DL to record abiotic stress responses is still hard because there aren't any good algorithms to separate images and analyze data from different sensors like RGB, hyperspectral, etc. It's also hard to keep the same level of stress and the same phenological stages when there is a lot of germplasm at different phenological stages for screening or GWAS analysis.

Despite the advancements in high-throughput crop phenotyping technologies and platforms, we still need to overcome several challenges to improve the accuracy of cereal phenotypic traits under abiotic stress conditions. In the future, we might combine HTPP image-based phenotypic traits with other metabolic and transcriptomic data to get a full picture of how plants react to abiotic stress. This information can then be used to find important genes using bioinformatic tools like GWAS/QTL, haplotype analysis, and gene network analysis. Cereal breeding can utilize a comprehensive understanding and genetic architecture of abiotic stress tolerance to enhance stress tolerance crops, either through gene editing with genome editing tools like CRISPR/Cas9 or molecular breeding techniques like genomic selection.

**TABLE 11.3**
**Overview of Key Advancement and Application of HTPPPs for Abiotic Stress Resilience/Tolerance in Cereals**

| S. No. | Cereal Crop | Stress | (GH/field) | HTPPP | Sensor | Advancement | Application | References |
|---|---|---|---|---|---|---|---|---|
| 1. | Maize | Drought | Green house | PHENOVISION | Hyper-spectral and RGB | Developed new protocol to extract Hyperspectral based traits | New spectral Indices to phenotype stress response | Mertens et al. (2021) |
| 2. | Wheat | Drought | Green house | LemnaTec Scanalyzer 3D | RGB and NIR | Developed indices from RGB and NIR images to calculate mean water content | Screen large genotypes for drought tolerance | Danzi et al. (2022) |
| 3. | Maize | Drought | Green house | RAP | RGB, Hyper Spectral and CT | RGB, HIS, CT imaging system derived-i-traits to identify significant MTAs by GWAS | Noval DT gene *ZmcPGM2* and *ZmFAB1A* and its function in DT validated | Wu et al. (2021) |
| 4. | Rice | Drought | Green house | RAP | RGB, | Combined RGB i-traits, leaf senescence scorer traits, yield traits under stress condition to identify significant MTAs by GWAS | Noval DR gene – *OsPP15*, was identified, and its function in DR validated | Guo et al. (2018) |
| 5. | Wheat | Drought | Green house | Henon university HTP | RGB, Hyperspectral | Combined RGB and HIS traits at different time series under drought stress and selected highly heritable i-traits to identify significant MTAs by GWAS | Noval DR gene – *TaPP2C6*, was identified, and its function in DR validated | Zhang et al. (2024) |
| 6. | Wheat | Drought | Green house | LemnaTec Scanalyzer 3D | Florescence imaging | demonstrates the full potential of ChlF image-based phenotyping for enhanced throughput, identifying stress tolerance RILs as well as sensor derived traits, and novel indices | RF ML model and cumulative difference index give higher accuracy in screen drought tolerance RILs | Arya et al. (2024) |
| 7. | Rice | Drought | Field | UAV | RGB | The structure from motion (SFM) algorithm, canopy height model and empirical linear regression model to predict Plant water status and CNN_LRS model to predict leaf scoring and calculated DRI | Locus 42 was associated with DRI for at least two time points, and OsSRO1c in this locus is significantly associated with drought resistance | Jiang et al. (2024) |

*(Continued)*

**TABLE 11.3 (*Continued*)**
**Overview of Key Advancement and Application of HTPPPs for Abiotic Stress Resilience/Tolerance in Cereals**

| S. No. | Cereal Crop | Stress | (GH/field) | HTPPP | Sensor | Advancement | Application | References |
|---|---|---|---|---|---|---|---|---|
| 8. | Sugarcane | Drought | Field | UAV | RGB, Thermal | Identified the tolerance genotypes from contrast germplasm based on NDVI and Canopy temperature difference using aerial phenotyping | This aerial phenotyping protocol can be used to screen large germplasm for drought tolerance in field | Hoffman et al. (2024) |
| 9. | Foxtail | Salinity | Green house | Inveon small animal PET scanner | PET | They demonstrated $^{22}Na$ PET imaging protocol to study sodium transport | This protocol can be used to screen large germplasm for identify candidate gene for tolerance | Ariño-Estrada et al. (2019) |
| 10. | Rice | Salinity | Green house | LemnaTec Scanalyzer 3D | RGB and Hyperspectral | HIS combined with PLSR model to measure leaf $Na^+$ and $K^+$ accumulation non-destructively | Distinguishes tolerant and sensitive genotypes in terms of $Na^+$ and $K^+$ accumulation in leaves | Pabuayon et al. (2022) |
| 11. | Rice | Salinity | Field | UAV | RGB and Hyperspectral | New indices and combined PLSR and principal component ML model to measure $Na^+$ and $K^+$ content and tolerance in field | This new indices and ML model can used to screen large germplasm for salinity stress tolerance in field | Das et al. (2020) |
| 12. | Maize | Salinity | Green house | HTP with PlantEye F500 with | RGB, 3D laser sensor (RNIR) | Using Digital biomass, Plant height and NDVI stress tolerance traits they found 3leaf period (T3) was identified as the key period for the phenotypic variation, distinguishing tolerance and susceptible germplasm | Using this this trait through GWAS analysis genes Zm00001d009708, Zm00001d018204 and Zm00001d042608 closely related to salt tolerance at seedling stages | Guo et al. (2023) |
| 13. | Barley | WL | Field | – | X-ray CT | Noval image analysis pipeline for phenotype aerenchyma in 2D and 3D and also timing of aerenchyma formation | This can be utilized for screen large germplasm for stress tolerance and also GWAS analysis for gene identification | Kehoe et al. (2022) |

(*Continued*)

**TABLE 11.3 (*Continued*)**
**Overview of Key Advancement and Application of HTPPPs for Abiotic Stress Resilience/Tolerance in Cereals**

| S. No. | Cereal Crop | Stress | (GH/field) | HTPPP | Sensor | Advancement | Application | References |
|---|---|---|---|---|---|---|---|---|
| 14. | Wheat | Heat | Plant incubation chamber | Semiautomated HTP | Mobile camera | Novel video/imaging and processing pipeline to phenotype wilting trait by measuring leaf angle under high temperature | Novel technique to tolerance and also GWAS analysis for gene identification | Rascio et al. (2020) |
| 15. | Wheat | D+H | Green house | PlantScreen™ Modular system | RGB, Thermal, NIR | Non-destructive imaging trait at later stage of stress episode effectively distinguish tolerance genotypes from sensitive compared to early stage of stress episodes | This technique can be further utilized to screen large germplasm for stress tolerance and also GWAS analysis for gene identification | Abdelhakim et al. (2021) |
| 16. | Wheat | D+H | Green house | PhenoLab | RGB, NIR, Multi spectral | Integration of cell physiological phenotyping, via the semi-high-throughput determination of enzyme activity signatures and metabolites, with high-throughput phenotyping methods | This methodology can be further utilized to screen large germplasm for stress tolerance | Correia, et al. (2022) |
| 17 | Sorghum | Chilling | GH | HTP | RGB | – | Using this traits 2,305 highly significant SNPs were identified nevertheless, *Sobic.007G033300. gene for cold tolerance* | Agnew et al. (2024) |
| 18 | Wheat | Frost | Field | UAV | Multispectral | Captured traits from multispectral such as biomass, chlorophyll and water content | Identified frost resistant genomic regions and novel TraesCS5B03G100850 genes which can further be utilized for stress tolerance improvement through molecular breeding or genetic engineering | Wu et al. (2024) |

## REFERENCES

Abdelhakim, L. O. A., Rosenqvist, E., Wollenweber, B., Spyroglou, I., Ottosen, C.-O., & Panzarová, K. (2021). Investigating combined drought- and heat stress effects in wheat under controlled conditions by dynamic image-based phenotyping. *Agronomy*, *11*(2), 364. https://doi.org/10.3390/agronomy11020364

Abiko, T., Kotula, L., Shiono, K., Malik, A. I., Colmer, T. D., & Nakazono, M. (2012). Enhanced formation of aerenchyma and induction of a barrier to radial oxygen loss in adventitious roots of *Zea nicaraguensis* contribute to its waterlogging tolerance as compared with maize (*Zea mays ssp.* mays). *Plant, Cell & Environment*, *35*(9), 1618–1630. https://doi.org/10.1111/j.1365-3040.2012.02513.x

Agnew, E., Ziegler, G., Lee, S., Lizárraga, C., Fahlgren, N., Baxter, I., Mockler, T. C., & Shakoor, N. (2024). Longitudinal genome-wide association study reveals early QTL that predict biomass accumulation under cold stress in sorghum. *Frontiers in Plant Science*, *15*. https://doi.org/10.3389/fpls.2024.1278802

Al-Tamimi, N., Langan, P., Bernád, V., Walsh, J., Mangina, E., & Negrão, S. (2022). Capturing crop adaptation to abiotic stress using image-based technologies. *Open Biology*, *12*(6). https://doi.org/10.1098/rsob.210353

Apel, K., & Hirt, H. (2004). REACTIVE OXYGEN SPECIES: Metabolism, oxidative stress, and signal transduction. *Annual Review of Plant Biology*, *55*(1), 373–399. https://doi.org/10.1146/annurev.arplant.55.031903.141701

Ariño-Estrada, G., Mitchell, G. S., Saha, P., Arzani, A., Cherry, S. R., Blumwald, E., & Kyme, A. Z. (2019). Imaging salt uptake dynamics in plants using PET. *Scientific Reports*, *9*(1). https://doi.org/10.1038/s41598-019-54781-z

Arya, S., Sahoo, R. N., Sehgal, V. K., Bandyopadhyay, K., Rejith, R. G., Chinnusamy, V., Kumar, S., Kumar, S., & Manjaiah, K. M. (2024). High-throughput chlorophyll fluorescence image-based phenotyping for water deficit stress tolerance in wheat. *Plant Physiology Reports*, *29*(2), 278–293. https://doi.org/10.1007/s40502-024-00783-7

Asaari, M. S. M., Mertens, S., Dhondt, S., Inzé, D., Wuyts, N., & Scheunders, P. (2019). Analysis of hyperspectral images for detection of drought stress and recovery in maize plants in a high-throughput phenotyping platform. *Computers and Electronics in Agriculture*, *162*, 749–758. https://doi.org/10.1016/j.compag.2019.05.018

Banerjee, A., & Roychoudhury, A. (2021). Role of phenomics in screening abiotic stress tolerance in plants. In *Plant Abiotic Stress Physiology* (pp. 47–59). Apple Academic Press. https://doi.org/10.1201/9781003180579-3

Beauchêne, K., Leroy, F., Fournier, A., Huet, C., Bonnefoy, M., Lorgeou, J., de Solan, B., Piquemal, B., Thomas, S., & Cohan, J.-P. (2019). Management and characterization of abiotic stress via phénofield®, a high-throughput field phenotyping platform. *Frontiers in Plant Science*, *10*. https://doi.org/10.3389/fpls.2019.00904

Bergsträsser, S., Fanourakis, D., Schmittgen, S., Cendrero-Mateo, M., Jansen, M., Scharr, H., & Rascher, U. (2015). HyperART: Non-invasive quantification of leaf traits using hyperspectral absorption-reflectance-transmittance imaging. *Plant Methods*, *11*(1), 1. https://doi.org/10.1186/s13007-015-0043-0

Borisjuk, L., Rolletschek, H., & Neuberger, T. (2012). Surveying the plant's world by magnetic resonance imaging. *The Plant Journal*, *70*(1), 129–146. https://doi.org/10.1111/j.1365-313x.2012.04927.x

Bracale, M., & Coraggio, I. (2003). Chilling and freezing stresses in plants: Cellular responses and molecular strategies for adaptation. In *Abiotic Stresses in Plants* (pp. 23–51). Springer Netherlands. https://doi.org/10.1007/978-94-017-0255-3_2

Busemeyer, L., Mentrup, D., Möller, K., Wunder, E., Alheit, K., Hahn, V., Maurer, H., Reif, J., Würschum, T., Müller, J., Rahe, F., & Ruckelshausen, A. (2013). BreedVision—A multi-sensor platform for non-destructive field-based phenotyping in plant breeding. *Sensors*, *13*(3), 2830–2847. https://doi.org/10.3390/s130302830

Cai, J., Okamoto, M., Atieno, J., Sutton, T., Li, Y., & Miklavcic, S. J. (2016). Quantifying the onset and progression of plant senescence by color image analysis for high throughput applications. *PLoS One*, *11*(6), e0157102. https://doi.org/10.1371/journal.pone.0157102

Caparas, M., Zobel, Z., Castanho, A. D. A., & Schwalm, C. R. (2021). Increasing risks of crop failure and water scarcity in global breadbaskets by 2030. *Environmental Research Letters*, *16*(10), 104013. https://doi.org/10.1088/1748-9326/ac22c1

Chiluwal, A., Bheemanahalli, R., Perumal, R., Asebedo, A. R., Bashir, E., Lamsal, A., Sebela, D., Shetty, N. J., & Krishna Jagadish, S. V. (2018). Integrated aerial and destructive phenotyping differentiate chilling stress tolerance during early seedling growth in sorghum. *Field Crops Research*, *227*, 1–10. https://doi.org/10.1016/j.fcr.2018.07.011

Chinnusamy, V., & Khanna-Chopra R. 2003. Effect of heat stress on grain starch content in diploid, tetraploid and hexaploid wheat species. *Journal of Agronomy and Crop Science* 189: 242–249.

Colmer, T. D., & Voesenek, L. A. C. J. (2009). Flooding tolerance: Suites of plant traits in variable environments. *Functional Plant Biology*, *36*(8), 665. https://doi.org/10.1071/fp09144

Correia, P. M. P., Cairo Westergaard, J., Bernardes da Silva, A., Roitsch, T., Carmo-Silva, E., & Marques da Silva, J. (2022). High-throughput phenotyping of physiological traits for wheat resilience to high temperature and drought stress. *Journal of Experimental Botany*, *73*(15), 5235–5251. https://doi.org/10.1093/jxb/erac160

Cossani, C. M., & Reynolds, M. P. (2012). Physiological traits for improving heat tolerance in wheat. *Plant Physiology*, *160*(4), 1710–1718.

Czedik-Eysenberg, A., Seitner, S., Güldener, U., Koemeda, S., Jez, J., Colombini, M., & Djamei, A. (2018). The 'PhenoBox', a flexible, automated, open-source plant phenotyping solution. *New Phytologist*, *219*(2), 808–823. https://doi.org/10.1111/nph.15129

Danzi, D., De Paola, D., Petrozza, A., Summerer, S., Cellini, F., Pignone, D., & Janni, M. (2022). The use of near-infrared imaging (NIR) as a fast non-destructive screening tool to identify drought-tolerant wheat genotypes. *Agriculture*, *12*(4), 537. https://doi.org/10.3390/agriculture12040537

Das, B., Manohara, K. K., Mahajan, G. R., & Sahoo, R. N. (2020). Spectroscopy based novel spectral indices, PCA- and PLSR-coupled machine learning models for salinity stress phenotyping of rice. *Spectrochimica Acta Part A: Molecular and Biomolecular Spectroscopy*, *229*, 117983. https://doi.org/10.1016/j.saa.2019.117983

Freeha Anjum, F. A., Abdul Wahid, A. W., Farrukh Javed, F. J., & Muhammad Arshad, M. A. (2008). Influence of foliar applied thiourea on flag leaf gas exchange and yield parameters of bread wheat (*Triticum aestivum*) cultivars under salinity and heat stresses. *International Journal of Agriculture and Biology*, *10*(6), 619–627.

Guo, S., Lv, L., Zhao, Y., Wang, J., Lu, X., Zhang, M., Wang, R., Zhang, Y., & Guo, X. (2023). Using high-throughput phenotyping analysis to decipher the phenotypic components and genetic architecture of maize seedling salt tolerance. *Genes*, *14*(9), 1771. https://doi.org/10.3390/genes14091771

Guo, Z., Yang, W., Chang, Y., Ma, X., Tu, H., Xiong, F., Jiang, N., Feng, H., Huang, C., Yang, P., Zhao, H., Chen, G., Liu, H., Luo, L., Hu, H., Liu, Q., & Xiong, L. (2018). Genome-Wide association studies of image traits reveal genetic architecture of drought resistance in rice. *Molecular Plant*, *11*(6), 789–805. https://doi.org/10.1016/j.molp.2018.03.018

Hairmansis, A., Berger, B., Tester, M., & Roy, S. J. (2014). Image-based phenotyping for non-destructive screening of different salinity tolerance traits in rice. *Rice*, *7*(1). https://doi.org/10.1186/s12284-014-0016-3

Hedhly, A., Hormaza, J. I., & Herrero, M. (2009). Global warming and sexual plant reproduction. *Trends in Plant Science*, *14*(1), 30–36. https://doi.org/10.1016/j.tplants.2008.11.001

Hoffman, N., Singels, A., & Joshi, S. (2024). Aerial phenotyping for sugarcane yield and drought tolerance. *Field Crops Research*, *308*, 109275. https://doi.org/10.1016/j.fcr.2024.109275

Holzman, M., Rivas, R., & Bayala, M. (2021). Relationship between TIR and NIR-SWIR as indicator of vegetation water availability. *Remote Sensing*, *13*(17), 3371. https://doi.org/10.3390/rs13173371

Jagadish, S. V. K., Bahuguna, R. N., Djanaguiraman, M., Gamuyao, R., Prasad, P. V. V., & Craufurd, P. Q. (2016). Implications of high temperature and elevated $CO_2$ on flowering time in plants. *Frontiers in Plant Science*, *7*. https://doi.org/10.3389/fpls.2016.00913

Jansen, M., Paulus, S., Nagel, K., & Dornbusch, T. (2018). Image processing for bioassays. In *Bioassays* (pp. 263–287). Elsevier. https://doi.org/10.1016/b978-0-12-811861-0.00013-9

Jayasinghe C, Badenhorst P, Jacobs J, Spangenberg G, Smith K. (2021). Image-based high-throughput phenotyping for the estimation of persistence of perennial ryegrass (*Lolium perenne* L.)—A review. *Grass and Forage Science*, *76*(3), 321–339. https://doi.org/10.1111/gfs.12520

Jiang, Z., Tu, H., Bai, B., Yang, C., Zhao, B., Guo, Z., Liu, Q., Zhao, H., Yang, W., Xiong, L., & Zhang, J. (2021). Combining UAV-RGB high-throughput field phenotyping and genome-wide association study to reveal genetic variation of rice germplasms in dynamic response to drought stress. *New Phytologist*, *232*(1), 440–455. https://doi.org/10.1111/nph.17580

Karwa, S., Bahuguna, R. N., Chaturvedi, A. K., Maurya, S., Arya, S. S., Chinnusamy, V., & Pal, M. (2020). Phenotyping and characterization of heat stress tolerance at reproductive stage in rice (*Oryza sativa* L.). *Acta Physiologiae Plantarum*, *42*(2). https://doi.org/10.1007/s11738-020-3016-5

Kaur, V., & Behl, R. (2010). Grain yield in wheat as affected by short periods of high temperature, drought and their interaction during pre- and post-anthesis stages. *Cereal Research Communications*, *38*(4), 514–520. https://doi.org/10.1556/crc.38.2010.4.8

Kehoe, S., Byrne, T., Spink, J., Barth, S., Ng, C. K. Y., & Tracy, S. (2022). A novel 3D X-ray computed tomography (CT) method for spatio-temporal evaluation of waterlogging-induced aerenchyma formation in barley. *The Plant Phenome Journal*, *5*(1). https://doi.org/10.1002/ppj2.20035

Khalil, A. M., Murchie, E. H., & Mooney, S. J. (2020). Quantifying the influence of water deficit on root and shoot growth in wheat using X-ray Computed Tomography. *AoB Plants*, *12*(5). https://doi.org/10.1093/aobpla/plaa036

Kim, S. L., Kim, N., Lee, H., Lee, E., Cheon, K.-S., Kim, M., Baek, J., Choi, I., Ji, H., Yoon, I. S., Jung, K.-H., Kwon, T.-R., & Kim, K.-H. (2020). High-throughput phenotyping platform for analyzing drought tolerance in rice. *Planta*, *252*(3). https://doi.org/10.1007/s00425-020-03436-9

Leveau, S., Parent, B., Zaka, S., & Martre, P. (2021). Differential sensitivity to temperature and evaporative demand in wheat relatives. *Journal of Experimental Botany*. *72*(21), 7580–7593 https://doi.org/10.1093/jxb/erab431

Mertens, S., Verbraeken, L., Sprenger, H., Demuynck, K., Maleux, K., Cannoot, B., De Block, J., Maere, S., Nelissen, H., Bonaventure, G., Crafts-Brandner, S. J., Vogel, J. T., Bruce, W., Inzé, D., & Wuyts, N. (2021). Proximal hyperspectral imaging detects diurnal and drought-induced changes in maize physiology. *Frontiers in Plant Science*, *12*. https://doi.org/10.3389/fpls.2021.640914

Nishiuchi, S., Yamauchi, T., Takahashi, H., Kotula, L., & Nakazono, M. (2012). Mechanisms for coping with submergence and waterlogging in rice. *Rice*, *5*(1). https://doi.org/10.1186/1939-8433-5-2

Pabuayon, I. C. M., Pabuayon, I. L. B., Singh, R. K., Ritchie, G. L., & de los Reyes, B. G. (2022). Applicability of hyperspectral imaging during salinity stress in rice for tracking $Na^+$ and $K^+$ levels in planta. *PLoS One*, *17*(7), e0270931. https://doi.org/10.1371/journal.pone.0270931

Pradhan, B., Chakraborty, K., Prusty, N., Deepa, Mukherjee, A. K., Chattopadhyay, K., & Sarkar, R. K. (2019). Distinction and characterisation of rice genotypes tolerant to combined stresses of salinity and partial submergence, proved by a high-resolution chlorophyll fluorescence imaging system. *Functional Plant Biology*, *46*(3), 248. https://doi.org/10.1071/fp18157.

Qin, W., Wang, J., Ma, L., Wang, F., Hu, N., Yang, X., Xiao, Y., Zhang, Y., Sun, Z., Wang, Z., & Yu, K. (2022). UAV-Based multi-temporal thermal imaging to evaluate wheat drought resistance in different deficit irrigation regimes. *Remote Sensing*, *14*(21), 5608. https://doi.org/10.3390/rs14215608

Rascio, A., De Santis, G., & Sorrentino, G. (2020). A low-cost method for phenotyping wilting and recovery of wheat leaves under heat stress using semi-automated image analysis. *Plants*, *9*(6), 718. https://doi.org/10.3390/plants9060718

Schmidt, J., Claussen, J., Wörlein, N., Eggert, A., Fleury, D., Garnett, T., & Gerth, S. (2020). Drought and heat stress tolerance screening in wheat using computed tomography. *Plant Methods*, *16*(1). https://doi.org/10.1186/s13007-020-00565-w

Sherstneva, O., Khlopkov, A., Gromova, E., Yudina, L., Vetrova, Y., Pecherina, A., Kuznetsova, D., Krutova, E., Sukhov, V., & Vodeneev, V. (2021). Analysis of chlorophyll fluorescence parameters as predictors of biomass accumulation and tolerance to heat and drought stress of wheat. *Functional Plant Biology*, 49(2), 155–169. https://doi.org/10.1071/fp21209

Siddiqui, Z. S., Cho, J.-I., Park, S.-H., Kwon, T.-R., Lee, G.-S., Jeong, M.-J., Kim, K.-W., Lee, S.-K., & Park, S.-C. (2014). Phenotyping of rice in salt stress environment using high-throughput infrared imaging. *Acta Botanica Croatica*, *73*(1), 312–321. https://doi.org/10.2478/botcro-2013-0027

Song, L., Guanter, L., Guan, K., You, L., Huete, A., Ju, W., & Zhang, Y. (2018). Satellite sun-induced chlorophyll fluorescence detects early response of winter wheat to heat stress in the Indian Indo-Gangetic Plains. *Global Change Biology*, *24*(9), 4023–4037.

Tang, C., Zhang, X., Zhao, Y., Li, L., Wei, J., Yang, Z., Chen, X., Zheng, H., Zhou, Z., Tang, W., & Xia, M. (2024). Response of chlorosis and growth to light intensity in rice seedlings. *ACS Agricultural Science & Technology*, *4*(2), 234–243. https://doi.org/10.1021/acsagscitech.3c00461

Virlet, N., Sabermanesh, K., Sadeghi-Tehran, P., & Hawkesford, M. J. (2017). Field Scanalyzer: An automated robotic field phenotyping platform for detailed crop monitoring. *Functional Plant Biology*, *44*(1), 143. https://doi.org/10.1071/fp16163

Viswanathan, C., & Khanna-Chopra, R. (2001). Effect of heat stress on grain growth, starch synthesis and protein synthesis in grains of wheat (*Triticum aestivum* L.) varieties differing in grain weight stability. *Journal of Agronomy and Crop Science*, *186*(1), 1–7. https://doi.org/10.1046/j.1439-037x.2001.00432.x

Wang, X., Zhang, J., Mao, W., Guan, P., Wang, Y., Chen, Y., Liu, W., Guo, W., Yao, Y., Hu, Z., Xin, M., Ni, Z., Sun, Q., & Peng, H. (2023). Association mapping identifies loci and candidate genes for grain-related traits in spring wheat in response to heat stress. *Plant Science*, *331*, 111676. https://doi.org/10.1016/j.plantsci.2023.111676

Wu, J., Yu, R., Liu, Y., Yuan, M., Jiang, P., Zhao, J., Zhang, C., Xu, X., Wang, Q., Wang, Y., Chen, T., Kang, Z., Zeng, Q., Zhao, Y., Su, B., & Han, D. (2024). *Genome-Wide Association Mapping and Candidate Genes Analysis of High-Throughput Image Descriptors for Wheat Frost Resistance*. Springer Science and Business Media LLC. https://doi.org/10.21203/rs.3.rs-4092348/v1

Wu, X., Feng, H., Wu, D., Yan, S., Zhang, P., Wang, W., Zhang, J., Ye, J., Dai, G., Fan, Y., Li, W., Song, B., Geng, Z., Yang, W., Chen, G., Qin, F., Terzaghi, W., Stitzer, M., Li, L., and Dai, M. (2021). Using high-throughput multiple optical phenotyping to decipher the genetic architecture of maize drought tolerance. *Genome Biology*, *22*(1). https://doi.org/10.1186/s13059-021-02377-0

Xiao, Q., Bai, X., Zhang, C., & He, Y. (2022). Advanced high-throughput plant phenotyping techniques for genome-wide association studies: A review. *Journal of Advanced Research*, *35*, 215–230.

Xie, C., Yang, C., & Moghimi, A. (2017). Detection of cold stressed maize seedlings for high throughput phenotyping using hyperspectral imagery. *Hyperspectral Imaging Sensors: Innovative Applications and Sensor Standards*, *10213*, 1021305.

Yang, W., Guo, Z., Huang, C., Duan, L., Chen, G., Jiang, N., Fang, W., Feng, H., Xie, W., Lian, X., Wang, G., Luo, Q., Zhang, Q., Liu, Q., & Xiong, L. (2014). Combining high-throughput phenotyping and genome-wide association studies to reveal natural genetic variation in rice. *Nature Communications*, *5*(1). https://doi.org/10.1038/ncomms6087

Zhang, H., Zhu, J., Gong, Z., & Zhu, J. K. (2022). Abiotic stress responses in plants. *Nature Reviews Genetics*, *23*(2), 104–119.

Zhang, L., Zhang, H., Niu, Y., & Han, W. (2019). Mapping maize water stress based on UAV Multispectral Remote Sensing. *Remote Sensing*, 11(6), 605. https://doi.org/10.3390/rs11060605

Zhang, X., Shabala, S., Koutoulis, A., Shabala, L., Johnson, P., Hayes, D., Nichols, D. S., & Zhou, M. (2015). Waterlogging tolerance in barley is associated with faster aerenchyma formation in adventitious roots. *Plant and Soil*, *394*(1–2), 355–372. https://doi.org/10.1007/s11104-015-2536-z

Zhang, Z., Qu, Y., Ma, F., Lv, Q., Zhu, X., Guo, G., Li, M., Yang, W., Que, B., Zhang, Y., He, T., Qiu, X., Deng, H., Song, J., Liu, Q., Wang, B., Ke, Y., Bai, S., Li, J. and Song, C. (2024). Integrating high-throughput phenotyping and genome-wide association studies for enhanced drought resistance and yield prediction in wheat. *New Phytologist*, *243*(5), 1758–1775. https://doi.org/10.1111/nph.19942

Zhao, Y., Han, Q., Ding, C., Huang, Y., Liao, J., Chen, T., Feng, S., Zhou, L., Zhang, Z., Chen, Y., Yuan, S., & Yuan, M. (2020). Effect of low temperature on chlorophyll biosynthesis and chloroplast biogenesis of rice seedlings during greening. *International Journal of Molecular Sciences*, *21*(4), 1390. https://doi.org/10.3390/ijms21041390

Zhu, F., Saluja, M., Dharni, J. S., Paul, P., Sattler, S. E., Staswick, P., Walia, H., & Yu, H. (2021). PhenoImage: An open-source graphical user interface for plant image analysis. *The Plant Phenome Journal*, *4*(1). https://doi.org/10.1002/ppj2.20015

# 12 High-Throughput Phenotyping and Phenomics for Climate-Smart Crop Breeding

*Arumugam Tamilselvan, Raman Rajasekar, Seethalaxmi, Sreehari V. Santhosh, Biswabiplab Singh, Ramesh Ramasamy, Sudhir Kumar, Madhurima Das, Mahesh Kumar, and Viswanathan Chinnusamy*

## 12.1 INTRODUCTION

The biggest challenge to agricultural production is the burgeoning population and the ongoing changes in the climate resulting from human action involves the widespread and careless use of natural resources, including the burning of hydrocarbons, cutting down of forests, continuous monoculture without proper residue management, industrial waste disposal, and landfills have greatly increased the levels of greenhouse gases (GHGs) like carbon dioxide ($CO_2$), nitrous oxide ($N_2O$), and methane (CH4) (Masson-Delmotte et al., 2021). Over the past fifty years (1972–2022), data indicate an average $CO_2$ concentration increase (ppm/year) by 155%, reaching its peak level recorded in 2022 at 3 ppm/year, alongside a global temperature rise of 1.0°C over the past hundred years (Masson-Delmotte et al., 2021). Alarmingly, from 2018 to 2023, the world experienced a 0.25°C rise in mean global temperature and a 20% surge in atmospheric $CO_2$ concentration compared to the period from 2013 to 2018 (Polya, 2023). Climate change is projected to markedly worsen water deficits on both regional and global levels. Currently, approximately two-thirds of the global population is experiencing severe water scarcity (Masson-Delmotte et al., 2021). Agriculture, highly vulnerable to the alterations in environmental conditions caused by global climate change, is facing significant threats. Climate change affects agricultural systems by modifying bloom periods, disrupting water resources, deteriorating soil quality, and increasing the prevalence of both biotic and abiotic stress factors. These effects collectively diminish agricultural output (Rosenzweig and Hillel, 1995; Raza et al., 2019), certainly, the quality of the food yield is also affected reducing the consumption of particular food crops. These susceptibilities present a major risk to worldwide food and nutritional stability. Hence, there is an urgent need to develop Climate-smart crops that are resilient varieties bred to endure climate change impacts, use water and nutrients efficiently, resist pests and diseases, reduce environmental footprint, and ensure productivity under extreme weather, thus supporting sustainable agriculture in the changing environment (FAO, 2022).

## 12.2 EFFECT CLIMATE CHANGE ON YIELD OF FIELD CROPS

Asia is facing severe environmental challenges due to climate change, with climate models predicting that if the current rate of temperature rise continues, the global mean temperature may increase by 1.5°C between 2030 and 2050 (Guntukula et al., 2020), leading to significant declines in crop yields, particularly for rice and wheat, as evidenced by a 9%–10% decrease in global cereal production from extreme heat and droughts between 1964 and 2007, with drought-related losses rising from 6.7% to 13.7% during this period, while models such as DSSAT predict reductions of 15.2% for rice and


DOI: 10.1201/9781003505631-12

14.1% for wheat, and APSIM estimates declines of 17.2% for rice and 12% for wheat, with research indicating that each degree Celsius increase in global temperature could result in a 6% decrease in wheat production, varying geographically, as mid and high-latitude regions may benefit from moderate warming, whereas equatorial regions will face more severe impacts due to already high temperatures exceeding optimal conditions for wheat cultivation, and these effects are expected to be more pronounced in developing countries with limited access to advanced technologies (Habib-ur-Rahman et al., 2022), despite significant opportunities for improvement in these areas, while current studies reveal that prevailing carbon dioxide levels have a generally negative impact on wheat production, with some mitigation in high-latitude regions but less in tropical and low-latitude areas, and climate-related stressors causing reductions in wheat grain starch content, with heat stress exacerbating decreases in protein, soluble sugar, and starch levels, thereby affecting the quality of wheat-based products (Malhi et al., 2021), and also altering pest dynamics with higher temperatures linked to increased aphid infestations and the prevalence of certain crop diseases (Tao et al., 2017), while in rice production, heat stress disrupts developmental stages and drought stress negatively affects yield components such as number of spikelets in panicle, grain size, and grain weight (Xiao et al., 2020), highlighting the need for adaptive strategies in crop management and breeding programs to develop resilient varieties that can maintain yield and quality, with recent advances in crop model ensembles providing more accurate projections and insights into agricultural trends essential for effective food security policies, and adaptation strategies including modifying management practices, adopting diverse approaches, utilizing climate-resilient varieties, employing advanced technologies, implementing smart farming, and promoting biodiversity to enhance ecosystem resilience, with climate change significantly affecting wheat grain quality by reducing protein, sugar, and starch content, as increased $CO_2$ and rising temperatures, while potentially stimulating growth, lead to decreased protein content and further reductions in soluble sugar and protein levels due to drought conditions.

## 12.3 CONVENTIONAL BREEDING APPROACHES

Plant breeding has been crucial in maintaining global food safety and security, with significant contributions to agricultural productivity over the past 10,000 years through the use of artificial selection to develop crops with desirable traits (Sufyan et al., 2023). However, as the global population is projected to grow by 25% to approximately 10 billion within the next 30 years, the yield improvements achieved through conventional breeding methods are proving insufficient to meet future food demands (Hickey et al., 2019). Although traditional breeding techniques have successfully produced high-yielding and nutritionally rich crops suitable for mechanical harvesting, these methods often result in reduced genetic diversity within crop populations. This lack of diversity increases the susceptibility of crops to environmental stresses and diseases, highlighting a significant challenge for future food security. In response to these limitations, there is a growing interest in leveraging advanced genetic approaches to enhance agro-ecological sustainability. These modern strategies aim to improve crop resilience, reduce reliance on chemical inputs, and mitigate environmental impacts, addressing the shortcomings of conventional breeding methods (Mangal et al., 2023; Saeed et al., 2023). Innovations in scientific research, such as trait discovery and the genetic dissection of complex traits, have provided new opportunities for identifying and deploying beneficial genes in crop varieties (Govindaraj et al. 2018). Despite these advancements, the integration of identified quantitative trait loci (QTLs) and genes into mainstream breeding programs has been limited, primarily due to challenges in the precise phenotyping of targeted traits and the subtle genetic differences among parental lines used for QTL mapping (Govindaraj et al. 2018). Many QTLs have not been effectively translated into breeding populations due to difficulties in accurately phenotyping traits and the marginal genetic variations between parental lines. Additionally, the relationship between QTLs/genes and phenotypic outcomes in real-world environments remains complex, with issues such as false-positive QTLs and context-specific functionality of QTLs, which may only be effective under certain genetic backgrounds or specific abiotic stress conditions. These challenges underscore the need for more robust, location-specific phenotyping approaches to accurately determine the impact of genetic factors on crop performance. Consequently, high-throughput phenotyping methods are essential for enhancing predictive accuracy regarding the relationships between genes and traits associated with various stresses. Such methods allow for the comprehensive evaluation of large numbers of plants under diverse environmental conditions, providing valuable insights into how genetic factors influence trait expression and resilience (Basavaraj and Rane, 2020).

## 12.4 HIGH-THROUGHPUT PHENOMICS

The laborious, time-consuming, potentially damaging, subjective, and expensive nature of manual phenotyping, along with its inefficiencies and inconsistencies in repeatability across evaluations (Araus et al., 2018), has highlighted the need for sophisticated, automated, and accurate technologies that can capture phenotypic data across all stages of growth and correlate it with genomic data (Song et al., 2021; Xu and Li, 2022). Yang et al. (2020) defined crop phenomics as a multidisciplinary approach to acquiring and analysis of high dimensional accurate data on an organism-wide scale through adopting advanced sensors, machine vision, and automation technology. Over a few decades, the rapid evolution of High-Throughput Phenotyping (HTP) technologies has revolutionized plant phenotyping, providing unparalleled opportunities to capture detailed phenotypic data across large plant populations. Recent advancements include the development of sophisticated imaging technologies, sensor-based systems, and automated platforms, which collectively enhance the precision and scale of phenotypic assessments. Automated platforms, including conveyor belts, drones, and field robots, facilitate the collection of phenotypic data under controlled and field conditions, significantly reducing labor and minimizing human error. In controlled high-throughput phenomics platforms (HTPP), hundreds to thousands of plants were grown in the greenhouse environment where a similar level of stress imposition was done to capture the real-time responses of the crop through various sensors that depend on the need of the experimental purposes. Each plant is mounted on a car having a specific RFID tag that is fed with pot details for storing the information that can retrieved for later use in analysis and interpretation. Through a conveyor belt, it is brought to the imaging chamber (Plant-to-sensor type) for recording the responses by capturing images with the high-resolution sensor. The list of sensors that are used in HTPPs is mentioned in the Table 12.1. In addition, it is also equipped with an automated weighing and watering facility by which a predefined volume of water is given to each plant and recorded to estimate the water use over a period of time (Rane et al., 2021). One of the limitations in the controlled environment is that the temporal frequency of observation and phenotypic scale is less but field-based Unmanned Aerial Vehicle (UAV) offer significant potential for assessing phenotypic trait variations among crops, enabling the rapid and cost-effective collection of diverse vegetation data across large areas without damaging the plants (Al-Tamimi et al., 2022). The integration of visible imaging (Vis) with UAVs is gaining considerable attention, with increasing interest from researchers in applying these technologies for high-throughput phenotyping (HTP) across a wide range of plant species. Sensors deployed on UAVs produce vast datasets of canopy spectral reflectance, represented as Vegetation Indices (VIs). These VIs act as UAV-HTP-derived traits, providing a comprehensive evaluation of canopy structure and photosynthetic performance by quantifying the light reflected from the crop canopy. Further LiDAR offers precise prediction of biomass in sorghum (Siebers et al., 2024).

## 12.5 SENSORS UTILIZED FOR SCREENING DIFFERENT STRESS

### 12.5.1 Drought

Selecting appropriate sensors is crucial for dissecting the component traits associated with the complex responses of crops, depending on the experimental focus, whether at the cellular or organ level. For studies involving different biotic and abiotic stresses, it is essential to have prior knowledge

**TABLE 12.1**
**Sensors Utilized in High-Throughput Phenomics**

| Sensors | Principle of Trait Capture | Traits Quantified | Reference |
|---|---|---|---|
| Visible light imaging | Reflectance captured between EMS of 400–700 nm when illuminated with visible light | Shoot/root growth, architecture, Greenness, leaf area, leaf rolling, senescence | Rane et al. (2021); Singh et al. (2023) |
| Fluorescence imaging camera | Chlorophyll molecules absorb that short wavelength is emitted as long wave fluorescence at red/far-red wavelength (680 & 735 nm) | PSII efficiency and NPQ | Pradhan et al. (2022); Sahranavard et al. (2024) |
| Thermal imaging | The image sensor produces an electrical signal by converting the infrared light (8–13 μm) produced by the object into tissue temperature. This temperature indicates the degree of evapotranspiration | Transpiration and stomatal conductance | Pradhan et al. (2022) |
| Near Infrared Imaging | The 900–1,700 nm range of the electromagnetic spectrum possesses very little energy. However, it is well absorbed by water molecules inside the plant system, making it a reliable indicator of tissue water content | Tissue water content, stomatal conductance | Rane et al. (2021) |
| Hyperspectral Reflectance imaging camera | The VIS-NIR (visible-near infrared, 400–1,000 nm) and SWIR (short wavelength infrared, 1,000–2,500 nm) bands were used to capture multidimensional spectral signatures. These signatures were then combined with machine learning (ML) and deep learning (DL) models to predict various biochemical and physiological changes at the cellular level | Chlorophyll content, tissue water content, Na and K content, Photosynthetic rate, secondary metabolites, etc. | Shi et al. (2024) |
| Light detecting & ranging (LIDAR) | A laser beam is projected onto plants and the energy scattered from the plant is captured for the computation | Biomass, morphological and geometrical features such as leaf angle | Siebers et al. (2024) |

of the specific impact of a particular stress to accurately record the responses, which is vital for the advancement of climate-smart crops. Under drought conditions, plant growth is stunted due to reduced cell division and enlargement, leading to smaller leaves, decreased shoot length, and reduced photosynthesis. This reduction is primarily caused by changes in enzyme activity associated with ATP synthesis, ultimately resulting in decreased crop yield and water use efficiency. In selecting drought-tolerant genotypes, one of the targets is to identify plants that exhibit higher biomass accumulation, greater tissue water content, increased fluorescence activity, and elevated osmolyte accumulation, even under drought conditions. Temporal biomass accumulation in various crop species, including rice (Kim et al., 2020), mung bean (Rane et al., 2021), and wheat (Zhang et al., 2024), has been measured using RGB sensors by analyzing pixel data from images captured from top and side views. Plants with superior root growth under drought conditions typically show higher tissue water content, which can be measured using near-infrared (NIR) imaging at the 1,450 nm water absorption band, as suggested by Seeling et al. (2009). This method has been applied in studies on mung bean (Rane et al., 2021) and rice (Kim et al., 2020). The intensity of NIR imaging is an excellent indicator of drought stress, comparable to color changes. Additionally, plants under drought stress often exhibit higher leaf surface temperatures due to stomatal closure, which reduces transpiration cooling. Infrared (IR) thermal imaging has been used to measure canopy temperature in various crops, such as mung bean (Rane et al., 2021), wheat (Ma et al., 2024), and chickpea (Pappula-Reddy et al., 2024), identifying genotypes that can maintain cooler canopies under drought conditions. Chlorophyll fluorescence has proven effective in detecting drought stress, and indicates the maximum quantum efficiency ($F_v/F_m$) in wheat (Arya et al., 2024).

### 12.5.2 High Temperature

A primary sign of heat-induced premature leaf senescence is reduced chlorophyll content, seen as leaf chlorosis, caused by heat-accelerated chlorophyll breakdown and inhibited biosynthesis (Jespersen and Huang, 2014). For characterizing stay green, genotypes with delayed senescence are estimated which can higher rate of retention of chlorophyll. Thus, Stay-Green genotypes offer valuable germplasm for improving crops to better withstand heat and drought stress (Kamal et al., 2019). Evaluating the genotypes by the conventional scoring leaf greenness method is tedious and error-prone. Hence, using vegetative indices (Vis) such as CCCI, RECI, and CVI derived from UAV-based multispectral images under field conditions identified 28 quantitative trait loci (QTL) for vegetative growth, senescence, and grain yield (Hassan et al., 2021). Under high temperatures, the light-harvesting complex PSII gets damaged and shows reduced quantum efficiency ($F_v/F_m$) which can be estimated using chlorophyll fluorescence imaging in sorghum (Pradhan et al., 2022).

### 12.5.3 Salinity

Salt stress is one of the major abiotic stresses that significantly affects the yield. High salt concentration in the shoot causes physiological drought destabilizing the cell membranes and causing protein degradation due to the toxicity of Na+ ions ultimately reducing the plant growth (Atieno et al., 2017) and accelerating the accumulation of ROS which disrupts photosynthetic machinery (PSII) and cellular integrity (Hasanuzzaman et al., 2021). Genotypic salinity tolerance in chickpeas was assessed using an RGB sensor, which tracked variations in growth over time. The results showed a 20% decline in growth rate, as observed through leaf expansion, accompanied by a 15% decrease in plant height and a 28% drop in shoot biomass. In addition to the growth rate, chlorophyll content in soybeans was also estimated using an RGB camera that shows a high correlation ($r$=0.89) with a handheld SPAD meter (Zhou et al., 2018). Such automated sensors can be adopted to screen the genotypic variation among germplasm which reduces the experimental time and that donors can be used for the breeding programs. To record the photosynthetic performance under Chlorophyll fluorescence imaging, salinity tolerance in Arabidopsis thaliana was studied to phenotype early and late salt stress response and found early qP, ΦP, and ΦNPQ responses to salt stress could identify salt-tolerant from salt-sensitive plants (Awlia et al., 2016). To study the internal activity of salt-affected plants using Hyperspectral imaging and record the dynamic response in okra plants. Using spectral information more accurate prediction of Physiological traits such as SPAD, Na concentration, $P_N$ and T (Feng et al., 2020). This could mitigate the bottleneck inherent in traditional approaches for assessing biochemical and physiological traits in plants.

## 12.6 BIG DATA GENERATION AND ANALYSIS

The recent Adoption of several high-throughput sensors in research experiments for monitoring at various timescales led to the generation of huge volumes of data causing issues with analysis and interpretation (Gill et al., 2022) causing hesitation in wide adaptation for monitoring of plant stresses. To overcome such limitations, current advancements in Machine learning (ML) and Deep Learning (DL) approaches paved the way for image feature extraction and huge data analysis. Machine learning integrates probability, statistics, and neural networks to link data with phenotypes (Samuel, 1959), enabling plant scientists to analyze multiple traits simultaneously, unlike traditional single-feature analysis methods. Machine learning identifies patterns in datasets using computerized models to make predictions without explicit instructions (Sandhu et al. 2021). It involves a calibration process where models are trained on large datasets, known as training sets, and validated on testing sets to assess accuracy and precision for future applications (LeCun et al., 2015). It involves two different approaches such as supervised learning where labeled data is used for training the model and unsupervised learning identify patterns independently. Several experiments adopted the Support Vector Machine (Behmann et al., 2014) for drought detection in barely, discriminant analysis, k-means clustering (Awlia et al., 2016) for extracting values in RGB images under salinity, neural networks, clustering, dimensional reduction, and least discriminant analysis (Gill et al., 2022). By utilizing machine learning algorithms to predict the biomass using image-derived parameters such as in wheat (Singh et al., 2023) used 16 ML models freely available in *predMod* and found 90% accuracy in predicting biomass with Bayesian LASSO model showed relatively stable and high in all the traits and experiments. Under water-stressed conditions in mungbean, vegetative stage biomass has been estimated using eight different models and the Partial Least Square model was selected and further utilized to calculate different growth rates (Rane et al., 2021). Leaf area is also estimated in *Dendrobium nobile* through ML techniques including multiple linear regression (MLR), artificial neural networks (ANNs), gradient boosting regression (GBR), support vector regression (SVR), and found that GBR was the best-fit model for leaf area prediction (Das et al., 2022)

Currently rapid adoption of Deep Learning tools for easier data analysis of a large number of images with high accuracy which does not require prior feature selection (Sandhu et al., 2021). The important DL models used for phenomics include but are not limited to, a convolutional neural network (CNN), multilayer perceptron (MLP), recurrent neural network (RNN), and generative adversarial networks (GAN) (Gill et al., 2022). Using the pix2pix deep learning network, the nutrient distribution of 402 accessions from all around the world was visualized using VIS-NIR (400–1,700 nm) hyperspectral imaging (Shi et al., 2024). A UAV imaging system combined with a CNN model was utilized to predict soybean yield across 972 breeding lines, accounting for 78% of the observed yield with a root mean square error of 391.0 kg/ha, demonstrating its potential to improve breeding efficiency (Zhou et al., 2022). With the advancement in computation tools, several sensors are utilized to record the response of the crop accurately and precisely paving the way for breeding climate-smart crops. The integration of High-throughput phenomics, artificial intelligence and genomics for the development of climate-smart crops is given in Figure 12.1.

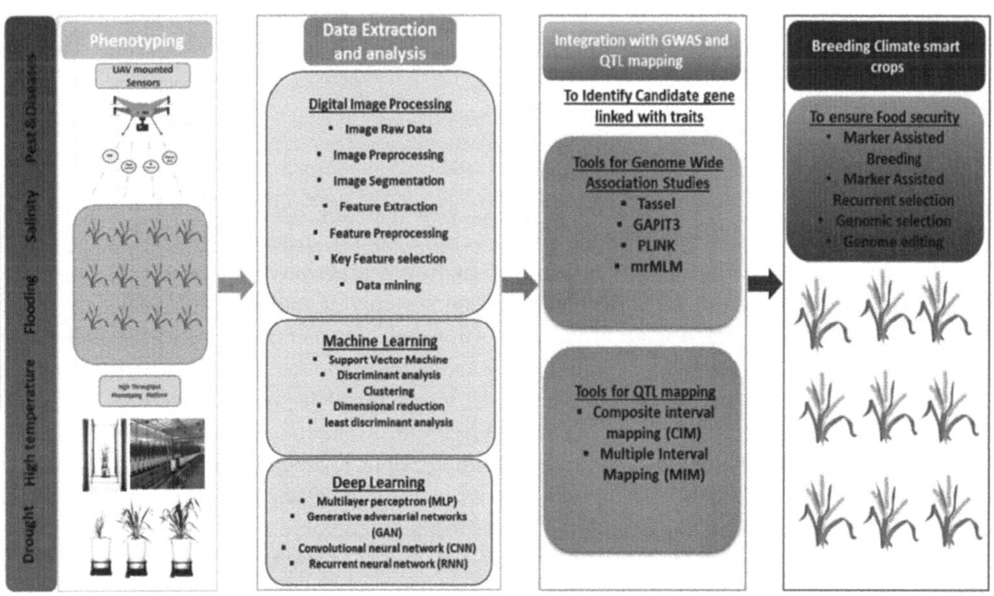


**FIGURE 12.1** The pipeline of integrating high-throughput phenotyping for breeding climate-smart crops.

## 12.7 INTEGRATION OF HTP FOR QTL IDENTIFICATION

Quantitative trait locus (QTL) analysis involves associating phenotypic traits with genotypic data to pinpoint genetic regions linked to specific phenotypes. Recent breakthroughs in next-generation DNA sequencing technologies have substantially advanced our understanding and decoding of crop genomes. Consequently, the research emphasis in plant science has transitioned from genomics to phenomics. This shift underscores the imperative for the development of high-throughput, non-invasive phenotyping technologies capable of systematically capturing phenotypic trait data. Such data can then be integrated with genomic information to facilitate crop improvement initiatives (Furbank and Tester, 2011). High-throughput phenotyping has the potential to effectively bridge this gap. This approach has facilitated the discovery of numerous genes and variants responsible for agriculturally important traits in crops, as highlighted by Zuo and Li (2014). Various studies have utilized high-throughput platforms to accurately phenotype mapping populations that are listed in the Table 12.2. The role of bioinformatics software and algorithms is crucial in deciphering and integrating this data to identify genomic variants and pathways associated with climate resilience phenotypes. These identified targets can be introduced into elite varieties using genome editing (GE) or genomic selection (GS) to enhance crop resilience.

## 12.8 INTEGRATING HIGH-THROUGHPUT PHENOMICS WITH GWAS

Genome-wide association Studies (GWAS) employ statistical techniques to identify links between genomic polymorphisms such as single nucleotide polymorphisms (SNPs), insertions, deletions, and structural variations and phenotypic variation. GWAS plays a pivotal role in elucidating the genetic architectures underlying complex traits (Rebolledo et al., 2016). Linkage mapping in plants, based on bi-parental crosses, limits recombination, resulting in QTLs confined to large chromosomal regions due to insufficient genome shuffling (Holland, 2007). GWAS outperforms traditional genetic mapping by using natural populations, which reduces research time and effort. It offers high efficiency and resolution, enabling genome-wide detection of multiple alleles at a locus. The genetic diversity in these populations allows for the analysis of various traits without repeatedly developing new populations, cutting sequencing costs. Given the complexity of traits like yield and disease resistance, which are controlled by multiple genes and affected by the environment, GWAS is a crucial tool for uncovering plant genetics (Xiao et al., 2022). Some of the GWAS studies conducted using HTP are given in Table 12.3.

## 12.9 CONCLUSION

Integrating High-Throughput Phenotyping (HTP), and artificial intelligence are being employed to capture stress responses nondestructively at cellular, plant, and canopy level under various stress. Nevertheless, still numerous challenges in trait extraction from captured biochemical and physiological stress responses through imaging systems in control and field settings due to the lack of powerful digital image processing. However, advancements in techniques such as ML algorithms & DL, and the use of multiple sensors could reduce the error during image trait extraction and comprehend the physiological and biochemical basis of image traits effectively conferring screening of diverse germplasm and helping in understanding the genetic architecture of complex stress tolerance traits through finding relationship between genotypic data. This accelerates the development of climate-resilient crops through either advanced molecular breeding approaches or genetic engineering techniques in a sustainable way. These innovations are crucial for ensuring sustainable agricultural productivity and safeguarding food and nutritional security in the face of a changing climate.

**TABLE 12.2**
**List of Some of the QTL Studies Done using High-Throughput Phenotyping**

| Crop | Mapping Population | Environment | Sensor Utilized | QTLs Identified | Reference |
|---|---|---|---|---|---|
| Arabidopsis | 62 RILs and 92 NILs were developed from crossing between Cape Verde Islands (Cvi) and Landsberg erecta (Ler) | To study the root gravitropism under a controlled environment | Digital images for 8 hours at 2 minutes intervals | Time-dependent QTL was detected on chromosome 1,3,4 for the root tip angle | Moore et al. (2013) |
| Barely | For the experiment, a total of forty-seven wild barley ILs from the S42IL collection and the recipient parent Scarlett were chosen | Controlled environment | RGB sensor | There are 44 Quantitative Trait Loci (QTLs) identified for 11 different phenotypes. Three quantitative trait loci (QTL) were discovered for the Absolute Growth Rate Integral (AGRI), whereas two QTL were revealed for Water Use Efficiency | Honsdorf et al. (2014) |
| Maize | A total of 167 recombinant lines, derived from the parents B73 and BY804, were planted in the RAP | Controlled environment | RGB sensor | 988 QTLs and three QTL hotspots for 106 traits across 16 growth stages have been identified | Zhang et al. (2017) |
| Maize | Three bi-parental mapping population, 237 RILs from Ki3xNC356,102 RILs from Tx740xNC356 and 178 RILs from LH82xLAMA-Y C | Field condition (Irrigated and non-irrigated/rainfed) | RGB images across two irrigation regimes using UAV-mounted | Nine QTL were identified across the three populations and two environments each explaining 5.1%–9.4% of genetic variance | Anderson et al. (2020) |
| Maize | 187 RILs from a cross between Zong3 and 87-1 | Greenhouse condition (control and water stress treatment) | Infrared thermography | 9 significant QTL were detected for LTD, RSFW and RSDW) | Liu et al. (2011) |
| Wheat | 198 RILs lines from a cross between Yangmai16/Zhongmai895 | Field condition | UAV-based multispectral | 22 QTL significant QTL were identified for senescence loci | Hassan et al., 2021 |
| Rice | 210 lines from a cross between, Zhenshan 97 and Minghui 63 | Greenhouse condition | RGB sensor | 3,586 time-specific QTLs were identified | Geng et al. (2024) |
| wheat | 120 RILs from crossing between Kohdasht (KHD) and Gonbad (GND) | Field condition | Chlorophyll fluorescence | Thirteen of the fourteen QTLs discovered in 2019 and four of the nine QTLs discovered in 2020 had an explanatory coefficient greater than 15% and were considered major QTLs | Sahranavad et al. (2024) |
| Wheat | 309 RILs from the cross of spring Worrakatta × Berkut | Field condition | Multispectral camera sensors mounted on DJI Phantom4 | Eight and three QTLs for LAI and chlorophyll content respectively were identified | Wang et al. (2022) |
| *S. pimpinellifolium* | In total, 274 *S. pimpinellifolium* accessions | Greenhouse and field conditions for salinity stress | Hyperspectral camera, RGB and Thermal infra-red camera | Three loci on chromosome 1, with eight, seven, and four significant SNPs, were linked to absolute growth under salt stress | Morton et al. (2023) |

**TABLE 12.3**
**List of Some of the GWAS Studies Done using High-Throughput Phenotyping**

| Crop | Population/Accession | Environment | Sensor Utilized | GWAS | Reference |
|---|---|---|---|---|---|
| Wheat | 565 diverse wheat accessions | Field condition | The MicaSense Altum sensor was integrated with DJI UAVs to acquire data across five narrow spectral bands: Green, Red, Red Edge, Near-Infrared, and Blue | An RSGS-based genome-wide study identified 47 QTL with 3,079 SNPs linked to SG and 1,085 candidate genes, including 15 known QTL | Yu et al. (2024) |
| Wheat | 155 wheat genotypes, mainly from Chinese wheat core collections and widely grown cultivars over several decades | Controlled condition (Drought) | Red-green-blue (RGB) and hyperspectral imaging (HSI) | A GWAS identified 5,320 drought-related significant SNPs and 27 clusters of these SNPs. One of the novel discoveries in this hotspot cluster was TaPP2C6 which showed a negative regulator of drought resistance | Zhang et al. (2024) |
| Rice | 187 KZ-RILs. | Field condition | Vis/NIR hyperspectral imaging | GWAS identified chromosomal regions and candidate genes linked to rice grain quality traits, revealing key genes involved in starch, sucrose metabolism, and grain chalkiness | Barnaby et al. (2020) |
| Maize | 228 maize inbred lines | Greenhouse condition | RGB sensor | An i-trait genome-wide association study uncovered 4,945 SNPs associated with traits, pinpointed 2,603 genetic loci, and identified 1,974 potential candidate genes | Wang et al. (2023) |

## REFERENCES


Al-Tamimi, N., Langan, P., Bernád, V., Walsh, J., Mangina, E., & Negrão, S. (2022). Capturing crop adaptation to abiotic stress using image-based technologies. *Open Biology*, 12(6), 210353.

Anderson, S. L., Murray, S. C., Chen, Y., Malambo, L., Chang, A., Popescu, S., ... & Jung, J. (2020). Unoccupied aerial system enabled functional modeling of maize height reveals dynamic expression of loci. *Plant Direct*, 4(5), e00223.

Araus, J. L., & Kefauver, S. C. (2018). Breeding to adapt agriculture to climate change: Affordable phenotyping solutions. *Current Opinion in Plant Biology*, *45*, 237–247.

Arya, S., Sahoo, R. N., Sehgal, V. K., Bandyopadhyay, K., Rejith, R. G., Chinnusamy, V., ... & Manjaiah, K. M. (2024). High-throughput chlorophyll fluorescence image-based phenotyping for water deficit stress tolerance in wheat. *Plant Physiology Reports*, *29*, 1–16.

Atieno, J., Li, Y., Langridge, P., Dowling, K., Brien, C., Berger, B., ... & Sutton, T. (2017). Exploring genetic variation for salinity tolerance in chickpea using image-based phenotyping. *Scientific Reports*, 7(1), 1300.

Awlia, M., Nigro, A., Fajkus, J., Schmoeckel, S. M., Negrão, S., Santelia, D., ... & Panzarová, K. (2016). High-throughput non-destructive phenotyping of traits that contribute to salinity tolerance in Arabidopsis thaliana. *Frontiers in Plant Science*, 7, 1414.

Barnaby, J. Y., Huggins, T. D., Lee, H., McClung, A. M., Pinson, S. R., Oh, M., ... & Edwards, J. D. (2020). Vis/NIR hyperspectral imaging distinguishes sub-population, production environment, and physicochemical grain properties in rice. *Scientific Reports*, *10*(1), 9284.

Basavaraj, P. S., & Rane, J. (2020). Avenues to realize potential of phenomics to accelerate crop breeding for heat tolerance. *Plant Physiology Reports*, 25(4), 594–610.

Behmann, J., Schmitter, P., Steinrücken, J., & Plümer, L. (2014). Ordinal classification for efficient plant stress prediction in hyperspectral data. *The International Archives of the Photogrammetry, Remote Sensing and Spatial Information Sciences*, *40*, 29–36.

Das, M., Deb, C. K., Pal, R., & Marwaha, S. (2022). A Machine Learning Approach for the Non-Destructive Estimation of Leaf Area in Medicinal Orchid Dendrobium nobile L. *Applied Sciences*, *12*(9), 4770.

FAO. (2022). *World Food and Agriculture Statistical Yearbook 2022*. FAO.

Feng, X., Zhan, Y., Wang, Q., Yang, X., Yu, C., Wang, H., ... & He, Y. (2020). Hyperspectral imaging combined with machine learning as a tool to obtain high-throughput plant salt-stress phenotyping. *The Plant Journal*, 101(6), 1448–1461.

Furbank, R. T., & Tester, M. (2011). Phenomics–technologies to relieve the phenotyping bottleneck. *Trends in Plant Science*, 16(12), 635–644.

Geng, Z., Lu, Y., Duan, L., Chen, H., Wang, Z., Zhang, J., ... & Yang, W. (2024). High-throughput phenotyping and deep learning to analyze dynamic panicle growth and dissect the genetic architecture of yield formation. *Crop and Environment*, 3(1), 1–11.

Gill, T., Gill, S. K., Saini, D. K., Chopra, Y., de Koff, J. P., & Sandhu, K. S. (2022). A comprehensive review of high throughput phenotyping and machine learning for plant stress phenotyping. *Phenomics*, 2(3), 156–183.

Govindaraj, M., Pattanashetti, S. K., Patne, N., Kanatti, A. A., & Ciftci, Y. O. (2018). Breeding cultivars for heat stress tolerance in staple food crops. *Next Generation Plant Breeding*, 76480(3).

Guntukula, R. (2020). Assessing the impact of climate change on Indian agriculture: Evidence from major crop yields. *Journal of Public Affairs*, *20*(1), e2040.

Habib-ur-Rahman, M., Ahmad, A., Raza, A., Hasnain, M. U., Alharby, H. F., Alzahrani, Y. M., ... & El Sabagh, A. (2022). Impact of climate change on agricultural production; Issues, challenges, and opportunities in Asia. *Frontiers in Plant Science*, *13*, 925548.

Hasanuzzaman, M., Raihan, M. R. H., Masud, A. A. C., Rahman, K., Nowroz, F., Rahman, M., ... & Fujita, M. (2021). Regulation of reactive oxygen species and antioxidant defense in plants under salinity. *International Journal of Molecular Sciences*, 22(17), 9326.

Hassan, M. A., Yang, M., Rasheed, A., Tian, X., Reynolds, M., Xia, X., ... & He, Z. (2021). Quantifying senescence in bread wheat using multispectral imaging from an unmanned aerial vehicle and QTL mapping. *Plant Physiology*, 187(4), 2623–2636.

Hickey, L. T., N. Hafeez, A., Robinson, H., Jackson, S. A., Leal-Bertioli, S. C., Tester, M., ... & Wulff, B. B. (2019). Breeding crops to feed 10 billion. *Nature Biotechnology*, 37(7), 744–754.

Holland, J. B. (2007). Genetic architecture of complex traits in plants. *Current Opinion in Plant Biology*, 10(2), 156–161.

Jespersen, D., & Huang, B. (2014). Physiological and biochemical mechanisms of plant tolerance to heat stress. In *Handbook of Plant and Crop Physiology*, Third Edition (pp. 389–404). Taylor and Francis.

Kamal, N. M., Gorafi, Y. S. A., Abdelrahman, M., Abdellatef, E., & Tsujimoto, H. (2019). Stay-green trait: A prospective approach for yield potential, and drought and heat stress adaptation in globally important cereals. *International Journal of Molecular Sciences*, 20(23), 5837.

Kim, S. L., Kim, N., Lee, H., Lee, E., Cheon, K. S., Kim, M., ... & Kim, K. H. (2020). High-throughput phenotyping platform for analyzing drought tolerance in rice. *Planta*, 252(3), 38.

LeCun, Y., Bengio, Y., & Hinton, G. (2015). Deep learning. *Nature*, *521*(7553), 436–444.

Liu, Y., Subhash, C., Yan, J., Song, C., Zhao, J., & Li, J. (2011). Maize leaf temperature responses to drought: Thermal imaging and quantitative trait loci (QTL) mapping. *Environmental and Experimental Botany*, 71(2), 158–165.

Ma, S., Liu, S., Gao, Z., Wang, X., Ma, S., & Wang, S. (2024). Water deficit diagnosis of winter wheat based on thermal infrared imaging. *Plants*, 13(3), 361.

Malhi, G. S., Kaur, M., & Kaushik, P. (2021). Impact of climate change on agriculture and its mitigation strategies: A review. *Sustainability*, *13*(3), 1318.

Mangal, V., Lal, M. K., Tiwari, R. K., Altaf, M. A., Sood, S., Gahlaut, V., ... & Kumar, D. (2023). A comprehensive and conceptual overview of omics-based approaches for enhancing the resilience of vegetable crops against abiotic stresses. *Planta*, *257*(4), 80.

Masson-Delmotte, V., Zhai, P., Pirani, A., Connors, S. L., Péan, C., Berger, S., ... & Zhou, B. (2021). Climate change 2021: The physical science basis. *Contribution of Working Group I to the Sixth Assessment Report of the Intergovernmental Panel on Climate Change*, *2*(1), 2391.

Moore, C. R., Johnson, L. S., Kwak, I. Y., Livny, M., Broman, K. W., & Spalding, E. P. (2013). High-throughput computer vision introduces the time axis to a quantitative trait map of a plant growth response. *Genetics*, *195*(3), 1077–1086.

Morton, M., Fiene, G., Ahmed, H. I., Rey, E., Abrouk, M., Angel, Y., ... & Julkowska, M. M. (2024). Deciphering salt stress responses in Solanum pimpinellifolium through high-throughput phenotyping. *The Plant Journal*, *119*(5), 2514–2537.

Pappula-Reddy, S. P., Kumar, S., Pang, J., Chellapilla, B., Pal, M., Millar, A. H., & Siddique, K. H. (2024). High-throughput phenotyping for terminal drought stress in chickpea (Cicer arietinum L.). *Plant Stress*, 11, 100386.

Polya, G. (2023). *WMO Warning: 1.5 Degree C Warming Breach Very Soon & With Increasing Frequency*. Act Now!.

Pradhan, A., Aher, L., Hegde, V., Jangid, K. K., & Rane, J. (2022). Cooler canopy leverages sorghum adaptation to drought and heat stress. *Scientific Reports*, 12(1), 4603.

Rane, J., Raina, S. K., Govindasamy, V., Bindumadhava, H., Hanjagi, P., Giri, R., ... & Nair, R. M. (2021). Use of phenomics for differentiation of mungbean (Vigna radiata L. Wilczek) genotypes varying in growth rates per unit of water. *Frontiers in Plant Science*, 12, 692564.

Raza, A., Razzaq, A., Mehmood, S. S., Zou, X., Zhang, X., Lv, Y., & Xu, J. (2019). Impact of climate change on crops adaptation and strategies to tackle its outcome: A review. *Plants*, *8*(2), 34.

Rebolledo, M. C., Peña, A. L., Duitama, J., Cruz, D. F., Dingkuhn, M., Grenier, C., & Tohme, J. (2016). Combining image analysis, genome wide association studies and different field trials to reveal stable genetic regions related to panicle architecture and the number of spikelets per panicle in rice. *Frontiers in Plant Science*, 7, 1384.

Rosenzweig, C., & Hillel, D. (1995). Potential impacts of climate change on agriculture and food supply. *Consequences*, *1*(2), 23–32.

Saeed, F., Chaudhry, U. K., Raza, A., Charagh. S., Bakhsh, A., Bohra, A., Ali, S. et al. (2023). Developing future heat-resilient vegetable crops. *Functional & Integrative Genomics, 23*(1), 47.

Sahranavard, N., Jorjani, E., Sabouri, H., Alegh, S. M., & Katouzi, M. (2024). Detecting QTLs controlling chlorophyll fluorescence parameters in Iranian wheat recombinant inbred lines. *Plant Gene*, 37, 100437.

Samuel, A. L. (1959). Some studies in machine learning using the game of checkers. *IBM Journal of Research and Development*, *3*(3), 210–229.

Sandhu, K., Patil, S. S., Pumphrey, M., & Carter, A. (2021). Multitrait machine-and deep-learning models for genomic selection using spectral information in a wheat breeding program. *The Plant Genome*, *14*(3), e20119.

Seelig, H. D., Hoehn, A., Stodieck, L. S., Klaus, D. M., Adams, W. W., & Emery, W. J. (2009). Plant water parameters and the remote sensing R 1300/R 1450 leaf water index: Controlled condition dynamics during the development of water deficit stress. *Irrigation Science*, 27, 357–365.

Shi, T., Gao, Y., Song, J., Ao, M., Hu, X., Yang, W., ... & Feng, H. (2024). Using VIS-NIR hyperspectral imaging and deep learning for non-destructive high-throughput quantification and visualization of nutrients in wheat grains. *Food Chemistry*, *461*, 140651.

Siebers, M. H., Fu, P., Blakely, B. J., Long, S. P., Bernacchi, C. J., & McGrath, J. M. (2024). Fast, nondestructive and precise biomass measurements are possible using lidar-based convex hull and voxelization algorithms. *Remote Sensing, 16*(12), 2191.

Singh, B., Kumar, S., Elangovan, A., Vasht, D., Arya, S., Duc, N. T., ... & Chinnusamy, V. (2023). Phenomics based prediction of plant biomass and leaf area in wheat using machine learning approaches. *Frontiers in Plant Science*, *14*, 1214801.

Song, P., Wang, J., Guo, X., Yang, W., & Zhao, C. (2021). High-throughput phenotyping: Breaking through the bottleneck in future crop breeding. *The Crop Journal*, *9*(3), 633–645.

Sufyan, M., Daraz, U., Hyder, S., Zulfiqar, U., Iqbal, R., Eldin, S. M., ... & Ali, I. (2023). An overview of genome engineering in plants, including its scope, technologies, progress and grand challenges. *Functional & Integrative Genomics*, 23(2), 119.

Tao, F., Rötter, R. P., Palosuo, T., Díaz-Ambrona, C. G. H., Mínguez, M. I., Semenov, M. A., ... & Schulman, A. H. (2017). Designing future barley ideotypes using a crop model ensemble. *European Journal of Agronomy*, *82*, 144–162.

Wang, W., Gao, X., Cheng, Y., Ren, Y., Zhang, Z., Wang, R., ... & Geng, H. (2022). QTL mapping of leaf area index and chlorophyll content based on UAV remote sensing in wheat. *Agriculture*, *12*(5), 595.

Wang, W., Guo, W., Le, L., Yu, J., Wu, Y., Li, D., ... & Pu, L. (2023). Integration of high-throughput phenotyping, GWAS, and predictive models reveals the genetic architecture of plant height in maize. *Molecular Plant*, *16*(2), 354–373.

Xiao, D., Li Liu, D., Wang, B., Feng, P., Bai, H., & Tang, J. (2020). Climate change impact on yields and water use of wheat and maize in the North China Plain under future climate change scenarios. *Agricultural Water Management*, *238*, 106238.

Xiao, Q., Bai, X., Zhang, C., & He, Y. (2022). Advanced high-throughput plant phenotyping techniques for genome-wide association studies: A review. *Journal of Advanced Research*, 35, 215–230.

Yang, W., Feng, H., Zhang, X., Zhang, J., Doonan, J. H., Batchelor, W. D., ... & Yan, J. (2020). Crop phenomics and high-throughput phenotyping: Past decades, current challenges, and future perspectives. *Molecular Plant*, *13*(2), 187–214.

Yu, R., Cao, X., Liu, J., Nie, R., Zhang, C., Yuan, M., ... & Wu, J. (2024). Using UAV-based temporal spectral indices to dissect changes in the stay-green trait in wheat. *Plant Phenomics*, *6*, 0171.

Zhang, X., Huang, C., Wu, D., Qiao, F., Li, W., Duan, L., ... & Yan, J. (2017). High-throughput phenotyping and QTL mapping reveals the genetic architecture of maize plant growth. *Plant Physiology*, 173(3), 1554–1564.

Zhang, Z., Qu, Y., Ma, F., Lv, Q., Zhu, X., Guo, G., ... & Song, C. P. (2024). Integrating high-throughput phenotyping and genome-wide association studies for enhanced drought resistance and yield prediction in wheat. *New Phytologist*, *243*(5), 1758–1775.

Zhou, J., Beche, E., Vieira, C. C., Yungbluth, D., Zhou, J., Scaboo, A., & Chen, P. (2022). Improve soybean variety selection accuracy using UAV-based high-throughput phenotyping technology. *Frontiers in Plant Science*, *12*, 768742.

Zhou, J., Chen, H., Zhou, J., Fu, X., Ye, H., & Nguyen, H. T. (2018). Development of an automated phenotyping platform for quantifying soybean dynamic responses to salinity stress in greenhouse environment. *Computers and Electronics in Agriculture*, 151, 319–330.

Zuo, J., & Li, J. (2014). Molecular dissection of complex agronomic traits of rice: A team effort by Chinese scientists in recent years. *National Science Review*, *1*(2), 253–276.

# 13 High-Throughput Plant Phenotyping and Phenomics

## *Advances and Prospects*

*Seethalaxmi, Ezhumalai Sivapragasam, Arumugam Tamilselvan, Raman Rajasekar, Sreehari V. Santhosh, Mahesh Kumar, Viswanathan Chinnusamy, and Sudhir Kumar*

## LIST OF ABBREVIATIONS

**AI**: Artificial Intelligence
**CNNs**: Convolutional Neural Networks
**CT**: Computed Tomography
**DL**: Deep Learning
**FIR**: Far Infra-Red
**GIS**: Global Information System
**GPS**: Global Positioning System
**HTP**: High-Throughput Phenotyping
**IR**: Infra-Red
**LAI**: Leaf Area Index
**LAD**: Leaf Area Duration
**LED**: Light Emitting Diode
**LiDAR**: Light Detection and Ranging
**LWIR**: Long Wavelength Infra-Red
**ML**: Machine Learning
**MRI**: Magnetic Resonance Imaging
**NDVI**: Normalized Difference Vegetation Index
**NIR**: Near Infra-Red
**RGB**: Red, Green and Blue
**RNNs**: Recurrent Neural Networks
**SPAD**: Soil Plant Analysis Development
**SWIR**: Shortwave Infra-Red
**TIR**: Thermal Infra-Red
**ToF**: Time of Flight
**UV**: Ultraviolet
**VNIR**: Visible-Near Infrared

## 13.1 INTRODUCTION

Global food security is predicted to be threatened over the next century by an expected increase in population growth, changing demography, and ecological variability. Efficient management of input resources is absolutely necessary for maintaining and enhancing crop output for food and fuel


DOI: 10.1201/9781003505631-13

production in the presence of these challenges (Fahlgren et al., 2015). The study of plant biology has undergone significant transformation in the last 20 years due to breakthroughs in "genomics. Researchers can sequence several genotypes of a particular crop for the purposes of association mapping and allele mining because the cost of genome sequencing is significantly lower. The public domain has the genomes of many crop plants, but phenotypic information on their genomes is still scarce because of environmental factors that make it difficult to characterize phenomes and elevate the possibility of measurement errors in traits. Ultimately, the outcome determines the phenotype, regardless of the exciting discoveries made in the field of molecular biology (Rahman et al., 2015). Enhancing our knowledge and understanding of structural and physiological features is essential (Awada et al., 2024). It is essential to comprehend how genes function and how the environment affects plants (Zhang et al., 2018). Understanding the phenotype—the intricate interaction between a plant's genetic information and its environmental surroundings—both on and off the ground during its growth and development—revolutionizes the field of plant breeding today (Awada et al., 2024). There has emerged a field called "Phenomics", the study of a wide range of phenotypic plant features. The Next Generation Phenotyping (NGP) method, known as phenomics, bridges the phenotype-genotype gap by illuminating the inner workings of living plants.

Phenomics is an interdisciplinary field that combines computer science, engineering, biology, bioinformatics, and statistics to provide extensive knowledge of the intricate molecular and physiological features of plants (Awada et al., 2024). The rapid advancement of genomics has led to the emergence of plenty of phenotyping platforms to look into multifaceted characteristics associated with growth, development, yield, and tolerance to biotic or abiotic stress (Zhang et al., 2018). High-throughput phenotyping (HTP) technology is expanding quickly and has completely changed the field of phenotyping. Automated sensing, data collection, and analysis, which depend on automated trait analysis to produce phenotypic data. It makes it possible for scientists to quickly and precisely gather phenotypic data from thousands of plants. These characteristics may include those pertaining to yield, growth, stress tolerance, nutrient status, and resistance to disease. Large-scale data collection on plant characteristics, including leaf area, plant height, biomass, blooming time, and disease resistance, is a requirement of HTP (Sheik et al., 2024). This chapter provides a detailed discussion of phenomics and high-throughput phenotyping, including the developments reported thus far.

## 13.2 ADVANTAGES OF PLANT PHENOMICS OVER CONVENTIONAL METHODS

Plant phenomics offers several key advantages over traditional phenotyping methods, as listed below:

### 13.2.1 High-Throughput and Non-Destructive

Phenomics techniques allow for rapid, automated, and non-destructive measurement of plant traits, enabling the screening of large populations in a short time without damaging plants. In contrast, conventional methods are often laborious, time-consuming, and destructive (Walter et al., 2015).

### 13.2.2 Quantitative and Objective Data

Phenomics provides quantitative, objective, and precise measurements of plant traits using advanced imaging and sensor technologies. This reduces the subjectivity and inaccuracies associated with manual measurements (Raju et al., 2020).

### 13.2.3 Temporal Resolution

Phenomics enables repeated measurements of the same plants over time, providing insights into dynamic processes like growth and development. Conventional methods often rely on destructive sampling at specific time points (Perez-Sanz et al., 2017).

### 13.2.4 Multi-Scale Analysis

Phenomics platforms can measure traits on multiple scales, from individual plants to whole canopies and fields. As a result, plant performance can be understood with greater depth (Yang et al., 2020).

### 13.2.5 Broad Trait Coverage

Modern phenotyping platforms can measure a wide range of structural, physiological, and biochemical traits. This includes traits related to growth, architecture, photosynthesis, water status, and nutrient content. Conventional methods typically focus on a limited set of traits (Perez-Sanz et al., 2017).

### 13.2.6 Abiotic Stress Phenotyping

Phenomics is particularly useful for noninvasive screening plants under abiotic stresses like drought, salinity, heat, cold, and nutrient deficiencies. This is crucial for crop improvement under climate change scenarios (Kumar et al., 2015).

### 13.2.7 Bridging Genotype to Phenotype

By providing high-dimensional phenotypic data, phenomics helps establish genotype-phenotype associations and enables genomics-assisted breeding. This is key to unlocking the potential of genetic resources for crop development (Pasala and Pandey, 2020).

This enables more efficient and comprehensive plant performance characterization, ultimately accelerating crop improvement.

## 13.3 FORWARD AND REVERSE PHENOMICS

Plant phenomics offers new opportunities through the use of high-throughput screens, numerous camera units, quantitative analysis, non-destructive measurements, growth dynamics monitoring, and stress assessment linked to genomes. Functional analysis of individual genes, forward and reverse genetic analyses, and the cultivation of elite plants all rely heavily on phenotyping. Forward phenomics selects the "best of the best" genotype/elite line using phenotyping methods to identify the most promising genotypes and desired attributes from a large collection (Pasala and Pandey, 2020). Using phenotyping instruments, forward phenomics "sieves" germplasm collections to identify salient and observable features. The function is employed to determine which germplasm is most suitable for a given characteristic. Using automated imaging technology, forward phenomics accelerates the phenotyping of several plants or germplasm lines, enabling the identification of intriguing traits or plants that are appropriate for a given circumstance. Using an automated imaging system, many plants are cultivated in pots that are barcode-labeled in advance and evaluated for unique characteristics. After being chosen, the plants with the desired characteristics can be allowed to mature and yield seeds for additional research and breeding (Rahman et al., 2015).

Reverse phenomics is the thorough analysis of the mechanisms behind particular traits that enables the exploitation of these mechanisms or the candidate gene(s) linked to the trait that can be used to introduce the trait into new varieties or, through the use of genetic transformation technology, to transfer the trait to other plant species. In reverse phenomics, desired characteristics, such as drought tolerance, are already present in particular germplasm. Then, researchers look for the genes that underlie the process and the mechanisms that govern the trait. Large-scale physiological and biochemical analysis is used to do this, and the data are then connected to genes involved in specific biochemical or physiological pathways. By using the reverse phenomics setup to identify the candidate gene(s), they intend to compare the expression pattern of these gene(s) with that of other genotypes (Rahman et al., 2015).

## 13.4 UTILIZATION OF SENSORS IN HIGH-THROUGHPUT PHENOTYPING

The integration of advanced technologies such as image processing (Li et al., 2014), machine learning (Pound et al., 2017), and communication methods (Al-Turjman, 2019) has significantly contributed to plant phenotyping. Studies have shown that both traditional and innovative sensors—like visible spectrum image sensors (Tsaftaris and Noutsos, 2009), thermal image sensors (Sagan et al., 2019), near-infrared image sensors (Chen et al., 2014), multispectral image sensors (Xu et al., 2019), hyperspectral image sensors (Behmann et al., 2018), LiDAR (Lin, 2015), ultrasonic sensors (Murray et al., 2016), fluorescence sensors (Hairmansis et al., 2014), and depth image sensors (Xia et al., 2015)—can be combined to assess various plant phenotyping parameters.

### 13.4.1 Visible-Spectrum Image Sensors

The visible spectrum, encompassing electromagnetic wavelengths typically between 400 and 780 nm, is what the human eye can perceive (Sliney, 2016). Image sensors designed for the visible spectrum capture light in the red, green, and blue (RGB) wavelengths to generate images (Zhu et al., 2019). These sensors can detect visible radiation from objects to create RGB images, which are useful in assessing plant health, biotic stress, and physical characteristics (Qiu et al., 2018). These sensors are popular due to their availability, cost-effectiveness, and ease of use. One approach in high-throughput systems involves capturing multiple images of plants from different angles using several sensors to ensure accurate and reliable phenotyping.

In one study, RGB images of barley plants were taken from three angles using a high-throughput platform, the LemnaTec 3D Scanalyzer (LemnaTec, GmbH, Wuerselen, Germany). The study found a strong correlation between biomass estimates from visible spectrum images and actual plant biomass, demonstrating the effectiveness of this sensor system in high-throughput plant phenotyping (Honsdorf et al., 2014). Similarly, Chen et al. (2014) developed an integrated framework for processing high-throughput plant phenotyping data, which was used to monitor plant growth and biomass changes. Other studies have applied similar methods to wheat (Golzarian et al., 2011) and sorghum (Neilson et al., 2015). However, a significant challenge with these sensors in plant phenotyping is distinguishing plants with similar colors from their backgrounds. Despite advancements in high-throughput platforms and image analysis techniques, limitations of digital image sensors persist.

### 13.4.2 Infrared Spectrum Image Sensors

Infrared wavelengths range from 780 to 1,000 μm, and the spectrum is divided into five subsets: near-infrared (NIR), mid-wave infrared (MWIR), longwave infrared (TIR/LWIR), and far-infrared (FIR). NIR image sensors are particularly useful in evaluating leaf water content and plant responses to drought, as plant tissues reflect more NIR and SWIR light than visible light, depending on leaf thickness and water content (Chen et al., 2014; Briglia et al., 2019; Seelig et al., 2008).

Thermal infrared (TIR) sensors detect longwave infrared (LWIR) spectra and visualize the distribution of infrared radiation on plant surfaces. These sensors convert the heat emitted by an object into visual images that map the surface temperature of leaves and other plants. By capturing thermal data within the 3–5 μm or 7–14 μm range, these sensors can assess plant physiological conditions such as canopy or leaf temperature (Ahmadi et al., 2017), transpiration, stomatal conductance (Prashar et al., 2013), and water status (Jangra et al., 2021; Li et al., 2014) in response to biotic and abiotic stress. However, environmental factors can easily affect thermal infrared images, making it difficult to detect small temperature changes (Aeberli et al., 2023).

### 13.4.3 Multispectral Image Sensors

Multispectral imaging, traditionally associated with space-based applications, has been effectively adapted for use in agriculture, national defense, and environmental monitoring. These sensors capture light across several broad spectral bands, making them particularly useful in agriculture for monitoring visible and near-infrared spectra. The data obtained from these sensors can be employed to assess various plant traits, such as water stress, disease detection, and the normalized difference vegetation index (NDVI) (Sankaran et al., 2015; Kitić et al., 2019). NDVI, derived from canopy reflectance in the infrared spectrum, has been shown to correlate strongly with ground-based measurements (Zaman-Allah et al., 2015). Despite their effectiveness, a significant limitation of multispectral sensors is their focus on a limited number of spectral bands, which may restrict their utility in certain applications.

### 13.4.4 Hyperspectral Image Sensors

Hyperspectral sensors capture hundreds of narrow spectral bands in the wavelength range of 250–2500 nm, including UV, VIS, NIR, and SWIR (Reynolds et al., 2019). The key distinction between multispectral and hyperspectral sensors lies in the number of wavebands that they can capture. Hyperspectral sensors have emerged as some of the most powerful tools in remote sensing, finding applications in estimating nutrient content, detecting diseases (Araus et al., 2014; Sambrekar et al., 2019; Pound et al., 2017), and monitoring fruit ripening and other physiological traits (Reynolds et al., 2019). Despite their potential for high-throughput plant phenotyping, hyperspectral sensors face challenges such as high costs and complex image processing and analysis (Fahlgren et al., 2015; Sankaran et al., 2015). Advanced techniques like image fusion, convolutional neural networks (Yang et al., 2019), artificial neural networks (Seiffert et al., 2010), and unsupervised learning (Römer et al., 2012) can help address these issues.

### 13.4.5 Fluorescence Imaging

Fluorescence occurs when an object absorbs light at a specific wavelength and then emits light at a shorter wavelength. In fluorescence imaging, plants are exposed to blue light (<500 nm), causing them to emit red-spectrum fluorescence (600–750 nm). The software is then used to analyze fluorescence variations during photosynthesis and convert them into false-color data. Fluorescence sensors typically use charge-coupled devices (CCDs) that are sensitive to fluorescence signals. This imaging technique measures traits like chlorophyll fluorescence, photosynthetic efficiency, leaf health, shoot architecture, and non-photochemical quenching. When external factors impact photosynthesis, plant metabolism is altered, making fluorescence sensors useful for detecting plant stress and disease (Li et al., 2014; Rousseau et al., 2013). However, these sensors have drawbacks, such as a limited field of view and a need for strong lighting (Qiu et al., 2018). Robust data processing software has been suggested for large-scale plant phenotyping tasks, and a uniform approach to fluorescence image processing is needed (Li et al., 2014).

### 13.4.6 Light Detection and Ranging (LiDAR)

LiDAR, also known as laser scanning, is a popular remote sensing method that measures distances by illuminating objects with laser light and recording reflections (Collis, 1970). LiDAR is widely used in high-throughput plant phenotyping to estimate plant characteristics and reconstruct plants in 3D (Lin, 2015). It has been employed to calculate wheat biomass and crop nitrogen, with LiDAR-derived data showing strong correlations with plant biomass and nitrogen content, enhancing crop management in wheat-growing areas (Eitel et al., 2014). LiDAR has also been used to monitor cotton plant growth, measuring parameters like plant height, canopy area, and volume, with findings showing excellent correlations between manual and LiDAR measurements (Sun et al., 2018). However, LiDAR's sensitivity to even minor changes in route length is both an advantage and a limitation, and highly precise LiDAR sensors can be costly (Qiu et al., 2018). Advances in LiDAR technology with high-resolution and long-range capabilities may address some of these issues (Zhao et al., 2019b).

### 13.4.7 Tomographic Imaging

Tomographic imaging techniques, such as positron emission tomography (PET), magnetic resonance imaging (MRI), and X-ray computed tomography (X-ray CT) are used to obtain high-resolution 3D images of plants or plant sections (Metzner et al., 2015). MRI enables the quantification of both static and dynamic properties, such as structural changes and metabolic processes over time. X-ray CT, on the other hand, is employed to visualize the internal and external three-dimensional architecture of plants at both macro and micro scales. As X-rays pass through the sample, part of the radiation is absorbed, while the remaining radiographs are detected and captured. Multiple 2D projections are then taken and reconstructed into 3D images (Piovesan et al., 2021). However, these imaging techniques are time-consuming, and unsuitable for large-scale sampling, and their equipment is often too large and heavy for aerial platforms (Sari et al., 2022).

### 13.4.8 Depth Image Sensors

Depth image sensors provide real-time depth information in images and are based on three principles: time of flight (ToF), stereo depth, and structured light. In the structured light approach, a projector shines patterned light onto an object, and an image sensor provides depth information based on the distortion of the pattern (Wu and Bainbridge-Smith, 2011). The ToF method emits light from the projector onto the target object, and the reflected light forms the depth image based on the time difference. Stereo depth uses two cameras to capture an object's image from two different angles, and depth is determined by triangulating the object's distance from each camera. This technique has been used in various studies, such as using an RGB-D sensor (a combination of RGB and depth sensors) for 3D reconstruction of plant canopy and architecture (Qiu et al., 2018). However, due to their susceptibility to environmental light, structured light sensors are less suitable for outdoor plant phenotyping (Xiang et al., 2019) (Figure 13.1).

## 13.5 AFFORDABLE PLANT PHENOTYPING DEVICES

Affordable phenotyping sensors have become increasingly important for monitoring plant health and growth in agricultural research. These sensors offer cost-effective solutions while maintaining the ability to gather vital data. Various low-cost sensors are available, each designed to measure specific aspects of plant physiology, such as chlorophyll content, thermal properties, and light reflectance. These tools are often integrated with platforms like Raspberry Pi, making them accessible and easy to use for large-scale plant phenotyping, thus enabling researchers to perform high-throughput analyses without the financial burden of expensive equipment.

### 13.5.1 AS7263 Spectral Sensor

The "ams multi-spectral sensor" package features affordable spectral sensors capable of identifying multiple channels in visible and near-infrared spectra. Within this package, the AS7263 spectral sensor stands out as a cost-effective tool for plant phenotyping, specifically operating in the near-infrared spectrum. This sensor is equipped with six distinct NIR channels, allowing it to detect wavelengths at 610, 680, 730, 760, 810, and 860 nm.

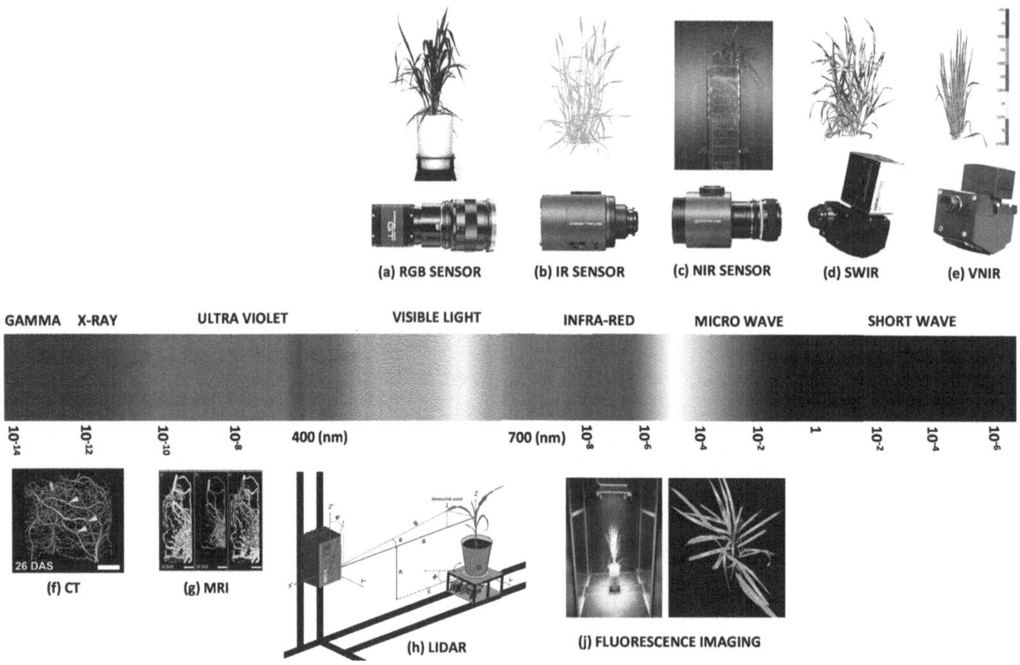


**FIGURE 13.1** Illustration of various imaging technologies and sensors across the electromagnetic spectrum, ranging from gamma and X-ray to visible light and infrared. (a) RGB, (b) IR image, (c) NIR image, (d) SWIR image, (e) VNIR image and (j) Fluorescence image were captured in our lab. (f) CT and (g) MRI (Metzner et al., 2015), (h) LIDAR (Thapa et al., 2018).

### 13.5.2 Low-Cost Chlorophyll Meter

Chlorophyll plays a crucial role in assessing plant health (Shaahan et al., 1999). However, the chlorophyll meters currently available on the market, such as MC-100, CL-01, and SPD-502, tend to be quite expensive, leading to a demand for more affordable alternatives. Addressing this need, Hutomo et al. (2016) developed a budget-friendly chlorophyll meter (LCCM) that operates on the light reflectance principle. This device utilizes a single red LED and a near-infrared LED as light sources, and a TSL250 photodiode converting light intensity into output voltage.

### 13.5.3 FLIR Lepton

FLIR Lepton sensors are cost-effective long-wave infrared (LIR) image sensors designed for capturing thermal images. These sensors are available in two easily integrable versions: Lepton 2.5 with 80×60 pixels and Lepton 3.5 with 160×120 pixels (Sangha et al., 2022). Bruno et al. (2017) utilized the FLIR Lepton 3.5, in combination with Raspberry Pi to produce thermal images of crops. Analyzing these thermal images may help predict water deficit stress. The results indicate that FLIR Lepton 3.5 serves as an affordable, non-invasive infrared imaging sensor for monitoring crop health.

### 13.5.4 Raspberry Pi-Series Cameras and Sensors

The low-cost Raspberry Pi camera family includes Raspberry Pi Camera Module 2 and the Raspberry Pi Camera Module 2 NoIR, which lacks an infrared filter. Both cameras have a Sony IMX219 8-megapixel image sensor. While the standard Module 2 camera captures RGB images, the NoIR version can take infrared images due to the absence of an infrared filter. Sangjan et al.

(2021) employed these sensor systems, specifically the Raspberry Pi Module 2 and Module 2 NoIR camera, for automated in-field phenotyping of wheat plants. They calculated the vegetation index, NDVI, and plant height through image analysis. This sensor system provides an affordable method for high-throughput plant phenotyping.

## 13.6 PHENOTYPING APPROACHES

The evaluation of crop phenotypic traits under controlled conditions and in the field has significantly increased over the last 20 years due to the rapid development of non-destructive sensing and imaging tools (Song et al., 2021). The study objective and the necessary level of information determine the phenotyping approach used. As a result of the combination of technologies with robotics, computers, automated control and aeronautics, a growing variety of high-throughput phenotyping devices for examining phenotypic features of crops emerged (Awada et al., 2024). For the fully automated tracking and analysis of adult plant growth, imaging, data gathering and assessment, a number of platforms have been developed (Rahman et al., 2015). Phenotyping platforms are categorized into three categories in this review according to imaging level: (1). Microscopic phenotyping (2). Ground-based phenotyping (3). Aerial phenotyping

These platforms provide an assessment of phenotypic characteristics at several levels, including tissue, organ, individual plant, plot and field systems (Song et al., 2021) (Figure 13.2).

### 13.6.1 Microscopic Phenotyping

The various tissue morphologies and micro-phenotypes observed in plants have attracted a great deal of attention. Rapid and non-destructive measurement of crop micro-phenotypes is desirable. A multitude of novel algorithms and tools have been presented for the observation and extraction of micro-phenotypes of crop organs or tissues, despite several obstacles such as the need for automated image analysis approaches, low microscale imaging resolution, and damaging and complex pre-treatment procedures. It is useful for phenotyping specific plant parts. Kernel features are important factors in determining final production, analyzing them is time-consuming and frequently necessitates damaging harvesting (Song et al., 2021). Creating a reliable, non-destructive kernel-trait analysis technique that can handle big populations is a current challenge. Hughes et al. (2017) developed a reliable technique for accurately extracting and measuring grain and spike morphometric parameters from X-ray micro-computed tomography (μCT) images have been developed. The created image analysis pipeline extracts morphometric data, performs image analysis, and automatically detects plant material of interest in μCT images. Based on micro-computed tomography (CT) scans, three-dimensional models of the kernel geometry of maize have been created. Additionally, a number of recently defined three-dimensional phenotypes have been highlighted, such as kernel aspect ratio, kernel-to-kernel distance, porosity and kernel angle (Zhang et al., 2018).

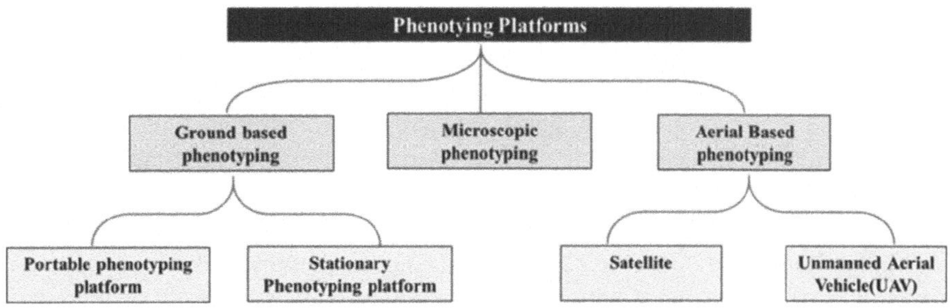


**FIGURE 13.2** Various platforms for High-throughput plant phenotyping.

The stomata are vital gateways for the exchange of gases between plants and their surroundings. The infrared gas exchange system has significantly sped up stomatal conductance research. Nevertheless, little is known about the activity of stomata themselves because there are no in-situ monitoring methods available, particularly in environments that are nocturnal. Sun et al. (2023) designed a deep learning-based stoma-monitoring pipeline that was utilized to continuously monitor stoma traits for the first time utilizing a comprehensive, constant and non-invasive video data set (Stomata Tracker). The anatomical characteristics of the stalk serve a significant part in plant performance and comprehending how plants have evolved to adapt to their surroundings provides evidence of evolutionary adaptation to the environment. Developing appropriate data processing methods for the identification and detection of the microscopic phenotypes of stalk tissue is hampered by the complicated microstructure of crop stalks. Zhang et al. (2013) devised a technique that uses a computerized color imaging analysis method for dyed maize stem (cross section) to measure the amount of lignification in maize tissues.

### 13.6.2 Ground-Based Phenotyping

#### 13.6.2.1 Portable Phenotyping Platform

Proximal phenotyping is the process of phenotyping plants using platforms positioned on the ground (Thakur et al., 2024). A useful tool for phenotyping at the individual plant and field levels is a ground-based phenotyping platform. Generally, depending on the application, this phenotyping platform can be categorized as moveable, fixed, or portable (Song et al., 2021). Modified cars and remote sensing sensors are examples of ground-based phenotyping platforms. These have great potential because they enable real-time, wide-area data acquisition for assessing plant conditions and contain practical tools like geographic information systems (GIS), global positioning systems (GPS), and remote sensing tools, which are referred to as phenotowers and are useful for examining spatial changeability. There have been several recent developments in phenotower technology recently (Omari et al., 2020). Lemna Tec, for instance, produced Scanalyzer3D systems, which are installed and utilized extensively worldwide. Conveyor belt systems and image capture equipment make up Scanalyzer3D systems. Every plant is cultivated in a container and placed in a carrier affixed to the conveyor belt. The conveyor belt mechanism autonomously collects and moves each plant via image capture devices. Large plant populations can be efficiently screened in this plant-to-sensor method (Jiang and Zhu, 2024). Another technique is sensor-to-plant, which describes active sensing systems that adjust sensors to predetermined locations to effectively gather phenotypic data. For instance, Du et al. (2020) used a planar scanning motion mechanism to create a greenhouse phenotyping platform. Using a path that included designated greenhouse zones, the imaging device was able to capture top-down images of thousands of plants.

#### 13.6.2.2 Stationary Phenotyping Platform

Static platforms are fixed phenotyping tools that can accommodate numerous sensors, including RGB, fluorescent, thermochromic, multispectral or hyperspectral and 3-dimensional imaging sensors to improve phenotyping's accuracy, resolution, and throughput both in the field and in controlled settings. A manual cart, a driven tractor, or a phenotyping robot platform can all be modified to become a mobile phenotyping platform by installing many sets of sensors on them. An operator traveling behind a cart propelled by an electric wheel may estimate canopy height, ground cover, and aboveground biomass quickly and nondestructively using light/laser radar (LiDAR) (Song et al., 2021). Some of them are mentioned here. The Field Phenotyping Platform (FIP) of the Eidgenössische Technische Hochschule (ETH) Zürich consists of hanging cameras placed from four 24-m-high poles throughout a $130 \times 100$ m area. Another platform developed was a basic cart with two meters of width and one meter of length made of steel, which was attached to two bicycle frames at a certain height that allowed for one meter of vertical clearance. The arms fastened with

U-bolts hold instruments like infrared thermometers and radiometers to the frame. Computers, batteries, and data recorders can all be mounted on a sizable horizontal surface. On flat ground, one person can easily maneuver the cart; on uneven ground, two people can maneuver it more easily due to berms, furrows, and other obstructions (Deery et al., 2014).

### 13.6.3 Aerial-Based Phenotyping

When ground-based phenotyping platforms are too challenging, aerial-based phenotyping platforms are often employed as a useful substitute. This platform provides quick measurements of several plants as well as plots in a few seconds. The most common aerial-based phenotyping platforms use small planes to securely attain the slow velocities required for low-altitude, superior-resolution photographs. Aerial-based phenotyping platforms now in use differ considerably in terms of payload and management cost (Omari et al., 2020). Robotics and computer vision technologies enable ground-level plant phenotyping screening. In the same way, aerial drones, satellite imaging systems, and field robots like BoniRob contribute significantly to remote sensing data collection (Feng et al., 2021).

#### 13.6.3.1 Satellite Imaging-Based Plant Phenotyping

Satellite images and images from diverse sources are utilized for assessing crop yields, including biomass (Chandra et al., 2020). Satellite imagery can be used to estimate soil pH, moisture, and other related parameters (Ghazali et al., 2020). Monitoring and historical analysis can be conducted regularly and consistently across extensive areas. GISs can extract precise results by utilizing satellite images. NASA, Geocento, and Google Earth Engine are among the leading providers of satellite imagery services, along with numerous others. In various nations, satellite imagery is employed for forest monitoring to detect changes in climate.

#### 13.6.3.2 Unmanned Aerial Vehicle (UAV) Based Phenotyping

Unmanned aerial vehicles (UAVs) are multi-mission capable aircraft that are powered, controlled by radio, or flow autonomously. UAVs can be deployed in various missions because they are easy to operate and control remotely. UAVs can be very useful in areas that are hard to reach by people. At the large-scale field level, UAV-based remote sensing can be used to gather phenotypic features within the necessary frequency of time and spatial resolution (Yang et al., 2018). Using communications and GPS technology, High-definition phenotyping information about plants throughout the field can be collected quickly and non-destructively by UAVs equipped with more than one sensor (Zheng et al., 2020; Zaman et al., 2014). Researchers now depend heavily on UAV-based phenotyping to rapidly identify plant phenotypes.

The sensor is the fundamental component of low-altitude remote sensing technology UAV data acquisition system. Auxiliary and phenotyping sensors are the two primary categories of sensors (Jing et al., 2017). Thermal infrared imaging cameras, RGB, multispectral, hyperspectral, fluorescence, and LIDAR sensors are possible phenotyping sensors that may be mounted on the UAV. UAVs most frequently utilize RGB cameras for phenotyping various crops due to their excellent resolution quality and affordable price in comparison with other sensors (Chee and Zhong, 2013). For instance, pictures captured by cameras installed on UAVs may be used to assess aboveground biomass (AGB), plant height, LAI, chlorophyll content, weed mapping, and yield (see Table 13.1).

Multispectral sensors have the ability to collect emittance, reflectance, or backscattered radiation in multiple bands from a region or object upon atmospheric calibration (Kerkech et al., 2020). Multispectral images may be used for disease identification, assessment of stem water potential, grain yield, nitrogen content, SPAD classification, and physiological stress. The wavelength range of 780–2526 nm was near-infrared. Hyperspectral cameras have a higher spectral resolution in the spectral area than multispectral cameras (Maes and Steppe, 2019). It is employed to identify plant stress and nutritional conditions. To obtain thermal infrared images of plants, an unmanned aerial vehicle (UAV) was equipped with a thermal infrared image sensor. Thermal infrared imaging

**TABLE 13.1**
**Overview of the Most Commonly used Imaging Sensors in High-Throughput Plant Phenotyping**

| Type of Sensor | Operation Principle | Phenotypic Traits | Advantages | Limitations | Applications | Reference |
|---|---|---|---|---|---|---|
| Visible Spectrum Image Sensor | Capture light within the 400–700 nm wavelength range to create an RGB image | Chlorophyll levels, phenological stages, leaf angle, structure and distribution, photosynthetic health, plant height and canopy coverage, leaf length and biomass, Yield characteristics such as total and filled spikelets, grain length and width, 1,000-grain weight, Panicle attributes such as length, number of branches, and grain count, Root system architecture | Affordable, high-resolution, widely available, user-friendly, inexpensive, simple to operate, easy to maintain and capable of rapid data collection | Restricted to three color bands, sensitive to lighting conditions, lacks spectral calibration, and only offers physiological data | Monitoring, stress reactions, and detection of pests and diseases | Zhu et al. (2019) |
| Infrared spectrum image sensor | Capture light within the 780 nm-1000 μm wavelength range to create an RGB image | Canopy temperature, Leaf greenness, leaf color, leaf chlorophyll content, phenological stages, photosynthetic activity, leaf water content, LAI, LAD, shoot temperature | This method provides precise temperature measurement, It is easy to handle and has reflectance properties that correlate with water content, It also offers a wide range of measurement | Sensor calibration and atmospheric correction are required, making it challenging to compare data over time. A reference is needed for accurate comparison, and it's difficult to distinguish between soil and plant temperatures in sparse canopies. Additionally, environmental conditions can easily affect the measurements | Growth monitoring and stress response | Ahmadi et al. (2017) |
| Multispectral image sensor | Capture light at few discrete wide spectral bands | Plant stress, leaf water content, NDVI, biomass, canopy cover and volume, plant diseases and pests, leaf greenness, photosynthetic status, chlorophyll content and leaf color | Simple in image processing, well-established technology, delivers multispectral data | Restricted to a few spectral bands; spectral data requires regular calibration using reference objects; data signal is influenced by camera geometry, lighting conditions and sun angle | Growth monitoring, phenology monitoring, morphological structure capture | Sankaran et al. (2015) |
| Hyperspectral image sensor | Capture light at hundreds of narrow spectral bands | Plant height, canopy coverage and volume, leaf greenness and nitrogen status, leaf/canopy water status, disease and pest presence, photosynthetic rate, nutrient stress, and metabolite levels | Offers high spectral resolution, includes extensive spectral information across multiple bands, can filter out background interference, well-suited for UAV use | High cost, low spatial resolution, and bulk data pose challenges for analysis and storage; affected by optimal lighting conditions; requires frequent sensor calibration; expensive with complex data interpretation | Growth monitoring, biomass estimation, biotic and abiotic stress detection, quality evaluation, and metabolite prediction. | Reynolds et al. (2019) |

*(Continued)*

**TABLE 13.1 (*Continued*)**
**Overview of the Most Commonly used Imaging Sensors in High-Throughput Plant Phenotyping**

| Type of Sensor | Operation Principle | Phenotypic Traits | Advantages | Limitations | Applications | Reference |
|---|---|---|---|---|---|---|
| Fluorescence sensor | It detects chlorophyll through fluorescence | Photosynthetic efficiency, Chlorophyll content, canopy coverage, photosynthetic status, water content, leaf growth and photosynthetic efficiency under multi stress conditions | Broad measurement range; capable of removing background interference; well-suited for UAV deployment | Challenges with fluorescence excitation, restricted field application, requires pre-acclimation conditions, tedious measurement of the canopy level due to low signal-to-noise ratio, needs intense illumination, not suitable for large-scale plants | Growth monitoring, plant stress detection, phenology monitoring | Qiu et al. (2018) |
| LiDAR | It creates a 3D model of the plant | Plant height, leaf length, canopy cover, biomass, canopy volume, LAI, LAD, canopy volume, chlorophyll content, phenology, leaf distribution, leaf angle, photosynthetic status | Captures three-dimensional shapes; ideal for UAV use | Difficult to detect minor variations; costly; some laser scanning instruments require specialized illumination; data processing is time-consuming | Structure capturé, monitoring growth stages | Zhao et al. (2019) |
| Depth Image sensors | It offers an image with spatial depth details | Geometric plant characteristics like shape, length, canopy structure and volume, plant height, leaf length, leaf angle, leaf width, biomass | Cost-effective and offers depth information | Limited resolution, short sensing range, and sensitive to optical interference | 3 D reconstruction of plants to estimate plant traits | Xiang et al. (2019) |
| AS7263 spectral sensors | It operates by measuring light intensity across multiple discrete wavelength bands. | Chlorophyll content, NDVI, plant stress | Portable and compact, high spectral resolution, multispectral data | Limited spectral bands, frequent calibration required, sensitivity to environmental conditions, complex data interpretation | Plant health monitoring, nutrient management, vegetation analysis, precision agriculture | Bäumker et al. (2021) |

(*Continued*)

**TABLE 13.1 (*Continued*)**
**Overview of the Most Commonly used Imaging Sensors in High-Throughput Plant Phenotyping**

| Type of Sensor | Operation Principle | Phenotypic Traits | Advantages | Limitations | Applications | Reference |
|---|---|---|---|---|---|---|
| Low-cost chlorophyll meter | It operates by shining light of specific wavelengths onto a leaf and measuring the amount of light reflected or transmitted | Chlorophyll content, leaf greenness | Affordable, simple and quick, portable, direct measurement | Limited to chlorophyll measurement, potential calibration issues, environmental sensitivity, not suitable for large-scale applications | Fertilization management, Crop monitoring | Hutomo et al. (2016) |
| FLIR lepton | It detects infrared radiation (heat) emitted by objects. | Canopy temperature, heat stress, plant water stress | Early stress detection, compact and lightweight, energy efficient | Low resolution, short sensing distance, sensitive to ambient light conditions, requires frequent calibration | Water stress detection, heat stress monitoring, canopy temperature assessment, early problem detection | Bruno et al. (2017) |
| Raspberry pi-series sensors | It captures visible light images or videos and convert light into electrical signals thus producing digital images | It captures visible light images or videos and convert light into electrical signals thus producing digital images | Versatile imaging, cost effective, customizable, portable | Low resolution, limited field of view processing and storage challenges, light and environmental sensitivity | Plant growth monitoring, visual phenotyping, time-lapse imaging, environmental monitoring | Sangjan et al. (2022) |

sensors help to measure canopy temperatures is possible through the use of highly focused infrared detectors, which are sensitive to certain environmental conditions (Sagan et al., 2019). A laser beam's period of flight provides the basis for a physical measurement known as light detection and ranging or LIDAR (Wallace et al., 2012). The primary applications of fluorescence sensors are the monitoring of plant metabolic data and the detection of agricultural diseases (Barbagallo et al., 2003).

## 13.7 HIGH-THROUGHPUT IMAGE DATA ANALYSIS

The image analysis process varies according to different sensors utilized for various experiments thus concluding a common pipeline is not possible. But for a proper image analysis, several steps to be followed (Souza and Yang, 2021), such as:

**Image Pre-processing:** The initial phase involves preparing raw images for further analysis. This includes clipping or merging the images as necessary, enhancing contrast to improve visibility, reducing noise to eliminate extraneous information, and transforming the spatial characteristics of the image to facilitate subsequent processing steps.

**Image Segmentation:** In this stage, techniques such as thresholding and morphological processing are employed to delineate the objects of interest from the background. This process isolates the relevant components of the image, making them ready for further analysis.

**Feature Extraction:** The objects of interest have been isolated and then proceed towards the extraction of relevant features based on the specific goals of the experiment. This process typically involves the identification of grayscale values, gradients, edges, contours, shapes, sizes, textures, corner points, and color characteristics, among others.

**Feature Pre-processing and Selection:** Given the potentially large volume of raw features generated, it is crucial to employ pre-processing techniques and feature selection methods. These steps are designed to filter and identify the most significant features, thereby refining the dataset for more focused analysis.

**Data Mining:** The final step involves the application of data mining techniques to uncover patterns, relationships, and insights from processed and refined data, leading to meaningful conclusions and knowledge discovery (Figure 13.3).

## 13.8 APPLICATIONS OF HTP

High-throughput phenotyping reduces the time taken to record experimental data in an automated manner, which reduces labor and cost. Additionally, high-throughput phenotyping tools (HTP) record temporal data to precisely analyze the dynamic response of crops to particular stresses and screen a large set of germplasm without risk of error under controlled and field conditions. Estimating biomass at each stage of the crop requires manual destruction of the crop and leads to repeated experimentation but in HTP, through imaging-based pixel calculation, the dynamic growth rate is calculated through modeling (Singh et al., 2023). With automated weighing and watering stations in high-throughput phenomics, daily water use is calculated (Rane et al., 2021), which

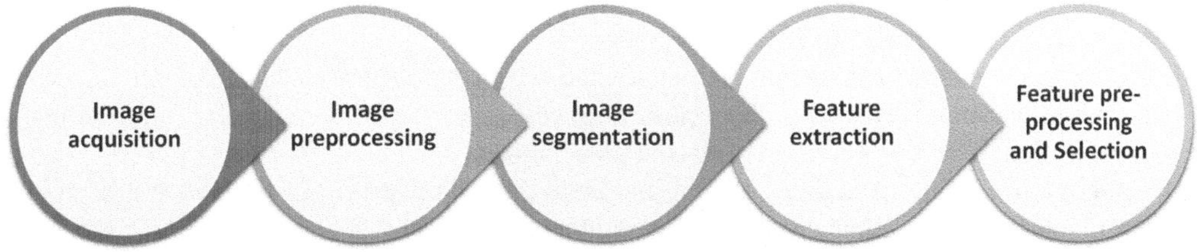


FIGURE 13.3 Sequential steps for high-throughput image data analysis.

helps estimate transpiration and water use efficiency. Which is one of the target traits for screening drought-tolerant genotypes. Using an imaging system helps with proper crop management and reduces yield loss (Dong et al., 2024). Counting spikelets was easier with the YOLOv8-seg model, which satisfies the practical demands for acquiring trait parameters of wheat spikes and significantly advances intelligent, noninvasive spike phenotyping approaches.

## 13.9 ADVANCES IN HIGH-THROUGHPUT PHENOTYPING

Advances in plant high-throughput phenotyping (HTP) have significantly transformed agricultural research. High-throughput phenotyping, Artificial Intelligence (AI) and advanced imaging technologies have become effective tools for crop development. These tools make it possible for scientists and plant breeders to quickly and precisely evaluate the characteristics and output of vast plant populations, resulting in more effective and focused breeding techniques. HTP platforms make use of machine vision algorithms and imaging technology to measure and analyze multiple traits across thousands of plants quickly. This makes it possible for scientists and breeders to carry out extensive screening tests, evaluate the effectiveness of various germplasm, target desired features, and quicken the breeding process (Sheikh et al., 2024).

### 13.9.1 Machine Learning

Machine Learning (ML) is the term for the collection of automated modeling techniques that can recognize patterns in data and use such patterns to automatically generate judgments without the need for explicit rule programming. The basic principle behind machine learning is to identify trends, patterns, or differences in data that can be utilized to correctly categorize or explain a new event using past experiences or examples. The capacity of ML technologies to extrapolate similarities or patterns from the provided information is one of its primary features. ML-enabled HTSP enables accurate, precise, and quick screening of various stresses, it will be beneficial to plant breeders, pathologists, physiologists, extension agents, entomologists, and farmers. This would also speed up the process of finding new genes and make it easier to introduce innovative selection strategies for complex quantitative parameters like yield, and biotic and abiotic stresses. An inherently multidisciplinary approach to data analysis that heavily integrates ideas from optimization, probability theory, statistics, decision theory, and visualization. When a large amount of data is available, machine learning techniques can help connect inputs (such as image data) to the desired number of outputs. By examining several aspects simultaneously, as opposed to studying each feature (trait) separately, physiologists, plant breeders, and biologists can examine enormous databases to identify patterns and influence discoveries. This is one of the primary advantages of applying ML techniques. This was formerly a significant bottleneck due to the high dimensionality of individual images and their large number, which makes it very challenging to evaluate them using traditional methods (Singh et al., 2016). The key roles of ML in HTP include.

**Data Processing:** ML algorithms can process extensive datasets, identifying relevant features that correlate with specific phenotypic traits. For example, regression models can predict traits like plant height based on spectral data, whereas classification algorithms can categorize plants based on growth stages (Mir et al., 2019).

**Pattern Recognition:** ML techniques excel at discovering patterns within complex datasets. Unsupervised learning methods, such as clustering, can group plants with similar traits, aiding in the identification of genetic variations and trait associations (Yang et al., 2020).

**Predictive Modeling:** Machine learning facilitates the development of predictive models that forecast plant performance under various environmental conditions. By training models on historical data, researchers can make informed decisions about breeding strategies and crop management practices (Gill et al., 2022)

### 13.9.2 Deep Learning

Deep learning (DL) is a branch of machine learning that employs several nonlinear transformations to create additional levels of abstraction in data using a complicated model. The word "deep" in "deep learning" highlights the many hidden layers of deep learning algorithms (i.e., significant credit assignment route or CAP depth) that constitute the basis of transformed data. The term "CAP depth" describes a series of operations applied to the input data to produce output. The key aspects of deep learning are

**Convolutional Neural Networks (CNNs):** CNNs function specifically well in HTP image analysis tasks. They can automatically extract features from images captured by drones or high-resolution cameras, enabling accurate identification and quantification of traits such as leaf area and disease symptoms. CNNs minimize the requirement for manual feature extraction by learning hierarchical features from raw data (Wang et al., 2019)

**Temporal Analysis with Recurrent Neural Networks (RNNs):** For handling sequential data, RNNs and Long Short-Term Memory (LSTM) networks are effective. In HTP, these networks can analyze time series data to capture dynamic growth patterns over time, which is essential for understanding plant responses to environmental changes (Arya et al., 2022).

**Transfer Learning:** Using this method, researchers can modify pre-trained models for certain phenotyping tasks by using them on huge datasets. Transfer learning reduces the time and resources needed to train models from scratch, making it particularly useful in scenarios with limited labeled data (Gill et al., 2022).

## 13.10 PROSPECTS

With the advancement of high-throughput technology, plant phenotyping is on the verge of the arrival of the big data era. Individual phenotype statistics alone are inadequate for conducting association studies. As a result, future research will rely heavily on comprehensive phenomics data and the interaction of multiple scales, including physiology, structure, omics, genetics, and environmental factors. To address emerging environmental challenges, future advancements in plant phenotyping need to focus on the development of innovative cost-effective technologies driven by artificial intelligence and remote sensing for image-based phenotyping. In the context of big data, there is a need for the development of processing software capable of handling effective data analysis. The integration of advanced technologies like machine learning and artificial intelligence, into the sensing and control of phenotyping platforms is essential for better comprehension and more effective extraction of phenotyping traits. To achieve high-quality real-time data processing, novel data processing algorithms should be employed to enhance the speed and accuracy of automated information processing in production environments with varying light conditions and backgrounds. The design of mobile platform structures must be innovative to improve stability, reduce vibration, and ensure precise data collection over challenging terrains. Flexible platforms equipped with automatic control devices or modular mechanisms can adapt to various environments and perform tasks related to phenotypic data collection for different crops, accommodating diverse planting systems, such as plant height, row spacing, and field layout. To acquire phenotypic information effectively, a multi-dimensional system that integrates ground-based and aerial platforms for data collection should be established. This could involve the use of swarm robots or other autonomous ground-based acquisition systems to enhance the throughput of multi-scale monitoring systems and ensure comprehensive spatial data coverage. Strengthening the collaboration between phenotyping platform developers and sensor manufacturers is crucial for establishing a unified, open platform and sensor standard and facilitating wider access to secondary development. The proposed collaboration will also support the integration of multiple sensors and intelligent data acquisition. Consequently, research is needed to support the rapid advancement of genetic improvement in breeding programs as well as the development of reliable, automated, comprehensible, and versatile high-throughput phenotyping procedures.

## 13.11 CONCLUSION

The transition from traditional phenotyping methods to high-throughput phenotyping (HTP) driven by innovative technologies like image processing, machine learning, and artificial intelligence has significantly revolutionized the field. The integration of these technologies has enabled researchers to collect large datasets with high precision and speed, enabling a deeper understanding of plant performance and the intricate interplay between genotype and phenotype. The multifaceted applications of HTP across diverse plant characteristics, including growth, yield, stress tolerance, and nutrient status. It discusses the various sensor technologies employed in HTP, from visible spectrum and infrared sensors to hyperspectral and fluorescence imaging. The document also delves into the specific advantages of each sensor technology, highlighting their unique contributions to phenotype analysis. Machine Learning algorithms streamline data processing, facilitate pattern recognition, and enable the development of predictive models for forecasting plant performance. The integration of deep learning techniques, such as CNNs and RNNs, further enhances the accuracy and efficiency of HTP data analysis.

## REFERENCES

Aeberli, A., Phinn, S., Johansen, K., Robson, A. & Lamb, D. W. (2023). Characterisation of banana plant growth using high-spatiotemporal-resolution multispectral UAV imagery. *Remote Sensing*, *15*(3), 679.

Ahmadi, S. H., Agharezaee, M., Kamgar-Haghighi, A. A. & Sepaskhah, A. R. (2017). Comparing canopy temperature and leaf water potential as irrigation scheduling criteria of potato in water-saving irrigation strategies. *International Journal of Plant Production*, *11*(2), 333–348.

Al-Turjman, F. (2019). The road towards plant phenotyping via WSNs: An overview. *Computers and Electronics in Agriculture*, *161*, 4–13.

Araus, J. L. & Cairns, J. E. (2014). Field high-throughput phenotyping: The new crop breeding frontier. *Trends in Plant Science*, *19*(1), 52–61.

Arya, S., Sandhu, K. S., Singh, J. & Kumar, S. (2022). Deep learning: As the new frontier in high-throughput plant phenotyping. *Euphytica*, *218*(4), 47.

Awada, L., Phillips, P. W. & Bodan, A. M. (2024). The evolution of plant phenomics: Global insights, trends, and collaborations (2000–2021). *Frontiers in Plant Science*, *15*, 1410738.

Barbagallo, R. P., Oxborough, K., Pallett, K. E. & Baker, N. R. (2003). Rapid, non invasive screening for perturbations of metabolism and plant growth using chlorophyll fluorescence imaging. *Plant Physiology*, *132*(2), 485–493.

Bäumker, E., Zimmermann, D., Schierle, S. & Woias, P. (2021). A novel approach to obtain PAR with a multi-channel spectral microsensor, suitable for sensor node integration. *Sensors*, *21*(10), 3390.

Behmann, J., Acebron, K., Emin, D., Bennertz, S., Matsubara, S., Thomas, S. & Rascher, U. (2018). Specim IQ: Evaluation of a new, miniaturized handheld hyperspectral camera and its application for plant phenotyping and disease detection. *Sensors*, *18*(2), 441.

Briglia, N., Montanaro, G., Petrozza, A., Summerer, S., Cellini, F. & Nuzzo, V. (2019). Drought phenotyping in Vitis vinifera using RGB and NIR imaging. *Scientia Horticulturae*, *256*, 108555.

Bruno, H., Narciso, M. G., Silva, G. O. & de Souza, M. A. (2017). Low-cost thermal imager for plant phenotyping. In *Latin-American Conference on Plant Phenotyping and Phenomics for Plant Breeding*. São Carlos: Embrapa Instrumentação.

Chandra, A. L., Desai, S. V., Guo, W. & Balasubramanian, V. N. (2020). *Computer Vision with Deep Learning for Plant Phenotyping in Agriculture: A Survey*, 2006.11391.

Chee, K. Y. & Zhong, Z. W. (2013). Control, navigation and collision avoidance for an unmanned aerial vehicle. *Sensors and Actuators A: Physical*, *190*, 66–76.

Chen, D., Neumann, K., Friedel, S., Kilian, B., Chen, M., Altmann, T. & Klukas, C. (2014). Dissecting the phenotypic components of crop plant growth and drought responses based on high-throughput image analysis. *The Plant Cell*, *26*(12), 4636–4655.

Collis, R. T. H. (1970). Lidar. *Applied Optics*, *9*(8), 1782–1788.

Deery, D., Jimenez-Berni, J., Jones, H., Sirault, X. & Furbank, R. (2014). Proximal remote sensing buggies and potential applications for field-based phenotyping. *Agronomy*, *4*(3), 349–379.

Eitel, J. U., Magney, T. S., Vierling, L. A., Brown, T. T. & Huggins, D. R. (2014). LiDAR based biomass and crop nitrogen estimates for rapid, non-destructive assessment of wheat nitrogen status. *Field Crops Research*, *159*, 21–32.

Fahlgren, N., Gehan, M. A. & Baxter, I. (2015). Lights, camera, action: High-throughput plant phenotyping is ready for a close-up. *Current Opinion in Plant Biology*, *24*, 93–99.

Feng, L., Chen, S., Zhang, C., Zhang, Y. & He, Y. (2021). A comprehensive review on recent applications of unmanned aerial vehicle remote sensing with various sensors for high-throughput plant phenotyping. *Computers and Electronics in Agriculture*, *182*, 106033.

Ghazali, M. F., Wikantika, K., Harto, A. B. & Kondoh, A. (2020). Generating soil salinity, soil moisture, soil pH from satellite imagery and its analysis. *Information Processing in Agriculture*, *7*(2), 294–306.

Gill, T., Gill, S. K., Saini, D. K., Chopra, Y., de Koff, J. P. & Sandhu, K. S. (2022). A comprehensive review of high throughput phenotyping and machine learning for plant stress phenotyping. *Phenomics*, *2*(3), 156–183

Golzarian, M. R., Frick, R. A., Rajendran, K., Berger, B., Roy, S., Tester, M. & Lun, D. S. (2011). Accurate inference of shoot biomass from high-throughput images of cereal plants. *Plant Methods*, *7*, 1–11.

Hairmansis, A., Berger, B., Tester, M. & Roy, S. J. (2014). Image-based phenotyping for non-destructive screening of different salinity tolerance traits in rice. *Rice*, *7*, 1–10.

Honsdorf, N., March, T. J., Berger, B., Tester, M. & Pillen, K. (2014). High-throughput phenotyping to detect drought tolerance QTL in wild barley introgression lines. *PLoS One*, *9*(5), e97047.

Hughes, A., Askew, K. & Scotson, C. P. (2017) Non-destructive, high-content analysis of wheat grain traits using X-ray micro computed tomography. *Plant Methods*, *13*, 76.

Jangra, S., Chaudhary, V., Yadav, R. C. & Yadav, N. R. (2021). High-throughput phenotyping: A platform to accelerate crop improvement. *Phenomics*, *1*(2), 31–53.

Jiang, N. & Zhu, X. G. (2024). Modern phenomics to empower holistic crop science, agronomy, and breeding research. *Journal of Genetics and Genomics.*

Jing, R., Gong, Z., Zhao, W., Pu, R. & Deng, L. (2017). Above-bottom biomass retrieval of aquatic plants with regression models and data acquired by a UAV platform – A case study in Wild Duck Lake Wetland, Beijing, China. *ISPRS Journal of Photogrammetry and Remote Sensing*, *134*, 122–134.

Kerkech, M., Hafiane, A. & Canals, R. (2020). Vine disease detection in UAV multispectral images using optimized image registration and deep learning segmentation approach. *Computers and Electronics in Agriculture*, *174*, 105446.

Kitić, G., Tagarakis, A., Cselyuszka, N., Panić, M., Birgermajer, S., Sakulski, D. & Matović, J. (2019). A new low-cost portable multispectral optical device for precise plant status assessment. *Computers and Electronics in Agriculture*, *162*, 300–308.

Kumar, J., Pratap, A. & Kumar, S. (2015). Plant phenomics: An overview. *Phenomics in Crop Plants: Trends, Options and Limitations*, 1–10.

Li, L., Zhang, Q. & Huang, D. (2014). A review of imaging techniques for plant phenotyping. *Sensors*, *14*(11), 20078–20111.

Lin, Y. (2015). LiDAR: An important tool for next-generation phenotyping technology of high potential for plant phenomics? *Computers and Electronics in Agriculture*, *119*, 61–73.

Maes, W. H. & Steppe, K. (2019). Perspectives for remote sensing with unmanned aerial vehicles in precision agriculture. *Trends in Plant Science*, *24*(2), 152–164.

Metzner, R., Eggert, A., van Dusschoten, D., Pflugfelder, D., Gerth, S., Schurr, U. & Jahnke, S. (2015). Direct comparison of MRI and X-ray CT technologies for 3D imaging of root systems in soil: Potential and challenges for root trait quantification. *Plant Methods*, *11*, 1–11.

Mir, R. R., Reynolds, M., Pinto, F., Khan, M. A. & Bhat, M. A. (2019). High-throughput phenotyping for crop improvement in the genomics era. *Plant Science*, *282*, 60–72.

Murray, S. C., Knox, L., Hartley, B., Méndez-Dorado, M. A., Richardson, G., Thomasson, J. A. & Rooney, W. L. (2016). High clearance phenotyping systems for season-long measurement of corn, sorghum and other row crops to complement unmanned aerial vehicle systems. *Autonomous Air and Ground Sensing Systems for Agricultural Optimization and Phenotyping*, *9866*, 14–21.

Neilson, E. H., Edwards, A. M., Blomstedt, C. K., Berger, B., Møller, B. L. & Gleadow, R. M. (2015). Utilization of a high-throughput shoot imaging system to examine the dynamic phenotypic responses of a C4 cereal crop plant to nitrogen and water deficiency over time. *Journal of Experimental Botany*, *66*(7), 1817–1832.

Omari, M. K., Lee, J., Faqeerzada, M. A., Joshi, R., Park, E. & Cho, B. K. (2020). Digital image-based plant phenotyping: A review. *Korean Journal of Agricultural Science*, *47*(1), 119–130.

Pasala, R. & Pandey, B. B. (2020). Plant phenomics: High-throughput technology for accelerating genomics. *Journal of Biosciences*, *45*(1), 111.

Perez-Sanz, F., Navarro, P. J. & Egea-Cortines, M. (2017). Plant phenomics: An overview of image acquisition technologies and image data analysis algorithms. *GigaScience*, *6*(11).

Piovesan, A., Vancauwenberghe, V., Van De Looverbosch, T., Verboven, P. & Nicolaï, B. (2021). X-ray computed tomography for 3D plant imaging. *Trends in Plant Science*, *26*(11), 1171–1185.

Pound, M. P., Atkinson, J. A., Townsend, A. J., Wilson, M. H., Griffiths, M., Jackson, A. S. & French, A. P. (2017). Deep machine learning provides state-of-the-art performance in image-based plant phenotyping. *Gigascience*, *6*(10), 1–10.

Prashar, A., Yildiz, J., McNicol, J. W., Bryan, G. J. & Jones, H. G. (2013). Infra-red thermography for high throughput field phenotyping in Solanum tuberosum. *PLoS One*, *8*(6), e65816. https://doi.org/10.1371/journal.pone.0065816

Qiu, R., Wei, S., Zhang, M., Li, H., Sun, H., Liu, G. & Li, M. (2018). Sensors for measuring plant phenotyping: A review. *International Journal of Agricultural and Biological Engineering*, *11*(2), 1–17.

Rahman, H., Ramanathan, V., Jagadeeshselvam, N., Ramasamy, S., Rajendran, S., Ramachandran, M. & Muthurajan, R. (2015). Phenomics: Technologies and applications in plant and agriculture. *PlantOmics: The Omics of Plant Science*, 385–411.

Raju, S. K. K., Thompson, A. M. & Schnable, J. C. (2020). Advances in plant phenomics: From data and algorithms to biological insights. *Applications in Plant Sciences*, *8*(8), e11386. https://doi.org/10.1002/aps3.11386

Reynolds, D., Ball, J., Bauer, A., Davey, R., Griffiths, S. & Zhou, J. (2019). CropSight: A scalable and open-source information management system for distributed plant phenotyping and IoT-based crop management. *Gigascience*, *8*(3), giz009.

Römer, C., Wahabzada, M., Ballvora, A., Pinto, F., Rossini, M., Panigada, C. & Plümer, L. (2012). Early drought stress detection in cereals: Simplex volume maximisation for hyperspectral image analysis. *Functional Plant Biology*, *39*(11), 878–890.

Rousseau, C., Belin, E., Bove, E., Rousseau, D., Fabre, F., Berruyer, R. & Boureau, T. (2013). High throughput quantitative phenotyping of plant resistance using chlorophyll fluorescence image analysis. *Plant Methods*, *9*, 1–13.

Sagan, V., Maimaitijiang, M., Sidike, P., Eblimit, K., Peterson, K. T., Hartling, S. & Mockler, T. (2019). UAV-based high resolution thermal imaging for vegetation monitoring, and plant phenotyping using ICI 8640 P, FLIR Vue Pro R 640, and thermomap cameras. *Remote Sensing*, *11*(3), 330.

Sambrekar, K. & Rajpurohit, V. S. (2019). Fast and efficient multiview access control mechanism for cloud based agriculture storage management system. *International Journal of Cloud Applications and Computing (IJCAC)*, *9*(1), 33–49.

Sangha, J. S. (2022). A review on sensing technologies for high-throughput plant phenotyping. *IEEE Open Journal of Instrumentation and Measurement,* 1, 1–21. https://doi.org/10.1109/OJIM.2022.3178468

Sangjan, W., Carter, A. H., Pumphrey, M. O., Jitkov, V. & Sankaran, S. (2021). Development of a raspberry pi-based sensor system for automated in-field monitoring to support crop breeding programs. *Inventions*, *6*(2), 42.

Sankaran, S., Khot, L. R., Espinoza, C. Z., Jarolmasjed, S., Sathuvalli, V. R., Vandemark, G. J. & Pavek, M. J. (2015). Low-altitude, high-resolution aerial imaging systems for row and field crop phenotyping: A review. *European Journal of Agronomy*, *70*, 112–123.

Seelig, H. D., Hoehn, A., Stodieck, L. S., Klaus, D. M., Adams Iii, W. W. & Emery, W. J. (2008). The assessment of leaf water content using leaf reflectance ratios in the visible, near-, and short-wave-infrared. *International Journal of Remote Sensing*, *29*(13), 3701–3713.

Seiffert, U., Bollenbeck, F., Mock, H. P. & Matros, A. (2010). Clustering of crop phenotypes by means of hyperspectral signatures using artificial neural networks. In *2010 2nd Workshop on Hyperspectral Image and Signal Processing: Evolution in Remote Sensing*, 1–4. Reykjavik, Iceland. https://doi.org/10.1109/WHISPERS.2010.5594947

Shaahan, M. M., El-Sayed, A. A. & Abou El-Nour, E. A. A. (1999). Predicting nitrogen, magnesium and iron nutritional status in some perennial crops using a portable chlorophyll meter. *Scientia Horticulturae*, *82*(3–4), 339–348.

Sheikh, M., Iqra, F., Ambreen, H., Pravin, K. A., Ikra, M. & Chung, Y. S. (2024). Integrating artificial intelligence and high-throughput phenotyping for crop improvement. *Journal of Integrative Agriculture*, *23*(6), 1787–1802.

Shippert, P. (2003). Introduction to hyperspectral image analysis. *Online Journal of Space Communication*, *2*(3), 8.

Singh, A., Ganapathysubramanian, B., Singh, A. K. & Sarkar, S. (2016). Machine learning for high-throughput stress phenotyping in plants. *Trends in Plant Science*, *21*(2), 110–124.

Singh, A., Kushwaha, S. K. P., Nandy, S., Padalia, H., Ghosh, S., Srivastava, A., Kumari, N. 2023. Aboveground forest biomass estimation by the integration of TLS and ALOS PALSAR data using machine learning. *Remote Sensing*, *15*, 1143.

Sliney, D. H. (2016). What is light? The visible spectrum and beyond. *Eye*, *30*(2), 222–229.

Song, P., Wang, J., Guo, X., Yang, W. & Zhao, C. (2021). High-throughput phenotyping: Breaking through the bottleneck in future crop breeding. *The Crop Journal*, *9*(3), 633–645.

Sun, S., Li, C., Paterson, A. H., Jiang, Y., Xu, R., Robertson, J. S. & Chee, P. W. (2018). In-field high throughput phenotyping and cotton plant growth analysis using LiDAR. *Frontiers in Plant Science*, *9*, 16.

Thakur, P., Dogra, P., Bairwa, S. K., Meena, S. K.,& Chandra, K. (2024). Modern phenotyping: A paradigm of present need in crop improvement. *International Journal of Plant & Soil Science*, *36*(1), 24–37.

Thapa, S., Zhu, F., Walia, H., Yu, H. & Ge, Y. (2018). A novel LiDAR-based instrument for high-throughput, 3D measurement of morphological traits in maize and sorghum. *Sensors*, *18*(4), 1187.

Tsaftaris, S. A. & Noutsos, C. (2009). Plant phenotyping with low-cost digital cameras and image analytics. In *Information Technologies in Environmental Engineering: Proceedings of the 4th International ICSC Symposium Thessaloniki,* Greece, May 28–29, 2009, 238–251.

Wallace, L., Lucieer, A., Watson, C. & Turner, D. (2012). Development of a UAV-LiDAR system with application to forest inventory. *Remote Sensing*, *4*(6), 1519–1543.

Walter, A., Liebisch, F. & Hund, A. (2015). Plant phenotyping: From bean weighing to image analysis. *Plant Methods*, *11*, 1–11.

Wang, X., Xuan, H., Evers, B., Shrestha, S., Pless, R. & Poland, J. (2019). High-throughput phenotyping with deep learning gives insight into the genetic architecture of flowering time in wheat. *GigaScience*, *8*(11), giz120.

Wu, H. H. & Bainbridge-Smith, A. (2011). *Advantages of Using a Kinect Camera in Various Applications.* University of Canterbury, 1–4.

Xia, C., Wang, L., Chung, B. K. & Lee, J. M. (2015). In situ 3D segmentation of individual plant leaves using an RGB-D camera for agricultural automation. *Sensors*, *15*(8), 20463–20479.

Xu, R., Li, C. & Paterson, A. H. (2019). Multispectral imaging and unmanned aerial systems for cotton plant phenotyping. *PLoS One*, *14*(2), e0205083.

Yang, S., Yang, X. & Mo, J. (2018). The application of unmanned aircraft systems to plant protection in China. *Precision Agriculture*, *19*, 278–292.

Yang, W., Feng, H., Zhang, X., Zhang, J., Doonan, J. H., Batchelor, W. D. & Yan, J. (2020). Crop phenomics and high-throughput phenotyping: Past decades, current challenges, and future perspectives. *Molecular Plant*, *13*(2), 187–214.

Yang, W., Yang, C., Hao, Z., Xie, C. & Li, M. (2019). Diagnosis of plant cold damage based on hyperspectral imaging and convolutional neural network. *IEEE Access*, *7*, 118239–118248.

Zaman, B., Jensen, A., Clemens, S. R. & McKee, M. (2014). Retrieval of spectral reflectance of high-resolution multispectral imagery acquired with an autonomous unmanned aerial vehicle. *Photogrammetric Engineering & Remote Sensing*, *80*(12), 1139–1150.

Zaman-Allah, M., Vergara, O., Araus, J. L., Tarekegne, A., Magorokosho, C., Zarco-Tejada, P. J. & Cairns, J. (2015). Unmanned aerial platform-based multi-spectral imaging for field phenotyping of maize. *Plant Methods*, *11*, 1–10.

Zhang, Y., Du, J., Ma, L., Pan, X., Wang, J. & Guo X. (2018). Three-dimensional segmentation, reconstruction and phenotyping analysis of maize kernel based on Micro-CT images, *Fresenius Environmental Bulletin*, *27*, 3965–3969.

Zhang, Y., Legay, S., Barrière, Y., Méchin, V., & Legland, D. (2013). Color quantification of stained maize stem section describes lignin spatial distribution within the whole stem. *Journal of Agricultural and Food Chemistry*, 61(13), 3186–3192.

Zhao, C., Zhang, Y., Du, J., Guo, X., Wen, W., Gu, S. & Fan, J. (2019a). Crop phenomics: Current status and perspectives. *Frontiers in Plant Science*, *10*, 714.

Zhao, F., Jiang, H. & Liu, Z. (2019b). Recent development of automotive LiDAR technology, industry and trends. In *Eleventh International Conference on Digital Image Processing (ICDIP 2019)* (Vol. *11179*, pp. 1132–1139). SPIE, Guangzhou, China.

Zheng, H., Zhou, X., He, J., Yao, X., Cheng, T., Zhu, Y. & Tian, Y. (2020). Early season detection of rice plants using RGB, NIR-GB and multispectral images from unmanned aerial vehicle (UAV). *Computers and Electronics in Agriculture*, *169*, 105223.

Zhu, Y., Lei, X., Wang, K. X. & Yu, Z. (2019). Compact CMOS spectral sensor for the visible spectrum. *Photonics Research*, *7*(9), 961–966.

# 14 Technical Advancements in Plant High-Throughput Phenotyping and Phenomics

## *An Overview*

*Sarfraz Ahmad, Maksud Hasan Shah, Mohammad Ghani Raghib, Saniya, Vikas Belwal, Shokat Ali, Rubina Khan, Shoumik Saha, and Sk Naim Aktar*

## 14.1 INTRODUCTION

Danish plant physiologist Wilhelm Johannsen coined the terms "genotype" and "phenotype" in 1911. The word phenotype is derived from the Greek phainein (meaning show) and typos (meaning type) (Johannsen, 1911), which means the external appearance of an individual. The precise methods and procedures used to measure the phenotypic features of plants are known as plant phenotyping. Plant phenomics is an interdisciplinary field that focuses on accurately acquiring and analyzing multidimensional phenotypes at high-throughput, organism-wide scale via crop growth (Yang et al., 2020). Phenomics, also known as phenometrics, was introduced as a new science in human genomic investigations to link clinical phenotypes with genotypes. In some literature, phenomics is also interchangeable with crop plant high-throughput phenotyping (HTP) technology. High-throughput phenotyping (HTP) refers to the rapid and accurate measurement of a large number of plant traits using automated and semi-automated systems. HTP enables the assessment of plant phenotypes on a scale with speed previously unattainable with traditional methods (Li et al., 2021).

High-throughput plant phenotyping technology, an interdisciplinary marvel, is essential for converting traditional crop breeding operations to modern breeding approaches. Climate-smart crops could enhance agricultural production and sustainability, but traditional breeding programs rely on time-consuming, labor-intensive, and subjective phenotypic observations. Plant HTP technology aims to bridge this gap by providing high-density genomic data from emerging sequencing technology and low-density phenomics data from traditional methods, reducing the development time of new crop cultivars. Plant HTP technology is projected to overcome significant crop breeding limits using an interdisciplinary approach that includes plant science, molecular genetics, engineering, bioinformatics, high-performance computing, and artificial intelligence technology.

The continuous growth of the global population and the corresponding demand for food have necessitated significant advancements in agricultural research. Traditional methods of crop improvement are no longer sufficient to meet these demands. Modern agricultural practices rely heavily on high-throughput phenotyping and phenomics to enhance crop yields, improve stress resistance, and ensure food security.

The rapid advancement of phenomics is necessary to maintain and increase crop yields with the advantages of recent genetics technologies. Modern phenomics tools aim to collect data


DOI: 10.1201/9781003505631-14

on traits like plant growth and development, plant stature, biomass, etc., for many plants in a very short duration. The advent of high-throughput phenotyping platforms has revolutionized the screening of many traits in large numbers of germplasms, coupled with reduced labor input made possible by automation and high-throughput data (image) analysis pipelines (Mir et al., 2019).

This chapter provides an overview of the technical advancements in plant high-throughput phenotyping (HTP) and phenomics, highlighting key technologies, methodologies, and their applications in contemporary agricultural research. Moreover, this chapter highlights the use of high-throughput phenomics for trait dissection in various crops and detecting trait-specific genes in different crop plants.

## 14.2 FORWARD AND REVERSE PHENOMICS

Forward and reverse phenomics are two powerful approaches to studying plant phenotypes. These approaches mirror the concepts of forward and reverse genetics and are used to explore the relationship between genotype and phenotype differently.

Forward phenomics involves starting with a specific phenotype or set of phenotypic traits and working towards identifying these traits' underlying genetic or molecular basis. The process begins with the observation of a phenotype of interest in a population, followed by extensive phenotypic characterization and, ultimately, the identification of the genetic factors that contribute to the observed traits (Figure 14.1). Reverse phenomics begins with a known genetic modification or variation, such as a specific gene or allele, and investigates the resulting phenotypic effects (Pasala & Pandey, 2020).

Forward phenomics is essential for discovering new genes associated with important agricultural traits, while reverse phenomics provides insights into the specific functions of genes and their contributions to plant biology (Chen et al., 2022). Together, these approaches enhance our understanding of plant genetics and contribute to the development of improved crop varieties.

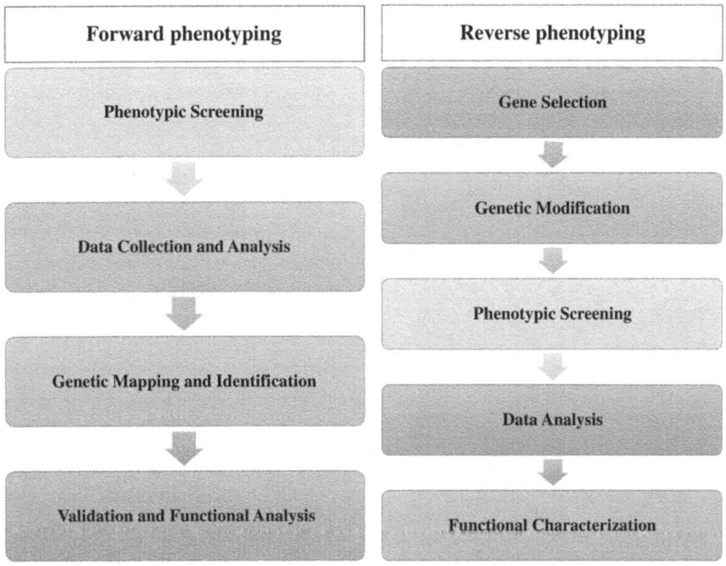


**FIGURE 14.1** The steps involved in the forward and reverse phenotyping process. In both approaches, high-throughput phenotyping (HTP) technologies are often employed at the phenotypic screening stage to measure and record phenotypic data across many individuals efficiently.

## 14.3 PRINCIPLES OF HIGH-THROUGHPUT PLANT PHENOTYPING AND PHENOMICS

A high-throughput plant phenotyping (HTP) system combines advanced sensor systems, automation platforms, and data analytics to collect high-resolution plant phenotypic data in various environments. It continuously measures a large population of plants using automated or robotic platforms. Automation systems like the robotic gantry, mobile, UAVs, and airplanes integrate plant phenotypic data with genomic and environmental data to identify genes associated with agronomic traits (Sarkar et al., 2023). Advanced data analytics technologies like machine learning and deep learning are used to develop predictive models (Atefi et al., 2021).

Plant HTP systems use sensors to collect crop phenotypic data, which can be classified as contact or noncontact sensors based on whether they contact plants. HTP systems' sensors must gather plant data rapidly, remotely, and without causing damage. Optical sensors, including digital, multispectral, and depth cameras, are the most often used sensors for measuring crop attributes at all sizes (Omia et al., 2023).

## 14.4 KEY TECHNOLOGIES IN HIGH-THROUGHPUT PHENOTYPING

### 14.4.1 Imaging Technologies

Imaging technologies are central to plant high-throughput phenotyping (HTP), allowing for the non-destructive, rapid, and accurate measurement of various plant traits across large populations. These technologies capture detailed plant morphology, physiology, and health information, providing insights essential for crop improvement, stress assessment, and functional genomics studies (Kaur et al., 2021). Pictures from widely used image sensors in high-throughput phenotyping are displayed in Figure 14.2. Below is an overview of the key imaging technologies used in plant HTP.

**FIGURE 14.2** Images obtained from commonly used imaging technologies in high-throughput phenotyping (a) Visible light imaging (RGB), (b) Multispectral imaging, (c) Hyperspectral imaging, (d) Thermal imaging, (e) Fluorescence imaging, (f) 3D imaging (LiDAR), (g) Tomographic imaging of root using fast neutron (Yu et al., 2014; Mahlein, 2016; Cen et al., 2017; Jin et al., 2021; Totzke et al., 2021).

#### 14.4.1.1 Visible Light Imaging

Visible light imaging is the most fundamental and widely used imaging technique in HTP. It detects light in the 400–700 nm wavelength spectrum, providing reflected RGB (red, green, and blue) values (Tang et al., 2023). It captures images of plants using standard digital cameras, producing high-resolution photographs that can be analyzed to extract various phenotypic traits. High-resolution RGB images accurately measure plant biomass, economic yield, root architecture, germination rate, growth rate, and biotic and abiotic stress detection (Abebe et al., 2023; Kolhar & Jagtap, 2023).

Recent advancements include the development of automated image analysis software that can process large datasets with minimal human intervention. Machine learning algorithms are increasingly being integrated to improve the accuracy of trait extraction and pattern recognition (Zhang et al., 2023).

#### 14.4.1.2 Multispectral Imaging

It extends beyond the visible spectrum, capturing data across a few discrete wavelengths, including visible and near-infrared (NIR) bands. This provides more detailed information about the physiological status of plants. Multispectral imaging can assess chlorophyll content, plant water content and nutrient levels, enabling plant health assessment (Koh et al., 2021). Moreover, early detection of stress factors such as drought, nutrient deficiency, or disease is possible by analyzing specific wavelength bands.

The development of portable multispectral cameras, including those mounted on UAVs, has increased data collection's spatial and temporal resolution. Enhanced image processing algorithms now allow for real-time analysis of multispectral data, facilitating quick decision-making in the field (Guo et al., 2021).

#### 14.4.1.3 Hyperspectral Imaging

Hyperspectral imaging captures data across a continuous range of wavelengths, often from the visible to the infrared spectrum. Hyperspectral imaging is a powerful tool for analyzing chemical composition, detecting stress and disease, and phenotyping for complex traits. It can detect subtle differences in chlorophyll, lignin, and protein content and identify early signs of biotic and abiotic stress before they become visible (Saric et al., 2022). Compared to multispectral imaging, which uses images from a few different spectra, hyperspectral imaging uses images from many different spectra.

Recent advancements in mini-sized and airborne and spaceborne hyperspectral sensors, such as Headwall Micro-Hyperspec and Cubert UHD 185-Firefly, make hyperspectral imaging more accessible for agricultural applications (Lu et al., 2020). Using open-source software and commercial hardware, a low-cost, lightweight, and deployable data-collecting device for the open-source hyperspectral imager (OpenHSI) was created (Mao et al., 2023). The device offered a low-cost hyperspectral data remote sensing method and was shown on a Matrice M600 drone in Sydney, Australia (Mao et al., 2023).

#### 14.4.1.4 Thermal Imaging

Thermal imaging measures the infrared radiation emitted by plants, which correlates with their temperature. Since plant temperature is closely related to transpiration and water status, thermal imaging is a powerful tool for assessing plant water use and detecting stress. It is crucial for detecting water stress, measuring transpiration rates, and monitoring plant responses to heat stress. It helps detect drought stress, understand water use efficiency, and identify tolerant varieties (Vieira & Ferrarezi, 2021; Jangra et al., 2021). Advanced thermal imaging technology has led to the development of more sensitive and higher-resolution cameras capable of detecting minute temperature differences.

#### 14.4.1.5 Fluorescence Imaging

It involves detecting light emitted by chlorophyll and other fluorescent molecules in plant tissues when excited by specific wavelengths of light. This technique provides information about the photosynthetic efficiency, stress detection, nutrient deficiency and health of plants (Adhikari et al., 2019; Kumar et al., 2021). High-throughput fluorescence imaging systems have been developed for laboratory and field conditions, enabling rapid screening of large plant populations.

#### 14.4.1.6 3D Imaging and LiDAR

3D imaging technologies, including LiDAR (Light Detection and Ranging), capture detailed three-dimensional representations of plants. LiDAR uses laser pulses to measure distances, generating precise 3D models of plant structures (Omasa et al., 2007). 3D imaging provides detailed information about canopy architecture, including plant height, leaf angle, and overall biomass. Traits such as leaf area, stem diameter, and branching patterns can be quantified using 3-D imaging data (Lin, 2015; Liu et al., 2020a).

Recent advancements in LiDAR integration with multispectral and hyperspectral imaging enhance plant trait assessment, while data processing and analysis software make 3D imaging more accessible for large-scale phenotyping studies (Jin et al., 2021).

#### 14.4.1.7 Tomographic Imaging

Imaging methods based on tomography, including positron emission tomography (PET), magnetic resonance imaging (MRI), and computed tomography (CT), provide non-invasive internal views of plant structures, allowing for detailed examination of internal morphology (Galieni et al., 2021; Jaiswal et al., 2022). X-ray CT can visualize and analyze root system architecture in soil, providing insights into root growth patterns and environmental interactions (Hou et al., 2022). MRI can study internal tissues, including water transport, vascular structure, and nutrient distribution within plants (Boulch et al., 2024). Tomographic imaging is also used to assess the internal qualities of seeds and fruits, such as void spaces, tissue density, and moisture content (Liu et al., 2020c).

Advancements in tomographic imaging have led to higher-resolution images and faster scanning times, making conducting detailed internal analyses on a larger scale feasible. The development of algorithms for automated image reconstruction and trait extraction has further enhanced the utility of this technology in phenotyping.

### 14.4.2 Platforms and Systems for High-Throughput Phenotyping

High-throughput phenotyping (HTP) platforms and systems are designed to automate and scale the process of measuring plant traits, enabling the analysis of large populations with speed and precision. These platforms can be broadly categorized into field-based and controlled environment systems, each offering unique capabilities and applications. Shared platforms and systems for high-throughput phenotyping are presented in Figure 14.3.

#### 14.4.2.1 Field-Based Phenotyping Platforms

Field-based phenotyping platforms are designed to operate in agricultural environments where plants are grown under natural or semi-natural conditions. These platforms are crucial for assessing traits influenced by environmental factors, such as stress resistance, yield, and growth dynamics.

##### *14.4.2.1.1 Ground-Based Platforms*

Phenomobiles and gantry systems are common ground-based platforms; each has unique functions and uses. *Phenomobiles* are mobile platforms, often resembling small vehicles or robots, with various sensors and imaging devices. They traverse through crop fields, capturing detailed phenotypic data from plants at varying stages of growth (Li et al., 2014).

**FIGURE 14.3** An overview of common platforms and systems for high-throughput phenotyping. Field-based phenotyping platforms use airborne (UAVs and satellites) and ground-based (phenomobiles and gantry systems) platforms. Controlled environment phenotyping systems encompass rhizotron, greenhouse, and growth chamber configurations (Qiao et al., 2022; Enyew et al., 2023).

*Gantry systems* consist of large structures spanning crop fields, supporting sensors and imaging devices on overhead rails. These systems can move back and forth along the rails, capturing high-resolution data from plants below. Gantry systems offer exact control over the movement of sensors, enabling detailed phenotypic measurements at specific points in the field (Roitsch et al., 2019). These systems benefit long-term studies where continuous plant growth and development monitoring is required. Phenomobiles and gantry systems use sensors like multispectral cameras, LiDAR sensors, thermal cameras, and visible light imaging systems to measure plant traits like height, leaf area, canopy temperature, and biomass. Advances in autonomous navigation allow them to operate without human intervention, and data integration allows them to collect vast amounts of real-time insights into plant performance.

#### *14.4.2.1.2 Aerial Platforms*

Unmanned aerial vehicles (UAVs) and satellites are the primary aerial platforms utilized in HTP. The UAVs, commonly known as drones, have become increasingly popular for field-based phenotyping due to their capacity to cover large areas rapidly and provide aerial views of crop fields. The ability to rapidly fly over large fields makes UAVs highly efficient for collecting data across extensive areas, making them ideal for large-scale breeding programs and environmental monitoring (Li et al., 2021).

Satellite-based phenotyping uses data from Earth observation satellites to assess plant traits on a regional or global scale. While less detailed than UAV or ground-based systems, satellite phenotyping provides valuable information on large-scale agricultural trends. Satellites can monitor vast agricultural regions, making them helpful in tracking crop health, drought conditions, and yield forecasts over entire seasons (Zhang et al., 2020). Satellites and UAVs can be equipped with multispectral and hyperspectral cameras to assess plant health, nutrient status, and stress responses from above.

##### *14.4.2.1.3 Controlled Environment Phenotyping Systems*

These systems are designed for use in growth chambers, greenhouses, or laboratories where environmental variables can be precisely controlled. *Growth chambers* provide highly controlled environments for studying plant responses to specific conditions, such as temperature, light, humidity, and $CO_2$ levels. Growth chambers are ideal for small-scale, detailed experiments where control over environmental conditions is critical. It enables researchers to create specific environmental scenarios like drought, heat stress, or nutrient deficiencies. It is equipped with automated imaging and sensing systems for continuously monitoring plant growth and physiological responses.

*Greenhouse phenotyping systems* offer a semi-controlled environment where plants can be grown under natural light with controlled temperature, humidity, and irrigation. Greenhouse phenotyping systems are often more automated and scalable than growth chambers. It often includes conveyor belts, automated irrigation systems, and robotic arms that move plants through different sensors for continuous monitoring. Greenhouses can accommodate larger plant populations than growth chambers, making them suitable for medium-scale experiments and early-stage breeding programs.

*Rhizotrons* are specialized systems designed to study root phenotypes under controlled conditions. They typically consist of transparent or semi-transparent containers where roots can grow, allowing for non-invasive imaging and monitoring. Rhizotrons are equipped with imaging systems that capture root architecture, growth patterns, and soil or other media interactions. These systems allow researchers to manipulate soil properties, such as moisture content, nutrient availability, and microbial activity, to study their effects on root development. Rhizotrons are particularly valuable for studying the interactions between roots and soil, including nutrient uptake and root exudation (Wendel et al., 2022).

### 14.4.3 Integration and Data Management

Integration and data management in plant phenotyping involve organizing, processing, and analyzing vast amounts of data collected from various phenotyping platforms (Figure 14.4).

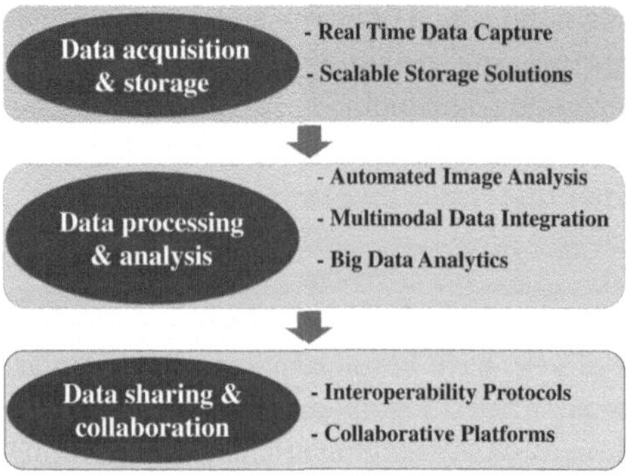


**FIGURE 14.4** Flow diagram of integration and data management system for high-throughput phenotyping.

#### 14.4.3.1 Data Acquisition and Storage

All HTP platforms generate large volumes of data, necessitating efficient acquisition and storage solutions. Modern phenotyping systems often include cloud-based platforms that enable real-time data uploading, storage, and access from remote locations (Kim, 2020).

#### 14.4.3.2 Data Processing and Analysis

The vast amounts of data generated by HTP platforms require sophisticated data processing and analysis tools. Advances in machine learning, artificial intelligence, and statistical modelling have significantly enhanced the ability to extract meaningful insights from phenotypic data. Machine learning algorithms can identify complex patterns and relationships in large datasets, enabling the discovery of novel phenotypic traits and their genetic associations. (Cheng et al. 2021) Applying deep learning techniques to image analysis has further improved the accuracy and robustness of phenotypic trait extraction. Multimodal data integration is essential for comprehensive plant phenotype analysis. Big data analytics techniques help identify patterns, correlations, and trends in large datasets, facilitating the discovery of new traits and their genetic associations (Singh et al., 2021; Sheikh et al., 2024).

#### 14.4.3.3 Data Sharing and Collaboration

Standardization and interoperability are critical for data sharing and collaboration across research institutions. HTP platforms use standardized data formats and protocols for researchers to share and use easily. These protocols ensure data comparability across different platforms and research projects. Cloud-based data management systems also feature collaborative tools, enabling simultaneous data access and analysis, promoting teamwork and data sharing.

## 14.5 APPLICATIONS OF HIGH-THROUGHPUT PHENOTYPING AND PHENOMICS

### 14.5.1 Crop Improvement and Breeding

HTP has revolutionized the field of crop improvement and breeding by enabling the rapid screening of large populations of different crop species (Table 14.1). This capability allows breeders to identify superior genotypes with desirable traits, such as increased yield, disease resistance, and stress tolerance. Integrating HTP with genomic selection and marker-assisted breeding has accelerated the development of improved crop varieties.

**Genomic Selection**: By integrating phenotypic data with genomic information, breeders can perform genomic selection, predicting the performance of plants based on their genetic makeup. Genome selection (GS) uses markers and statistical modelling to select complex traits controlled by multiple alleles, predicting crop performance before field tests (Merrick et al., 2022). High-throughput phenotyping platforms, such as UAVs, have improved genetic prediction (GS) in crops like sorghum (Watanabe et al., 2017). These platforms, equipped with remote-sensing units with RGB or NIR-GB cameras, are cost-effective and easy to operate. The data collected from these platforms can be used in wheat genetics to improve grain yield prediction accuracy (Rutkoski et al., 2016).

**Accelerating Breeding Cycles**: HTP allows breeders to rapidly screen large populations of plants for desirable traits, such as yield, disease resistance, drought tolerance and quality improvement (Shakoor et al., 2017; Reynolds et al., 2020). This accelerates selection and shortens breeding cycles.

**Trait Discovery**: HTP facilitates the identification of new traits or the detailed characterization of known traits, helping breeders discover novel genetic variations linked to specific plant characteristics. Similarly, large-scale genetic screening, such as CRISPR-Cas9 libraries, can be phenotypically assessed using HTP to identify genes responsible for specific traits (Liu et al., 2020b).

**Nutritional Quality**: HTP can be used to assess traits related to nutritional quality, such as protein content, micronutrient levels, and biofortification. This helps in breeding crops that have higher yields and offer better nutrition.

**TABLE 14.1**
**High-Throughput Phenotyping (HTP) Application in Crop Improvement and Functional Genomics (Yang et al., 2020; Li et al., 2021)**

| Crop | Imaging Technology/Platform | Trait Studied | References |
|---|---|---|---|
| Arabidopsis | Phenovator | For photosynthesis and growth | Flood et al. (2016) |
| Wheat | Unmanned Aerial Vehicle (UAV) linked with a terrestrial platform in Italy | Leaf rolling, dry biomass, phenology score, vegetation index, leaf chlorophyll content, and a chosen drought-adaptive indicator | Condorelli et al. (2018) |
| | High-throughput plant phenotyping system in National Plant Phenomics Centre, UK | Plant area, height, water consumption, senescence, and fitted parameters | Camargo et al. (2018) |
| | Tractor-based semiautomatic phenotyping system, China | Sensor-based traits, yield-related canopy architecture | Jiang et al. (2019) |
| | UAV-Ebee/Multispectral | Yield and yield attributing traits | Hu et al. (2020) |
| Rice | Tractor-based high-throughput phenotyping, USA | Canopy height, canopy temperature reduction, and three reflectance ratios | Tanger et al. (2017) |
| | Hyperspectral Imaging System, China | Hyperspectral indices, chlorophyll content | Feng et al. (2017) |
| | Micro-CTRGB imaging system | Tiller traits, tiller growth traits, biomass, shoot morphological and shoot growth traits | Wu et al. (2019) |
| | UAV- Self-developed Octorotor/RGB, multispectral | Canopy height, VIs, canopy Coverage | Wan et al. (2020) |
| Maize | High-throughput rice phenotyping facility, China | Plant morphological traits, leaf architecture, colour, growth, and biomass-related characteristics | Zhang et al. (2017) |
| | PhenoArch phenotyping platform (INRA, France) | Transpiration rate, fresh biomass, and plant leaf area | Prado et al. (2018) |
| | Unmanned aerial vehicle, China | Plant height and growth rate at four growth stages | Wang et al. (2019) |
| | UAV: 3DR Solo Quadcopter/ Multispectral | hemispherical-conical reflectance factors (HCRF), the chlorophyll red-edge index (CHL) and the NDVI | Fawcett et al. (2020) |
| Sorghum | Phenobot 1.0, an auto-steered and self-propelled field-based HTP platform equipped with stereo RGB cameras, USA | Plant height and stem diameter | Salas Fernandez et al. (2017) |
| | RGB imaging box and semiautomated imaging analysis pipeline: Toolkit for inflorescence, USA | Panicle size and shape | Zhou et al. (2019) |
| | UAV-Anaconda/RGB, multispectral | Plant height, VIs | Shi et al. (2016) |
| Barley | UAV-Customized Hyperspectral | Normalized vegetation index (NDVI), yield | Oehlschläger et al. (2018) |
| Cotton | UAV-Matrice 600 Pro/RGB, multispectral camera, infrared thermal | Yield and yield-related traits | Feng et al. (2020) |

### 14.5.2 Functional Genomics and Systems Biology

High-resolution phenotypic data, combined with genomic and transcriptomic information, facilitate the identification of genes and regulatory networks underlying complex traits. Phenomics allows researchers to link gene function with phenotypic traits (Gene function analysis). By systematically knocking out or overexpressing genes and analyzing the resulting phenotypes, scientists can infer the roles of specific genes in plant development and stress responses. Further, HTP data can be used in Quantitative trait loci (QTL) mapping studies to identify genomic regions associated with important traits.

### 14.5.3 Stress Phenotyping

Understanding plant responses to biotic and abiotic stresses is critical for developing resilient crop varieties. HTP enables the detailed characterization of stress responses under controlled and field conditions. For example, thermal imaging can detect early signs of drought stress, while hyperspectral imaging can identify nutrient deficiencies and disease symptoms (Gerhards et al., 2019). Further, HTP is crucial for developing crops that can adapt to changing climate conditions by simulating future scenarios and phenotyping plant responses, enabling researchers to select or engineer suitable crops.

### 14.5.4 Precision Agriculture

Applying HTP in precision agriculture enables site-specific management of crops, optimizing resource use and maximizing yield. High-resolution phenotypic data can inform variable rate applications of fertilizers, pesticides, and irrigation, reducing inputs and minimizing environmental impact (Bai & Ge, 2021). Integrating HTP with decision support systems and remote sensing technologies further enhances the precision and efficiency of agricultural practices. Data from HTP can be used to create detailed maps of field variability, guiding the precise application of inputs like water, fertilizers, and pesticides, which enhances efficiency and sustainability.

### 14.5.5 Crop Phenology and Growth Modelling

HTP is utilized for phenological studies, monitoring and modelling key developmental stages in crops like flowering and maturation (Bandaru et al., 2020), which aids in optimizing planting schedules and predicting harvest times, and for growth modelling, improving crop performance predictions under various environmental conditions.

## 14.6 CHALLENGES AND FUTURE DIRECTIONS

Despite the significant advancements in HTP, several technical challenges remain. Developing robust and scalable data integration frameworks is essential for managing the vast amounts of phenotypic data generated by modern HTP systems. Additionally, improving the accuracy and reliability of sensor technologies under field conditions is critical for broader adoption. Further, the lack of standardization and interoperability among different HTP platforms and data formats hinders data sharing and collaboration. Establishing standardized protocols and data formats is crucial for promoting interoperability and facilitating large-scale phenotypic studies. Addressing these challenges requires ongoing innovation and interdisciplinary collaboration. Moreover, the widespread adoption of HTP and phenomics raises important ethical and social considerations. Ensuring equitable access to advanced phenotyping technologies and data is essential for promoting global food security and reducing disparities in agricultural research. Additionally, addressing data privacy and ownership concerns is critical for building trust and fostering responsible innovation.

The future of HTP and phenomics lies in the continued integration of advanced technologies and methodologies. The convergence of high-resolution imaging, sensor technologies, and big data analytics will enable more comprehensive and precise phenotypic analyses. Furthermore, integrating HTP with other omics technologies, such as genomics and metabolomics, will provide deeper insights into plant biology and support the development of next-generation crop varieties.

## 14.7 CONCLUSION

High-throughput phenotyping and phenomics represent transformative advancements in agricultural research, offering unprecedented opportunities for crop improvement and sustainable agriculture. The integration of cutting-edge technologies, such as imaging, spectroscopy, and machine learning, has revolutionized our ability to measure and analyze plant phenotypes on a large scale. While challenges remain, ongoing innovation and interdisciplinary collaboration will continue to drive progress in this field, ultimately contributing to global food security and agricultural sustainability.

## REFERENCES


Abebe, A. M., Kim, Y., Kim, J., Kim, S. L., & Baek, J. (2023). Image-based high-throughput phenotyping in horticultural crops. *Plants*, *12*(10), 2061.

Adhikari, N. D., Simko, I., & Mou, B. (2019). Phenomic and physiological analysis of salinity effects on lettuce. *Sensors*, *19*(21), 4814.

Atefi, A., Ge, Y., Pitla, S., & Schnable, J. (2021). Robotic technologies for high-throughput plant phenotyping: Contemporary reviews and future perspectives. *Frontiers in Plant Science*, *12*, 611940.

Bai, G., & Ge, Y. (2021). Crop sensing and its application in precision agriculture and crop phenotyping. *Fundamentals of Agricultural and Field Robotics*, 137–155.

Bandaru, V., Yaramasu, R., Koutilya, P. N. V. R., He, J., Fernando, S., Sahajpal, R., Wardlow, B. D., Suyker, A., & Justice, C. (2020). PhenoCrop: An integrated satellite-based framework to estimate physiological growth stages of corn and soybeans. *International Journal of Applied Earth Observation and Geoinformation*, *92*, 102188.

Boulch, P. N., Collewet, G., Guillon, B., Quellec, S., Leport, L., & Musse, M. (2024). Quantitative MRI imaging of parenchyma and venation networks in Brassica napus leaves: Effects of development and dehydration. *Plant Methods*, *20*(1), 69.

Camargo, A. V., Mackay, I., Mott, R., Han, J., Doonan, J. H., Askew, K., Corke, F., Williams, K., & Bentley, A. R. (2018). Functional mapping of quantitative trait loci (QTLs) associated with plant performance in a wheat MAGIC mapping population. *Frontiers in Plant Science*, *9*, 887.

Cen, H., Weng, H., Yao, J., He, M., Lv, J., Hua, S., Li, H., & He, Y. (2017). Chlorophyll fluorescence imaging uncovers photosynthetic fingerprint of citrus Huanglongbing. *Frontiers in Plant Science*, *8*, 1509.

Chen, C. J., Rutkoski, J., Schnable, J. C., Murray, S. C., Wang, L., Jin, X., Stich, B., Crossa, J., Hayes, B. J., & Zhang, Z. (2022). Role of the genomics–phenomics–agronomy paradigm in plant breeding. *Plant Breeding Reviews*, *46*, 627–673.

Cheng, C. Y., Li, Y., Varala, K., Bubert, J., Huang, J., Kim, G. J., Halim, J., Arp, J., Shih, H. J. S., Levinson, G., & Park, S. H. (2021). Evolutionarily informed machine learning enhances the power of predictive gene-to-phenotype relationships. *Nature Communications*, *12*(1), 5627.

Condorelli, G. E., Maccaferri, M., Newcomb, M., Andrade-Sanchez, P., White, J. W., French, A. N., Sciara, G., Ward, R., & Tuberosa, R. (2018). Comparative aerial and ground based high throughput phenotyping for the genetic dissection of NDVI as a proxy for drought adaptive traits in durum wheat. *Frontiers in Plant Science*, *9*, 349736.

Enyew, M., Geleta, M., Feyissa, T., Hammenhag, C., Tesfaye, K., Seyoum, A., & Carlsson, A. S. (2023). Sorghum genotypes grown in simple rhizotrons display wide variation in root system architecture traits. *Plant and Soil*, *502*, 1–16.

Fawcett, D., Panigada, C., Tagliabue, G., Boschetti, M., Celesti, M., Evdokimov, A., Biriukova, K., Colombo, R., Miglietta, F., Rascher, U., & Anderson, K. (2020). Multi-scale evaluation of drone-based multispectral surface reflectance and vegetation indices in operational conditions. *Remote Sensing*, *12*(3), 514.

Feng, A., Zhou, J., Vories, E. D., Sudduth, K. A., & Zhang, M. (2020). Yield estimation in cotton using UAV-based multi-sensor imagery. *Biosystems Engineering*, *193*, 101–114.

Feng, H., Guo, Z., Yang, W., Huang, C., Chen, G., Fang, W., Xiong, X., Zhang, H., Wang, G., Xiong, L., & Liu, Q. (2017). An integrated hyperspectral imaging and genome-wide association analysis platform provides spectral and genetic insights into the natural variation in rice. *Scientific Reports*, *7*(1), 4401.

Flood, P. J., Kruijer, W., Schnabel, S. K., van der Schoor, R., Jalink, H., Snel, J. F., Harbinson, J., & Aarts, M. G. (2016). Phenomics for photosynthesis, growth and reflectance in Arabidopsis thaliana reveals circadian and long-term fluctuations in heritability. *Plant Methods*, *12*, 1–14.

Galieni, A., D'Ascenzo, N., Stagnari, F., Pagnani, G., Xie, Q., & Pisante, M. (2021). Past and future of plant stress detection: An overview from remote sensing to positron emission tomography. *Frontiers in Plant Science*, *11*, 609155.

Gerhards, M., Schlerf, M., Mallick, K., & Udelhoven, T. (2019). Challenges and future perspectives of multi-/Hyperspectral thermal infrared remote sensing for crop water-stress detection: A review. *Remote Sensing*, *11*(10), 1240.

Guo, W., Carroll, M. E., Singh, A., Swetnam, T. L., Merchant, N., Sarkar, S., Singh, A. K., & Ganapathysubramanian, B. (2021). UAS-based plant phenotyping for research and breeding applications. *Plant Phenomics*, *2021*, 1–21.

Hou, L. H., Gao, W., Weng, Z. H., Doolette, C. L., Maksimenko, A., Hausermann, D., Zheng, Y., Tang, C., Lombi, E., & Kopittke, P. M. (2022). Use of X-ray tomography for examining root architecture in soils. *Geoderma*, *405*, p.115405.

Hu, Y., Knapp, S., & Schmidhalter, U. (2020). Advancing high-throughput phenotyping of wheat in early selection cycles. *Remote Sensing*, *12*(3), 574.

Jaiswal, Y. S., Xue, Y., Xiao, T., & Williams, L. L. (2022). Use of microcomputed tomography and image processing tools in medicinal and aromatic plants. In A. B. Sharangi & K. V. Peter (Eds.), *Medicinal Plants* (pp. 361–386). Apple Academic Press.

Jangra, S., Chaudhary, V., Yadav, R. C., & Yadav, N. R. (2021). High-throughput phenotyping: A platform to accelerate crop improvement. *Phenomics*, *1*(2), 31–53.

Jiang, L., Sun, L., Ye, M., Wang, J., Wang, Y., Bogard, M., Lacaze, X., Fournier, A., Beauchêne, K., Gouache, D., & Wu, R. (2019). Functional mapping of N deficiency-induced response in wheat yield-component traits by implementing high-throughput phenotyping. *The Plant Journal*, *97*(6), 1105–1119.

Jin, S., Sun, X., Wu, F., Su, Y., Li, Y., Song, S., Xu, K., Ma, Q., Baret, F., Jiang, D., & Ding, Y. (2021). Lidar sheds new light on plant phenomics for plant breeding and management: Recent advances and future prospects. *ISPRS Journal of Photogrammetry and Remote Sensing*, *171*, 202–223.

Johannsen, W. (1911). The genotype conception of heredity. *American Naturalist*, *45*, 129–159.

Kaur, B., Sandhu, K. S., Kamal, R., Kaur, K., Singh, J., Röder, M. S., & Muqaddasi, Q. H. (2021). Omics for the improvement of abiotic, biotic, and agronomic traits in major cereal crops: Applications, challenges, and prospects. *Plants*, *10*(10), 1989.

Kim, J. Y. (2020). Roadmap to high throughput phenotyping for plant breeding. *Journal of Biosystems Engineering*, *45*, 43–55.

Koh, J. C., Spangenberg, G., & Kant, S. (2021). Automated machine learning for high-throughput image-based plant phenotyping. *Remote Sensing*, *13*(5), 858.

Kolhar, S., & Jagtap, J. (2023). Plant trait estimation and classification studies in plant phenotyping using machine vision–A review. *Information Processing in Agriculture*, *10*(1), 114–135.

Kumar, P., Eriksen, R. L., Simko, I., & Mou, B. (2021). Molecular mapping of water-stress responsive genomic loci in lettuce (Lactuca spp.) using kinetics chlorophyll fluorescence, hyperspectral imaging and machine learning. *Frontiers in Genetics*, *12*, 634554.

Li, D., Quan, C., Song, Z., Li, X., Yu, G., Li, C., & Muhammad, A. (2021). High-throughput plant phenotyping platform (HT3P) as a novel tool for estimating agronomic traits from the lab to the field. *Frontiers in Bioengineering and Biotechnology*, *8*, 623705.

Li, L., Zhang, Q., & Huang, D. (2014). A review of imaging techniques for plant phenotyping. *Sensors*, *14*(11), 20078–20111.

Lin, Y. (2015). LiDAR: An important tool for next-generation phenotyping technology of high potential for plant phenomics? *Computers and Electronics in Agriculture*, *119*, 61–73.

Liu, H., Bruning, B., Garnett, T., & Berger, B. (2020a). Hyperspectral imaging and 3D technologies for plant phenotyping: From satellite to close-range sensing. *Computers and Electronics in Agriculture*, *175*, 105621.

Liu, H.J., Jian, L., Xu, J., Zhang, Q., Zhang, M., Jin, M., Peng, Y., Yan, J., Han, B., Liu, J. and Gao, F., 2020b. High-throughput CRISPR/Cas9 mutagenesis streamlines trait gene identification in maize. *The Plant Cell*, *32*(5), 1397–1413.

Liu, W., Liu, C., Jin, J., Li, D., Fu, Y., & Yuan, X. (2020c). High-throughput phenotyping of morphological seed and fruit characteristics using X-ray computed tomography. *Frontiers in Plant Science*, *11*, 601475.

Lu, B., Dao, P. D., Liu, J., He, Y., & Shang, J. (2020). Recent advances of hyperspectral imaging technology and applications in agriculture. *Remote Sensing*, *12*(16), 2659.

Mahlein, A. K. (2016). Plant disease detection by imaging sensors–parallels and specific demands for precision agriculture and plant phenotyping. *Plant Disease*, *100*(2), 241–251.

Mao, Y., Betters, C. H., Garske, S., Randle, J., Wong, K. C., Cairns, I. H., & Evans, B. J. (2023). A customisable data acquisition system for open-source hyperspectral imaging. *Sensors*, *23*(20), 8622.

Merrick, L. F., Herr, A. W., Sandhu, K. S., Lozada, D. N., & Carter, A. H. (2022). Optimizing plant breeding programs for genomic selection. *Agronomy*, *12*(3), 714.

Mir, R. R., Reynolds, M., Pinto, F., Khan, M. A., & Bhat, M. A. (2019). High-throughput phenotyping for crop improvement in the genomics era. *Plant Science*, *282*, 60–72.

Oehlschläger, J., Schmidhalter, U., & Noack, P. O. (2018). UAV-based hyperspectral sensing for yield prediction in winter barley. In *2018 9th Workshop on Hyperspectral Image and Signal Processing: Evolution in Remote Sensing (WHISPERS)* (pp. 1–4). Netherlands: IEEE.

Omasa, K., Hosoi, F., & Konishi, A. (2007). 3D lidar imaging for detecting and understanding plant responses and canopy structure. *Journal of Experimental Botany*, *58*(4), 881–898.

Omia, E., Bae, H., Park, E., Kim, M. S., Baek, I., Kabenge, I., & Cho, B. K. (2023). Remote sensing in field crop monitoring: A comprehensive review of sensor systems, data analyses and recent advances. *Remote Sensing*, *15*(2), 354.

Pasala, R., & Pandey, B. B. (2020). Plant phenomics: High-throughput technology for accelerating genomics. *Journal of Biosciences*, *45*(1), 111.

Prado, S. A., Cabrera-Bosquet, L., Grau, A., Coupel-Ledru, A., Millet, E. J., Welcker, C., & Tardieu, F. (2018). Phenomics allows identification of genomic regions affecting maize stomatal conductance with conditional effects of water deficit and evaporative demand. *Plant, Cell & Environment*, *41*(2), 314–326.

Qiao, Y., Valente, J., Su, D., Zhang, Z., & He, D. (2022). AI, sensors and robotics in plant phenotyping and precision agriculture. *Frontiers in Plant Science*, *13*, 1064219.

Reynolds, M., Chapman, S., Crespo-Herrera, L., Molero, G., Mondal, S., Pequeno, D. N., Pinto, F., Pinera-Chavez, F. J., Poland, J., Rivera-Amado, C., & Saint Pierre, C. (2020). Breeder friendly phenotyping. *Plant Science*, *295*, 110396.

Roitsch, T., Cabrera-Bosquet, L., Fournier, A., Ghamkhar, K., Jiménez-Berni, J., Pinto, F., & Ober, E. S. (2019). New sensors and data-driven approaches—A path to next generation phenomics. *Plant Science*, *282*, 2–10.

Rutkoski, J., Poland, J., Mondal, S., Autrique, E., Pérez, L. G., Crossa, J., Reynolds, M., & Singh, R. (2016). Canopy temperature and vegetation indices from high-throughput phenotyping improve accuracy of pedigree and genomic selection for grain yield in wheat. *G3: Genes, Genomes, Genetics*, *6*(9), 2799–2808.

Salas Fernandez, M. G., Bao, Y., Tang, L., & Schnable, P. S. (2017). A high-throughput, field-based phenotyping technology for tall biomass crops. *Plant Physiology*, *174*(4), 2008–2022.

Saric, R., Nguyen, V. D., Burge, T., Berkowitz, O., Trtílek, M., Whelan, J., Lewsey, M. G., & Čustović, E. (2022). Applications of hyperspectral imaging in plant phenotyping. *Trends in Plant Science*, *27*(3), 301–315.

Sarkar, S., Ganapathysubramanian, B., Singh, A., Fotouhi, F., Kar, S., Nagasubramanian, K., Chowdhary, G., Das, S .K., Kantor, G., Krishnamurthy, A., & Merchant, N. (2023). Cyber-agricultural systems for crop breeding and sustainable production. *Trends in Plant Science*. 29(2), 130–149.

Shakoor, N., Lee, S., & Mockler, T. C. (2017). High throughput phenotyping to accelerate crop breeding and monitoring of diseases in the field. *Current Opinion in Plant Biology*, *38*, 184–192.

Sheikh, M., Iqra, F., Ambreen, H., Pravin, K. A., Ikra, M., & Chung, Y. S. (2024). Integrating artificial intelligence and high-throughput phenotyping for crop improvement. *Journal of Integrative Agriculture*, *23*(6), 1787–1802.

Shi, Y., Murray, S. C., Rooney, W. L., Valasek, J., Olsenholler, J., Pugh, N. A., Henrickson, J., Bowden, E., Zhang, D., & Thomasson, J. A. (2016). Corn and sorghum phenotyping using a fixed-wing UAV-based remote sensing system. *Autonomous Air and Ground Sensing Systems for Agricultural Optimization and Phenotyping*, *9866*, 46–53.

Singh, A., Jones, S., Ganapathysubramanian, B., Sarkar, S., Mueller, D., Sandhu, K., & Nagasubramanian, K. (2021). Challenges and opportunities in machine-augmented plant stress phenotyping. *Trends in Plant Science*, *26*(1), 53–69.

Tang, Y., Song, S., Gui, S., Chao, W., Cheng, C., & Qin, R. (2023). Active and low-cost hyperspectral imaging for the spectral analysis of a low-light environment. *Sensors*, *23*(3), 1437.

Tanger, P., Klassen, S., Mojica, J. P., Lovell, J. T., Moyers, B. T., Baraoidan, M., Naredo, M. E. B., McNally, K. L., Poland, J., Bush, D. R., & Leung, H. (2017). Field-based high throughput phenotyping rapidly identifies genomic regions controlling yield components in rice. *Scientific Reports*, *7*(1), 42839.

Totzke, C., Kardjilov, N., Hilger, A., Rudolph-Mohr, N., Manke, I., & Oswald, S. E. (2021). Three-dimensional in vivo analysis of water uptake and translocation in maize roots by fast neutron tomography. *Scientific Reports*, *11*(1), 10578.

Vieira, G. H. S., & Ferrarezi, R. S. (2021). Use of thermal imaging to assess water status in citrus plants in greenhouses. *Horticulturae*, *7*(8), 249.

Wan, L., Cen, H., Zhu, J., Zhang, J., Zhu, Y., Sun, D., Du, X., Zhai, L., Weng, H., Li, Y., & Li, X. (2020). Grain yield prediction of rice using multi-temporal UAV-based RGB and multispectral images and model transfer–a case study of small farmlands in the South of China. *Agricultural and Forest Meteorology*, *291*, 108096.

Wang, X., Zhang, R., Song, W., Han, L., Liu, X., Sun, X., Luo, M., Chen, K., Zhang, Y., Yang, H., & Yang, G. (2019). Dynamic plant height QTL revealed in maize through remote sensing phenotyping using a high-throughput unmanned aerial vehicle (UAV). *Scientific Reports*, *9*(1), 3458.

Watanabe, K., Guo, W., Arai, K., Takanashi, H., Kajiya-Kanegae, H., Kobayashi, M., Yano, K., Tokunaga, T., Fujiwara, T., Tsutsumi, N., & Iwata, H. (2017). High-throughput phenotyping of sorghum plant height using an unmanned aerial vehicle and its application to genomic prediction modeling. *Frontiers in Plant Science*, *8*, 421.

Wendel, A. S., Bauke, S. L., Amelung, W., & Knief, C. (2022). Root-rhizosphere-soil interactions in biopores. *Plant and Soil*, *475*(1), 253–277.

Wu, D., Guo, Z., Ye, J., Feng, H., Liu, J., Chen, G., Zheng, J., Yan, D., Yang, X., Xiong, X., & Liu, Q. (2019). Combining high-throughput micro-CT-RGB phenotyping and genome-wide association study to dissect the genetic architecture of tiller growth in rice. *Journal of Experimental Botany*, *70*(2), 545–561.

Yang, W., Feng, H., Zhang, X., Zhang, J., Doonan, J. H., Batchelor, W. D., Xiong, L., & Yan, J. (2020). Crop phenomics and high-throughput phenotyping: Past decades, current challenges, and future perspectives. *Molecular Plant*, *13*(2), 187–214.

Yu, K. Q., Zhao, Y. R., Li, X. L., Shao, Y. N., Liu, F., & He, Y. (2014). Hyperspectral imaging for mapping of total nitrogen spatial distribution in pepper plant. *PloS One*, *9*(12), e116205.

Zhang, C., Marzougui, A., & Sankaran, S. (2020). High-resolution satellite imagery applications in crop phenotyping: An overview. *Computers and Electronics in Agriculture*, *175*, 105584.

Zhang, H., Wang, L., Jin, X., Bian, L., & Ge, Y. (2023). High-throughput phenotyping of plant leaf morphological, physiological, and biochemical traits on multiple scales using optical sensing. *The Crop Journal*, 1303–1318.

Zhang, X., Huang, C., Wu, D., Qiao, F., Li, W., Duan, L., Wang, K., Xiao, Y., Chen, G., Liu, Q., & Xiong, L. (2017). High-throughput phenotyping and QTL mapping reveals the genetic architecture of maize plant growth. *Plant Physiology*, *173*(3), 1554–1564.

Zhou, Y., Srinivasan, S., Mirnezami, S. V., Kusmec, A., Fu, Q., Attigala, L., Salas Fernandez, M. G., Ganapathysubramanian, B., & Schnable, P. S. (2019). Semiautomated feature extraction from RGB images for sorghum panicle architecture GWAS. *Plant Physiology*, *179*(1), 24–37.

# 15 Statistical Tools and Data Analysis for Organizing Plant High-Throughput Phenomics

*Mahankali Sravanalakshmi, Mani Manoj, Priti Kumari, Macharla Srinidhi, Kannan Vijayarani, Shanmugam Velayuthaprabhu, Muhukrishnan Saradhadevi, and Arumugam Vijaya Anand*

## LIST OF ABBREVIATIONS

**3D**: Three Dimensional
**AI**: Artificial Intelligence
**ANNs**: Artificial Neural Networks
**ANOVA**: Analysis of Variance
**DIVA-GIS**: Digital Vegetation Analysis-Geographic Information System
**GIS**: Geographic Information System
**GRASS GIS**: Geographical Resources Analysis Support System Geographic Information System
**HTP**: High-Throughput Phenotyping
**IMAP**: Image Mapping and Analytics for Phenotyping
**IoT**: Internet of Things
**LMA**: Leaf Mass per Area
**MANOVA**: Multivariate Analysis of Variance
**ML**: Machine Language
**OCR**: Optical Character Recognition
**PCA**: Principal Component Analysis
**Plant CV**: Plant Computer Vision
**RGB**: Red, Green and Blue
**SVM**: Support Vector Machine Learning Algorithm
**UAS**: Unmanned Aircraft Systems
**UAV**: Unmanned Aerial Vehicle

## 15.1 INTRODUCTION

High-throughput phenotyping (HTP) is an innovative approach that enhances the efficiency and accuracy of plant phenotype evaluation through non-destructive methods. It leverages advanced technologies such as computer vision, machine learning, and various sensors to collect extensive data on plant traits across multiple samples and time scales, significantly surpassing traditional phenotyping methods in throughput and precision (Singhvi et al., 2024; Jiang and Zhu, 2024). HTP facilitates the extraction of specific traits, enabling applications in genetic dissection and genomic-assisted breeding, which are crucial for improving crop varieties (Xu and Li, 2022). Moreover, it plays a vital role in assessing plant responses to biotic stress, such as insect damage, by quantifying damage and insect behavior in a high-throughput manner (Kundu et al., 2024).


DOI: 10.1201/9781003505631-15

### 15.1.1 Definition and Importance of Phenomics in Plant Research

Phenomics is an interdisciplinary field focused on understanding the relationships between plant traits and their genetic and environmental contexts. It is crucial in advancing plant research, particularly in breeding and crop management. By utilizing high-throughput technologies and multi-modal data collection, phenomics enables the assessment of diverse plant traits across various conditions, significantly enhancing the efficiency of traditional measurement methods (Jiang and Zhu, 2024). This is particularly important in addressing challenges posed by abiotic stresses, such as drought and salinity, which threaten crop production and food security (Kumar et al., 2020). Moreover, the development of platforms like Plant PAD illustrates the importance of phenomics in disease management, facilitating large-scale image analysis for effective disease diagnosis and control (Dong et al., 2024). Overall, the evolution of plant phenomics underscores its vital role in fostering global collaboration and innovation in agricultural practices, ultimately contributing to the development of climate-resilient crop varieties (Awada et al., 2024).

It enables the collection of extensive data on diverse plant traits across various time scales, significantly enhancing the throughput of data collection compared to traditional methods (Jiang and Zhu, 2024). HTP techniques, such as drone-based imaging, allow researchers to analyze temporal changes in plant growth dynamics, revealing how genetic and environmental factors interact over time (Adak and Murray, 2023). Additionally, the use of sophisticated sensor technologies and data analysis pipelines facilitates accurate measurement of physiological traits, which are essential for optimizing crop management and breeding practices (Kumar et al., 2015).

### 15.1.2 Overview of High-Throughput Phenotyping Platforms

HTP platforms have emerged as essential tools for the non-destructive and efficient evaluation of plant phenotypes, particularly in agricultural research. These platforms leverage advanced technologies such as computer vision, machine learning, and automation to enhance the accuracy and speed of phenotypic assessments across various crops, including vegetables and root systems (Singhvi et al., 2024). Recent innovations, such as the RootBot, facilitate the study of root growth under drought conditions by simulating environmental stressors, thus enabling the evaluation of root traits that are typically challenging to assess (Ilakiya et al., 2020). However, the implementation of HTP systems is not without challenges; issues such as the need for precise calibration, the potential for measurement errors due to environmental variability, and the necessity for comprehensive datasets must be carefully managed to optimize their utility (Hendrik et al., 2023). Overall, while HTP platforms present significant advantages for plant phenotyping, their effectiveness hinges on thoughtful design and application tailored to specific research objectives.

### 15.1.3 Common Challenges in Managing and Analyzing Phenomics Data

Managing and analyzing phenomics data presents several common challenges. Firstly, there is a significant deficiency in analytical and modelling approaches tailored for the multi-dimensional nature of phenomics data, which complicates the extraction of meaningful insights from complex datasets (Ying, 2023). Additionally, the high dimensionality of data, particularly in electronic health records, poses difficulties in predictive modelling and interpretation, often requiring advanced dimension reduction techniques that can obscure clinical relevance (Allen et al., 2021). Furthermore, the integration of HTP data with existing genetic frameworks necessitates robust statistical methods to enhance prediction accuracy, which remains an ongoing challenge (Morota et al., 2022). Lastly, the lack of universal standards for data acquisition and management, along with the need for effective tools for exploratory analysis, such as Pheno-Mapper, highlights the necessity for improved methodologies in handling these vast datasets (Zhou et al., 2021).

## 15.2 DATA ACQUISITION AND PRE-PROCESSING

### 15.2.1 Types of Data Generated in Plant Phenomics

Plant phenomics generates a diverse array of data types that enhance our understanding of plant traits and their interactions with environmental factors. This data is crucial for advancing agricultural practices and crop breeding. HTP data utilizes sensors to collect reflective, emitted, and fluorescence signals from plants, enabling the assessment of various traits across large sample sizes and time scales (Jiang and Zhu, 2024). Temporal phenotype data captured through repetitive drone flights, this data provides insights into growth dynamics and the influence of environmental factors on different genotypes over time (Adak and Murray, 2023). Three-dimensional (3D) data advanced techniques like 3D reconstruction from multi-view images allow for detailed monitoring of plant growth and structure, facilitating accurate leaf area estimation (Daiki and Noshita, 2024). Genomic and environmental data integration combining phenomic data with genomic insights enhances predictive capabilities for complex traits, demonstrating the synergistic potential of these data types (Adak and Murray, 2023). While the focus on high-dimensional data collection is promising, challenges remain in standardizing data management and ensuring effective application in agricultural settings.

### 15.2.2 Data Quality Control and Cleaning

Data quality control and cleaning are critical components in organizing plant high-throughput phenomics, particularly given the complexity and volume of data generated by modern phenotyping technologies. High-throughput phenomics involves the acquisition of extensive phenotypic data through automated imaging and sensor technologies, which can produce vast datasets that require rigorous management and analysis to ensure their reliability and utility in research and breeding programs. The integration of diverse data types, including genomic, phenotypic, and environmental data, poses significant challenges in data quality control. Emphasize the necessity of combining heterogeneous datasets in plant phenomics, particularly in the context of climate change and food security, where accurate data is vital for effective decision-making (Pradal et al., 2017). Similarly, highlights the importance of methodological advancements in data acquisition and processing, underscoring that the focus should not only be on sensor development but also the handling of the resulting data (Tardieu et al., 2017). This is echoed by, those who discuss the need for robust data management frameworks to facilitate the integration of image data with genomic information, which is essential for machine learning applications in phenomics (Zhou et al., 2018). Moreover, the development of optimized phenomic predictors, as described, illustrates how high-throughput phenomics can enhance breeding decisions by providing reliable data that correlates well with traditional measurements (Parmley et al., 2019). This correlation is crucial for ensuring that the data collected through automated systems can be trusted for further analysis and application in breeding programs. Also provides insights into the non-invasive methods used in HTP, which generate comprehensive datasets that require careful documentation and quality control to maintain their integrity (Arend et al., 2016).

### 15.2.3 Data Normalization and Transformation Techniques

Data normalization and transformation techniques are crucial for organizing plant high-throughput phenomics data, ensuring accuracy and comparability across various measurements. The digital adjustment of plant development method automates the normalization of time-series data by referencing early developmental stages, reducing variance in phenotyping data by up to 2.5 times compared to traditional methods (Lozano-Claros et al., 2020). This approach enhances the uniformity of measurements, allowing for more informative comparisons between individual plants. Image standardization techniques, such as automated adjustments of colour profiles based on reference panels, address variations in image quality that can bias results (Berry et al., 2018). These adjustments, including brightness and contrast corrections, improve the accuracy of morphological measurements essential for effective phenotyping. Additionally, a statistical framework that integrates data mining and machine learning has been developed to analyze complex phenomics datasets. This method enhances prediction accuracy and reduces computational time, facilitating better classification of plant traits (Rahaman et al., 2019). Despite these advancements, challenges persist in ensuring consistency across diverse experimental conditions and plant species, highlighting the need for further research to refine these techniques and broaden their applicability.

## 15.3 EXPERIMENTAL DESIGN IN PHENOMICS STUDIES

### 15.3.1 Randomization and Replication in Phenomics Experiments

Randomization and replication are fundamental to the reliability and validity of high-throughput phenomics experiments, helping to minimize confounding variables and enhance the reproducibility of findings, crucial for translating research into practical applications. In plant high-throughput phenomics, block randomization plays a key role in preventing imbalances in sample allocation by addressing both known and unknown confounders, similar to its importance in proteomics where sample conditions can significantly impact outcomes (Burger et al., 2020). In preclinical animal studies, randomization enhances scientific rigour and reduces bias, with tools like the 'Mouse Randomization' app aiding in proper sample size estimation and ethical research practices (da Fonseca Cardoso et al., 2024). Replication also plays a critical role, as demonstrated by Mendelian randomization, replicates findings across different traits and diseases to establish causal relationships between proteins and phenotypes, exemplified by the identification of 38 proteins linked to human diseases (Bretherick et al., 2020). However, the complexities of replication are underscored by discrepancies between randomized controlled trials and real-world evidence, making it essential to understand these differences for accurate interpretation of findings (Thompson, 2021).

### 15.3.2 Blocking and Factorial Design for Field and Greenhouse Studies

HTP in plant studies requires robust experimental designs, such as blocking and factorial arrangements, to enhance data accuracy and reliability by mitigating errors from environmental variability and measurement inaccuracies in both greenhouse and field settings. Proper experimental design is vital for reliable phenotypic data collection as it minimizes confounding factors and increases statistical power (Clarke et al., 2022). In greenhouse experiments, blocking experimental units effectively reduces residual variance, thereby improving the precision of the results (Hartung et al., 2019). High-throughput plant phenotyping platforms equipped with advanced sensors allow for non-destructive monitoring and extensive data collection across various environments (Li et al., 2021). These platforms can integrate with software tools like image mapping and analytics for phenotyping (IMAP), efficiently processing the image data to provide plot-level metrics essential for large-scale studies (Kim, 2021). Comparative research has demonstrated that fixed-position arrangements are more effective than mobile pot re-arrangements in reducing variance, underscoring the value of strategic blocking in greenhouse experiments (Hartung et al., 2019). Additionally, the use of miniature greenhouses with controlled environmental factors enhances the ability to accurately derive phenomic and genomic correlations (Dong et al., 2019).

### 15.3.3 Sampling Strategies for High-Throughput Phenotyping

The integration of machine learning and computer vision in HTP has automated the phenotyping process, significantly improving efficiency in handling large datasets (Zhang et al., 2023). For instance, optical character recognition (OCR) techniques have been applied to read plant labels with 94.31% accuracy, streamlining data extraction. Additionally, image segmentation combined with machine learning algorithms has facilitated precise morphological classifications, enabling accurate trait identification (Singhvi et al., 2024). Sensor technologies, including multispectral and thermal imaging, play a pivotal role in assessing plant responses to environmental stresses like drought and nutrient deficiency, allowing for rapid, non-invasive evaluations of crop health. Ground and aerial platforms equipped with these sensors have been used to differentiate rice genotypes based on spectral reflectance data, deepening the understanding of plant responses under varying conditions (Sahoo et al., 2023). Furthermore, anatomical phenotyping, through the Anatomics approach, focuses on high-throughput imaging of plant anatomy, providing insights into genetic and physiological traits by quantifying anatomical features (Strock et al., 2022).

## 15.4 STATISTICAL METHODS FOR DATA ANALYSIS

### 15.4.1 Descriptive Statistics

Statistics plays an important role in the proper application of experiments (Raudonius, 2017). In any research project data description is very important before we start testing the hypothesis. Mean is the average value and is calculated by summing up all the values in a dataset and dividing it by the total number of values. To reduce the variance and bias sometimes methods like analysis of covariance is preferred (Frison et al., 1992). The central value of all the observations is called the Median (Whitley and Ball, 2001). The median statistics is used to obtain the estimates of two constants of nature for example Newton's gravitational and Hubble constant (Bethapudi and Desai, 2017). Standard deviation is a measure of the degree how which the given data set has deviated from the mean value. Methods like standard deviation are used to analyze the genetic specificity of crop plants at intraspecific levels (Mohammadi and Prasanna, 2003).

Measuring distribution functions and sampling probabilities are one of very important tools in molecular stimulation. At a given set of stimulations, the macroscopic states of system frequencies can be easily measured. Histograms can be easily employed to reconstruct the probability reconstructions (Shell et al., 2007). For example, to classify plants into appropriate taxonomies, different features like width and length, perimeter, a distance map and importantly a color histogram is also used (Cardina et al., 1995). Batches of Data are displayed in Boxplots. Five sets of data like extremes, the upper and lower hinges and the median from a set of data are used. These are widely used tools in preparing summaries which are visual for statisticians and non-statisticians and in exploratory data analysis (McGill et al., 1978). In the field of root biology, the generalized boxplot concept is used to present the visual summary of root ensembles by aggregating the images of roots with identical genetic characters (Vad et al., 2017).

### 15.4.2 Multivariate Analysis

Principal component analysis (PCA) is a commonly utilized method for lowering the dimensionality of the datasets and improving interpretability while minimizing information loss (Jolliffe and Cadima, 2016). To preserveg as much variance as feasible, PCA transfers high-dimensional data to a lower-dimensional space. Developed by Pearson and Hotelling. To identify the principal components, PCA first computes the dataset's covariance matrix before calculating and sorting the eigenvectors and eigenvalues (Sewell, 2008). In domains such as exploratory biomedicine, where data-driven analyses can advance more quickly than hypothesis-based research, PCA is very helpful (Hasan and Abdulazeez, 2021). For example, if your data set consists of a thousand 100-dimensional points, PCA can reduce its dimensionality to 10 without sacrificing too much information, meaning 90% compression. In certain instances, the minor variance directions also hold significance. To remove correlations, PCA can also be used to normalize data so that it has unit variance in all directions. This procedure is known as whitening or sphering. However, the bias of PCA is not always appropriate; depending on the application, features with low variance may potentially have significant predictive value (Sewell, 2008).

To find trends in plant phenotypic data and obtain a variety of parent sets for breeding programs, cluster analysis is a useful method. Although hierarchical clustering has been the conventional approach, it has significant processing complexity and poor accuracy. Researchers have suggested many strategies to deal with these problems. For soybean phenotypic data, spectral clustering with Pivotal Sampling has demonstrated increased accuracy and decreased computational cost in comparison to hierarchical clustering (Shastri et al., 2021). Non-parametric clustering is another method that works well for assessing high-dimensional, complicated phenotypic data without making assumptions about the data distribution. This technique has proven effective in identifying unique phenotypic patterns in mutant lines of Arabidopsis chloroplasts (Gao et al., 2015). In a similar vein, time-series photos of maize plants were taken using unmanned aerial vehicle (UAV) remote sensing, and plant height data was extracted for phenotypic analysis. To find typical dynamic patterns in plant height, growth rate, and height contribution rate, Han et al. created a fuzzy c-means clustering technique. This technique was effective in identifying genotype-to-genotype variability in phenotype. These studies show how HTP and sophisticated clustering algorithms can improve plant breeding operations by making it easier to choose different parent sets and improve the analysis of phenotypic features (Han et al., 2019).

### 15.4.3 Linear and Non-Linear Modelling

Linear regression models are widely used for trait prediction in plants, with genomic best linear unbiased prediction being a common approach (Gianola et al., 2018). To increase forecast accuracy, however, academics have looked into several different approaches. Multiple linked features have been modeled simultaneously by the application of multitask learning and multiple output regression approaches, which may improve prediction accuracy (He et al., 2016). Research comparing machine learning and classical approaches has demonstrated that while complicated neural network architectures do not always outperform well-established methods, Bayes B and linear regression models with sparsity restrictions frequently do well (John et al., 2022). In some circumstances, robust alternatives to BLUP, including linear models with t or Laplace distributions, have shown encouraging results, especially for variables like milk yield in cows and flowering time in Arabidopsis (Gianola et al., 2018). For genome-enabled prediction of continuous features, the Bayesian linear regression package in R implements a variety of Bayesian regression models, such as the Bayesian LASSO (de los Campos et al., 2013). Ridge regression, or best linear unbiased prediction, works with the assumption of additive genetic effects. On the other hand, non-additive kernels, such as the Gaussian kernel, can take epistatic effects into account. Non-additive kernels have been shown through cross-validation experiments to dramatically increase prediction accuracy for specific features, including wheat grain yield (Endelman, 2011).

In statistical analysis, linear mixed models are widely used to represent a wide range of data types, including multilevel, clustered, repeated measurements, and longitudinal data. These models are also known as Analysis of Variance (ANOVA) models (Kuran and Özkale, 2020). As a result, the machine language (ML) concept can be applied to parameter estimation for data that satisfies the normality condition (Agostinelli and Yohai, 2016). Henderson (1984) created a thorough explanation of the mixed model methodology, including the mixed model equations, their characteristics, and methods for estimating both the random and fixed effects (Henderson, 1984). The within- and between-subject variabilities can both be accounted for via a linear mixed model (Aghamohammadi and Meshkani, 2017). Incorporating random effects into linear models allows for the consideration of both within-subject correlation and response variability across subjects (Martinez and Holian, 2014).

Non-linear models, such as artificial neural networks (ANNs) and logistic regression, have been applied to various plant-related studies. ANNs outperform binary logistic regression models in the prediction of insect distribution (Peacock et al., 2007). For modeling wild oat field emergence, a bivariate ANN approach was demonstrated (Chantre et al., 2013). The area under the receiver-operating characteristic curve and the area under the curve indicate that neural networks performed better in tree mortality modeling than in logistic regression (King, 2003). Non-linear models' efficacy in transcriptome prediction tasks isn't always evident, though. Although neural networks are capable of detecting non-linear patterns in gene expression data, they might not always outperform logistic regression (Heil et al., 2022).

### 15.4.4 ANOVA and MANOVA

Presenting the statistical comparison of plant growth rate faces difficulties, in which analysis of variance of primary data could guide the interpretation of results. Heterogeneity of variance is seen in the analysis of variance of data from different harvests. ANOVA of separate harvests can be performed, even though analysis of variance of overall data improves the apprehension of plant growth by linking the interaction between time and treatment under study. In analysis of variance experimental errors are random, independent and normally distributed. ANOVA model includes I, II and III in which model I include fixed effects, model II random effects and model III includes both random and fixed effects. Fixed and mixed ANOVA models have identical sum of squares and they differ in expected mean square. Analysis of variance performed as three crossed factors. Classical growth analysis was performed in experiment P on bean cultivars in pots as well as in fields with only three harvests and was considered a fixed effect and ANOVA model I was used. Analysis was performed as two–crossed factors in experiments in rice cultivars in which harvest has a random effect and ANOVA model III is used. Error mean square is an appropriate term for testing the hypothesis in a fixed model whereas in a mixed model, the choice of a suitable error term is more difficult when main effects are tested (Araújo, 2003). The computation of variances, sums of squares, and F-ratios as well as the least squares technique are explained (Hector, 2021) point-wise ANOVA can yield discontinuous growth curves that may not accurately reflect plant growth dynamics. Functional ANOVA is proposed as an alternative, allowing for nonparametric estimation of trait changes over time without assumptions about curve shape (Xu et al., 2018) enhanced statistical power and direct genetic parameter estimation are two benefits of this modified ANOVA method for assessing quantitative traits based on single-locus genotypes.

In biological and agricultural research, Multivariate Analysis of Variance (MANOVA) is a potent statistical technique for examining several connected features. Limiting Type I error and taking into account correlations between dependent variables, provides advantages over numerous individual ANOVAs. It has been successfully used in studies on plant growth responses to acoustic treatments (Abdullah et al., 2018) and tomato fruit ripening proteomics (Pratta et al., 2010). A variety of experimental designs, including fully randomized designs with factorial layouts, can be used using MANOVA (Din and Hayat, 2021). However, in some circumstances, such as when a genetic mutation impacts all attributes equally, MANOVA may lose its power to overcome this drawback, and unified score-based association tests have been developed. These tests combine the advantages of marginal models and MANOVA to provide greater detection power in genome-wide association studies that involve many correlated traits (Ray et al., 2014).

### 15.4.5 Time-Series Analysis

The sudden advancements in remote sensing agronomic research enable the more accurate characterization of the dynamic nature of longitudinal traits such as the potential genetic improvements in crop efficiency (Novotny et al., 1991). The use of appropriate statistical methods to assess phenotypes and traits like light, biomass accumulation, and response to stresses was necessitated by the data generated by methods like HTP (Moreira et al., 2020). Researchers analyzed the Heritability and genetic correlation by measuring trees multiple times over some time and collected the longitudinal data. The longitudinal data can also be analyzed with the help of Kirkpatrick's approach using covariance functions (Apiolaza et al., 2001). The longitudinal family studies which involve repeated measurements are difficult. A method used for this study is the Gibbs-sampling-based approach which focuses on cross-sectional family studies, pedigrees on arbitrary complexity, continuous traits and repeated binary disease states (Burton et al., 2005).

The study of biogeochemical cycles and ecosystem processes depended on the seasonal litter inputs in forests. The amount of litter decomposability is higher in the wet season whereas it is very low in the dry season. This is because the phenolic concentration is found high in the dry season (Parsons et al., 2014). In certain plants such as *sphagnum* species the litter decomposition is found higher at the end of the summer season which may be related to the presence of the N15 isotope which occurred due to the accumulation of microbial organic compounds rich in N15, whereas the litter's of *Eriophorum vaginatum* and *Carex rostrata* had a decreased amount of N15 as compared to the initial amount at the summer endings (Bragazza et al., 2010).

## 15.5 ADVANCED COMPUTATIONAL TOOLS

### 15.5.1 Machine Learning Approaches

Supervised learning techniques were employed for variable selection (identifying the genetic and clinical features most associated or predictive of outcome) and prediction, with the outcome being case-control status or quantitative trait value (Figure 15.1). These techniques included regression-based techniques (e.g., generalized linear models, logic regression, and regularized regression) and tree-based techniques (e.g., decision trees and random forests) (Dasgupta et al., 2011). A comparison of multinomial logistic regression with support vector machines, random forests, and ANNs for predicting appearance traits found similar performance across methods (Katsara et al., 2021). The Support Vector Machine learning algorithm (SVM), is based on structural risk minimization and statistical learning theory established by Vapnik and associates. SVMs were successfully utilized were those in which the primary goal was to locate hidden patterns in the data concerning label classes (Duan, 2011). This algorithm can effectively differentiate various classes using both linear and non-linear discrimination functions (Amarappa and Sathyanarayana, 2014).

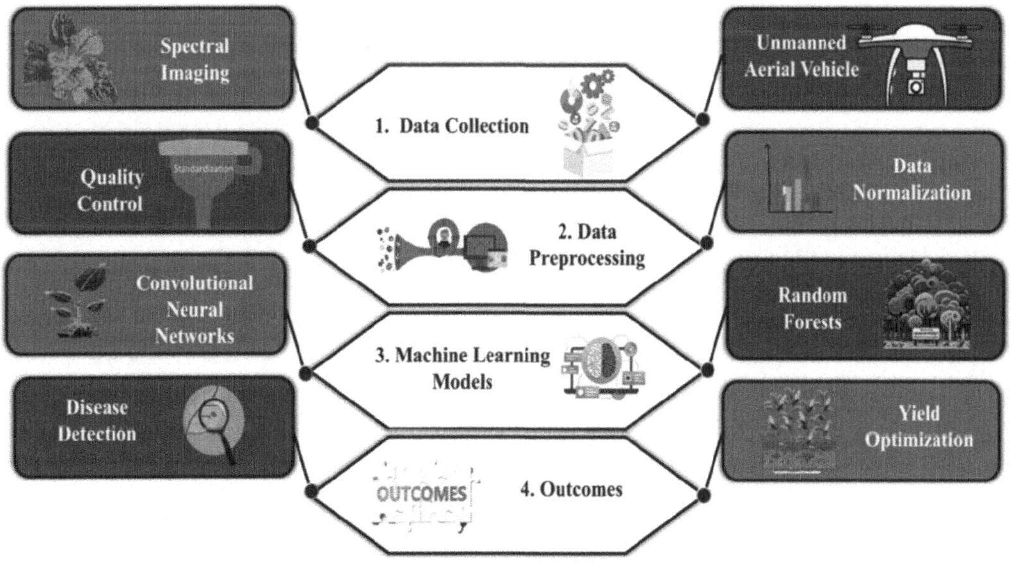


**FIGURE 15.1** AI and machine learning workflow in plant phenomics.

In plant phenotypic categorization and trait analysis, unsupervised learning techniques, particularly K-means clustering have shown promise. The plant shoot length, total plant height, blooming percentage, number of pods per plant, pod length, number of seeds per plant, and seed length variations between germplasm are six significant morphological variables that can be usefully analyzed using the K-means clustering algorithm (Amal et al., 2020). Using K-means clustering on photos of greenhouses, the KAT4IA pipeline creates training data for the extraction and analysis of plant attributes from an image-based field phenotyping system. The following are the key steps in the KAT4IA pipeline: building training sets under supervision, segmenting plants from photos of plants in the field, automatically separating target plants, calculating plant characteristics, and fitting functional curves to the extracted characteristics (Guo et al., 2021). K-Means, neural gas, growing neural gas, self-organizing map, and adaptive resonance theory were assessed in Arabidopsis using one task-specific index in addition to standard validity indices. A low agreement was established between task restrictions and traditional validity metrics, with K-Means exhibiting the greatest results (Backhaus et al., 2010). In the Big Data era, these approaches provide promising means of evaluating large-scale plant phenotypic data and enhancing our knowledge of plant biology.

Deep learning algorithms now drive innovations in recognizing, profiling, and predicting cellular phenotypes, addressing critical biological questions. As imaging assays become increasingly complex and large-scale, deep learning provides essential computational tools to uncover and analyze previously unexplored details of cellular phenotypes, enhancing our understanding of cell biology. Deep learning methods use numerous hidden layers in the network, each of which boosts the network's capacity for discrimination and prediction by sequentially applying a basic operation to the images (Pound et al., 2017). In phenotyping, deep learning architectures have been applied to leaf counting, plant morphology classification (Wang et al., 2019), plant recognition and identification (Ghosal et al., 2019), identification of root and shoot features, and identification and classification of plant stress.

### 15.5.2 Genome-Wide Association Studies

The use of Genome-wide association studies to examine complex features has grown to be widespread and effective. This was largely driven by developments in genomic technology, which allowed us to analyze genetic variants throughout the genome in a variety of genetic materials (Tibbs et al., 2020). With the use of Red, Green, and Blue (RGB) cameras, hyperspectral sensors, and computed tomography, HTP technology platforms may track phenotypic variation in plants. These sensors can be connected to environmental and genotypic data. The next-generation sequencing and whole-genome sequencing of numerous crops have made it possible to separate functional genomic studies of crops into two categories: the high-throughput period and the main data (Ahmed et al., 2023). Therefore, integrating data necessitates a great deal of scientific work, including the creation of appropriate infrastructures, analytical tools, standards, procedures, and models. Maureen O'Malley and Orkun Soyer have conducted a recent analysis of integration in molecular systems biology (O'Malley and Soyer, 2012). They highlight the significance of the scientific work related to data integration and distinguish it from other forms of integration, like methodological and explanatory integration, which have garnered the interest of science philosophers more readily (Leonelli, 2013).

### 15.5.3 Data Integration and Multi-Omics Analysis

These techniques such as proteomics, metabolomics, and physionomics have been gradual, particularly in their application to studying plants, including agriculturally important crops (Figure 15.2). However, utilizing these omics platforms can offer valuable insights into the molecular functioning

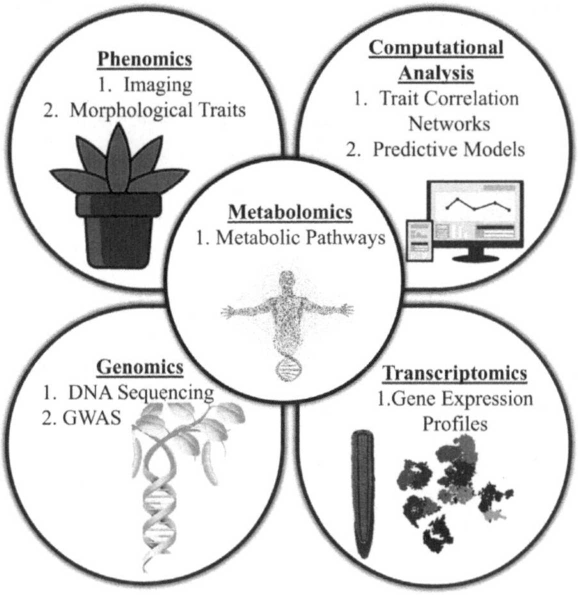


**FIGURE 15.2** Integration of multi-omics data for trait analysis.

of plants, even when employing just a single omics approach (Maroli et al., 2018). The OPTIMAS data warehouse (OPTIMAS-DW) is a comprehensive data collection for maize and integrates data from different data domains such as transcriptomics, metabolomics, genomics, proteomics and phenomics. In the OPTIMAS project, a 44K oligo chip was designed and annotated to describe the functions of the selected unigenes (Colmsee et al., 2012). In the model plant species Arabidopsis, the senescence of organs like flowers, fruits, and roots is comparatively under-researched, as is senescence in perennials like trees. There is a pressing need to integrate multi-omics techniques and physiological phenotyping into comprehensive phenomics approaches to better understand the complex process of senescence (Großkinsky et al., 2018) applications of metabolomics in plant genetics and breeding efforts offer efficient and effective solutions to challenges in our agricultural systems. Metabolites are an indispensable component of plant metabolism owing to their influence on plant biomass and architecture. In recent years, metabolomics has established itself as one of the breakthroughs in science, paving the way for accurate metabolite profiling in microbes, plants and animals (Yandeau-Nelson et al., 2016; Kumar et al., 2017).

Correlations among plant traits often represent significant trade-offs or allometric relationships in key biological functions such as carbon gain, structural support, water uptake, and reproduction, which are linked to different plant organs. These correlations can be visualized as a trait network, where traits are represented as network nodes and strong correlations between traits are depicted as edges, connecting the nodes within the network (Kleyer et al., 2019). Spurious correlations can arise even when random variables $X$, $Y$, and $Z$ are uncorrelated with each other. For instance, the ratio $X/Z$ might positively correlate with $Y/Z$ simply because $Z$, the denominator, can vary widely. When $Z$ is small, both ratios can take on large values, and when $Z$ is large, both ratios may be small, creating an apparent correlation. In the context of the leaf economic spectrum, the photosynthetic rate per unit leaf area may show little to no correlation with the nitrogen content per unit leaf area. However, if both variables are adjusted by dividing by leaf mass per area (LMA) or multiplying by specific leaf area, which is the inverse of LMA, a significant correlation is likely to emerge between photosynthetic capacity per unit mass and nitrogen concentration per unit mass (Poorter et al., 2014).

## 15.6 VISUALIZATION TECHNIQUES

### 15.6.1 Basic Plotting

HTP relies on advanced visualization techniques, such as basic plotting, to enhance the analysis of plant data, particularly for assessing traits under varying environmental conditions. Basic plotting serves as a critical tool for organizing and interpreting complex datasets by providing visual insights into plant traits and growth patterns. Software like IMAP allows for rapid image processing and visualization, generating plot-level metrics from aerial images within seconds, which is crucial for large-scale phenotyping studies (Kim, 2021). Additionally, tools like the "implant" package in R integrate image processing with statistical analysis, enabling robust feature extraction and functional growth curve analysis to better understand plant dynamics and developmental stages (Wang et al., 2020). Furthermore, plot-level models that utilize P-splines and mixed models capture essential growth stages and quantify traits through repeated measurements, thereby enhancing data accuracy and reliability (Roth et al., 2021). Non-invasive imaging techniques, such as thermal and multispectral sensors, provide valuable data on plant responses to water and nitrogen stress, enriching the phenotyping process and enabling more precise monitoring of plant health and performance (Sahoo et al., 2023). These visualizations and plotting techniques are essential for extracting meaningful insights from high-throughput data, ultimately advancing plant research and breeding programs.

### 15.6.2 Advanced Visualization

Interactive dashboards are used for exploring complex phenotypic data in biomedical research. These platforms enable to visualization and analysis of multidimensional datasets, providing the discovery of patterns and associations. Phenome-wide association Studies-Mendelian randomization extension allows for the exploration of multi-morbidity patterns in phenome-wide association studies (Strayer et al., 2019), while Pheno-Mapper utilizes topological analysis for interactive visualization of large-scale phenomics data (Zhou et al., 2021). Hyppo-X employs algebraic topology and graph theory to characterize environmental influences on phenotypic traits (Kamruzzaman et al., 2017). PEAX integrates visual phenotype model definition with statistical testing of expression data, enabling clinical domain experts to explore relationships between multivariate phenotype models and gene expression (Hinterberg et al., 2014). These tools demonstrate the potential of interactive dashboards to enhance the exploration and interpretation of complex phenotypic data, supporting hypothesis generation and advancing personalized medicine research.

3D visualization of plant architecture has emerged as an important tool for analyzing plant growth and adaptation to environmental factors and morphology. Compared to other methods like 2D methods, 3D visualization offers more effectiveness in detecting plant architectural characteristics (Van et al., 2019). Various approaches have been developed for 3D plant modeling, including fractal geometry, L-systems and stochastic processes (Hu et al., 2001). Recent advancements include a non-destructive, inexpensive 3D reconstruction system for Arabidopsis thaliana using visual volume intersection, enabling time-sequential scanning and quantitative evaluation of plant architectures (Kunita et al., 2021). For plants with sparse canopies, such as soybean and maize in early growth stages, computer vision-based 3D reconstruction using structure from motion and multiple view stereo techniques have proven to be an accurate and cost-effective alternative to manual measurements or expensive laser scanning (Santos and Oliveira, 2012). Functional-structural plant models integrate 3D representations with physiological and environmental processes to study complex plant interactions across scales (Godin and Sinoquet, 2005). Various techniques have been developed for 3D plant modeling and reconstruction, including optical projection tomography for capturing high-resolution 3D data of plant specimens at different scales (Lee et al., 2006). These methods can be categorized into creating non-existent plants (modeling) and reconstructing models from real-world plants (Okura, 2022). Visualization approaches for 3D plant models have been systematically analyzed, with static model visualization accounting for 83% of research and a focus on plant phenotypic characteristics and growth processes (Di et al., 2021).

### 15.6.3 Geospatial Analysis

A Geographic Information System (GIS) is typically defined as a collection of workers, computer hardware, software, and data used for collecting, managing, analyzing, and presenting all types of geographically linked data. Plant scientists can examine, analyze, comprehend, interpret, and visualize data in any way that reveals relationships, patterns, and trends in the form of maps, globes, reports, and charts by combining the visual elements of an application with the analytical capabilities of a database (Morgan, 2011). Developments in GIS technologies have made it possible to integrate open-source Python modules that perform various geospatial tasks (Grippa et al., 2017). The creators of GRASS GIS (Geographical resources analysis support system), which is now a key component of open-source geographic software, first employed this tactic in the 1980s (Neteler et al., 2012). The centerpiece of the Open-Source geographic information system, GRASS integrates network analytic features, SQL-based attribute management, and raster, vector, and image processing capabilities. Plant distribution patterns can be revealed through genetic, ecological, and geographic patterns through the examination of genebank and herbarium databases using DIVA-GIS (Digital vegetation analysis), a specialist software for genetic resources work (Hijmans et al., 2001). ArcPlantPattern is an ArcGIS extension that helps with planning planting schemes and analyzing plant distribution patterns for a range of uses, such as landscape architecture and land reclamation (Zhang et al., 2011).

## 15.7 SOFTWARE AND TOOLS FOR PHENOMICS DATA ANALYSIS

### 15.7.1 Overview of Popular Software Packages

The Pliman R package is a set of functions intended for use in plant image analysis, among other purposes. The package will assist researchers in the following tasks: (1) manipulating, segmenting, and computing image indexes based on Red, Green, Blue, Red-Edge, and Near-Infrared bands; (2) measuring the area and shape of leaves; (3) estimating the severity of plant diseases; (4) counting and extracting features (such as area, perimeter, radius, circularity, and eccentricity) of objects like grains, leaves, pods, pollen, and cells; and (5) computing RGB indexes for each item in an image (Olivoto, 2022). Similar to this, the FIELDimageR software makes it easier to analyze orthomosaic images from field trials in agriculture, allowing for the computation of vegetation indices, canopy cover, and plant count at the plot level (Matias et al., 2020). Easy Leaf Area can be easily customized to meet particular experimental needs because it was created in Python, a free and open-source programming language with mathematics and image processing capabilities. This program offers a quick, easy, and accurate way to calculate the leaf area from digital photos. Without masking soil or depending on infrared pictures to identify leaf area from background soil, this tool's capacity to separate leaf pixels from different backdrops also makes it useful for measuring leaf area and crop canopy (Easlon et al., 2014). To assess the genetic influences and heritabilities of quantitative traits in plant populations, specialized software programs like SASQuant have been created (Gusmini et al., 2007).

### 15.7.2 Dedicated Tools for Plant Phenomics

PhenoImage is an open-source, cross-platform, graphical user interface-based solution for HTP image processing that aims to enable image analysis for people with little to no programming experience. Because it is open-source, it can be made more user-friendly to accommodate individual needs. It is flexible to examine photos from a broad range of plant species and platforms thanks to the availability of numerous functions and filtering factors. High-performance computing clusters and individual computers can both run PhenoImage (Zhu et al., 2021). Plant CV (Plant computer vision), another open-source software package, is a toolset for image processing used in plant phenotyping analysis. The objective of the Plant CV project is to create an open-source, community-developed suite of modular, reusable, and repurposable tools for plant image analysis (Gehan et al., 2017). Open-source web-based image analysis platform that analyzes field data in the form of photographs using open CV (open-source computer vision). One of the outcomes would be the computation of vegetative indicators to forecast plant performance in actual field settings (Guzman et al., 2015). Researchers have also used ImageJ and other computer languages to create tools for 3D segmentation, co-localization, measuring confocal cells, and measuring roots.

### 15.7.3 Workflow Automation and Pipeline Integration

Workflow automation and pipeline integration are essential for enhancing the efficiency and accuracy of high-throughput phenomics in plant research, driven by recent technological advancements that enable streamlined data collection and analysis. Automated pipelines, such as those designed for multi-modal integration, play a critical role in extracting and analyzing large datasets from various sensors, significantly reducing the reliance on manual assessment (Hernandez et al., 2024). These systems are capable of processing extensive data, and linking phenotypic traits with environmental factors, thereby deepening the understanding of plant responses (Zhang et al., 2023). Additionally, high-throughput robotic platforms are increasingly employed to manage experimental workflows, using directed acyclic graphs to track processes and ensure reproducibility and data provenance. Automation tools like Apache Airflow further optimize operations, from sample transport to data collection, improving resource scheduling and experimental design (Kaspersetz et al., 2024). The integration of advanced technologies, including AI and computer vision, into phenomics, enables real-time monitoring and analysis, which is crucial for unravelling complex biological systems (Kumar et al., 2015). These automated systems also support the cultivation of diverse plant models, enhancing the scalability of phenotyping efforts (Cao et al., 2023). Despite these advancements, challenges such as standardizing protocols and ensuring data quality across diverse platforms persist, highlighting the need for continued innovation to fully realize the potential of high-throughput phenomics in agricultural research.

## 15.8 CASE STUDIES AND APPLICATIONS

### 15.8.1 Examples of Successful Phenomics Studies

High-throughput phenomics studies employ advanced statistical tools and data analysis techniques, such as machine learning and computer vision, to enhance the evaluation of plant traits, as demonstrated by several successful case studies. One notable study on *Populus trichocarpa* utilized a dataset of 1,672 images from Oak Ridge National Laboratory, applying OCR and machine learning for trait classification. The OCR model achieved a 94.31% accuracy in extracting treatment labels, and morphological classifications reached an average accuracy of 62.82%, showcasing the potential of automated data extraction and analysis in phenomics (Singhvi et al., 2024). Another significant example is the PlantPAD platform, which hosts over 421,000 images spanning 63 crops and 310 diseases, enabling intelligent disease diagnosis through pre-trained deep-learning models. This platform illustrates the effective integration of extensive image data and machine learning to enhance disease management in agricultural settings (Dong et al., 2024). Additionally, a systematic review of HTP techniques underscored the critical role of both hardware and software, noting that deep learning algorithms were employed in 69% of the reviewed methods, highlighting their importance in analyzing large-scale phenotypic data (Solimani et al., 2023).

### 15.8.2 Translational Applications in Crop Improvement and Breeding

Translational applications in crop improvement and breeding are increasingly leveraging statistical tools and data analysis to optimize high-throughput phenomics, significantly enhancing the efficiency of trait assessment and selection. The integration of AI with HTP technologies has revolutionized crop breeding by enabling the rapid analysis of large datasets to identify patterns between phenotypic traits and genetic factors, thereby improving the efficiency of trait selection (Sheikh et al., 2024). Machine learning techniques are particularly valuable in extracting meaningful insights from extensive phenotypic data, addressing a major bottleneck in traditional breeding programs by automating the identification of complex trait correlations (Gill et al., 2022). Additionally, the use of vegetation indices linked with AI and remote sensing technologies provides automated evaluations of crop health and productivity, which are crucial for precision agriculture. These indices enhance the accuracy of phenotypic assessments by enabling the prediction of various crop traits based on spectral data (Tayade et al., 2022). Software innovations, such as IMAP, further streamline the processing and visualization of image data, delivering rapid plot-level metrics that are essential for informed breeding decisions. IMAP's capability to process aerial imagery in real time exemplifies the potential of high-throughput data analysis in accelerating breeding cycles and improving crop varieties (Kim, 2021). Despite these advancements, challenges in data management and the integration of diverse phenotyping platforms persist, underscoring the need for continued innovation to fully harness the potential of HTP in crop improvement and breeding programs.

### 15.8.3 Challenges and Lessons Learned from Large-Scale Phenomics Projects

Large-scale phenomics projects in plant high-throughput phenomics face considerable challenges in data organization and analysis, particularly due to issues with data interoperability, statistical modeling, and the integration of diverse datasets. One major obstacle is the lack of standardized formats and descriptions for large-scale image data, which hampers data sharing and collaborative research across disciplines, as highlighted by the limited connection between phenomics and modelling communities (Dong et al., 2024; Saint Cast et al., 2022). Furthermore, the vast amounts of data generated by HTP require robust statistical frameworks to enhance the accuracy of genomic selection, particularly in integrating high-dimensional data. The complexity of multi-trait and genotype-environment interactions adds another layer of difficulty in accurately predicting phenotypic outcomes, necessitating advanced models that can handle these intricacies (Morota, 2022). Lessons from successful projects underline the importance of adopting standardized data formats and ontologies to improve interoperability and foster collaborative research (Saint Cast et al., 2022). Moreover, the integration of advanced imaging technologies and machine learning has proven effective in improving the analysis of phenomic data, offering deeper insights into plant traits and responses (Mansoor et al., 2024). Addressing these challenges is crucial for maximizing the potential of high-throughput phenomics in crop improvement and breeding programs.

## 15.9 FUTURE TRENDS IN PHENOMICS DATA ANALYSIS

### 15.9.1 Emerging Technologies

UAVs and the Internet of Things (IoT) are two popular technologies being used in agriculture that are ushering in a new era of precision agriculture by replacing old farming methods. The application of unmanned aircraft systems (UAS) as platforms for sensing and/or communication is another innovative technology that has great promise for precision agriculture (Ray, 2018). It was first presented as an affordable substitute method for high spatial and temporal resolution imaging acquisition, environmental monitoring, and imaging acquisition. UAVs are increasingly being used in agriculture these days to help farmers with decision-making and monitoring (Atzori et al., 2017). UAS are used in many agricultural operations, including weed control, fertilizer, irrigation, and insecticides. Additionally, the integration of innovative 3D reconstruction modeling techniques with UAS technology has made it possible to monitor. The fundamental idea is that various physical objects can communicate with one another through the use of particular addressing methods to be connected to the Internet. Several vertical areas, such as industry, transportation, healthcare, automobiles, smart homes, and agriculture, can benefit from the application of IoT technology (Wang et al., 2006). IoT devices in an agricultural setting offer helpful data on a variety of physical characteristics to improve cultivation techniques (Freeman et al., 2015). Since wireless data transmission is the foundation for the great majority of IoT applications across several markets, wireless sensor networks play a critical role in IoT technology.

### 15.9.2 Big Data Challenges and Solutions

HTP in plant research presents substantial challenges in managing the vast datasets generated, particularly in terms of data volume, complexity, and integration. The immense volume of data collected from various sensors complicates analysis and interpretation due to its heterogeneity and the complexity of extracting meaningful insights (Gill et al., 2022). Moreover, integrating data from diverse phenotyping platforms and sensor types remains a significant hurdle, limiting comprehensive analysis and reducing the effectiveness of data-driven decision-making in plant breeding (Kumar et al., 2022). To address these issues, innovative solutions are being developed, such as leveraging machine learning and deep learning algorithms, which enhance the accuracy of trait identification by extracting meaningful patterns from large, complex datasets (Singhvi et al., 2024; Gill et al., 2022). Additionally, automated systems, including drones and multispectral cameras, facilitate real-time monitoring and data collection, significantly improving the efficiency and scalability of phenotyping efforts. Collaborative, interdisciplinary research efforts are also crucial, fostering the development of user-friendly platforms and databases that seamlessly integrate genomics and phenomics data (Panda et al., 2022; Kumar et al., 2022). Despite these advancements, ongoing challenges in data integration and platform accessibility must be addressed to fully harness the potential of HTP in plant phenomics.

### 15.9.3 The Role of Artificial Intelligence and Automation in Phenomics

Artificial Intelligence (AI) components such as computer vision, machine learning, and deep learning have been effectively incorporated into non-invasive imaging methods. With the use of deep learning and machine learning for reliable picture analysis, this integration is progressively increasing the effectiveness of data gathering and analysis. Furthermore, the development of instruments and software used in field phenotyping for data management and gathering has been facilitated by AI (Nabwire et al., 2021). These technologies are essential to precision agriculture because they make it easier to identify complicated plant features and environmental data (Tripodi et al., 2022). Plant breeding and pathology operations are being improved by AI-based image analysis approaches, such as feature extraction, identification, classification, and prediction. Acquisition of images, pre-processing, segmentation, representation, description, and recognition are usually steps in the process (Hilli, 2022). Though AI has great promise for phenomics, it necessitates advanced programming knowledge and an awareness of different algorithms (Nakhle and Harfouche, 2021).

## 15.10 CONCLUSION

High-throughput phenomics platforms have revolutionized plant phenotyping by offering unprecedented precision and scalability. However, these platforms still face challenges in data management, analysis, and standardization, which must be addressed to fully realize their potential in breeding and crop management. Effective data acquisition, pre-processing, and quality control are critical to unlocking the potential of HTP in crop breeding. Refining transformation techniques will further enhance the accuracy and utility of large datasets, particularly when used under diverse environmental conditions. Integrating advanced experimental designs, such as randomization, blocking, and innovative sampling strategies, is essential for ensuring the validity of the data and enabling more accurate phenotypic analysis. Statistical approaches, such as PCA, clustering, linear regression, ANOVA, and time-series analysis, play a crucial role in enhancing the efficiency and accuracy of phenotypic data interpretation. As these methods evolve, they will continue to optimize plant breeding and data-driven agricultural research. Furthermore, machine learning and multi-omics platforms allow researchers to delve deeper into complex plant traits and genomic data, driving innovation in plant breeding, genetics, and phenotypic analysis, ultimately improving agricultural outcomes. Advanced visualization techniques and geospatial tools provide critical insights into plant trait dynamics, supporting informed decision-making in plant breeding and research. At the same time, specialized software and automated pipelines have significantly enhanced the precision and efficiency of HTP, accelerating discoveries in plant science. Nonetheless, persistent challenges in standardizing workflows and ensuring data consistency across platforms remain. Therefore, ongoing innovation and standardization efforts are necessary to fully harness the potential of AI, automation, and emerging technologies in crop improvement and precision agriculture.

## ACKNOWLEDGEMENT

All the authors are thankful to their respective Universities and Institutions for their valuable supports.

## DATA AVAILABILITY

The authors confirm that the data supporting the findings of this study are available within the book/chapter.

## REFERENCES

Abdullah, N. A., Noor, A. M., Rahiman, M. H., & Rani, K. A. (2018). Multivariate analysis of variance followed by descriptive discriminant analysis: An analysis of the acoustic treatments effect on mung beans' growth. In *AIP Conference Proceedings, 2013*(1). Kuala Lumpur: AIP Publishing.

Adak, A., & Murray, S. C. (2023). Crop Science Futurology: A Data-Driven Approach through Phenomic, Genomic and Enviromic Insights. *Authorea Preprints*

Aghamohammadi, A., and Meshkani, M. R. (2017). Bayesian quantile regression for skew-normal linear mixed models. *Communications in Statistics – Theory and Methods, 46*(22), 10953–10972.

Agostinelli, C., and Yohai, V. J. (2016). Composite robust estimators for linear mixed models. *Journal of the American Statistical Association, 111*(516), 1764–1774.

Ahmed, H. G. M. D., Zeng, Y., Fiaz, S., Rashid, A. R. (2023). Applications of high-throughput phenotypic phenomics. In: Prakash, C. S., Fiaz, S., Nadeem, M. A., Baloch, F. S., Qayyum, A. (Eds.), *Sustainable Agriculture in the Era of the OMICs Revolution* (pp. 119–134). Springer International Publishing.

Allen, C., Hu, J., Kumar, V., Ahmad, M. A., & Teredesai, A. (2021). Interpretable phenotyping for electronic health records. In *2021 IEEE 9th International Conference on Healthcare Informatics (ICHI)* (pp. 161–170). Victoria, Canada.

Amal, T. C., Thottathil, A. T., Veerakumari, K. P., Rakkiyappan, R., & Vasanth, K. (2020). Morphological traits of drought tolerant horse gram germplasm: Classification through machine learning. *Journal of the Science of Food and Agriculture, 100*(13), 4959–4967.

Amarappa, S., & Sathyanarayana, S. V. (2014). Data classification using support vector machine (SVM), a simplified approach. *International Journal of Electronics and Computer Science Engineering, 3*(4), 435–445.

Apiolaza, L. A., & Garrick, D. J. (2001). Analysis of longitudinal data from progeny tests: Some multivariate approaches. *Forest Science, 47*(2), 129–140.

Araújo, A. P. (2003). Analysis of variance of primary data on plant growth analysis. *Pesquisa Agropecuária Brasileira, 38*, 1–10.

Arend, D., Lange, M., Pape, J. M., Weigelt-Fischer, K., Arana-Ceballos, F., Mücke, I., & Junker, A. (2016). Quantitative monitoring of Arabidopsis thaliana growth and development using high-throughput plant phenotyping. *Scientific Data*, *3*(1), 1–13.

Atzori, L., Iera, A., & Morabito, G. (2017). Understanding the Internet of Things: Definition, potentials, and societal role of a fast-evolving paradigm. *Ad Hoc Networks*, *56*, 122–140.

Awada, L., Phillips, P. W., & Bodan, A. M. (2024). The evolution of plant phenomics: Global insights, trends, and collaborations (2000–2021). *Frontiers in Plant Science*, *15*, 1410738.

Backhaus, A., Kuwabara, A., Fleming, A., & Seiffert, U. (2010). Validation of unsupervised clustering methods for leaf phenotype screening. In *ESANN*, Bruges, Belgium.

Berry, J. C., Fahlgren, N., Pokorny, A. A., Bart, R. S., & Veley, K. M. (2018). An automated, high-throughput method for standardizing image color profiles to improve image-based plant phenotyping. *PeerJ*, *6*, e5727.

Bethapudi, S., & Desai, S. (2017). Median statistics estimates of Hubble and Newton's constants. *The European Physical Journal Plus*, *132*, 1–8.

Bragazza, L., Iacumin, P., Siffi, C., & Gerdol, R. (2010). Seasonal variation in nitrogen isotopic composition of bog plant litter during 3 years of field decomposition. *Biology and Fertility of Soils, 46*, 877–881.

Bretherick, A. D., Canela-Xandri, O., Joshi, P. K., Clark, D. W., Rawlik, K., Boutin, T. S., & Haley, C. (2020). Linking protein to phenotype with Mendelian randomization detects 38 proteins with causal roles in human diseases and traits. *PLoS Genetics*, *16*(7), e1008785.

Burger, B., Vaudel, M., & Barsnes, H. (2020). Importance of block randomization when designing proteomics experiments. *Journal of Proteome Research*, *20*(1), 122–128.

Burton, P. R., Scurrah, K. J., Tobin, M. D., & Palmer, L. J. (2005). Covariance components models for longitudinal family data. *International Journal of Epidemiology*, *34*(5), 1063–1077.

Cao, R., Li, N. T., Latour, S., Cadavid, J. L., Tan, C. M., Forman, A., & McGuigan, A. P. (2023). An automation workflow for high-throughput manufacturing and analysis of scaffold-supported 3D tissue arrays. *Advanced Healthcare Materials*, *12*(19), 220242.

Cardina, J., Sparrow, D. H., &Mc COY, E. L. (1995). Analysis of spatial distribution of common lambsquarters (Chenopodium album) in no-till soybean (Glycine max). *Weed Science*, *43*(2), 258–268.

Chantre, G. R., Blanco, A. M., Forcella, F., van Acker, R. C., Sabbatini, M. R., & González-Andújar, J. L. (2013). A comparative study between non-linear regression and artificial neural network approaches for modelling wild oat (Avena fatua) field emergence. *The Journal of Agricultural Science, 152*, 254–262.

Clarke, J. L., Qiu, Y., & Schnable, J. C. (2022). Experimental design for controlled environment high-throughput plant phenotyping. In A. Lorence & K. M. Jimenez (Eds.), *High-Throughput Plant Phenotyping: Methods and Protocols* (pp. 57–68). Springer US.

Colmsee, C., Mascher, M., Czauderna, T., Hartmann, A., Schlüter, U., Zellerhoff, N., & Sonnewald, U. (2012). OPTIMAS-DW: A comprehensive transcriptomics, metabolomics, ionomics, proteomics and phenomics data resource for maize. *BMC Plant Biology*, *12*, 1–10.

da Fonseca Cardoso, L. M., da Silva Ferreira, N. C., de Araripe Lopes Correa, M., da Silva, S. A., & Alves, L. A. (2024). Software for animal randomization: A tool for increasing the reproducibility of science. *Laboratory Animals*, *58*(2), 164–169.

Daiki, S., & Noshita, K. (2024). A comparative study of plant phenotyping workflows based on three-dimensional reconstruction from multi-view images. In *8th International Plant Phenotyping Symposium*, Fukoka, Japan.

Dasgupta, A., Sun, Y. V., König, I. R., Bailey-Wilson, J. E., & Malley, J. D. (2011). Brief review of regression-based and machine learning methods in genetic epidemiology: The Genetic Analysis Workshop 17 experience. *Genetic Epidemiology*, *35*(S1), S5–S11.

de los Campos, G., Pérez, P., Vazquez, A. I., & Crossa, J. (2013). Genome-enabled prediction using the BLR (Bayesian Linear Regression) R-package. *Methods in Molecular Biology, 1019*, 299–320.

Di, S., Cao, L., & Ge, Y. (2021). Visualization methods of 3D plant models: A systematic mapping study. *Journal of Electrical and Computer Engineering*, *2021*(1), 2754343.

Din, I. U., & Hayat, Y. (2021). ANOVA or MANOVA for correlated traits in agricultural experiments. *Sarhad Journal of Agriculture*, *37*(4), 1250–1259.

Dong, L., Jiang, H., & Aluru, M. (2019). *U.S. Patent No. 10,470,379*. Washington, DC: U.S. Patent and Trademark Office.

Dong, X., Zhao, K., Wang, Q., Wu, X., Huang, Y., Wu, X., & Hao, G. (2024). PlantPAD: A platform for large-scale image phenomics analysis of disease in plant science. *Nucleic Acids Research*, *52*(D1), D1556–D1568.

Duan, J. C., Härdle, W. K., & Gentle, J. E. (Eds.). (2011). *Handbook of Computational Finance*. Springer Science & Business Media.

Easlon, H. M., & Bloom, A. J. (2014). Easy Leaf Area: Automated digital image analysis for rapid and accurate measurement of leaf area. *Applications in Plant Sciences*, *2*(7), 1400033.

Endelman, J. B. (2011). Ridge regression and other Kernels for genomic selection with R package rrBLUP. *The Plant Genome, 4*(3). DOI: 10.3835/plantgenome2011.08.0024

Freeman, P. K., & Freeland, R. S. (2015). Agricultural UAVs in the US: Potential, policy, and hype. *Remote Sensing Applications: Society and Environment*, *2*, 35–43.

Frison, L., & Pocock, S. J. (1992). Repeated measures in clinical trials: Analysis using mean summary statistics and its implications for design. *Statistics in Medicine*, *11*(13), 1685–1704.

Gao, Q., Ostendorf, E., Cruz, J. A., Jin, R., Kramer, D. M., & Chen, J. (2015). Inter-functional analysis of high-throughput phenotype data by non-parametric clustering and its application to photosynthesis. *Bioinformatics, 32*(1), 67–76.

Gehan, M. A., Fahlgren, N., Abbasi, A., Berry, J. C., Callen, S. T., Chavez, L., & Sax, T. (2017). PlantCV v2: Image analysis software for high-throughput plant phenotyping. *PeerJ*, *5*, e4088.

Ghosal, S., Zheng, B., Chapman, S. C., Potgieter, A. B., Jordan, D. R., Wang, X., & Guo, W. (2019). A weakly supervised deep learning framework for sorghum head detection and counting. *Plant Phenomics, 2019*, 1525874.

Gianola, D., Cecchinato, A., Naya, H., & Schön, C. (2018). Prediction of complex traits: Robust alternatives to best linear unbiased prediction. *Frontiers in Genetics, 9, 195*

Gill, T., Gill, S. K., Saini, D. K., Chopra, Y., de Koff, J. P., & Sandhu, K. S. (2022). A comprehensive review of high throughput phenotyping and machine learning for plant stress phenotyping. *Phenomics*, *2*(3), 156–183.

Godin, C., & Sinoquet, H. (2005). Functional-structural plant modelling. *The New Phytologist*, *166 3*, 705–8.

Grippa, T., Lennert, M., Beaumont, B., Vanhuysse, S., Stephenne, N., & Wolff, E. (2017). An open-source semi-automated processing chain for urban object-based classification. *Remote Sensing*, *9*(4), 358.

Großkinsky, D. K., Syaifullah, S. J., &Roitsch, T. (2018). Integration of multi-omics techniques and physiological phenotyping within a holistic phenomics approach to study senescence in model and crop plants. *Journal of Experimental Botany, 69*(4), 825–844.

Guo, X., Qiu, Y., Nettleton, D., Yeh, C. T., Zheng, Z., Hey, S., & Schnable, P. S. (2021). KAT4IA: K-means assisted training for image analysis of field-grown plant phenotypes. *Plant Phenomics, 2021*, 9805489.

Gusmini, G., Wehner, T. C., & Donaghy, S. B. (2007). SASQuant: A SAS software program to estimate genetic effects and heritabilities of quantitative traits in populations consisting of 6 related generations. *Journal of Heredity*, *98*(4), 345–350.

Guzman, H. A. R., González-Navarro, F. F., Selvaraj-Gomez, M., Valencia, M., & Delgado, A. (2015). Field phenomics: A web-based image analysis platform using open-source tools. In *2015 International Conference on Computational Science and Computational Intelligence (CSCI)* (pp. 851–852). IEEE, Las Vegas, USA.

Han, L., Yang, G., Dai, H., Yang, H., Xu, B., Feng, H., Li, Z., & Yang, X. (2019). Fuzzy Clustering of Maize Plant-Height Patterns Using Time Series of UAV Remote-Sensing Images and Variety Traits. *Frontiers in Plant Science, 10*, 926.

Hartung, J., Wagener, J., Ruser, R., & Piepho, H. P. (2019). Blocking and re-arrangement of pots in greenhouse experiments: Which approach is more effective? *Plant Methods*, *15*, 1–11.

Hasan, B. M., & Abdulazeez, A. M. (2021). A review of principal component analysis algorithm for dimensionality reduction. *Journal of Soft Computing and Data Mining*, *2*(1), 20–30.

He, D., Kuhn, D. N., & Parida, L. (2016). Novel applications of multitask learning and multiple output regression to multiple genetic trait prediction. *Bioinformatics, 32*(12), i37- i43.

Hector, A. (2021). *The New Statistics with R: An Introduction for Biologists*. Oxford University Press.

Heil, B. J., Crawford, J., & Greene, C. S. (2022). The effects of nonlinear signal on expression-based prediction performance. *bioRxiv*.2022-06.

Henderson, C. R. (1984). *Applications of Linear Models in Animal Breeding 462*. University of Guelph.

Hendrik, P., Fabio, F., Olivia, V., Ulrich, S., & Anika, W. K. (2023). Pitfalls and potential of high-throughput plant phenotyping platforms. *Frontiers in Plant Science*, *14,* 1233794.

Hernandez, K. J., Ciardi, T. G., Yamamoto, R., Lu, M., Nihar, A., Jimenez, J. C., & Bruckman, L. S. (2024). L-PBF high-throughput data pipeline approach for multi-modal integration. *Integrating Materials and Manufacturing Innovation*, *13*, 758–772.

Hijmans, R. J., Guarino, L., Cruz, M. M., & Rojas, E. (2001). *Computer Tools for Spatial Analysis of Plant Genetic Resources Data: 1*. DIVA-GIS.

Hilli, J. H. (2022). An overview on phenomics applications in different agriculture disciplines. *International Journal of Plant & Soil Science*, *34*(22), 631–637.

Hinterberg, M. A., Kao, D. P., Bristow, M. R., Hunter, L. E., Port, J. D., & Görg, C. (2014). PEAX: Interactive visual analysis and exploration of complex clinical phenotype and gene expression association. In *Pacific Symposium on Biocomputing. Pacific Symposium on Biocomputing* (pp. 419–430), Aurora, USA.

Hu, B. G., Zhao, X., Yan, H. P., De Reffye, P., Blaise, F., Xiong, F. L., & Wang, Y. M. (2001). Plant growth modeling and visualization-review and perspective. *Acta Automatica Sinica*, *27*(6), 816–835.

Ilakiya, T., Parameswari, E., Davamani, V., Swetha, D., & Prakash, E. (2020). High-throughput crop phenotyping in vegetable crops. *Pharma Innovation*, *9*, 184–191.

Jiang, N., & Zhu, X. G. (2024). Modern phenomics to empower holistic crop science, agronomy, and breeding research. *Journal of Genetics and Genomics*. *51*(8), 790–800.

John, M., Haselbeck, F., Dass, R., Malisi, C., Ricca, P., Dreischer, C., Schultheiss, S. J., & Grimm, D. G. (2022). A comparison of classical and machine learning-based phenotype prediction methods on simulated data and three plant species. *Frontiers in Plant Science, 13*, 932512

Jolliffe, I. T., & Cadima, J. (2016) Principal component analysis: A review and recent developments. *Philosophical Transactions of the Royal Society A*, *374*, 20150202.

Kamruzzaman, M., Kalyanaraman, A., Krishnamoorthy, B., Hey, S., & Schnable, P. S. (2017). Hyppo-X: A scalable exploratory framework for analyzing complex phenomics data. *IEEE/ACM Transactions on Computational Biology and Bioinformatics, 18*, 1535–1548.

Kaspersetz, L., Schroeder-Kleeberg, F., Mione, F. M., Martinez, E. C., Neubauer, P., & Cruz-Bournazou, M. N. (2024). Automation of experimental workflows for high throughput robotic cultivations. In F. Manenti & V. Gintaras (Eds.), *Computer Aided Chemical Engineering 53* (pp. 2971–2976). Elsevier.

Katsara, M. A., Branicki, W., Walsh, S., Kayser, M., Nothnagel, M., & VISAGE Consortium. (2021). Evaluation of supervised machine-learning methods for predicting appearance traits from DNA. *Forensic Science International: Genetics*, *53*, 102507.

Kim, J. Y. (2021). Software design for image mapping and analytics for high throughput phenotyping. *Computers and Electronics in Agriculture*, *191*, 106550.

King, S. L. (2003). Using ROC curves to compare neural networks and logistic regression for modeling individual noncatastrophic tree mortality. *United States Department of Agriculture Forest Service General Technical Report nc* (pp. 349–358).

Kleyer, M., Trinogga, J., Cebrián-Piqueras, M. A., Trenkamp, A., Fløjgaard, C., Ejrnæs, R., & Blasius, B. (2019). Trait correlation network analysis identifies biomass allocation traits and stem specific length as hub traits in herbaceous perennial plants. *Journal of Ecology*, *107*(2), 829–842.

Kumar, J., Pratap, A., & Kumar, S. (2015). Plant phenomics: An overview. *Phenomics in Crop Plants: Trends, Options and Limitations*, *1*, 1–10.

Kumar, M., Mahato, A., Kumar, S., & Mishra, V. K. (2020). Phenomics-assisted breeding: An emerging way for stress management. *New Frontiers in Stress Management for Durable Agriculture*, 295–310.

Kumar, R., Bohra, A., Pandey, A. K., Pandey, M. K., & Kumar, A. (2017). Metabolomics for plant improvement: Status and prospects. *Frontiers in Plant Science,* 8, 1302.

Kundu, P., Puri, H., Mohanty, D., Shinde, S., & Grover, S. (2024). High-throughput phenotyping and its importance in host plant resistance. In S. Kumar & M. Furlong (Eds.), *Plant Resistance to Insects in Major Field Crops* (pp. 301–329). Springer Nature Singapore.

Kunita, I., Morita, M. T., Toda, M., & Higaki, T. (2021). A three-dimensional scanning system for digital archiving and quantitative evaluation of arabidopsis plant architectures. *Plant and Cell Physiology, 62*, 1975–1982.

Kuran, Ö., and Özkale, M. R. (2021). Improvement of mixed predictors in linear mixed models. *Journal of Applied Statistics*, *48*(5), 924–942.

Lee, K., Avondo, J., Morrison, H., Blot, L., Stark, M., Sharpe, J., Bangham, A., & Coen, E. (2006). Visualizing plant development and gene expression in three dimensions using optical projection tomography [W]. *The Plant Cell Online, 18*, 2145–2156.

Leonelli, S. (2013). Integrating data to acquire new knowledge: Three modes of integration in plant science. *Studies in History and Philosophy of Science Part C: Studies in History and Philosophy of Biological and Biomedical Sciences*, *44*(4), 503–514.

Li, D., Quan, C., Song, Z., Li, X., Yu, G., Li, C., & Muhammad, A. (2021). High-throughput plant phenotyping platform (HT3P) as a novel tool for estimating agronomic traits from the lab to the field. *Frontiers in Bioengineering and Biotechnology*, *8*, 623705.

Lozano-Claros, D., Meng, X., Custovic, E., Deng, G., Berkowitz, O., Whelan, J., & Lewsey, M. G. (2020). Developmental normalization of phenomics data generated by high throughput plant phenotyping systems. *Plant Methods*, *16*, 1–11.

Mansoor, S., Karunathilake, E. M. B. M., Tuan, T. T., & Chung, Y. S. (2024). Genomics, Phenomics, and Machine Learning in Transforming Plant Research: Advancements and Challenges. *Horticultural Plant Journal*. DOI: 10.1016/j.hpj.2023.09.005

Maroli, A. S., Gaines, T. A., Foley, M. E., Duke, S. O., Doğramacı, M., Anderson, J. V., & Tharayil, N. (2018). Omics in weed science: A perspective from genomics, transcriptomics, and metabolomics approaches. *Weed Science*, *66*(6), 681–695.

Martinez, M.-J., and Holian, E. (2014). An alternative estimation approach for the heterogeneity linear mixed model. *Communications in Statistics – Simulation and Computation, 43*(10), 2628–2638.

Matias, F. I., Caraza-Harter, M. V., & Endelman, J. B. (2020). FIELDimageR: An R package to analyze orthomosaic images from agricultural field trials. *The Plant Phenome Journal*, *3*(1), e20005.

McGill, R., Tukey, J. W., & Larsen, W. A. (1978). Variations of box plots. *The American Statistician*, *32*(1), 12–16.

Mohammadi, S. A., & Prasanna, B. M. (2003). Analysis of genetic diversity in crop plants—salient statistical tools and considerations. *Crop Science*, *43*(4), 1235–1248.

Moreira, F. F., Oliveira, H. R., Volenec, J. J., Rainey, K. M., & Brito, L. F. (2020). Integrating high-throughput phenotyping and statistical genomic methods to genetically improve longitudinal traits in crops. *Frontiers in Plant Science, 11,* 681.

Morgan, B. J. (2011). Geographic information systems for the plant sciences. *Arnoldia*, *69*(1), 14–22.

Morota, G., Jarquin, D., Campbell, M. T., & Iwata, H. (2022). Statistical methods for the quantitative genetic analysis of high-throughput phenotyping data. In A. Lorence & K. M. Jimenez (Eds.), *High-Throughput Plant Phenotyping: Methods and Protocols* (pp. 269–296). Springer US.

Nabwire, S., Suh, H. K., Kim, M. S., Baek, I., & Cho, B. K. (2021). Application of artificial intelligence in phenomics. *Sensors*, *21*(13), 4363.

Nakhle, F., & Harfouche, A. L. (2021). Ready, steady, go AI: A practical tutorial on fundamentals of artificial intelligence and its applications in phenomics image analysis. *Patterns*, *2*(9), 100323.

Neteler, M., Bowman, M. H., Landa, M., & Metz, M. (2012). GRASS GIS: A multi-purpose open-source GIS. *Environmental Modelling & Software*, *31*, 124–130.

Novotny, V., Jones, H., Feng, X., & Capodaglio, A. (1991). Time series analysis models of activated sludge plants. *Water Science and Technology*, *23*(4–6), 1107–1116.

O'malley, M. A., & Soyer, O. S. (2012). The roles of integration in molecular systems biology. *Studies in History and Philosophy of Science Part C: Studies in History and Philosophy of Biological and Biomedical Sciences*, *43*(1), 58–68.

Okura, F. (2022). 3D modeling and reconstruction of plants and trees: A cross-cutting review across computer graphics, vision, and plant phenotyping. *Breeding Science, 72*, 31–47.

Olivoto, T. (2022). Lights, camera, pliman! An R package for plant image analysis. *Methods in Ecology and Evolution*, *13*(4), 789–798.

Panda, D., Kumar, M., Mahalingam, L., & Raveendran, M. (2022). Artificial intelligence-aided phenomics in high-throughput stress phenotyping of plants. In U. Kose, V. B. S. Prasath, M. R. H. Mondal, P. Podder, & S. Bharati (Eds.), *Artificial Intelligence and Smart Agriculture Applications* (pp. 185–212). Auerbach Publications.

Parmley, K., Nagasubramanian, K., Sarkar, S., Ganapathysubramanian, B., & Singh, A. K. (2019). Development of optimized phenomic predictors for efficient plant breeding decisions using phenomic-assisted selection in soybean. *Plant Phenomics,* 2019, 5809404.

Parsons, S. A., Valdez-Ramirez, V., Congdon, R. A., & Williams, S. E. (2014). Contrasting patterns of litterfall seasonality and seasonal changes in litter decomposability in a tropical rainforest region. *Biogeosciences*, *11*(18), 5047–5056.

Peacock, L., Worner, S. P., & Pitt, J. P. (2007). The application of artificial neural networks in plant protection. *Eppo Bulletin, 37*, 277–282.

Poorter, H., Lambers, H., & Evans, J. R. (2014). Trait correlation networks: A whole-plant perspective on the recently criticized leaf economic spectrum. *New Phytologist*, *201*(2), 378–382.

Pound, M. P., Atkinson, J. A., Townsend, A. J., Wilson, M. H., Griffiths, M., Jackson, A. S., & French, A. P. (2017). Deep machine learning provides state-of-the-art performance in image-based plant phenotyping. *Gigascience*, *6*(10), gix083.

Pradal, C., Artzet, S., Chopard, J., Dupuis, D., Fournier, C., Mielewczik, M., & Cohen-Boulakia, S. (2017). InfraPhenoGrid: A scientific workflow infrastructure for plant phenomics on the grid. *Future Generation Computer Systems*, *67*, 341–353.

Pratta, N. N., Quaglino, M. B., Rodríguez, G. R., Zorzoli, R., & Pratta, G. R. (2010). A multivariate approach to the proteomics of tomato fruit ripening. *Genes, Genomes and Genomics*, *4*, 48–51.

Rahaman, M. M., Ahsan, M. A., & Chen, M. (2019). Data-mining techniques for image-based plant phenotypic traits identification and classification. *Scientific Reports*, *9*(1), 19526.

Raudonius, S. (2017). Application of statistics in plant and crop research: Important issues. *Zemdirbyste-Agriculture*, *104*(4), 377–382.

Ray, D., Pankow, J. S., & Basu, S. (2014). USAT: A unified score-based association test for multiple phenotype-genotype analysis. *Genetic Epidemiology, 40*(1), 20–34.

Ray, P. P. (2018). A survey on internet of things architectures. *Journal of King Saud University-Computer and Information Sciences*, *30*(3), 291–319.

Roth, L., Rodríguez-Álvarez, M. X., van Eeuwijk, F., Piepho, H. P., & Hund, A. (2021). Phenomics data processing: A plot-level model for repeated measurements to extract the timing of key stages and quantities at defined time points. *Field Crops Research*, *274*, 108314.

Sahoo, R. N., Viswanathan, C., Kumar, M., Bhugra, S., Karwa, S., Misra, T., & Kumar, S. (2023). High-throughput phenomics of crops for water and nitrogen stress. In M. H. Masthigowda, K. Gopalareddy, R. Khobra, G. Singh, & G. P. Singh (Eds.), *Translating Physiological Tools to Augment Crop Breeding* (pp. 291–310). Springer Nature Singapore.

Saint Cast, C., Lobet, G., Cabrera-Bosquet, L., Couvreur, V., Pradal, C., Tardieu, F., & Draye, X. (2022). Connecting plant phenotyping and modelling communities: Lessons from science mapping and operational perspectives. *Silico Plants*, *4*(1), diac005.

Santos, T. T., & De Oliveira, A. A. (2012). Image-based 3D digitizing for plant architecture analysis and phenotyping. In *Conference on Graphics, Patterns and Images*, 25, 2012, Ouro Preto. Workshop on Industry Applications. [Sl]. Conference Publishing Series, 2012.

Sewell, M. (2008). *Principal Component Analysis*. University College London.

Shastri, A. A., Ahuja, K., Ratnaparkhe, M. B., & Busnel, Y. (2021). Probabilistically sampled and spectrally clustered plant species using phenotypic characteristics. *PeerJ*, *9*, e11927.

Sheikh, M., Iqra, F., Ambreen, H., Pravin, K. A., Ikra, M., & Chung, Y. S. (2024). Integrating artificial intelligence and high-throughput phenotyping for crop improvement. *Journal of Integrative Agriculture*, *23*(6), 1787–1802.

Shell, M. S., Panagiotopoulos, A., &Pohorille, A. (2007). Methods based on probability distributions and histograms. In C. Chipot & A. Pohorille (Eds.), *Free Energy Calculations: Theory and Applications in Chemistry and Biology* (pp. 77–118). Springer.

Singhvi, V., Lunga, L., Nidhi, P., Keum, C., & Prakash, V. (2024). *High-Throughput Phenotyping using Computer Vision and Machine Learning. arXiv preprint arXiv:2407.06354.*

Solimani, F., Cardellicchio, A., Nitti, M., Lako, A., Dimauro, G., & Renò, V. (2023). A systematic review of effective hardware and software factors affecting high-throughput plant phenotyping. *Information*, *14*(4), 214.

Strayer, N., Shirey-Rice, J. K., Shyr, Y., Denny, J. C., Pulley, J. M., & Xu, Y. (2021). PheWAS-ME: A web-app for interactive exploration of multimorbidity patterns in PheWAS. *Bioinformatics*, *37*(12), 1778–1780.

Strock, C. F., Schneider, H. M., & Lynch, J. P. (2022). Anatomics: High-throughput phenotyping of plant anatomy. *Trends in Plant Science*, *27*(6), 520–523.

Tardieu, F., Cabrera-Bosquet, L., Pridmore, T., & Bennett, M. (2017). Plant phenomics, from sensors to knowledge. *Current Biology*, *27*(15), R770–R783.

Tayade, R., Yoon, J., Lay, L., Khan, A. L., Yoon, Y., & Kim, Y. (2022). Utilization of spectral indices for high-throughput phenotyping. *Plants*, *11*(13), 1712.

Thompson, D. (2021). Replication of randomized, controlled trials using real-world data: What could go wrong? *Value in Health*, *24*(1), 112–115.

Tibbs Cortes, L., Zhang, Z., & Yu, J. (2021). Status and prospects of genome-wide association studies in plants. *The Plant Genome*, *14*(1), e20077.

Tripodi, P., Nicastro, N., Pane, C., & Cammarano, D. (2022). Digital applications and artificial intelligence in agriculture toward next-generation plant phenotyping. *Crop and Pasture Science, 74*, 597–614.

Vad, V., Cedrim, D., Busch, W., Filzmoser, P., & Viola, I. (2017). Generalized box-plot for root growth ensembles. *BMC Bioinformatics*, *18*, 1–15.

Van Pham, T., Lee, B. C., & King, S. A. (2019). A comparison of effectiveness between 2-dimensional and 3-dimensional data visualization in detecting plant architectural characteristics. In *Human Interface and the Management of Information. Visual Information and Knowledge Management: Thematic Area, HIMI 2019, Held as Part of the 21st HCI International Conference, HCII 2019*, July 26–31, 2019, Orlando, FL, USA, Proceedings, Part I 21 (pp. 223–236). Springer International Publishing.

Wang, N., Zhang, N., & Wang, M. (2006). Wireless sensors in agriculture and food industry—Recent development and future perspective. *Computers and Electronics in Agriculture*, *50*(1), 1–14.

Wang, R., Qiu, Y., Zhou, Y., Liang, Z., & Schnable, J. C. (2020). A high-throughput phenotyping pipeline for image processing and functional growth curve analysis. *Plant Phenomics,* 2020, 7481687.

Wang, X., Xuan, H., Evers, B., Shrestha, S., Pless, R., & Poland, J. (2019). High-throughput phenotyping with deep learning gives insight into the genetic architecture of flowering time in wheat. *GigaScience*, *8*(11), giz120.

Whitley, E., & Ball, J. (2001). Statistics review 1: Presenting and summarising data. *Critical Care*, *6*, 1–6.

Xu, R., & Li, C. (2022). A review of high-throughput field phenotyping systems: Focusing on ground robots. *Plant Phenomics*, 2022, 9760269.

Xu, Y., Qiu, Y., & Schnable, J. C. (2018). Functional modeling of plant growth dynamics. *The Plant Phenome Journal*, *1*(1), 1–10.

Yandeau-Nelson, M. D., Lauter, N., &Zabotina, O. A. (2016). Advances in metabolomic applications in plant genetics and breeding. *CABI Reviews,* 2015, 1–15.

Ying, W. (2023). Phenomic studies on diseases: Potential and challenges. *Phenomics*, *3*(3), 285–299.

Zhang, M., Xu, S., Han, Y., Li, D., Yang, S., & Huang, Y. (2023). High-throughput horticultural phenomics: The history, recent advances and new prospects. *Computers and Electronics in Agriculture*, *213*, 108265.

Zhang, Y., Woodward, N. T., Unger, D., Hung, I. K., Oswald, B. P., & Farrish, K. W. (2011). A GIS tool for plant spatial pattern analysis. *Environmental Modelling & Software*, *26*(10), 1251–1254.

Zhou, N., Siegel, Z. D., Zarecor, S., Lee, N., Campbell, D. A., Andorf, C. M., & Friedberg, I. (2018). Crowdsourcing image analysis for plant phenomics to generate ground truth data for machine learning. *PLoS Computational Biology*, *14*(7), e1006337.

Zhou, Y., Kamruzzaman, M., Schnable, P., Krishnamoorthy, B., Kalyanaraman, A., & Wang, B. (2021). Pheno-Mapper: An interactive toolbox for the visual exploration of phenomics data. In *Proceedings of the 12th ACM Conference on Bioinformatics, Computational Biology, and Health Informatics* (pp. 1–10). Florida, USA.

Zhu, F., Saluja, M., Dharni, J. S., Paul, P., Sattler, S. E., Staswick, P., & Yu, H. (2021). PhenoImage: An open-source graphical user interface for plant image analysis. *The Plant Phenome Journal*, *4*(1), e20015.

# 16 Robotics and Aerial Vehicle for High-Throughput Phenotyping
## *Technologies and Prospects*

*Priti Kumari, Mani Manoj, Mahankali Sravanalakshmi, Macharla Srinidhi, Marudhachalam Kamalesh, Shanmugam Gavaskar, Alagarsamy Sumithra, and Arumugam Vijaya Anand*

## LIST OF ABBREVIATIONS

**2D:** Two Dimensional
**3D:** Three Dimensional
**AI:** Artificial Intelligence
**GIS:** Geographic Information Systems
**GNSS:** Global Navigation Satellite System
**GPS:** Global Positioning System
**HIS:** Hyperspectral Imaging
**HTP:** High-Throughput Phenotyping
**IoT:** Internet of Things
**LiDAR:** Light Detection and Ranging
**ML:** Machine Learning
**MSI:** Multispectral Imaging
**RFID:** Radio Frequency Identification
**RGB:** Red, Green and Blue
**SQL:** Structured Query Language
**UAVs:** Unmanned Aerial Vehicles
**UGVs:** Unmanned Ground Vehicles
**VIs:** Vegetation Indices
**VTOL:** Vertical Take-Off & Landing

## 16.1 INTRODUCTION

### 16.1.1 Overview of High-Throughput Phenotyping

High-throughput phenotyping (HTP) is defined as the non-destructive and efficient evaluation of plant phenotypes, significantly enhancing the ability to analyze complex traits in plant breeding and research. This approach integrates advanced technologies such as computer vision and machine learning (ML), allowing for the rapid processing of large datasets to extract specific traits with high accuracy (Singhvi et al., 2024). HTP is particularly important in areas like plant-insect interactions, where it facilitates the quantification of plant damage and insect behavior, thus aiding in the development of insect-resistant varieties (Kundu et al., 2024). Additionally, HTP plays a crucial

 DOI: 10.1201/9781003505631-16

role in breeding programs by enabling precise measurements of traits like grain yield, which are essential for improving crop performance under varying environmental conditions (Resende et al., 2024). The non-destructive nature of HTP allows for continuous monitoring of plants, making it a vital tool for modern crop improvement strategies aimed at enhancing stress resilience and yield potential (Thakur et al., 2024).

HTP is increasingly vital in agriculture, particularly for enhancing crop breeding and evaluating product efficacy. In maize breeding, HTP facilitates the rapid assessment of complex traits, such as grain yield, through the use of UAVs to measure Vegetation indices (VIs) and correlate them with yield outcomes. Studies have shown significant correlations between VIs and grain yield, with accuracy levels ranging from 79.07% to 95.94% across various locations (Resende et al., 2024). Moreover, HTP applications extend to evaluating herbicide and adjuvant efficacy, where multispectral imaging is employed to assess plant responses under different stress conditions. This approach has demonstrated improved pesticide effectiveness when adjuvants are used, highlighting the potential for HTP to enhance agricultural product development (Ramírez-Rojas et al., 2022). While these advancements show promise, challenges remain, such as the need for comprehensive datasets and the integration of ML techniques to optimize trait extraction and analysis (Singhvi et al., 2024).

HTP has significant applications in forestry and environmental monitoring, leveraging advanced technologies for enhanced data collection and analysis (Figure 16.1). For instance, drone-based remote sensing has emerged as a powerful tool for tracking phenological changes in conifer seedlings, utilizing carotenoid-sensitive spectral indices to monitor photosynthetic activity effectively across large populations (D'Odorico et al., 2020). This method allows for rapid assessment of tree responses to environmental variations, crucial for understanding climate change impacts. Additionally, precision forestry techniques, including the use of modern sensors and cloud-based analytics, facilitate real-time monitoring of forest environments, enabling stakeholders to manage resources sustainably (Haile et al., 2023). These technologies can assess various factors, such as soil traffic ability and worker safety, while also providing critical data for ecological assessments through high-throughput molecular phenotyping (Marie et al., 2020).

### 16.1.2 Role of Robotics and Aerial Vehicles in HTP

The evolution and integration of robotics and aerial vehicles in high-throughput processing reflect significant advancements in design and functionality. Recent studies highlight the diversification of aerial robotic platforms, such as rotorcraft and hybrid designs, which are tailored to meet various mission

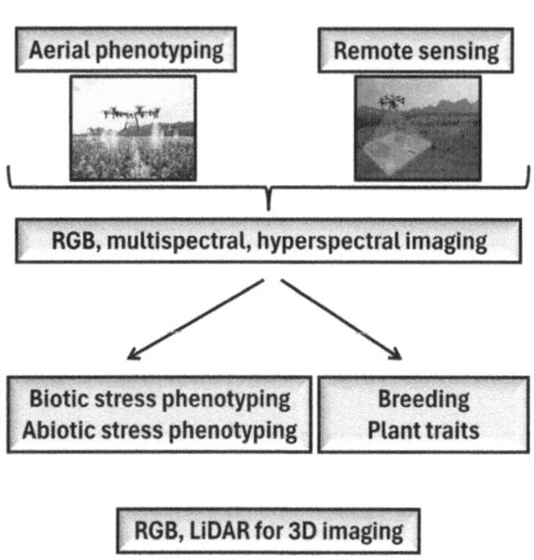


**FIGURE 16.1** Overview of high-throughput phenotyping.

profiles, including maintenance tasks in high-altitude environments (Butler and Costello, 2021). The integration of robotic manipulators into aerial systems enhances their operational capabilities, allowing for cooperative manipulation and improved interaction with the environment (Ollero et al., 2021). Furthermore, hybrid vehicles that combine aerial and ground functionalities are emerging, offering increased payload capacity and robustness against collisions, which is crucial for navigating challenging terrains (Kalantari et al., 2020). In the context of homeland security, unmanned aerial vehicles (UAVs) are being recognized for their strategic advantages in surveillance and reconnaissance, necessitating a cohesive integration strategy to avoid operational inefficiencies (Myers, 2005).

UAVs offer significant advantages over traditional methods in various applications, particularly in agriculture, surgery, and construction. UAVs enhance pesticide application efficiency by improving maneuverability and reducing labor intensity, which is crucial for effective pest control while minimizing environmental impact (Zhang et al., 2023). In bariatric surgery, robotic systems provide precision and a three-dimensional view, leading to reduced complication rates and improved surgical outcomes compared to conventional techniques (Ozmen et al., 2021). Furthermore, aerial robots facilitate complex construction tasks by operating autonomously in three-dimensional spaces, allowing for innovative building forms that traditional ground-based methods cannot achieve (Ammar, 2016). Additionally, aerial manipulation technologies enable efficient inspections and maintenance of high-altitude structures, reducing labor intensity and accident risks associated with manual operations (Hengjuan et al., 2018).

## 16.2 ROBOTICS IN HIGH-THROUGHPUT PHENOTYPING

### 16.2.1 Types of Robots Used in HTP

Unmanned ground vehicles (UGVs) have evolved significantly since their inception, with applications ranging from autonomous passenger cars to planetary exploration (Hebert et al., 2012). UGVs can be classified based on their control systems: teleoperated, autonomous, or using supervisory control (Gage, 1995). Recent advancements in UGV technology include the integration of advanced sensors like LiDAR (Light detection and ranging) cameras for 3D (Three dimensional) spatial data collection and the development of nature-inspired designs such as Satellite Pour I 'Observation de la Terre and Digit the Humanoid robots (Keshinro, 2023). These autonomous systems are increasingly being used to assist humans in various tasks, from food delivery to medical operations. UGV research has focused on improving mobility, navigation, sensing techniques, and route planning algorithms (Hebert et al., 2012). Despite the concept's long history, fully functioning UGVs capable of complex tasks have only recently become a reality, marking a new era in robotics (Tilbury and Ulsoy, 2011).

Autonomous and semi-autonomous robots are designed to operate in partially unknown environments, adapting to dynamic situations that require rapid behavioral modifications (Floreano et al., 1999). While fully autonomous systems are becoming more prevalent, there are still barriers to their widespread deployment, leading to the development of semi-autonomous systems that require some human intervention (Zilberstein, 2015). These systems combine autonomous operation with operator control, allowing for flexibility in complex scenarios (Sheh et al., 2009). Semi-autonomy is particularly relevant in personal robotics, where acceptability and adaptability to user needs are crucial (Laschi et al., 2001). Researchers are exploring various design choices and planning techniques to account for human involvement in semi-autonomous systems (Zilberstein, 2015). Implementing adaptable semi-autonomy through user modeling and explanation modules for human-robot communication can enhance the robot's ability to meet user preferences and improve overall performance in personal robotics applications (Laschi et al., 2001).

Field robots and drones (UAVs) are gaining popularity in various civil applications, including agriculture, surveillance, and disaster management (Ligon et al., 2018). In agriculture, UAVs are increasingly used for crop and animal monitoring, with the potential for controlling field robots and performing management actions (Vroegindeweij et al., 2014). The agricultural drone market is expected to grow significantly, as UAVs offer advantages over satellites in terms of frequency, detail, and cloud obstruction avoidance (Stehr, 2015). However, challenges remain, such as limited processing capabilities, low onboard storage, and communication range constraints (Ligon et al., 2018). Autonomous operation in complex agricultural environments requires further research, particularly in decision-making (Vroegindeweij et al., 2014). Cooperative systems combining UAVs and UGVs show promise for field surveillance missions, leveraging various sensors and mission processors to respond to dynamic environments (Zhang et al., 2018). As technology improves and costs decrease, drones are expected to see wider adoption in agriculture (Stehr, 2015).

### 16.2.2 Sensors and Data Collection Technologies

Imaging sensors play a crucial role in remote sensing and biodiversity monitoring. RGB (Red, Green and Blue) sensors, with their three broad spectral bands, are widely available and capable of broad habitat classification (Tait et al., 2019). However, multispectral and hyperspectral sensors offer superior spectral discrimination and accuracy for complex applications (Vagni, 2007; Tait et al., 2019). Hyperspectral imaging (HIS) systems typically collect over 100 narrow spectral bands, while multispectral systems capture fewer than 20 non-contiguous bands. These advanced sensors enhance target discrimination, camouflage detection, and other military applications (Vagni, 2007). Recent developments in compressive single-pixel HSI using RGB sensors offer a low-cost alternative to traditional hyperspectral systems (Tao et al., 2021). When comparing RGB and hyperspectral cameras for color discrimination, RGB devices are limited by metamerism and spectral variability, while hyperspectral cameras can reconstruct unique spectra for each pixel, providing insights into complex aspects of color vision (Garcia et al., 2015).

Non-imaging sensors like LiDAR, ultrasonics, and infrared are widely used in unmanned systems and everyday life (Bi et al., 2021). These sensors have proven effective for canopy characterization in vineyards and tree crops, with ultrasonic sensors providing average canopy measurements and LiDAR offering more detailed information (Llorens et al., 2011). Both technologies have shown good accuracy in measuring complex structures, with LiDAR slightly outperforming ultrasonic sensors (Chandan and Akhil, 2022). These sensors can accurately estimate geometrical and structural parameters of tree canopies, making them valuable for HTP and precision horticulture. Applications include real-time variable-rate solutions for adjusting plant protection product or fertilizer dose rates. While LiDAR technology shows great potential for tree phenotyping through highly accurate 3D modeling, its adoption by researchers for evaluating new plant varieties or management practices remains limited (Colaço et al., 2018).

Sensor fusion techniques combine data from multiple sensors to improve accuracy and reliability in various applications shown in Table 16.1. In mobile robotics, these techniques are crucial for tasks such as navigation, localization, and obstacle avoidance (Kam et al., 1997; Sharma et al., 2014). Common fusion methods include Bayesian, Dempster-Shafer, and fuzzy logic approaches, which can yield results differing by up to 20% (Chatterjee and Huntsberger, 1989). For mobile devices, sensor fusion helps identify users' daily activities but faces constraints like low memory and processing power (Pires et al., 2016). In robotics, sensor fusion is categorized into low-level fusion for direct integration of sensory data and high-level fusion for indirect integration in hierarchical architectures (Kam et al., 1997). These techniques address challenges such as over-dependency on individual sensors and unreliable data estimation (Sharma et al., 2014). Overall, sensor fusion enhances the interpretation of complex environments and improves decision-making in various technological applications.

### 16.2.3 Data Processing and Analysis

ML and artificial intelligence (AI) are revolutionizing data interpretation in various fields. In seismic interpretation, ML-assisted workflows can significantly reduce interpretation time from months to days, enabling multiple structural framework scenarios and improving decision-making

**TABLE 16.1**
**Phenotyping Robots are Categorized by the Drive Mechanism**

| Robot | Phenotyping Sensor | Perception Sensor | Crop | Advantages | Disadvantages | Applications |
|---|---|---|---|---|---|---|
| **Wheeled Robot (Skid Steering)** | | | | | | |
| VinBot | RGB-D camera<br>2D LiDAR | RTK-GNSSIMU | Grape | 1. Simple mechanical structure<br>2. Simple motion control<br>3. Low cost | 1. Low power efficiency in turning<br>2. High wear and tear of the tires can cause damage to the soil | 1. Yield estimation<br>2. Vineyard monitoring |
| Shrimp | 3D LiDAR RGB camera<br>Hyperspectral camera | RTK-GNSSIMU | Horticultural crop | 1. High maneuverability<br>2. Low Noise | 1. Limited payload capacity<br>2. Limited communication range | 1. Mango fruit detection<br>2. Almond mapping of flower, fruits, and yield<br>3. Almond fruit detection |
| Robotanist | Stereo camera<br>RGB camera | RTK-GNSSIMU<br>2D LiDAR<br>Stereo camera | Sorghum | 1. Scalability for large operations<br>2. Early detection of plant stress | 1. Technical complexity<br>2. Limited adaptability | 1. Corn stalk count and width estimation |
| Vinobot | Trinocular camera | GNSS<br>2D LiDAR | Maize, Sorghum | 1. Early disease detection<br>2. Real-time data for decision making | 1. High initial cost<br>2. Dependency on stable connectivity | 1. Simulation of Vinobot |
| TerraSentia | Multispectral camera<br>2D LiDAR<br>Hyperspectral camera<br>RGB camera<br>RGB-D camera | GNSS<br>Gyroscope | Maize | 1. Precise crop monitoring<br>2. Compact and lightweight | 1. Limited field coverage<br>2. Limited versatility | 1. Corn stem width estimation<br>2. Under canopy navigation |
| MARRIA | 2D LiDAR RGB-D camera | RTK-GNSSIMU | Not specified | 1. Autonomous operation<br>2. Data-driven decision making | 1. Technical complexity<br>2. Limited adaptability in certain environments | 1. Simulation of an agricultural robot |

*(Continued)*

**TABLE 16.1 (*Continued*)**
**Phenotyping Robots are Categorized by the Drive Mechanism**

| Robot | Phenotyping Sensor | Perception Sensor | Crop | Advantages | Disadvantages | Applications |
|---|---|---|---|---|---|---|
| | RGB-D camera | RTK-GNSS 2D LiDAR | Maize | 1. Reduction in human error<br>2. Support for precision planting and harvesting | 1. Maintenance cost<br>2. Data management challenges | 1. Corn stalk diameter estimation |
| **Wheeled Robot (Differential Drive)** | | | | | | |
| RobHortic | Thermal camera<br>Multispectral camera<br>RGB camera<br>NIR camera<br>Hyperspectral camera | RTK-GNSS | Horticultural crop | 1. Simple mechanical structure<br>2. Simple motion control<br>3. Low cost | 1. Less precise in steering control | 1. Carrot disease detection |
| **Wheeled Robot (2WD2WS and Ackerman Steering)** | | | | | | |
| AgBotll | RGB camera | RTK-GNSS | Row crop | 1. Precise steering control | 1. Need coordination between the drive wheel and steering wheel | 1. Weed detection |
| Phenobot 1.0 | Stereo camera | RTK-GNSS | Sorghum | 1. Automated data collection<br>2. Real-time data analysis | 1.Technical complexity<br>2.Limited adaptability to different environments | 1. Sorghum plant height and stalk diameter estimation<br>2. Sorghum plant architecture |
| AgriRover-01 | 3D LiDAR | RTK-GNSS | Corn | 1. Autonomous operation<br>2. Enhanced precision | 1. Technical complexity<br>2. Limited adaptability in certain environments | 1. Plant height and row spacing estimation |
| DairyBiobot | 2D LiDAR NDVI sensor | RTK-GNSS | Ryegrass | 1. Data collection and analysis<br>2. Reduced labor costs | 1. Maintenance and repair cost<br>2. Dependence on technology | 1. Ryegrass biomass estimation |

(*Continued*)

**TABLE 16.1 (*Continued*)**
**Phenotyping Robots are Categorized by the Drive Mechanism**

| Robot | Phenotyping Sensor | Perception Sensor | Crop | Advantages | Disadvantages | Applications |
|---|---|---|---|---|---|---|
| | | | **Wheeled Robot (Articulated Steering) Sorghum** | | | |
| Phenobot 3.0 | Stereo camera | RTK-GNSS | Sorghum | 1. Good mobility for rough terrain<br>2. Precise steering control | 1. Increased complexity in mechanical structure | 1. Development of a strawberry harvesting robot<br>2. Development of row-following algorithm in polytunnels |
| | | | **Wheeled Robot (4WD4WS)** | | | |
| Thorvald II | Application dependent | GPSIMU 2D LiDAR | Row crop Horticulture crop | 1. High maneuverability<br>2. Good terrain adaptation | 1. Complex mechanical structure<br>2. Complex motion control<br>High cost | 1. Weed detection<br>2. Dataset collection |
| Ladybird | RGB camera<br>2D LiDAR<br>Hyperspectral camera<br>Stereo camera<br>Thermal camera | 2D LiDAR<br>RTK-GNSSIMU | Row crop | 1. Miniaturized design<br>2. Advanced sensing capabilities | 1. Limited payload capacity<br>2. Limited operational range | 1. Estimation of plant height from LiDAR measurement<br>2. Estimation of wheat green area index from LiDAR measurement |
| Phenomobile V1 | 2D LiDAR<br>RGB camera<br>Multispectral camera | RTK-GNSS | Row crop Wheat Mixed crop | 1. Automated data collection<br>2. Real-time monitoring | 1. Maintenance and repair cost<br>2. Data management challenges | 1. Plant phenotyping<br>2. Plant breeding |

(*Continued*)

**TABLE 16.1 (*Continued*)**
**Phenotyping Robots are Categorized by the Drive Mechanism**

| Robot | Phenotyping Sensor | Perception Sensor | Crop | Advantages | Disadvantages | Applications |
|---|---|---|---|---|---|---|
| AgRover | Not specified | RTK-GNSS | Row crop | 1. Automated field operations<br>2. Reduced labor costs | 1. Maintenance and repair cost<br>2. Data management and analysis | 1. Crop monitoring<br>2. Weed and pest control |
| Flex-Ro | RGB camera<br>Ultrasonic sensor<br>Infrared thermometer<br>Spectrometer | GNSS Obstacle detector | Row crop | 1. Precision agriculture<br>2. Automated operations | 1. Technical complexity<br>2. Limited adaptability in certain conditions | 1. Soil analysis<br>2. Crop monitoring |
| MARS X | Application dependent | RTK-GNSS | Row crop | 1. High precision<br>2. Real-time data collection | 1. Technical complexity<br>2. Limited adaptability in certain conditions | 1. Weed and pest management<br>2. Field mapping |
| **Tracked (Differential Drive)** | | | | | | |
| Armadillo Scout | Application dependent | GNSS 2D LiDAR | Not specified | 1. Good mobility for rough terrain<br>2. Low ground pressure | 1. Complex mechanical structure<br>2. Low power efficiency in turning<br>3. Can damage the soil | 1. Advanced navigation and mobility<br>2. Precision agriculture |
| PHENObot | RGB camera | RTK-GNSS | Grape | 1. Advanced monitoring<br>2. Weed and pest management | 1. High initial cost<br>2. Technical complexity | 1. Grape bunch and berry detection |
| | RGB camera | RTK-GNSS | Soybean | 1. Precision agriculture<br>2. Automated operations | 1. Maintenance and repair cost<br>2. Limited adaptability in extreme conditions | 1. Crop monitoring<br>2. Weed and pest management |
| | RGB camera | RTK-GNSS | Not specified | 1. Soil analysis<br>2. Field mapping | 1. Battery life and power requirements<br>2. Data management challenges | 1. Environmental monitoring<br>2. Sustainable farming |

(*Continued*)

**TABLE 16.1 (*Continued*)**
**Phenotyping Robots are Categorized by the Drive Mechanism**

| Robot | Phenotyping Sensor | Perception Sensor | Crop | Advantages | Disadvantages | Applications |
|---|---|---|---|---|---|---|
| TERRA-MEPP | Stereo camera<br>Depth camera<br>RGB camera | RTK-GNSS<br>Wheel encoder<br>Gyroscope | Sorghum | 1. Advanced navigation and mobility<br>2. Real-time data collection | 1. Technical complexity<br>2. Limited adaptability in extreme conditions | 1. Crop monitoring<br>2. Weed and pest management |
| Phenomobile V2 | 2D LiDAR<br>RGB camera<br>Multispectral camera | RTK-GNSSIMU | Wheat Cotton<br>Sunflower<br>Mixed crop | 1. Multi-sensor capability<br>2. Automated and efficient | 1. Maintenance and repair<br>2. Limited use in extreme conditions | 1. Crop phenotyping<br>2. Disease and pest detection |
| **Wheel-Legged (4WD4WS)** | | | | | | |
| BoniRob | Application dependent | RTK-GNSS<br>Inertial sensor<br>3D LiDAR | Row crop | 1. High maneuverability<br>2. Good terrain adaptation<br>3. Physical dimension can be changed | 1. Complex mechanical structure<br>2. Complex motion control<br>High cost | 1. Soil compaction and moisture measurement<br>2. Weed image dataset collection<br>3. Image dataset collection<br>4. Ground and aerial robot collaboration |

4WD4WS: four-wheel driving and four-wheel steering: 2WD2WS: two-wheel driving and two-wheel steering.

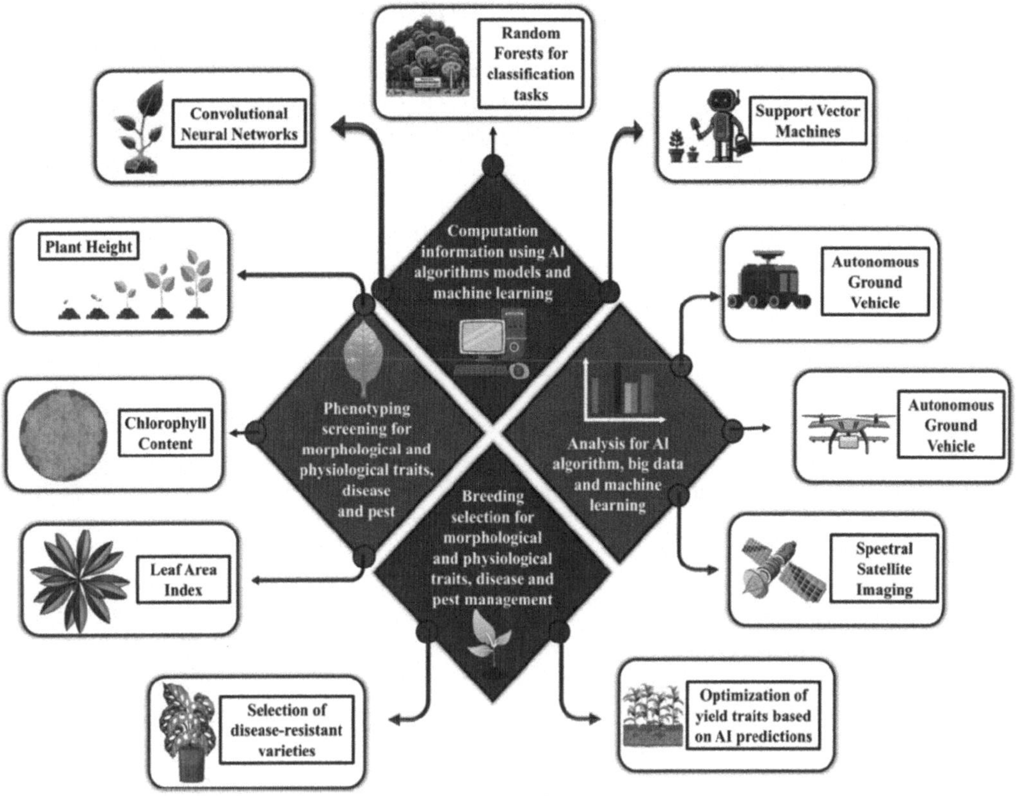


**FIGURE 16.2** Integration of artificial intelligence in high-throughput phenotyping.

(Dyer and Manral, 2021). Unsupervised neural networks, such as self-organizing maps, can reveal previously unidentified details in seismic data by analyzing multiple attributes simultaneously at the sample level (Sacrey and Roden, 2018). In formation testing, AI-based automation of the entire workflow, from protest to multiwall analysis, can reduce turnaround time from weeks to minutes while generating new insights (Ramatullayev and Mustapha, 2022). Gazprom Neft has developed ML algorithms to automate typical processes in seismic data interpretation, aiming to reduce the duration of processing and interpretation steps and ultimately shorten the overall exploration period (Semin et al., 2021). These advancements demonstrate the potential of ML and AI to transform geoscientists' roles and improve efficiency in data interpretation (Figure 16.2).

Feature extraction algorithms play a crucial role in computer vision and image processing. Early methods focused on pattern synthesis using Boolean matrices and matrix factorization (Abdali, 1971). Recent advances have led to more sophisticated algorithms, with the scale-invariant feature transform and maximally stable extremal regions being among the most effective (Salahat and Qasaimeh, 2017). These algorithms are computationally intensive but essential for enabling machines to interpret visual information. In multispectral imagery, features are extracted based on attributes such as color, texture, and intensity, with texture-based methods being particularly prominent (Vijayalakshmi et al., 2021). Common algorithms include mean shift, Principal Component transformation, and local binary patterns. For hyperspectral images, dimensionality reduction is a critical consideration. The choice of algorithm depends on the specific application and input type, as different methods may perform better under certain conditions (Salahat and Qasaimeh, 2017).

Big Data Analytics integration with enterprise systems offers significant opportunities for organizations to derive value from large-scale data processing and analysis. The integration process involves strategic planning, organizational alignment, and data governance (Loshin, 2013).

Scalable platforms built on technologies like Apache Hadoop enable data collection from diverse sources and support predictive analytics for process optimization (Sarnovský et al., 2017). The combination of big data and Hadoop has revolutionized various industries, enhancing decision-making and competitive advantage (Gandhi, 2023). For effective big data analytics, integrating R programming language with Hadoop provides a robust solution. R excels in data analysis, while Hadoop offers distributed computing capabilities for handling large datasets (Singh and Garg, 2021). Key components of big data ecosystems include high-performance analytics appliances, NoSQL (Structured Query Language) databases, and graph analytics tools (Loshin, 2013). Successful integration requires addressing challenges in data collection, processing, and analysis across different production environments (Sarnovský et al., 2017).

### 16.2.4 Applications of Robotics in HTP

Crop monitoring using remote sensing techniques has become increasingly important for food security and agricultural management. Multiyear polarimetric radar data can effectively track crop growth cycles and phenological stages, with an overall classification accuracy of 85% (Liu et al., 2012). Multimodal remote sensing approaches, combining data from multiple sources, offer more comprehensive and accurate crop monitoring by leveraging the strengths of different modalities (Karmakar et al., 2023). Various crop growth models, such as growth and yield simulators, coalition for environmentally responsible economies, and agricultural production systems simulators, have been developed to simulate crop development. Integration of remote sensing data with crop growth models can enhance simulation accuracy through "forced law" and "assimilation of law" methods. Future trends in crop monitoring are expected to combine moderate resolution satellite data (e.g., FengYun-3A, moderate resolution imaging spectroradiometer) with crop growth models, "3S" systems, and in situ observations for refined dynamic monitoring and quantitative assessment of crop growth (Zhang et al., 2010).

Pest and disease detection in crops is crucial for maintaining agricultural productivity. Various technologies and methods have been employed to address this challenge. Hyperspectral sensing has shown promise in detecting symptoms of pests and diseases in vegetable crops, with specific spectral bands identified as significant predictors (Apan et al., 2005). Image processing techniques have been utilized for pest identification and disease detection, involving image capture, pre-processing, and classification (Rajan et al., 2016). Detection systems have incorporated diverse technologies such as web-based, mobile-based, and Internet of Things (IoT) approaches, along with methods like expert systems, deep learning, and various ML algorithms (Anamisa et al., 2019). Artificial intelligence and deep learning have also been applied to develop effective monitoring systems for crop health. These technologies offer advantages over traditional methods like naked eye inspection or lab sampling, which can be time-consuming and inefficient for large-scale monitoring (Deepika and Kaliraj, 2021).

Recent research has focused on developing wireless sensor networks for measuring soil and environmental parameters. These systems typically employ multiple sensors to measure factors like soil moisture, temperature, and electrical conductivity (Plotog et al., 2014; Valente et al., 2007; Wang et al., 2010). Innovations include using three electrodes for improved soil moisture measurement and reducing electrode polarization (Plotog et al., 2014). Some designs incorporate solar-powered devices and ZigBee technology for wireless communication (Valente et al., 2007). Underground sensor nodes connected to above-ground relay nodes have been proposed for efficient data collection and transmission (Wang et al., 2010). These networks aim to provide comprehensive environmental monitoring with low power consumption and long-term deployment capabilities. Beyond soil parameters, modern methods and instruments can measure various atmospheric, hydrographic, and edaphic factors, including air quality, water properties, and climatic conditions (Posudin, 2014). These advancements contribute to more effective and widespread environmental monitoring and control.

### 16.2.5 Challenges and Limitations

Navigation and positioning systems face several technical challenges, including complexity, context, ambiguity, and environmental data handling in multisensor integration (Groves et al., 2014). Indoor location-based services struggle with reliability, availability of positioning technologies, up-to-date maps, and privacy concerns (Bassiri et al., 2017). Underwater vehicle navigation is advancing through improved sensor technology and algorithms, enabling novel missions and vehicles. Future challenges include near-bottom navigation, state estimation, and multi-vehicle navigation (Kinsey et al., 2006). Non-GNSS (Global Navigation Satellite System) radio frequency navigation, particularly using signals of opportunity, offers alternatives when GNSS performance is inadequate, such as indoors or in urban canyons. However, these systems face their own technical challenges that must be addressed for practical implementation (Raquet and Martin, 2008). Overcoming these obstacles is crucial for developing robust, reliable, and cost-effective navigation solutions across various environments and applications.

Environmental challenges, particularly weather and terrain variability, significantly impact military operations and exploration missions. Terrain and weather effects are fundamental to battlefield awareness and decision-making in net-centric operations (Hieb et al., 2007). For subterranean operations, terrain analysis is crucial for locating systems and developing courses of action, while weather conditions can affect system accessibility and detection (Bartolomeu, 2020). In exploration rover development, mission and environmental challenges include terrain effects, extreme temperatures, dust issues, and radiation protection (Zakrajsek et al., 2005). Modularity is identified as a key factor in developing viable surface mobility vehicles for Moon and Mars missions. To address these challenges in virtual warfare simulators, a multi-agent approach using neuro-fuzzy hybridization techniques has been proposed to model terrain and weather operations and predict their impact on air-tasking operations (Vijay et al., 2012). This approach aims to replicate real-world scenarios and assess weapon-induced damage in military training simulations.

Data overload is a significant challenge in modern decision-making processes. Managers across various levels experience both data and information overload, impacting their ability to make effective decisions (Letsholo and Pretorius, 2016). The problem can be characterized in three ways: as clutter, a workload bottleneck, or difficulty in finding significance, each leading to different solution approaches (Woods et al., 1999; Woods et al., 2002). The context-sensitivity of data significance poses a particular challenge, as it's difficult to determine a priori what information will be relevant (Woods et al., 2002). In the hospitality sector, managers cope with information overload using personal strategies like filtering, withdrawal, and summarizing, while organizations develop interactive dashboards (Saxena and Lamest, 2018). To address data overload effectively, solutions should focus on helping people recognize and explore relevant portions of data fields, rather than relying solely on machine processing, which can be brittle in unanticipated conditions (Woods et al., 2002).

## 16.3 AERIAL VEHICLES IN HIGH-THROUGHPUT PHENOTYPING

### 16.3.1 Types of Aerial Vehicles

UAVs, commonly known as drones, have gained widespread use in civilian and military applications. They can be broadly categorized into fixed-wing and multirotor designs, each with distinct advantages and disadvantages (Garg, 2022). Fixed-wing UAVs are suitable for surveying large areas, while multi rotors excel in maneuverability for close-range imaging (Atole et al., 2017). UAVs are equipped with various sensors, including cameras, inertial measurement systems, LiDAR, and GPS (Global positioning system), enabling real-time data collection and transmission. Applications range from agriculture and wildlife surveillance to security and construction (Atole et al., 2017). Recent technological advancements have improved UAV capabilities, allowing for high-resolution imagery and 3D mapping (Garg, 2022). However, challenges persist, including collision avoidance, battery life, and security concerns (Elmeseiry et al., 2021). The design and control of UAVs depend on their specific type and intended use, with ongoing research focusing on enhancing flight performance and addressing current limitations (Aber et al., 2019; Elmeseiry et al., 2021).

UAVs have gained significant attention in recent years, with applications spanning both civilian and military domains (Valavanis, 2007). UAVs range from insect-sized to commercial airliner dimensions, utilizing various propulsion types (Tsach et al., 2010). The U.S. military operates several major UAV systems, including the Predator, Global Hawk, and Hunter. Recent innovations include arming UAVs and teaming them with manned aircraft for improved targeting (Bone and Bolkcom, 2003). Vertical take-off and landing (VTOL) capabilities are desirable for both UAVs and manned aerial vehicles, offering advantages such as reduced dependency on airports. However, challenges remain in VTOL aircraft development, including technological constraints and performance optimization (Zhou et al., 2020). As UAV technology continues to advance, these systems present promising solutions for various airborne applications, potentially surpassing manned aircraft in performance and cost-effectiveness for certain missions (Tsach et al., 2010).

Recent advancements in satellite-based platforms have led to innovative solutions across various domains. Konakas and colleagues developed a multi-mission service platform for wide-area surveillance, integrating data from diverse sensors for applications like maritime monitoring (Konakas et al., 2014). Petcu and colleagues introduced a Grid-based platform specifically designed for satellite image processing, addressing the challenges of data intensity in remote sensing, particularly for environmental issues (Petcu et al., 2008). Roman-Gonzalez and Vargas-Cuentas proposed a user-centric web platform paradigm that adapts label priorities based on individual user needs, enhancing the management of satellite imagery through ML techniques. Lastly, (Zhao et al., 2003) presented a hierarchical three-dimensional grid architecture for satellite-based networks, demonstrating superior performance in routing algorithms through simulations on the optimized network engineering tools platform. Collectively, these studies highlight the versatility and potential of satellite-based platforms in addressing complex challenges in surveillance, environmental monitoring, and data management.

### 16.3.2 Sensor Technologies on Aerial Platforms

Recent advancements in HSI have addressed key challenges in spatial resolution and compactness. The mixel camera concept offers improved light collection and signal-to-noise ratio while eliminating keystone errors through a combination of hardware and software solutions (Hoye and Fridman, 2013). A compact voice information service near-infrared spectroscopy hyperspectral imager achieves high performance with minimal size and weight, suitable for UAV applications (Hyvärinen et al., 2011). Another approach utilizes a linear variable filter near the focal plane array, maintaining high spatial resolution through chromatic focusing corrections (Renhorn et al., 2016). To overcome resolution limitations, a matrix factorization technique reconstructs high-resolution hyperspectral images by combining lower-resolution hyperspectral data with high-resolution RGB images (Kawakami et al., 2011). These innovations collectively address the demand for improved spatial and spectral resolution in HSI across various applications, including geosensing and cultural heritage preservation.

LiDAR and radar technologies are crucial for autonomous vehicles and advanced driver-assistance systems, offering complementary capabilities. LiDAR provides high angular resolution and spatiotemporal consistency, while radar excels in adverse weather conditions and long-range detection (Bilik, 2022). Combining these technologies can enhance positioning accuracy and robustness in dynamic environments (Mounier et al., 2023). Research has shown that integrating radar with LiDAR can significantly reduce backscatter clutter, improving underwater target detection (Mullen et al., 1994). Both sensors employ different modulation schemes and imaging systems, with ongoing developments aimed at improving their performance (Giuffrida et al., 2023). Future advancements are expected to yield low-cost, long-range LiDAR systems and digital radars with LiDAR-like angular resolution (Bilik, 2022). As these technologies evolve, their integration and optimization for real-time processing will be crucial for meeting the demanding requirements of autonomous vehicle applications (Giuffrida et al., 2023).

The integration of GPS, GIS, and remote sensing technologies has revolutionized precision mapping and spatial data acquisition (Zeiler, 1992; Gao, 2002). This integration offers enhanced accuracy, cost-effectiveness, and rapid data collection for various applications, including electrical distribution mapping, ecological studies, and resource management (Zeiler, 1992; Rutter, 2007; Gao, 2002). Four integration models have been proposed: linear, interactive, hierarchical, and complex, with linear being the most common. Hierarchical integration is utilized in precision farming and environmental modeling, while complex integration is valuable for disaster mitigation and mobile mapping (Gao, 2002). Sub-meter precision GPS systems have extended the application range from landscape to patch-scale studies, enabling detailed analysis of animal behavior and social interactions (Rutter, 2007). Prospects include distributed mobile GIS and location-aware multimedia digital personal assistants, with full integration expected to find applications in new fields as mobile communications technologies improve (Gao, 2002).

### 16.3.3 Data Collection and Processing

Image stitching and 3D reconstruction are critical techniques in various fields, including forensic science and medical imaging. Iqbal and colleagues propose a method that utilizes multilayer information to convert 2D (Two dimensional) images into 3D representations, enhancing abnormal activity detection through the stitching of images captured from multiple viewpoints (Iqbal et al., 2016). Arya reviews various image stitching methods, emphasizing their application in creating high-resolution panoramas and reconstructing torn documents (Arya, 2015). Bergen and Wittenberg focus on endoscopic image stitching, highlighting its role in expanding the limited field of view during minimally invasive procedures (Bergen and Wittenberg, 2016). Shashank and colleagues provide a survey of image alignment and stitching methods, discussing the challenges and constraints involved in producing seamless images, particularly in forensic applications. Collectively, these studies illustrate the diverse applications and ongoing advancements in image stitching and 3D reconstruction techniques (Shashank et al., 2014).

HSI and multispectral imaging (MSI) are emerging technologies with applications across various fields, including pathology, agriculture, and remote sensing (Amigo, 2019; Ortega et al., 2020; Nalepa, 2021). These techniques capture spectral information beyond the visible range, complementing spatial data and potentially improving disease detection and clinical practice compared to traditional RGB imaging (Ortega et al., 2020). Recent advancements in sensor technology and data mining have expanded the possibilities for HSI and MSI analysis (Amigo, 2019; Nalepa, 2021). However, processing large volumes of hyperspectral data presents challenges, necessitating robust algorithms for target detection, background modeling, and dimensionality reduction (Kwon et al., 2013). Researchers are developing various image processing and ML pipelines to efficiently extract value from these highly dimensional datasets, enabling faster adoption of the technology in practice, including in hardware-constrained environments (Nalepa, 2021). The field continues to evolve, with ongoing research into non-Gaussian distribution modeling and comparative evaluations of anomaly detection methods (Kwon et al., 2013).

Temporal data collection and monitoring are crucial for various applications, including wildlife tracking and RFID (radio frequency identification) systems. In wildlife monitoring, wireless sensor networks offer new opportunities for collecting spatio-temporal data from remote locations, enabling the analysis of animal movement behavior (Baratchi et al., 2013). For RFID systems, temporal data management is essential due to the time-dependent, dynamic nature of RFID data, requiring specialized data models and query support for tracking and monitoring (Wang and Liu, 2005). The relational model has been proposed as an appropriate formalization for monitoring distributed systems, treating entities and relationships as time-varying relations (Snodgrass, 1982).

In the medical domain, temporal data collection and interpretation are particularly important for cardiovascular diagnoses, as the temporal characteristics of symptoms provide valuable diagnostic clues (Long et al., 1997). These diverse applications highlight the significance of effective temporal data collection and monitoring across multiple fields.

### 16.3.4 Applications of Aerial Vehicles in HTP

Recent advancements in large-scale crop phenotyping are crucial for enhancing crop breeding and functional genomics. Yang and colleagues highlight the bottleneck of acquiring large-scale phenotypic data and propose that technological innovations can bridge the phenotype-genotype gap, accelerating plant genetic improvements (Yang et al., 2020). Deery and colleagues discuss the potential of proximal remote sensing buggies equipped with various sensors for field-based phenotyping, emphasizing the need for a multidisciplinary approach to develop high-throughput techniques (Deery et al., 2014). Peshlov and colleagues describe the implementation of high-clearance mobile platforms (rovers) that automate data collection, significantly increasing the number of research plots analyzed per season (Peshlov et al., 2017). Shakoor and colleagues review the importance of high-resolution imaging and environmental sensors in HTP, which is essential for accelerating crop improvement and monitoring diseases. Collectively, these studies underscore the transformative potential of advanced phenotyping technologies in modern agriculture (Shakoor et al., 2017).

Environmental and climate monitoring has become crucial for understanding and addressing global challenges. Remote sensing technologies and robotic systems are increasingly used to gather data on various environmental factors, including temperature, rainfall, vegetation, and atmospheric conditions (Ceccato et al., 2014; Dunbabin and Marques, 2012). These monitoring efforts support decision-making in sectors such as agriculture, health, and disaster management by providing real-time, high-resolution data over large areas. The integration of satellite-derived products and in-situ measurements into early warning systems enables better response to climate-sensitive issues like malnutrition, infectious diseases, and agricultural pests (Ceccato et al., 2014). To ensure the accuracy and comparability of measurements across different sensors and time scales, stringent calibration and characterization processes are essential, particularly for climate monitoring applications (Cooksey et al., 2014). These advancements in environmental monitoring technologies and practices are critical for informed decision-making and long-term ecosystem management.

Forest biodiversity assessment is crucial for sustainable management and conservation. Key components include ecosystem structure, composition, and function (Spanos and Feest, 2007). Remote sensing technologies play a vital role in evaluating biodiversity across various spatial scales, from landscape to stand level, providing information on ecosystem diversity, structural attributes, and temporal changes (Innes and Koch, 1998). Important factors for assessment include processes generating and maintaining biodiversity, as well as silvicultural treatments that mimic natural disturbances (Spanos et al., 2010). International cooperation is essential for achieving global and regional perspectives on forest biodiversity, with various organizations contributing to assessment efforts (Bachmann et al., 2013). Monitoring methods range from line transect sampling to satellite remote sensing, allowing for comprehensive biodiversity evaluation (Bachmann et al., 1998). Integrating biodiversity information into forest management and policy-making is crucial for effective conservation strategies and sustainable forestry practices (Bachmann et al., 2013; Spanos et al., 2010).

### 16.3.5 Challenges and Limitations

Regulatory and safety issues in biotechnology and pharmaceuticals are complex and evolving. Stem cell therapies, while promising, face regulatory challenges due to safety concerns, despite growing evidence of their safety in clinical use (Centeno and Bashir, 2015). New biotechnological products, such as those produced by recombinant DNA and hybridoma techniques, present novel safety issues for regulatory agencies, requiring careful consideration of production methods and safety testing protocols (Petricciani, 1983). In the food industry, nanomaterials pose potential risks due to their ability to cross biological barriers, necessitating thorough toxicity assessments and regulatory frameworks (Tarhan, 2020). Responding to drug safety issues involves a systematic process of identification, investigation, and risk-benefit analysis, with the ultimate goal of protecting patients. This process requires proactive monitoring, evidence gathering, and effective communication of any necessary changes in drug use recommendations (Waller and Lee, 1999).

Data accuracy and validation are crucial for ensuring the reliability of research findings and decision-making processes. Validation studies of public health datasets, such as the Victorian Perinatal Data Collection, have shown high levels of accuracy for most data items, with some newer items requiring improvement (Flood et al., 2017). The quality of both primary and secondary data sources is essential, necessitating careful planning, thorough data entry procedures, and post-collection validation (Emanuelson and Egenvall, 2014). In the field of remote sensing, accuracy assessment has become an integral component of mapping projects, with error matrices commonly used to represent the accuracy of individual map categories (Congalton, 2001). The advent of big data has introduced new challenges in data quality, as the velocity and variety of data generation can lead to imperfections. Data validation processes and tools are crucial for recognizing and improving the quality of big data to prevent erroneous analysis results (Xie et al., 2017).

Cost and scalability are critical considerations in modern data systems. NoSQL databases have emerged to address the scalability challenges of traditional SQL systems, offering improved availability and scalability at the expense of certain consistency guarantees (Cattell, 2011). Large-scale data centers, viewed as warehouse-scale computers, require holistic design approaches to efficiently deliver Internet services (Barroso and Clidaras, 2022). However, the pursuit of scalability may come at a significant cost. The Configuration that Outperforms a Single Thread metric has been proposed to evaluate big data platforms, revealing that many systems require hundreds of cores before outperforming a single-threaded implementation (McSherry et al., 2015). This highlights the importance of considering both scalability and overhead when designing and selecting data systems. As data processing continues to move to the cloud, understanding these trade-offs becomes increasingly crucial for optimizing performance and cost-effectiveness in large-scale computing environments (Tao et al., 2013).

## 16.4 INTEGRATION OF ROBOTICS AND AERIAL VEHICLES

### 16.4.1 Collaborative Systems

Research demonstrates significant synergy between UAVs and UGVs in various applications. UAVs excel at covering large areas and providing aerial imagery, while UGVs offer accurate ground-level target localization (Grocholsky et al., 2006). This cooperation enhances surveillance operations, with UAVs autonomously following UGVs to assist in trajectory planning and navigation in challenging environments (Cantelli et al., 2013). In disaster response scenarios, opportunistic cooperation between UAVs and UGVs can emerge, either mediated by incident command staff or through direct coordination between robot teams (Perkins and Murphy, 2013). To extend UAV endurance, UGVs can serve as mobile docking stations, providing landing and recharging capabilities for UAVs during missions (Cocchioni et al., 2014). These collaborative approaches leverage the strengths of both platforms, enabling more efficient and effective operations in areas such as humanitarian demining, hazardous materials management, and extended indoor/outdoor missions.

Coordinated data collection strategies have been shown to significantly improve performance in large-scale data transfer scenarios compared to direct and non-coordinated methods (Cheng et al., 2004; Cheng et al., 2003; Chou et al., 2008). These approaches utilize multiple paths between source and destination hosts, considering the transfer demands of all sources to maximize network resource utilization (Cheng et al., 2004). Coordinated methods are more robust under inaccurate network condition information and various congestion types (Chou et al., 2008). In clinical trials, successful data collection strategies involve flexible, personnel-intensive approaches with multiple collection modes. High return rates (90% for healthcare service utilization calendars and 82% for annual surveys) were achieved by offering various collection methods (mail, email, telephone, and texting) and introducing incentives. Adapting to family preferences and providing flexible options are crucial for maximizing data collection response rates in studies involving remote data gathering (Finkelstein et al., 2016).

Real-time communication and data-sharing systems have become increasingly important with the rise of mobile devices and participatory sensing (Bagale et al., 2014). Web Socket technology has shown promise in supporting real-time communication, particularly for screen-sharing applications (Rahmatulloh et al., 2019). To address security concerns, a capability for sharing protected data in real-time has been developed, based on extending the mathematical foundations of need-to-know information security systems (James et al., 2012). In multiprocessor real-time systems, non-blocking protocols have been proposed to allow concurrent read and write operations on shared data, optimizing space requirements and reducing overhead compared to lock-based protocols (Tsigas and Zhang, 1999). These advancements collectively contribute to more efficient and secure real-time data sharing across various platforms, from mobile devices to multiprocessor systems, addressing challenges such as bandwidth limitations, scalability, and security while improving overall performance.

### 16.4.2 System Architectures

Research on centralized versus decentralized systems reveals a complex trade-off between efficiency and resilience. Centralized systems generally coordinate faster but are more vulnerable to systemic failures, while decentralized systems are more robust to errors but slower to coordinate (Veetil, 2017). This risk-return trade-off is particularly evident in decision-making processes, where centralization pools expertise but risks global consequences, whereas decentralization confines errors locally but increases their frequency (Arcuri and Dari-Mattiacci, 2010). In supply chain management, decentralized inventory systems reduce cost variance through risk diversification when supply disruptions occur, contrasting with the traditional preference for centralization in demand-driven scenarios (Schmitt et al., 2015). The optimal choice between centralization and decentralization depends on various factors, including communication error likelihood, urgency, and the need to prevent system collapse (Veetil, 2017). Historically, organizations have employed different approaches, highlighting the need for improved theories and business-oriented information technology management in this decision-making process (Hugoson, 2009).

Autonomous decision-making frameworks for unmanned vehicles are crucial for safe and efficient operation in dynamic environments. These frameworks typically involve multiple components, including motion planning, obstacle avoidance, and risk assessment (Widyotriatmo and Hong, 2008; Sankararaman, 2017). They must handle various uncertainties, such as sensor noise and unpredictable obstacles, often utilizing probabilistic methods like Markov decision processes (Widyotriatmo and Hong, 2008). Risk-informed decision-making is essential, with trajectories classified based on safety levels and potential outcomes (Sankararaman, 2017). The Beliefs, Desires, Intentions paradigm has been proposed as a basis for tactical decision-making systems, supported by tools for structural modeling and code generation (Evertsz et al., 2015). Adaptive Decision-Making Frameworks have shown promise in improving performance and robustness compared to static or random decision-making policies in multi-agent systems (Barber et al., 2003). These frameworks aim to balance reactivity with goal-oriented behavior in unpredictable environments.

Edge computing is emerging as a complementary paradigm to cloud computing, addressing challenges like latency, bandwidth, and real-time processing in IoT environments (Mittal et al., 2017; Bhalla and Kaur, 2024). By processing data closer to its source, edge computing enhances scalability and efficiency, particularly for applications requiring low-latency responses (Singh, 2017). The integration of edge and cloud computing creates a hybrid approach that leverages the strengths of both paradigms, enabling faster processing at the edge while utilizing cloud resources for more complex computations and storage. This convergence supports various industries, including manufacturing, transportation, and healthcare, by offering improved performance and real-time analytics capabilities (Sapra, 2023). However, challenges such as security, connectivity, and complexity need to be addressed for successful implementation (Bhalla and Kaur, 2024; Sapra, 2023). As the field evolves, future developments may include intelligent edge devices, enhanced security measures, and optimized integration with 5G networks (Sapra, 2023).

Head2Multi-modal phenotyping in agriculture utilizes various sensors and platforms to collect high-throughput data on crop traits. UAVs equipped with RGB cameras, LiDAR, and hyperspectral sensors enable rapid field coverage and data collection (Bartlett et al., 2017; Ravi et al., 2019). These systems can measure traits such as canopy cover, biomass, and plant stress, supporting both breeding programs and precision agriculture (Chawade et al., 2019). Integrating data from multiple sensors presents challenges in georeferencing and calibration, but can achieve accuracies up to 3 cm (Ravi et al., 2019). Modern phenomics accelerates crop physiology research, optimizes agronomic practices, and enhances breeding efforts. However, the field faces challenges in developing standards for data management, measuring physiological parameters, and enabling practical application by farmers and breeders (Jiang and Zhu, 2024). Overall, multi-modal phenotyping offers powerful tools for advancing crop science and agricultural production.

Integrated systems for environmental monitoring and disaster response combine various technologies to enhance data collection, analysis, and decision-making. These systems utilize remote sensing, satellite imagery, and in-situ sensors to gather real-time environmental data (Montgomery and Mundt, 2010; Hall et al., 2006). IoT and cloud computing enable efficient data management and processing (Fang et al., 2014). Geographic Information Systems (GIS) and web portals facilitate data visualization and collaboration among stakeholders (Montgomery and Mundt, 2010; Hürster et al., 2010). Such integrated approaches support various applications, including climate change monitoring, ecosystem assessment, and disaster preparedness (Fang et al., 2014; Hall et al., 2006). Early warning and risk management systems benefit from this interdisciplinary approach, incorporating multiple data sources and analytical tools (Hürster et al., 2012). These integrated systems aim to improve environmental protection, resource management, and disaster response by providing comprehensive, accessible information to researchers, policymakers, and emergency responders (Montgomery and Mundt, 2010; Hall et al., 2006).

Research and development projects are crucial for organizational competitiveness and innovation. Effective management of these projects requires advanced skills, tools, and methodologies, including competence management, talent and knowledge management, and modern information technology (Kisielnicki, 2014). Project managers play a multifunctional role in research and development projects, balancing research conditions, team dynamics, and project timelines (Jolović and Bobera, 2019). Various quantitative modeling approaches have been developed for research and development project selection and resource allocation, addressing the critical need for proper project choice to avoid resource waste (Heidenberger and Stummer, 1999). One proposed method combines Data Envelopment Analysis (DEA) for initial project ranking with the Value Creation Model for further evaluation of complex decisions (Linton et al., 2002). This approach aligns with managerial decision-making practices by automating obvious choices and focusing on nuanced considerations. Ultimately, successful research and development project management is essential for corporate renewal and market success (Heidenberger and Stummer, 1999).

## 16.5 FUTURE PROSPECTS

### 16.5.1 Technological Advancement

Recent advancements in AI and ML have led to significant improvements in precision and efficiency across various domains. Deep learning, reinforcement learning, and transfer learning techniques have been applied to enhance model performance (Patel, 2023). In medical imaging, AI and ML have shown promise in analyzing neuroimaging data, with the potential to uncover patterns that may have eluded human detection. ML enhancements have also been applied to fault-tolerant computing, improving real-time classification of non-identical inputs and environmental threats (Yin and Aroush, 2023). However, challenges remain, including the need for more transparent and explainable algorithms, addressing bias and fairness issues (Patel, 2023), and developing efficient learning methods suitable for real-time applications (Yin and Aroush, 2023). As AI and ML continue to evolve, researchers emphasize the importance of considering ethical implications and exploring the synergies between ML and fault-tolerant computing (Patel, 2023; Yin and Aroush, 2023).

Recent advances in sensor technologies have been driven by developments in electronic circuits, signal processing, and manufacturing techniques (Kanoun and Tränkler, 2004; Trankler and Kanoun, 2001). These advancements have led to innovative sensor structures with self-monitoring capabilities and improved performance. Miniaturization, multi-sensor systems, and wireless sensors are emerging trends in the field (Trankler and Kanoun, 2001). Various sensing technologies, including fiber optics, laser vibrometry, and microelectromechanical systems, have been developed for non-destructive testing and structural health monitoring applications (Hassani and Dackermann, 2023). The progress in sensor technology has expanded its applications in the medical, industrial, and automotive sectors (Krewski et al., 2007). These advancements have resulted in sensors with enhanced quality, reliability, and economic efficiency (Kanoun and Tränkler, 2004). The synergistic interaction of developments in related fields continues to drive innovation in sensor technology, promising further improvements in performance and capabilities.

Autonomous navigation and decision-making for unmanned vehicles have seen significant advancements in recent years. Research has focused on developing systems that can operate effectively with minimal human supervision in outdoor environments (Giesbrecht et al., 2007). These systems incorporate various modules, including path tracking, obstacle avoidance, and path planning, to enable point-to-point navigation in unknown terrain. Agent-oriented approaches have become increasingly popular for autonomous vehicle control, with architectures ranging from reactive to more advanced belief-goal-intention models (Veres et al., 2011). Recent methodologies employ techniques such as convolutional neural networks, model predictive control, and hidden Markov models for improved decision-making and path planning (Ramanathan and Kartik, 2021). Despite these advancements, challenges remain in ensuring safe execution across all driving situations. Standardization of decision-making through agent architectures has been proposed as a potential solution for future developments and legal certification of intelligent vehicles (Veres et al., 2011).

### 16.5.2 Potential Applications

Plant phenotyping, the identification of genotype and environmental effects on plant appearance and performance, is crucial for addressing future agricultural challenges (Minervini et al., 2015). However, phenotyping has become a bottleneck in crop improvement due to the rapid advances in genotyping (Fiorani and Schurr, 2013). To overcome this, researchers are developing high-throughput, non-invasive imaging techniques for both controlled and field environments (Araus et al., 2014; Fiorani and Schurr, 2013). These methods aim to provide quantitative analyses of plant structure and function, focusing on traits related to resource use efficiency and stress tolerance (Fiorani and Schurr, 2013). Field-based platforms combining remote sensing and automated environmental data collection are being developed to address spatial variability in field conditions (Araus et al., 2015). These advancements in phenotyping technologies are expected to accelerate breeding progress, enabling the development of crops with higher yield potential and improved stress resistance to meet future food production demands (Araus et al., 2014).

IoT integration in smart farming offers promising solutions for optimizing agricultural practices and increasing productivity. IoT-enabled systems can monitor various environmental parameters like temperature, humidity, soil moisture, and UV index in real-time, allowing farmers to make informed decisions (Doshi et al., 2019). These technologies facilitate process automation, waste reduction, and cost-efficiency in farming operations (Kanchana, 2018). The integration of IoT with

UAVs further enhances smart farming capabilities, enabling automated operations with minimal supervision (Islam et al., 2021). IoT-based smart farming incorporates expert intelligent techniques, robotics, and artificial intelligence algorithms to streamline data management, monitoring, and decision-making processes. While IoT in agriculture presents numerous opportunities, challenges such as connectivity limitations in remote areas and the need for skilled professionals to develop and utilize automated agricultural techniques must be addressed for successful implementation (Islam et al., 2021; Jaiswal and Rawat, 2021).

Precision farming is an innovative approach to agriculture that utilizes technology and data to optimize crop production while minimizing environmental impact (Hakkim et al., 2016; Finger et al., 2019). This site-specific management technique allows farmers to treat fields as heterogeneous entities, tailoring inputs spatially and temporally (Finger et al., 2019). By combining sensors, information systems, and enhanced machinery, precision agriculture enables better resource utilization and improved food supply sustainability (Gebbers and Adamchuk, 2010). The concept emerged as a response to the challenges of increasing food demand, limited land availability, and the need for eco-friendly agricultural practices (Mandal and Ghosh, 2000). While currently more prevalent in large farms in developed countries, precision farming has the potential to benefit small-scale farming systems in developing nations (Finger et al., 2019). As technology advances, precision farming tools are becoming more connected, accurate, and widely applicable, offering promising solutions for future food security challenges (Finger et al., 2019; Gebbers and Adamchuk, 2010).

### 16.5.3 Ethical and Societal Implications

Privacy and security concerns have emerged as significant barriers to e-commerce adoption and the effective utilization of big data. These concerns are the primary reason users hesitate to make online purchases (Udo, 2001). In the era of big data, protecting privacy and enhancing security has become crucial, with challenges arising from new technologies like IoT and cloud computing (Asasfeh et al., 2024; Bertino, 2016). The increasing data attack surface and the need to reconcile privacy with data used for security applications present additional challenges (Bertino, 2016). In the audit sector, the implementation of data analytics has raised concerns about client data privacy and security (Yunis et al., 2021). Addressing these concerns is essential for organizations to leverage the full potential of data analytics and maintain user trust. Future research should focus on developing comprehensive solutions to evolving data security and privacy challenges in various sectors (Bertino, 2016; Yunis et al., 2021).

Research on the impact of trade, automation, and immigration on employment reveals complex dynamics. Trade generally leads to short-term labor reallocation within sectors, with wage effects outweighing employment changes (Hoekman and Winters, 2005). Automation's overall impact on job numbers appears negligible, but it creates higher-skilled positions while potentially displacing low-skilled workers (Katz et al., 2023). Immigration's effect on average wages and employment is typically neutral or slightly positive, though it can have distributional consequences based on skill composition (Edo, 2019). Labor market institutions, including minimum wage, employment protection, unions, active labor market policies, and unemployment benefits, play crucial roles in shaping employment outcomes, with effects varying across countries and economic contexts (Giotis, 2024). These findings underscore the need for targeted policies to address potential negative impacts on vulnerable groups and to adapt to changing labor market dynamics.

Regulatory policy for emerging technologies like IoT and e-cigarettes requires balancing innovation with safety, security, and privacy concerns (Sicker and Redl, 2018; Bold and Krishnan-Sarin, 2019). Policymakers face challenges in promoting adoption while ensuring appropriate oversight, necessitating industry-led best practices and standards (Sicker and Redl, 2018). For nanomaterial-containing medical products, regulatory considerations include addressing the lack of toxicological data and standardized approaches for safety assessment (Faunce, 2008). Over the past two decades, OECD countries have developed various tools and institutions to improve regulatory quality, though implementation varies across nations (Malyshev, 2006). Ongoing research is crucial to inform evidence-based regulations, particularly for rapidly evolving technologies like e-cigarettes (Bold and Krishnan-Sarin, 2019). Effective regulatory policies must also consider cost-effectiveness and ensure that industry involvement in standard-setting does not hinder competition (Faunce, 2008).

## 16.6 CONCLUSION

Robotics and aerial vehicles have importantly transformed the landscape of HTP, and propose unprecedented accuracy and potency in agrarian research. These technologies enable the quick collection of large-scale phenotypic data, essential for improving crop varieties and addressing global food security challenges. The consolidation of advanced sensors, ML, and autonomous systems has enhanced the accuracy of data collection, reduced labor-intensive tasks, and increased the scalability of phenotyping operations. Furthermore, the continuous advancements in drone technology, coupled with robotics, present promising prospects for the future, such as real-time data processing, multi-modal sensing, and large-scale implementation. However, challenges remain, including the need for standardization, cost-effectiveness, and user-friendly interfaces to ensure broader adoption. As these technologies continue to evolve, they hold the potential to revolutionize agricultural practices, making HTP more accessible and impactful, ultimately contributing to sustainable food production and agricultural innovation.

Recent advancements in robotics and aerial vehicles have revolutionized high-throughput plant phenotyping. UAVs equipped with diverse sensors can expeditiously collect data on plant traits over large areas, providing essential insights into plant growth and stress responses. These technologies enable quick, accurate, and non-destructive measurements of morphological, chemical, and physiological properties of plants in both controlled environments and field conditions. While challenges remain in operating these systems in dynamic agricultural environments, the integration of robotics and remote sensing promises to importantly enhance plant breeding efforts and crop productivity assessment. As research continues, autonomous phenotyping systems are expected to advance rapidly, ushering in a new era of plant science and agricultural innovation.

## ACKNOWLEDGEMENT

All the authors are thankful to their respective Universities and Institutions for their valuable supports.

## DATA AVAILABILITY

The authors confirm that the data supporting the findings of this study are available within the book/chapter.

## REFERENCES

Abdali, S. K. (1971). Feature extraction algorithms. *Pattern Recognition*, *3*(1), 3–21.

Aber, J. S., Marzolff, I., Ries, J., & Aber, S. E. W. (2019). *Small-Format Aerial Photography and UAS Imagery: Principles, Techniques and Geoscience Applications*. Academic Press.

Amigo, J. M. (2019). Hyperspectral and multispectral imaging: Setting the scene. *Data Handling in Science and Technology, 32*, 3–16.

Ammar, M. (2016). *Aerial Construction: Robotic Fabrication of Tensile Structures with Flying Machines* (Doctoral dissertation, ETH Zurich).

Anamisa, D. R., Yusuf, M., Agustiono, W., & Svarief, M. (2019). Technologies. Methods, and approaches on detection system of plant pests and diseases. In *2019 6th International Conference on Electrical Engineering, Computer Science and Informatics (EECSI),* 163–170. IEEE, Bandung, Indonesia.

Apan, A., Datt, B., & Kelly, R. (2005). Detection of pests and diseases in vegetable crops using hyperspectral sensing: A comparison of reflectance data for different sets of symptoms. In *Proceedings of the 2005 Spatial Sciences Institute Biennial Conference 2005: Spatial Intelligence, Innovation and Praxis (SSC2005)*.

Araus, J. L., Elazab, A., Vergara, O., Cabrera-Bosquet, L., Serret, M. D., Zaman-Allah, M., & Cairns, J. E. (2015). *New Technologies for Phenotyping. Phenomics: How Next-Generation Phenotyping is Revolutionizing Plant Breeding*, 1–14. Springer Nature.

Araus, J. L., Li, J., Parry, M. A., & Wang, J. (2014). Phenotyping and other breeding approaches for a New Green Revolution. *Journal of Integrative Plant Biology*, *56*(5), 422–424.

Arcuri, A., & Dari-Mattiacci, G. (2010). Centralization versus decentralization as a risk-return trade-off. *The Journal of Law and Economics*, *53*(2), 359–378.

Arya, S. (2015). A review on image stitching and its different methods. *International Journal of Advanced Research in Computer Science and Software Engineering*, *5*(5), 299–303.

Asasfeh, A., Tarawneh, L., Pradhan, M. R., & Al-Dmour, N. A. (2024). Privacy and security concerns in big data. In *2024 2nd International Conference on Cyber Resilience (ICCR)*, 1–5. IEEE, Dubai, UAE.

Atole, R. R., Bello, L. C. S., & Lirag, J. R. S. (2017). Eyes in the sky: A review of civilian unmanned aerial vehicles (UAVs). *International Journal of Computer Applications*, *173*, 36–41.

Bachmann, P., Köhl, M., & Päivinen, R. (Eds.). (2013). *Assessment of Biodiversity for Improved Forest Planning, 51*. Springer Science & Business Media.

Bagale, J. N., Shiyanbola, A., Moore, J. P., & Kheirkhahzadeh, A. D. (2014). Towards a real-time data sharing system for mobile devices. In *2014 Eighth International Conference on Next Generation Mobile Apps, Services and Technologies*, 147–152. IEEE, Oxford, UK.

Baratchi, M., Meratnia, N., Havinga, P. J., Skidmore, A. K., & Toxopeus, B. A. (2013). Sensing solutions for collecting spatio-temporal data for wildlife monitoring applications: A review. *Sensors*, *13*(5), 6054–6088.

Barber, K. S., Gamba, I. M., & Martin, C. E. (2003). Representing and analyzing adaptive decision-making frameworks. *Agent Autonomy*, 29–54.

Barroso, L. A., & Clidaras, J. (2022). *The Datacenter as a Computer: An Introduction to the Design of Warehouse-Scale Machines*. Springer Nature.

Bartlett, P. W., Coblenz, L., Sherwin, G., Stambler, A., & van der Meer, A. (2017). A custom multi-modal sensor suite and data analysis pipeline for aerial field phenotyping. In J. A. Thomasson, M. McKee, & R. J. Moorhead (Eds.), *Autonomous Air and Ground Sensing Systems for Agricultural Optimization and Phenotyping II, 10218*, 19–28. SPIE.

Bartolomeu, J. P. S. (2020). Operations in subterranean systems: Terrain and weather variable. *Security and Defence Quarterly*, *29*(2), 39–60.

Basiri, A., Lohan, E. S., Moore, T., Winstanley, A., Peltola, P., Hill, C., & e Silva, P. F. (2017). Indoor location based services challenges, requirements and usability of current solutions. *Computer Science Review*, *24*, 1–12.

Bertino, E. (2016). Data security and privacy: Concepts, approaches, and research directions. In *2016 IEEE 40th Annual Computer Software and Applications Conference (COMPSAC), 1*, 400–407. IEEE, Atlanta, USA.

Bhalla, T., & Kaur, R. (2024). Convergence of Internet of Things (IoT) and Edge Computing: Enhancing Scalability and Efficiency through Cloud Integration: A Review. *International Journal of Engineering Science and Humanities*, *14*(1), 233–239.

Bi, X., & Bi, X. (2021). Infrared sensors and ultrasonic sensors. In X. Bi (Ed.), In *Environmental Perception Technology for Unmanned Systems*, 143–168.

Bilik, I. (2022). Comparative analysis of radar and lidar technologies for automotive applications. *IEEE Intelligent Transportation Systems Magazine*, *15*(1), 244–269.

Bold, K. W., & Krishnan-Sarin, S. (2019). E-cigarettes: Tobacco policy and regulation. *Current Addiction Reports*, *6*, 75–85.

Bone, E., & Bolkcom, C. (2003). *Unmanned Aerial Vehicles: Background and Issues for Congress*. Congressional Research Service, Library of Congress.

Butler, C., & Costello, M. (2021). Vehicle design in aerial robotics. *Current Robotics Reports*, *2*, 415–426.

Cantelli, L., Mangiameli, M., Melita, C. D., & Muscato, G. (2013). UAV/UGV cooperation for surveying operations in humanitarian demining. In *2013 IEEE International Symposium on Safety, Security, and Rescue Robotics (SSRR)*, 1–6. IEEE, Linkoping, Sweden.

Cattell, R. (2011). Scalable SQL and NoSQL data stores. *ACM Sigmod Record*, *39*(4), 12–27.

Ceccato, P., Fernandes, K., Ruiz, D., & Allis, E. (2014). Climate and environmental monitoring for decision making. *Earth Perspectives*, *1*, 1–22.

Centeno, C. J., & Bashir, J. (2015). Safety and regulatory issues regarding stem cell therapies: One clinic's perspective. *PM&R*, *7*(4), S4–S7.

Chandan, K. J., & Akhil, V. M. (2022). Investigation on accuracy of Ultrasonic and LiDAR for complex structure area measurement. In *2022 6th International Conference on Trends in Electronics and Informatics (ICOEI)*, 134–139. IEEE, Tirunelveli, India.

Chatterjee, P. S., & Huntsberger, T. L. (1989). Comparison of techniques for sensor fusion under uncertain conditions. In P. S. Schenker (Ed.), In *Sensor Fusion: Spatial Reasoning and Scene Interpretation*, *1003*, 194–201. SPIE.

Chawade, A., van Ham, J., Blomquist, H., Bagge, O., Alexandersson, E., & Ortiz, R. (2019). High-throughput field-phenotyping tools for plant breeding and precision agriculture. *Agronomy*, *9*(5), 258.

Cheng, W. C., Chou, C. F., Golubchik, L., Khuller, S., & Wan, Y. C. (2004). A coordinated data collection approach: Design, evaluation, and comparison. *IEEE Journal on Selected Areas in Communications*, *22*(10), 2004–2018.

Cheng, W. C., Chou, C. F., Golubchik, L., Khuller, S., & Wan, Y. C. (2003). Large-scale data collection: A coordinated approach. In *IEEE INFOCOM 2003. Twenty-Second Annual Joint Conference of the IEEE Computer and Communications Societies (IEEE Cat. No. 03CH3742), 1,* 218–228. IEEE, San Francisco, USA.

Chou, C. F., Wan, Y. C., Cheng, W. C., Golubchik, L., & Khuller, S. (2008). *A Performance Study of a Large-Scale Data Collection Problem*. Digital Repository at the University of Maryland.

Cocchioni, F., Pierfelice, V., Benini, A., Mancini, A., Frontoni, E., Zingaretti, P., & Longhi, S. (2014). Unmanned ground and aerial vehicles in extended range indoor and outdoor missions. In *2014 International Conference on Unmanned Aircraft Systems (ICUAS),* 374–382. IEEE, Orlando, USA.

Colaço, A. F., Molin, J. P., Rosell-Polo, J. R., & Escolà, A. (2018). Application of light detection and ranging and ultrasonic sensors to high-throughput phenotyping and precision horticulture: Current status and challenges. *Horticulture Research*, *5*, 35.

Congalton, R. G. (2001). Accuracy assessment and validation of remotely sensed and other spatial information. *International Journal of Wildland Fire*, *10*(4), 321–328.

Cooksey, C. C., Fraser, G. T., & Yoon, H. W. (2014). Sensor calibration and characterization to meet climate monitoring requirements. In *2014 IEEE Ninth International Conference on Intelligent Sensors, Sensor Networks and Information Processing (ISSNIP)*, 1–5. IEEE, Singapore.

Deepika, P., & Kaliraj, S. (2021). A survey on pest and disease monitoring of crops. In *2021 3rd International Conference on Signal Processing and Communication (ICPSC)*, 156–160. IEEE.

Deery, D., Jimenez-Berni, J., Jones, H., Sirault, X., & Furbank, R. (2014). Proximal remote sensing buggies and potential applications for field-based phenotyping. *Agronomy*, *4*(3), 349–379.

D'Odorico, P., Besik, A., Wong, C. Y., Isabel, N., & Ensminger, I. (2020). High-throughput drone-based remote sensing reliably tracks phenology in thousands of conifer seedlings. *New Phytologist*, *226*(6), 1667–1681.

Doshi, J., Patel, T., & Kumar Bharti, S. (2019). Smart Farming using IoT, a solution for optimally monitoring farming conditions. *Procedia Computer Science*, *160*, 746–751.

Dunbabin, M., & Marques, L. (2012). Robots for environmental monitoring: Significant advancements and applications. *IEEE Robotics & Automation Magazine*, *19*(1), 24–39.

Dyer, L., & Manral, S. (2021). Accelerating the Seismic Interpretation Workflow using Machine Learning. In *82nd EAGE Annual Conference & Exhibition, 2021*(1), 1–5. European Association of Geoscientists & Engineers, Amsterdam, The Netherlands.

Edo, A. (2019). The impact of immigration on the labor market. *Journal of Economic Surveys*, *33*(3), 922–948.

Elmeseiry, N., Alshaer, N., & Ismail, T. (2021). A detailed survey and future directions of unmanned aerial vehicles (UAVs) with potential applications. *Aerospace*, *8*(12), 363.

Emanuelson, U., & Egenvall, A. (2014). The data–sources and validation. *Preventive Veterinary Medicine*, *113*(3), 298–303.

Evertsz, R., Thangarajah, J., Yadav, N., & Ly, T. (2015). A framework for modelling tactical decision-making in autonomous systems. *Journal of Systems and Software*, *110*, 222–238.

Fang, S., Da Xu, L., Zhu, Y., Ahati, J., Pei, H., Yan, J., & Liu, Z. (2014). An integrated system for regional environmental monitoring and management based on internet of things. *IEEE Transactions on Industrial Informatics*, *10*(2), 1596–1605.

Faunce, T. A. (2008). Toxicological and public good considerations for the regulation of nanomaterial-containing medical products. *Expert Opinion on Drug Safety*, *7*(2), 103–106.

Finger, R., Swinton, S. M., El Benni, N., & Walter, A. (2019). Precision farming at the nexus of agricultural production and the environment. *Annual Review of Resource Economics*, *11*(1), 313–335.

Finkelstein, S. M., Celebrezze, M., Cady, R., Lunos, S., & Looman, W. S. (2016). Strategies to maximize data collection response rates in a randomized control trial focused on children with medical complexity. *Telemedicine and e-Health*, *22*(4), 295–301.

Fiorani, F., & Schurr, U. (2013). Future scenarios for plant phenotyping. *Annual Review of Plant Biology*, *64*(1), 267–291.

Flood, M. M., McDonald, S. J., Pollock, W. E., & Davey, M. A. (2017). Data accuracy in the Victorian perinatal data collection: Results of a validation study of 2011 data. *Health Information Management Journal*, *46*(3), 113–126.

Floreano, D., Godjevac, J., Martinoli, A., Mondada, F., & Nicoud, J. D. (1999). Design, control, and applications of autonomous mobile robots. *Advances in Intelligent Autonomous Systems*, 159–186.

Gage, D. W. (1995). UGV history 101: A brief history of Unmanned Ground Vehicle (UGV) development efforts. In *Unmanned Systems*, *13*(3), 9.

Gandhi, P. (2023). Advanced application development in agriculture—issues and challenges. In *International Conference on Information and Communication Technology for Intelligent Systems,* 625–635. Springer Nature Singapore, Ahmedabad, India.

Gao, J. (2002). Integration of GPS with remote sensing and GIS: Reality and prospect. *Photogrammetric Engineering and Remote Sensing*, *68*(5), 447–454.

Garcia, J. E., Girard, M. B., Kasumovic, M., Petersen, P., Wilksch, P. A., & Dyer, A. G. (2015). Differentiating biological colours with few and many sensors: Spectral reconstruction with RGB and hyperspectral cameras. *PLoS One*, *10*(5), e0125817.

Garg, P. K. (2022). Characterisation of fixed-wing versus multirotors UAVs/drones. *Journal of Geomatics*, *16*(2), 152–159.

Gebbers, R., & Adamchuk, V. I. (2010). Precision agriculture and food security. *Science*, *327*(5967), 828–831.

Giesbrecht, J., Collier, J., Broten, G., Monckton, S., & Mackay, D. (2007). A navigation and decision making architecture for unmanned ground vehicles. *DRDC Suffield TM*, *300*, 2007.

Giotis, G. (2024). Labor market institutions and employment. *Encyclopedia*, *4*(1), 273–294.

Giuffrida, L., Masera, G., & Martina, M. (2023). A survey of automotive radar and lidar signal processing and architectures. *Chips*, *2*(4), 243–261.

Grocholsky, B., Keller, J., Kumar, V., & Pappas, G. (2006). Cooperative air and ground surveillance. *IEEE Robotics & Automation Magazine*, *13*(3), 16–25.

Groves, P. D., Wang, L., Walter, D., Martin, H., Voutsis, K., & Jiang, Z. (2014). The four key challenges of advanced multisensor navigation and positioning. In *2014 IEEE/ION Position, Location and Navigation Symposium-PLANS 2014,* 773–792. IEEE.

Haile, M. A., Haile, D. T., & Zerihun, D. (2023). Real-time sensor data analytics and visualization in cloud-based systems for forest environment monitoring. *International Journal of Advances in Signal and Image Sciences*, *9*(1), 29–39.

Hakkim, V. A., Joseph, E. A., Gokul, A. A., & Mufeedha, K. (2016). Precision farming: The future of Indian agriculture. *Journal of Applied Biology & Biotechnology*, *4*(6), 68–72.

Hall, S. W., Carolla, M. A., & Deason, J. P. (2006). The use of imagery in environmental disaster preparedness and response. *Federal Facilities Environmental Journal*, *16*(4), 65–72.

Hassani, S., & Dackermann, U. (2023). A systematic review of advanced sensor technologies for non-destructive testing and structural health monitoring. *Sensors*, *23*(4), 2204.

Hebert, M. H., Thorpe, C. E., & Stentz, A. (Eds.). (2012). *Intelligent Unmanned Ground Vehicles: Autonomous Navigation Research at Carnegie Mellon*, *388*. Springer Science & Business Media.

Heidenberger, K., & Stummer, C. (1999). Research and development project selection and resource allocation: A review of quantitative modelling approaches. *International Journal of Management Reviews*, *1*(2), 197–224.

Hengjuan, L., Fengqin, S., & Junqing, W. (2018). Design of the Aerial Work Robot with Self-propelled. In *2018 2nd IEEE Advanced Information Management, Communicates, Electronic and Automation Control Conference (IMCEC)*, 1155–1158. IEEE, Xi'an, China.

Hieb, M. R., Mackay, S., Powers, M. W., Yu, H., Kleiner, M., & Pullen, J. M. (2007). Geospatial challenges in a net centric environment: Actionable information technology, design, and implementation. In *Defense Transformation and Net-Centric Systems 2007, 6578,* 369–380. SPIE.

Hoekman, B., & Winters, L. A. (2005). Trade and Employment: Stylized Facts and Research Findings. *World Bank Policy Research Working Paper,* 3676.

Høye, G., & Fridman, A. (2013). Mixel camera—a new push-broom camera concept for high spatial resolution keystone-free hyperspectral imaging. *Optics Express*, *21*(9), 11057–11077.

Hugoson, M. Å. (2009). Centralized versus decentralized information systems: A historical flashback. In *History of Nordic Computing 2: Second IFIP WG 9.7 Conference,* HiNC2, Turku, Finland, August 21–23, 2007, Revised Selected Papers 2, 106–115. Springer Berlin Heidelberg.

Hürster, W., Wilbois, T., & Chaves, F. (2012). An integrated systems approach for early warning and risk management systems. In F. Stowell (Ed.), *Systems Approach Applications for Developments in Information Technology*, 55–65. IGI Global.

Hyvärinen, T., Herrala, E., Procino, W., & Weatherbee, O. (2011). Compact high-resolution VIS/NIR hyperspectral sensor. In *Next-Generation Spectroscopic Technologies IV, 8032*, 237–244. SPIE.

Innes, J. L., & Koch, B. (1998). Forest biodiversity and its assessment by remote sensing. *Global Ecology & Biogeography Letters*, *7*(6), 397–419.

Iqbal, J. M., Parveen, M. S., & Arun, S. (2016). Image stitching and 2D to 3D Image Reconstruction for Abnormal Activity Detection. *International Journal of Computer Applications*, *133*(17), 1–7.

Islam, N., Rashid, M. M., Pasandideh, F., Ray, B., Moore, S., & Kadel, R. (2021). A review of applications and communication technologies for internet of things (Iot) and unmanned aerial vehicle (UAV) based sustainable smart farming. *Sustainability*, *13*(4), 1821.

Jaiswal, S., & Rawat, G. (2021). IoT-enabled smart farming: Challenges and opportunities. In A. Choudhury, A. Biswas, T. P. Singh, & S.K. Ghosh (Eds.), *Smart Agriculture Automation Using Advanced Technologies: Data Analytics and Machine Learning, Cloud Architecture, Automation and IoT*, 123–139. Springer Nature.

James, J. R., Mabry, F., & Huggins, K. (2012). Seeing the real world: Sharing protected data in real time. In *2012 45th Hawaii International Conference on System Sciences*, 2357–2365. IEEE.

Jiang, N., & Zhu, X. G. (2024). Modern phenomics to empower holistic crop science, agronomy, and breeding research. *Journal of Genetics and Genomics*.

Jolovic, I., & Bobera, D. (2019). Analysis of the project manager's role in the research and development projects. *Oditor*, 38.

Kalantari, A., Touma, T., Kim, L., Jitosho, R., Strickland, K., Lopez, B. T., & Agha-Mohammadi, A. A. (2020). Drivocopter: A concept hybrid aerial/ground vehicle for long-endurance mobility. In *2020 IEEE Aerospace Conference*, 1–10. IEEE, Big Sky, USA.

Kam, M., Zhu, X., & Kalata, P. (1997). Sensor fusion for mobile robot navigation. *Proceedings of the IEEE*, *85*(1), 108–119.

Kanchana, S. (2018). IoT in agriculture: Smart farming. *International Journal of Scientific Research in Computer Science, Engineering and Information Technology*, *3*, 181–184.

Kanoun, O., & Trankler, H. R. (2004). Sensor technology advances and future trends. *IEEE Transactions on Instrumentation and Measurement*, *53*(6), 1497–1501.

Karmakar, P., Teng, S. W., Murshed, M., Pang, S., Li, Y., & Lin, H. (2023). Crop monitoring by multimodal remote sensing: A review. *Remote Sensing Applications: Society and Environment*, *33*, 101093.

Katz, R., Callorda, F., & Jung, J. (2023). The impact of automation on employment and its social implications: Evidence from Chile. *Economics of Innovation and New Technology*, *32*(5), 646–662.

Kawakami, R., Matsushita, Y., Wright, J., Ben-Ezra, M., Tai, Y. W., & Ikeuchi, K. (2011). High-resolution hyperspectral imaging via matrix factorization. In *CVPR 2011*, 2329–2336. IEEE, Colorado Springs, USA.

Keshinro, B. I. (2023). *Human Activity Recognition Using Deep Learning Methods for Human-Robot Interaction* (Doctoral dissertation, North Carolina Agricultural and Technical State University).

Kinsey, J. C., Eustice, R. M., & Whitcomb, L. L. (2006). A survey of underwater vehicle navigation: Recent advances and new challenges. In *IFAC Conference of Manoeuvering and Control of Marine Craft*, *88*, 1–12. Lisbon, Portugal.

Kisielnicki, J. (2014). Project management in research and development. *Foundations of Management*, *6*(3), 57–70.

Konakas, S., Pantazis, S., Gardikis, G., Costicoglou, S., Andrikopoulos, I., Chaves, J. M., & Niebla, C. P. (2014). A multi-mission service platform for satellite-based wide area surveillance. In *2014 International Conference on Telecommunications and Multimedia (TEMU),* 19–23. IEEE, Heraklion, Greece.

Krewski, D., Glickman, B. W., Habash, R. W., Habbick, B., Lotz, W. G., Mandeville, R., & Weaver, D. F. (2007). Recent advances in research on radiofrequency fields and health: 2001–2003. *Journal of Toxicology and Environmental Health, Part B*, 10(4), 287–318.

Kundu, P., Puri, H., Mohanty, D., Shinde, S., & Grover, S. (2024). High-throughput phenotyping and its importance in host plant resistance. In *Plant Resistance to Insects in Major Field Crops*, 301–329. Springer Nature Singapore.

Kwon, H., Hu, X., Theiler, J., Zare, A., & Gurram, P. (2013). Algorithms for multispectral and hyperspectral image analysis. *Journal of Electrical and Computer Engineering*, *2013*, 908906-1.

Laschi, C., Teti, G., Tamburrini, G., Datteri, E., & Dario, P. (2001). Adaptable semi-autonomy in personal robots. In *Proceedings 10th IEEE International Workshop on Robot and Human Interactive Communication. ROMAN 2001 (Cat. No. 01TH8591)*, 152–157. IEEE, Bordeaux, France.

Letsholo, R. G., & Pretorius, M. P. (2016). Investigating managerial practices for data and information overload in decision making. *Journal of Contemporary Management*, *13*(1), 767–792.

Ligon, J., Bein, D., Ly, P., & Onesto, B. (2018). 3D point cloud processing using spin images for object detection. In *2018 IEEE 8th Annual Computing and Communication Workshop and Conference (CCWC),* 731–736. IEEE, Las Vegas, USA.

Linton, J. D., Walsh, S. T., & Morabito, J. (2002). Analysis, ranking and selection of R&D projects in a portfolio. *R&D Management, 32*(2), 139–148.

Liu, C., Shang, J., Vachon, P. W., & McNairn, H. (2012). Multiyear crop monitoring using polarimetric RADARSAT-2 data. *IEEE Transactions on Geoscience and Remote sensing, 51*(4), 2227–2240.

Liu, P., & Wang, F. (2005). Temporal management of RFID data. In *Proceedings of the 31st International Conference on Very Large Data Bases*, 1128–1139. Trondheim, Norway.

Llorens, J., Gil, E., Llop, J., & Escolà, A. (2011). Ultrasonic and LIDAR sensors for electronic canopy characterization in vineyards: Advances to improve pesticide application methods. *Sensors, 11*(2), 2177–2194.

Long, W. J., Fraser, H., Naimi, S., & Stahl, J. (1997). Collecting and Interpreting Temporal Data in an Expert System. In *Proceedings of the AMIA Annual Fall Symposium,* 872. American Medical Informatics Association, Nashville, USA.

Loshin, D. (2013). *Big Data Analytics: From Strategic Planning to Enterprise Integration with Tools, Techniques, NoSQL, and Graph.* Elsevier.

Malyshev, N. (2006). Regulatory policy: OECD experience and evidence. *Oxford Review of Economic Policy, 22*(2), 274–299.

Mandal, D., & Ghosh, S. K. (2000). Precision farming–The emerging concept of agriculture for today and tomorrow. *Current Science, 79*(12), 1644–1647.

Marie, B., Leloup, J., Bonin, A., Gallet, A., & Mouchet, M. (2020). Promising potential of high-throughput molecular phenotyping of freshwater fishes for environmental assessment. *bioRxiv*, 2020-12.

McSherry, F., Isard, M., & Murray, D. G. (2015). Scalability! but at what {COST}?. In *15th Workshop on Hot Topics in Operating Systems (HotOS XV)*, kartause Ittingen, Switzerland.

Minervini, M., Scharr, H., & Tsaftaris, S. A. (2015). Image analysis: The new bottleneck in plant phenotyping [applications corner]. *IEEE Signal Processing Magazine, 32*(4), 126–131.

Mittal, S., Negi, N., & Chauhan, R. (2017). Integration of edge computing with cloud computing. In *2017 International Conference on Emerging Trends in Computing and Communication Technologies (ICETCCT),* 1–6. IEEE, Dehradun, India.

Montgomery, K., & Mundt, C. (2010). A new paradigm for integrated environmental monitoring. In *Proceedings of the 1st International Conference and Exhibition on Computing for Geospatial Research & Application*, 1–7. Washington, USA.

Mounier, E., Dawson, E., Elhabiby, M., Korenberg, M., & Noureldin, A. (2023). Investigating the Complementary Use of Radar and LIDAR for Positioning Applications. *The International Archives of the Photogrammetry, Remote Sensing and Spatial Information Sciences, 48*, 909–914.

Mullen, L., & Herczfeld, P. R. (1994). Application of radar technology to aerial LIDAR systems. In *1994 IEEE MTT-S International Microwave Symposium Digest (Cat. No. 94CH3389-4)*, 175–178. IEEE, San Diego, USA.

Myers, M. K. (2005). *How Can Unmanned Aerial Vehicles Be Best Integrated into Homeland Security*. Naval War Coll Newport RI Joint Military Operations Dept.

Nalepa, J. (2021). Recent advances in multi-and hyperspectral image analysis. *Sensors, 21*(18), 6002.

Ollero, A., Tognon, M., Suarez, A., Lee, D., & Franchi, A. (2021). Past, present, and future of aerial robotic manipulators. *IEEE Transactions on Robotics, 38*(1), 626–645.

Ortega, S., Halicek, M., Fabelo, H., Callico, G. M., & Fei, B. (2020). Hyperspectral and multispectral imaging in digital and computational pathology: A systematic review. *Biomedical Optics Express, 11*(6), 3195–3233.

Ozmen, M., Gundogdu, E., & Guldogan, E. (2021). The current role of robotics in bariatric surgery. *Minerva Surgery, 76*(1), 24–32.

Patel, J. (2023). A comprehensive review of innovations in artificial intelligence and enhancing machine learning model precision and efficiency. In *2023 14th International Conference on Computing Communication and Networking Technologies (ICCCNT),* 1–4. IEEE, Delhi, India.

Perkins, T., & Murphy, R. R. (2013). Active and mediated opportunistic cooperation between an unmanned aerial vehicle and an unmanned ground vehicle. In *2013 IEEE International Symposium on Safety, Security, and Rescue Robotics (SSRR)*, 1–8. IEEE, Linkoping, Sweden.

Peshlov, B., Nakarmi, A., Baldwin, S., Essner, S., & French, J. (2017). Scaling up high throughput field phenotyping of corn and soy research plots using ground rovers. In J. A. Thomasson, M. McKee, & R. J. Moorhead (Eds.), *Autonomous Air and Ground Sensing Systems for Agricultural Optimization and Phenotyping II, 10218,* 1021802. SPIE.

Petcu, D., Gorgan, D., Pop, F., Tudor, D., & Zaharie, D. (2008). Satellite image processing on a Grid-based platform. *International Scientific Journal of Computing*, *7*(2), 51–58.
Petricciani, J. C. (1983). An overview of safety and regulatory aspects of the new biotechnology. *Regulatory Toxicology and Pharmacology*, *3*(4), 428–433.
Pires, I. M., Garcia, N. M., Pombo, N., & Flórez-Revuelta, F. (2016). From data acquisition to data fusion: A comprehensive review and a roadmap for the identification of activities of daily living using mobile devices. *Sensors*, *16*(2), 184.
Plotog, I., Varzaru, G., & Mihailescu, B. (2014). Device for multiple environment parameters measurements for agriculture use. In *2014 IEEE 20th International Symposium for Design and Technology in Electronic Packaging (SIITME)*, 207–210. IEEE, Bucharest, Romania.
Posudin, Y. (2014). *Methods of Measuring Environmental Parameters*. John Wiley & Sons.
Rahmatulloh, A., Darmawan, I., & Gunawan, R. (2019). Performance analysis of data transmission on WebSocket for real-time communication. In *2019 16th International Conference on Quality in Research (QIR): International Symposium on Electrical and Computer Engineering*, 1–5. IEEE, Padang, Indonesia.
Rajan, P., & Radhakrishnan, B. (2016). A survey on different image processing techniques for pest identification and plant disease detection. *International Journal of Computer Science and Network (IJCSN)*, 137–141.
Ramanathan, P., & Kartik. (2021). Autonomous driving cars: Decision-making. *Internet of Vehicles and its Applications in Autonomous Driving*, 31–39.
Ramatullayev, S., & Mustapha, H. (2022). Machine-Learning-Based Automation of Formation Testing Data Interpretation and Analysis. In *Abu Dhabi International Petroleum Exhibition and Conference*. Abu Dhabi, UAE, D032S170R004. SPE.
Ramírez-Rojas, C., Peña-Valdivia, C. B., García-Esteva, A., & Padilla-Chacón, D. (2022). Phenotyping of corn plants with effect of mesotrione herbicide. *Revista Mexicana De Ciencias Agrícolas*, *13*(8), 1399–1410.
Raquet, J., & Martin, R. K. (2008). Non-GNSS radio frequency navigation. In *2008 IEEE International Conference on Acoustics, Speech and Signal Processing*, 5308–5311. IEEE, Las Vegas, USA.
Ravi, R., Hasheminasab, S. M., Zhou, T., Masjedi, A., Quijano, K., Flatt, J. E., & Habib, A. (2019). UAV-based multi-sensor multi-platform integration for high throughput phenotyping. In *Autonomous Air and Ground Sensing Systems for Agricultural Optimization and Phenotyping IV, 11008,* 106–120. SPIE.
Renhorn, I. G., Bergström, D., Hedborg, J., Letalick, D., & Möller, S. (2016). High spatial resolution hyperspectral camera based on a linear variable filter. *Optical Engineering*, *55*(11), 114105–114105.
Resende, E. L., Bruzi, A. T., Cardoso, E. D. S., Carneiro, V. Q., Pereira de Souza, V. A., Frois Correa Barros, P. H., & Pereira, R. R. (2024). High-throughput phenotyping: Application in maize breeding. *Agri Engineering*, *6*(2), 1078–1092.
Rutter, S. M. (2007). The integration of GPS, vegetation mapping and GIS in ecological and behavioural studies. *Revista Brasileira de Zootecnia*, *36*, 63–70.
Sacrey, D., & Roden, R. (2018). Solving exploration problems with machine learning. *First Break*, *36*(6), 67–72.
Salahat, E., & Qasaimeh, M. (2017). Recent advances in features extraction and description algorithms: A comprehensive survey. In *2017 IEEE International Conference on Industrial Technology (ICIT)*, 1059–1063. IEEE, Toronto, Canada.
Sankararaman, S. (2017). Towards a computational framework for autonomous decision-making in unmanned aerial vehicles. In *Aiaa Information Systems-Aiaa Infotech@ Aerospace,* 0446. American Institute of Aeronautics and Astronautics.
Sapra, L. (2023). Cloud computing and edge computing for scalability and performance. *Turkish Online Journal of Qualitative Inquiry*.
Sarnovsky, M., Bednár, P., & Smatana, M. (2017). Data integration in scalable data analytics platform for process industries. In *2017 IEEE 21st International Conference on Intelligent Engineering Systems (INES),* 000187–000192. IEEE, Larnaca, Cyprus.
Saxena, D., & Lamest, M. (2018). Information overload and coping strategies in the big data context: Evidence from the hospitality sector. *Journal of Information Science*, *44*(3), 287–297.
Schmitt, A. J., Sun, S. A., Snyder, L. V., & Shen, Z. J. M. (2015). Centralization versus decentralization: Risk pooling, risk diversification, and supply chain disruptions. *Omega*, *52*, 201–212.
Semin, D., Zaytsev, S., Karpov, I., & Bondarev, E. (2021). Application of machine learning algorithms for seismic data interpretation: "Gazprom Neft" case stadies. In *ProGREss' 21*, *2021*(1), 1–4. European Association of Geoscientists & Engineers.
Shakoor, N., Lee, S., & Mockler, T. C. (2017). High throughput phenotyping to accelerate crop breeding and monitoring of diseases in the field. *Current Opinion in Plant Biology*, *38*, 184–192.

Sharma, R., Daniel, H., & Dušek, F. (2014). Sensor fusion: An application to localization and obstacle avoidance in robotics using multiple IR sensors. In *Nostradamus 2014: Prediction, Modeling and Analysis of Complex Systems*, 385–392. Springer International Publishing.

Shashank, K., SivaChaitanya, N., Manikanta, G., Balaji, C. N., & Murthy, V. V. S. (2014). A survey and review over image alignment and stitching methods. *Journal of Communications Technology and Electronics*, *5*, 50–52.

Sheh, R., Milstein, A., McGill, M., Salleh, R., Hengst, B., & Sammut, C. (2009). Semi-autonomous robots for robocup rescue. In *Proceedings of ACRA, 9*, 1–10. Sydney, Australia.

Sicker, D. C., & Redl, D. J. (2018). Policy and regulatory issues. *IEEE Internet of Things Magazine*, *1*(1), 4–5.

Singh, D., & Garg, R. (2021). R and Hadoop integration for big data analytics. In *Proceedings of 3rd International Conference on Computing Informatics and Networks: ICCIN 2020*, 13–22. Springer Singapore, Delhi, India.

Singh, S. (2017). Optimize cloud computations using edge computing. In *2017 International Conference on Big Data, IoT and Data Science (BID)*, 49–53. IEEE, Pune, India.

Singhvi, V., Lunga, L., Nidhi, P., Keum, C., & Prakash, V. (2024). High-Throughput Phenotyping using Computer Vision and Machine Learning. *arXiv preprint arXiv:2407.06354*.

Snodgrass, R. (1982). *Monitoring Distributed Systems: A Relational Approach* (Doctoral dissertation, Carnegie-Mellon University).

Spanos, K. A., & Feest, A. (2007). A review of the assessment of biodiversity in forest ecosystems. *Management of Environmental Quality: An International Journal*, *18*(4), 475–486.

Spanos, K. A., Feest, A., Petrakis, P. V., & Daskalakou, E. N. (2010). Salient points on the assessment and monitoring of forest biodiversity. *Bioremediation, Biodiversity and Bioavailability*, *4*(1), 1–7.

Stehr, N. J. (2015). Drones: The newest technology for precision agriculture. *Natural Sciences Education*, *44*(1), 89–91.

Tait, L., Bind, J., Charan-Dixon, H., Hawes, I., Pirker, J., & Schiel, D. (2019). Unmanned aerial vehicles (UAVs) for monitoring macroalgal biodiversity: Comparison of RGB and multispectral imaging sensors for biodiversity assessments. *Remote Sensing*, *11*(19), 2332.

Tao, C., Zhu, H., Wang, X., Zheng, S., Xie, Q., Wang, C., & Zheng, Z. (2021). Compressive single-pixel hyperspectral imaging using RGB sensors. *Optics Express*, *29*(7), 11207–11220.

Tao, D., Zhou, X., Lau, A., & Li, Y. (2013). Personalised information gathering and recommender systems: Techniques and trends. *EAI Endorsed Transactions on Scalable Information Systems*, *13*(1–3), 1–17.

Tarhan, Ö. (2020). Safety and regulatory issues of nanomaterials in foods. In S. M. Jafari (Ed.), *Handbook of Food Nanotechnology*, 655–703. Academic Press.

Thakur, P., Dogra, P., Bairwa, S. K., Meena, S. K., & Chandra, K. (2024). Modern phenotyping: A paradigm of present need in crop improvement. *International Journal of Plant & Soil Science*, *36*(1), 24–37.

Tilbury, D., & Ulsoy, A. G. (2011). A new breed of robots that drive themselves. *Mechanical Engineering*, *133*(02), 28–33.

Trankler, H. R., & Kanoun, O. (2001). Recent advances in sensor technology. In *IMTC 2001. Proceedings of the 18th IEEE Instrumentation and Measurement Technology Conference. Rediscovering Measurement in the Age of Informatics (Cat. No. 01CH 37188)*, *1*, 309–316. IEEE, Budapest, Hungary.

Tsach, S., Tatievsky, A., & London, L. (2010). Unmanned aerial vehicles (UAVs). *Encyclopedia of Aerospace Engineering*.

Tsigas, P., & Zhang, Y. (1999). Non-blocking data sharing in multiprocessor real-time systems. In *Proceedings Sixth International Conference on Real-Time Computing Systems and Applications. RTCSA'99 (Cat. No. PR00306)*, 247–254. IEEE, Hong Kong, China.

Udo, G. J. (2001). Privacy and security concerns as major barriers for e-commerce: A survey study. *Information Management & Computer Security*, *9*(4), 165–1.

Vagni, F. (2007). Survey of hyperspectral and multispectral imaging technologies (etude sur les technologies d'imagerie hyperspectrale et multispectrale). *Nato Research and Technology Organization Neuilly-Sur-Seine (France), Nato Re-Search and Technology Organization Neuilly-Sur-Seine (France), Tech. Rep. Rto-Tr-Set-065-P3*.

Valavanis, K. P. (2007). Advances in unmanned aerial vehicles: State of the art and the road to autonomy. In K. P. Valavanis (Ed.), *Intelligent Systems, Control and Automation. Science and Engineering*, *33*. Springer.

Valente, A., Morais, R., Serodio, C., Mestre, P., Pinto, S., & Cabral, M. (2007). A ZigBee sensor element for distributed monitoring of soil parameters in environmental monitoring. In *SENSORS, 2007 IEEE*, 135–138, IEEE, Atlanta, USA.

Veetil, V. P. (2017). Coordination in centralized and decentralized systems. *Available at SSRN 2600735*.

Veres, S. M., Molnar, L., Lincoln, N. K., & Morice, C. P. (2011). Autonomous vehicle control systems—a review of decision making. *Proceedings of the Institution of Mechanical Engineers, Part I: Journal of Systems and Control Engineering*, *225*(2), 155–195.

Vijay Rao, D., Iliadis, L., Papaleonidas, A., & Spartalis, S. (2012). Modelling environmental factors and effects in virtual warfare simulators by using a multi agent approach. *International Journal of Artificial Intelligence (to appear, 2012)* (ISSN 0974-0635).

Vijayalakshmi, V., Kumar, D. M., Kumar, S. P., & Thejaswini, P. (2021). Ndvi computation of Liss III images using Qgis. *Information Technology in Industry*, *9*(1), 1451–1454.

Vroegindeweij, B. A., van Wijk, S. W., & Van Henten, E. (2014). Autonomous unmanned aerial vehicles for agricultural applications. In *Abstract from AgEng 2014*, Zurich, Switzerland.

Waller, P. C., & Lee, E. H. (1999). Responding to drug safety issues. *Pharmacoepidemiology and Drug Safety*, *8*(7), 535–552.

Wang, Q., Terzis, A., & Szalay, A. (2010). A novel soil measuring wireless sensor network. In *2010 IEEE Instrumentation & Measurement Technology Conference Proceedings*, 412–415. IEEE, Austin, USA.

Widyotriatmo, A., & Hong, K. S. (2008). Decision making framework for autonomous vehicle navigation. In *2008 SICE Annual Conference,* 1002–1007. IEEE, Chofu, Japan.

Woods, D. D., Patterson, E. S., & Roth, E. M. (2002). Can we ever escape from data overload? A cognitive systems diagnosis. *Cognition, Technology & Work*, *4*, 22–36.

Woods, D. D., Patterson, E. S., Roth, E. M., & Christoffersen, K. (1999). Can we ever escape from data overload? A cognitive systems diagnosis. In *Proceedings of the Human Factors and Ergonomics Society Annual Meeting*, *43*(3), 174–178. SAGE Publications, San Diego, California.

Xie, C., Gao, J., & Tao, C. (2017). Big data validation case study. In *2017 IEEE Third International Conference on Big Data Computing Service and Applications (BigDataService),* 281–286. IEEE, Redwood City, USA.

Yang, W., Feng, H., Zhang, X., Zhang, J., Doonan, J. H., Batchelor, W. D., & Yan, J. (2020). Crop phenomics and high-throughput phenotyping: Past decades, current challenges, and future perspectives. *Molecular Plant*, *13*(2), 187–214.

Yin, M. L., & Aroush, H. (2023). Real-time fault-tolerant computing with machine learning enhancements. In *2023 Annual Reliability and Maintainability Symposium (RAMS)*, 1–7. IEEE, Orlando, USA.

Yunis, M. M., El-Khalil, R., & Ghanem, M. (2021). Towards a conceptual framework on the importance of privacy and security concerns in audit data analytics. In *Proceedings of the International Conference on Industrial Engineering and Operations Management*, Sao Paulo, Brazil, 5–8.

Zakrajsek, J., McKissock, D., Woytach, J., Zakrajsek, J., Oswald, F., McEntire, K., & Goodnight, T. (2005). Exploration rover concepts and development challenges. In *1st Space Exploration Conference: Continuing the Voyage of Discovery*, Orlando, USA, 2525.

Zeiler, M. N. (1992). Integrating GIS and GPS for mapping and analysis of electric distribution circuits. In *[Proceedings] 1992 Rural Electric Power Conference. Papers presented at the 36th Annual Conference*, D2-1. IEEE, New Orleans, USA.

Zhang, H. W., Chen, H. L., Zou, C. H., & Yu, W. D. (2010). The review of dynamic monitoring technology for crop growth. In C. M. U. Neale & A. Maltese (Eds.), *Remote Sensing for Agriculture, Ecosystems, and Hydrology XII*, *7824*, 545–557. SPIE.

Zhang, R., Hewitt, A. J., Chen, L., Li, L., & Tang, Q. (2023). Challenges and opportunities of unmanned aerial vehicles as a new tool for crop pest control. *Pest Management Science*, *79*(11), 4123–4131.

Zhang, S., Wang, H., He, S., Zhang, C., & Liu, J. (2018). An autonomous air-ground cooperative field surveillance system with quadrotor UAV and unmanned atv robots. In *2018 IEEE 8th Annual International Conference on CYBER Technology in Automation, Control, and Intelligent Systems (CYBER),* 1527–1532. IEEE, Tianjin, China.

Zhao, Z. W., Zhang, J., Sun, F., Li, L., & Chen, I. M. (2003). Routing strategy for satellite-based networks with a hierarchical space grid architecture. In *Fourth International Conference on Information, Communications and Signal Processing, 2003 and the Fourth Pacific Rim Conference on Multimedia. Proceedings of the 2003 Joint*, *3*, 1939–1943. IEEE, Singapore.

Zhou, Y., Zhao, H., & Liu, Y. (2020). An evaluative review of the VTOL technologies for unmanned and manned aerial vehicles. *Computer Communications*, *149*, 356–369.

Zilberstein, S. (2015). Building strong semi-autonomous systems. In *Proceedings of the AAAI Conference on Artificial Intelligence, 29*(1), 4088–4092. Austin, USA.

# 17 Technical Advancements in Plant Metabolic Phenotyping Based on High-Throughput and High-Resolution Analytical Platforms

*Kuttikattuparambil Sudheer Aditi, Manoharan Rajesh, Natarajan Pushpa, Chenniappan Anchana Devi, Kandasamy Vijayalakshmi, and Arumugam Vijaya Anand*

## LIST OF ABBREVIATIONS

| | |
|---|---|
| **GC-HR-TOF-MS**: | High-Resolution Time-of-Flight Mass Spectrometry |
| **GC-MS**: | Gas Chromatography-Mass Spectrometry |
| **HTP**: | High-Throughput Phenotyping |
| **IAP**: | Integrated Analysis Platform |
| **IMS**: | Imaging Mass Spectrometry |
| **LC-MS**: | Liquid Chromatography-Mass Spectrometry |
| **MALDI-FTICR-IMS**: | Matrix-Assisted Laser Desorption/Ionization-Fourier Transform Ion Cyclotron Resonance-Imaging Mass Spectrometry |
| **MS**: | Mass Spectrometry |
| **NIR**: | Near-Infrared |
| **NMR**: | Nuclear Magnetic Resonance |
| **SPPs**: | Superficially Porous Particles |
| **UAPs**: | Unmanned Aerial Platforms |
| **UAV**: | Unmanned Aerial Vehicle |
| **UHPLC**: | Ultrahigh-Pressure Liquid Chromatography |

## 17.1 INTRODUCTION

Enhancing the productivity of significant food and fodder crops has long been a major priority, and this challenge is being tackled globally. United Nations and CGIAR institutes are aware that current climatic conditions and continuous breeding efforts may not be sufficient to meet future food demands. The entire world is working with various advanced techniques such as genetic engineering, genome editing, and plant metabolic phenotyping for improved food productivity to meet future demands. The international research institutes tie up national partners for crop improvement and productivity. Genomic information of several plant species is available in the biological databases, however, limited phenotypic information poses challenges due to environmental factors and potential errors in trait measurements (Pieruschka & Schurr, 2019).

DOI: 10.1201/9781003505631-17 

Researchers are now making efforts to understand the uniqueness of individuals, considering both their genetic makeup and physical traits. To achieve this, high-throughput phenotyping (HTP) discovery methods are being utilized to conduct numerous experiments simultaneously and gain a deeper insight into these variations. The establishment of field trials at different locations is quite challenging and takes a lot of time. Therefore, we need an accurate, and rapid method on a large scale to forecast the interested traits. Recently, there has been a focus on creating high-throughput analysis to simplify the process of studying plant phenotyping. HTP can be applied across various stages, from individual cells to large-scale growth. Researchers can identify the precise gene of interest by beginning at the most fundamental level leading to more efficient and higher-yielding growth to achieve the desired phenotype.

This technology utilizes automated analysis of traits which swiftly produce clear data about the physical characteristics of plants using automated sensors, tools for data collection, and analysis instruments (Jangra et al., 2021). Its application has expedited plant breeding by efficiently screening numerous plants at various growth stages, enabling the rapid identification of desired traits without waiting for full plant maturation in the field. The importance of phenotypic data remains, as it plays a vital role in foreseeing different biological results like the health of plants, traits related to diseases, and mortality. This technology substantially enhances the capabilities of plant breeders and agronomists, offering a more efficient and precise approach to advancing plant research and breeding programs. This chapter discusses the methods and applications of HTP analysis of plant phenotyping.

## 17.2 HIGH-THROUGHPUT ANALYTICAL PLATFORMS FOR PLANT METABOLIC PHENOTYPING

Conventional field plant phenotyping, which was traditionally dependent on breeders' visual inspection, has been transformed by recent advances in Unmanned Aerial Vehicle (UAV), robotic, and powerful computing technology. These innovations have facilitated the adoption of HTP technology in breeding programs, allowing for the comprehensive collection of high-resolution temporal and geographical information on plant phenotypes. Utilized systems encompass robotic mobile platforms, UAV methods, and in-field sensing structures such as towers, gantries, or cable systems. Notably, UAV-based imaging technology has emerged as a promising HTP system for plant phenotyping (Ludovisi et al., 2017; Volpato et al., 2021).

**UAV-Established Imaging Technology for Field HTP involves**:

- **System Overview**: UAV platforms, provided with an array of cameras including visible, multispectral, and thermal, capture successive images of crops at various phases of growth within breeding fields.
- **Image Processing Pipelines**: Customized image processing pipelines are employed to analyze collected images based on image type and crop traits of interest.
- **Image Pre-processing**: During this stage, radiometric, temperature, and color calibration are conducted, followed by stitching and geo-referencing processes. These steps aim to produce orthomosaic images of fields.
- **Data Processing and Analysis**: Orthomosaic images undergo segmentation to isolate target areas, such as breeding plots, and to extract image features essential for assessing crop features (Bongomin et al., 2024; Ludovisi et al., 2017).

**Advanced-Data Analytics for Crop Trait Estimation**:

- **Machine Learning**: Sophisticated data analytic techniques, including machine learning and deep learning, are employed to investigate segmented images and the extracted image characteristics for each plot.
- **Predictive Models**: By leveraging these methods, predictive and prescriptive models are developed to quantify intricate crop qualities like yield and stress resilience.

Incorporating developed machine learning techniques has demonstrated significant prospects in uncovering concealed information from images, leading to precise estimations of complex crop traits. The use of UAV-based imaging technology in conjunction with sophisticated data analytics is poised to revolutionize field plant phenotyping by providing non-destructive, efficient, and detailed insights into plant physiology and agronomic traits (Zhang, 2023). UAVs, armed with cameras and sensors, facilitate swift and non-destructive measurements, presenting faster turnaround times compared to satellites at competitive costs. These UAVs provide high-resolution images that allow for the acquisition of detailed information about plant traits, overcoming the limitations of traditional ground-based techniques. Furthermore, the utilization of UAV-based hyperspectral imaging has emerged as a promising technique in detecting and monitoring crop diseases like Fusarium head blight in winter wheat (Ludovisi et al., 2017; Ma et al., 2021).

The use of unmanned aerial platforms (UAPs) furnished with sensors for multispectral imaging has been instrumental in assessing spatial area variability and crop phenotyping. Aerial spectral imaging provides a fast and lower-cost approach for testing field description, allowing for the identification of spatial variability which is crucial for crop improvement programs (Zaman-Allah et al., 2015). HTP using UAV-based technologies has been effective in estimating plant height accurately. Studies have shown that advancements in UAV platforms and image processing techniques have improved the accuracy of crop height estimations, making UAV-based phenotyping a beneficial tool for agricultural investigation and breeding programs (Volpato et al., 2021).

HTP analytical platforms, such as gas chromatography-mass spectrometry (GC-MS), nuclear magnetic resonance (NMR) spectroscopy, liquid chromatography-mass spectrometry (LC-MS), and infrared spectroscopy, are essential for plant metabolic phenotyping. These platforms offer insights into cellular and organismal metabolic status by analyzing primary metabolism and determining metabolites related to various physiological and agronomic traits. Because of the varied properties, no single platform can effectively profile all metabolites. The selection of the platform depends on the classes of metabolites under analysis. GC-MS, LC-MS, and NMR are commonly utilized methods, each with unique advantages such as reproducibility, sensitivity and comprehensive analysis (Wase et al., 2022).

GC-MS is a liable and HTP platform used for primary metabolism analysis in plants, providing reproducibility and a wide dynamic range for metabolite profiling. Its strength lies in reproducible GC retention times, MS fragmentation patterns, and the ability to quantify compounds with molecular weights below 600 a.u., making it convenient for HTP metabolite analysis. High-resolution time-of-flight mass spectrometry (GC-HR-TOF-MS) offers benefits over established GC-quadrupole (Q)-MS by providing high full-scan rates, mass accuracy, resolution, and enhanced detection limits. GC-HR-TOF-MS excels in separating ion signals at diverse exact masses, reducing incorrect peak annotations and eradicating the impact of co-eluting metabolites on estimation (Wase et al., 2022). Plant traits that can be measured using GC-MS for HTP phenotyping include the emission of volatile organic compounds, transpiration rates, and photosynthetic gas exchange. These traits are necessary for studying plant maturation, stress tolerance, and disease resistance, making them useful in plant breeding to increase drought resistance and resource efficiency (Jud et al., 2018; Li et al., 2021).

**NMR**: Plant metabolites can be identified and quantitatively analyzed *in vivo* or tissue extracts using NMR spectroscopy. NMR is extremely useful for analyzing main metabolites such as sugars and amino acids as well as secondary metabolites including phenolic chemicals, which are found in large quantities in plants (Kim et al., 2010). NMR complements MS in metabolomics, providing non-destructive, comprehensive metabolic information for plant phenotyping and functional genomics studies. NMR allows for metabolite fingerprinting and profiling in plants, aiding in discriminating between sample groups and identifying significant differences in metabolite concentrations. Different probe head designs influence sensitivity, with microprobes enabling analysis of small sample quantities and cryogenic probe heads enhancing sensitivity for improved signal detection. Most metabolic tests are performed in the 300-600MHz region, however NMR spectrometers with field strengths upto 21 Tesla can yield high resolution spectra (Krishnan, 2004).

**IR**: Plant samples can have their chemical makeup and geographical origin identified using infrared spectroscopy, which provides both qualitative and quantitative data on heterogeneous samples. Compared to traditional methods, infrared spectroscopy is cost-effective, needs little to no sample preparation, and does not involve the use of chemicals, making it environmentally friendly. It enables the representation showing the geographical distribution of active components in plant tissues, providing valuable insights into the molecular components at a cellular level. This technique allows for rapid data acquisition and analysis, making it suitable for HTP applications in plant research (Huck, 2015; Sohn et al., 2021).

High-throughput analytical systems in plants enable accurate measures of preferred characteristics among thousands of field-grown plants and thereby significantly speed up crop development efforts. These platforms aid in the selection of higher-yielding, more resilient to disease, and stress-tolerant lines (Jangra et al., 2021). The automation and HTP core of these systems permit the screening of hundreds of genotypes and thousands of respective plants or field plots, leading to increased efficiency and standardized evaluations (Poorter et al., 2023). Additionally, HTPs enable non-invasive and fast assessments of plant size, development, and physiological variables, contributing to a more thorough knowledge of plant traits and responses to surrounding circumstances (Poorter et al., 2023). However, there are limitations associated with high-throughput analytical platforms in plants. One challenge is the need for significant investments in infrastructure, operation, and maintenance, which can pose financial constraints for research groups considering adopting these technologies (Poorter et al., 2023). Another limitation is the potential deviation in plant size estimates caused by factors like diurnal changes in leaf angle when using top-view cameras for measurements, highlighting the importance of careful interpretation of data and calibration curve generation. Additionally, the phenotyping bottleneck is evident when analysing characteristics influenced by multiple genes, as plants of identical genotypes can show diverse phenotypes under varying environmental circumstances (Poorter et al., 2023). While high-throughput analytical platforms offer valuable tools for plant research and crop improvement programs by enhancing efficiency and providing detailed insights into plant traits, users must consider the financial investments, data interpretation challenges, and limitations associated with these systems to ensure accurate and meaningful results in plant phenotyping studies.

## 17.3 INTEGRATING HIGH-THROUGHPUT AND HIGH-RESOLUTION ANALYTICAL PLATFORMS

Integrating HTP and high-resolution analytical platforms in plant research can offer a comprehensive approach to phenotyping studies, combining the efficiency of HTP systems with the detailed insights provided by high-resolution imaging techniques. High-throughput systems, such as those used in drug discovery, enable rapid analysis of large sample sizes and can be applied even when detailed structural information of target proteins is not available (Szymański et al., 2011). These systems reduce manufacturing costs, require minimal space, and permit the screening of a considerable number of compounds in a short time, enhancing efficiency in data collection and analysis (Szymański et al., 2011). Alternatively, high-resolution analytical platforms provide detailed assessments of plant traits like size, development, and physiological variables through non-invasive imaging techniques. These platforms offer precise measurements of plant responses to environmental conditions, contributing to a more in-depth knowledge of plant phenotypes (Szymański et al., 2011). By incorporating high-resolution imaging with high-throughput screening methods, researchers can obtain both quantitative data from large sample sizes and detailed qualitative information on specific plant characteristics. This integration can direct more accurate and complete evaluations of plant responses to various stimuli, ultimately enhancing the efficiency and effectiveness of plant research endeavors.

The integration of high-throughput and high-resolution analytical platforms in plant research presents a promising approach to advancing phenotyping studies by combining the benefits of rapid data collection with detailed trait assessments. By leveraging the strengths of both types of analytical systems, researchers can acquire a more holistic experience of plant responses and traits, leading to improved crop improvement programs and scientific discoveries in the field of plant biology.

Recent developments in high-throughput analytical platforms have introduced several promising techniques for plant analysis, enhancing speed, efficiency, and data generation capabilities. In the field of plant research, liquid chromatography has been a key technique used in many HTP analytical workflows. One of the most significant advancements in HTP analytical platforms is ultrahigh-pressure liquid chromatography (UHPLC). By utilizing custom-made tools with short bed lengths and optimized geometries, researchers have achieved separation speeds down to the sub-second time limit, approaching sensor-like throughput. This advancement in UHPLC technology has significantly improved the speed and efficiency of liquid chromatography analyses for high-throughput applications. Moreover, the use of superficially porous particles (SPPs) in chromatography has shown promise in enhancing speed and efficiency. SPP columns have demonstrated the ability to achieve sub-minute separations with lower pressure drops, enabling faster analysis times and increased throughput. Additionally, open tubular columns have been explored as an alternative to packed-bed formats, offering potential advantages in efficiency and kinetic performance.

Moreover, MS has gained widespread use in HTP analysis workflows due to its ability to combine high sample throughput with high selectivity. MS-based tools offer rapid compound identification and quantification capabilities, making them invaluable for hit and lead identification in chemical reaction screening. Automation of sample preparation steps and the use of optimized short UHPLC methods have further enhanced the speed and efficiency of MS-based analyses in high-throughput settings (Vervoort et al., 2021).

## 17.4 APPLICATIONS OF HIGH-THROUGHPUT AND HIGH-RESOLUTION ANALYTICAL PLATFORMS IN PLANT METABOLIC PHENOTYPING

High-throughput and high-resolution analytical platforms play a crucial role in plant metabolic phenotyping by addressing challenges related to chemical complexity, efficiency of extraction, analytical dynamic range, and measure throughput. These platforms enable the comprehensive phenotyping of plants, ranging from ground-based imaging to aerial phenotyping across various plant species like rice, wheat, maize, cotton, Arabidopsis, and more using specialized tools such as RootReader3D, GROWSCREEN-Rhizo, Zeppelin NT aircraft, Phenocart, and others (Gill et al., 2022).

Modern phenomics centers that employ sensor platforms for plant phenotyping include the Leibniz Institute of Plant Genetics and Crop Plant Research in Germany, the Plant Accelerator and High-Resolution Plant Phenomics Center in Australia and other renowned organizations worldwide. Additionally, private companies like ‘LemnaTec’, ‘PhenoSpex’, and ‘We Provide Solutions’ offer wide-ranging HTP platforms for controlled and field environments. These advancements have led to the development of specific software applications like HYPOTrace, HTPheno, LAMINA, and others designed to extract a broad spectrum of measurements crucial for plant phenotyping like projected area, height/length, shape, width, digital volume, compactness, relative growth rate, and colorimetric analysis (Klukas et al., 2014). The Integrated Analysis Platform (IAP) is an open-source framework that facilitates HTP data processing by handling diverse image sources and organizing phenotypic data efficiently.

HTP platforms, such as those utilizing GC-MS metabolite profiling, are effective tools for noticing metabolic reactions in plants (Wase et al., 2022). These platforms help researchers identify variations in metabolites under different growth conditions or stressors. High-throughput platforms when coupled with metabolomics technologies play an important role in bridging the genotype-phenotype gap (Hall et al., 2022). By rapidly analysing a large number of samples, these platforms aid in understanding how genetic variations influence plant metabolic pathways. High-resolution analytical platforms offer a more increased resolution for capturing time-related phenotypic differences in plants compared to traditional experimental designs (Poorter et al., 2023). This capability allows researchers to monitor dynamic metabolic responses over time without the need for repeated planting and harvesting. HTP platforms provide non-destructive and adequate field-based plant phenotyping solutions (Gill et al., 2022). These platforms, whether manual, autonomous, or semi-autonomous, are equipped with HTP capabilities and offer efficient ways to collect plant phenotype data without damaging the plants.

### 17.4.1 Spatial Metabolomics and Imaging

Imaging mass spectrometry (IMS) is a potent analytical method that enables the visualization and localization of metabolites within tissue sections (Nakabayashi et al., 2021). By combining MS with spatial information, IMS allows researchers to map the distribution of various compounds in biological samples with high spatial resolution. Understanding how different molecules are located within plant tissues and how they relate to genes can help identify which genes are responsible for producing specific compounds in plants. By combining this information with data on where genes are active in the tissue, researchers can pinpoint which genes are involved in creating certain plant compounds. To further understand the role of a specific molecule, scientists can compare genetically modified plants which lack that molecule using advanced gene-editing techniques. This process allows researchers to determine the function of a molecule by observing how plants change when that molecule is absent (Nakabayashi et al., 2021).

Matrix-assisted laser desorption/ionization-Fourier transform ion cyclotron resonance-imaging mass spectrometry (MALDI-FTICR-IMS) is an analytical technique to visualize site-specific metabolite signal powers on longitudinal or cross-sections of tissue, with ultra-high resolution and exactness of ion peaks (Dong et al., 2020; Yamamoto et al., 2019; Nakabayashi et al., 2021).

### 17.4.2 Non-Destructive Techniques in Metabolic Phenotyping

Non-destructive techniques are essential in metabolic phenotyping for plants as they allow researchers to gather valuable data on plant metabolism and traits without altering or damaging the samples. Healthy and infected plants interact differently with electromagnetic radiation due to their distinct optical properties. Imaging techniques aid in detecting these differences, crucial for plant phenotyping by analysing light interactions with plant tissues, revealing unseen properties non-destructively. Imaging at distinct wavelengths is employed for diverse characteristics of plant phenotyping.

**Visible Imaging**: It is primarily employed to estimate characteristics of plant architecture like image-based projected biomass, color, growth dynamics, seedling vigor, root architecture, seed morphology, leaf area, leaf disease severity assessments, yield, fruit distribution and numbers. The silicon sensors, like charge-coupled devices or complementary metal-oxide-semiconductor arrays, are exposed to a wavelength range of 400–750 nm, enabling 2D imaging. These sensors form the foundation of critical applications for visible images and represent the easiest imaging approach for plant phenotyping. They create spatial matrices of intensity values corresponding to photon fluctuations in the visible light spectrum, with blue light around 450 nm, green light around 550 nm, and red light around 600 nm. Typically, conventional digital or RCB/CIR cameras are utilized in visible imaging due to their rapid data acquisition and effective solutions for plant phenotyping applications (Jangra et al., 2021). However, challenges arise in processing visible images to extract phenotypic data like biomass and leaf characteristics due to difficulties in accurately segmenting overlapping leaves. This limitation hinders precise analysis of plant traits using visible imaging techniques (Li et al., 2014).

**Fluorescence Imaging**: Fluorescence imaging, focusing on the chlorophyll complex, is key in disease detection in plants. It tracks metabolic changes from photosynthesis to respiration, aiding in monitoring plant health responses before visible growth declines. This technique estimates photosynthesis efficiency, screens for disease resistance, and assesses stress tolerance in plant phenotyping, offering valuable insights into physiological processes affected by biotic and abiotic stresses.

**Thermal Infrared Imaging**: Thermal imaging visualizes infrared radiation, reflecting surface temperatures. It measures leaf temperatures to explore plant water relations, especially stomatal conductance linked to evaporation rates. Combined with spectral imaging, it enhances leaf water content estimation, offering insights into plant hydration levels and stress responses.

**Imaging Spectroscopy** can provide an understanding of the drivers of growth dynamics by means like estimating spatiotemporal growth patterns during investigations and also for collecting plant spectroscopy information to quantify vegetation indices, water ranges, the composition parameters of seeds and pigment composition in yield possible investigations (Li et al., 2014). Imaging spectroscopy is utilized in phenomics for remote sensing of vegetation, where solar radiation deliberated by plants is analyzed. In the visible spectral range (400–700nm), single leaves or canopies exhibit low reflectance, with a reflectance peak around 550nm indicating minimal absorption by leaf pigments like chlorophyll. A substantial gain in reflectance occurs at the change from visible to near-infrared (NIR) wavelengths (red edge), as NIR radiation is predominantly reflected by leaves due to scattering within mesophyll cells. Factors influencing reflectance patterns include leaf and canopy architecture, like leaf density and growth patterns. Reflectance decreases gradually beyond 700nm due to increased water absorption inside leaves. Calibration models linking reference information and spectral data are crucial for plant spectral reflectance and NIR spectroscopy, involving sub-samples for calibration development and mathematical algorithms for prediction models. Successful calibration models enable the forecast of phenotypic values using spectral data for decision-making in crop advancement programs, integrating genotypic and environmental information (Jangra et al., 2021).

### 17.4.3 Bioinformatics and Data Analysis

Bioinformatics helps to analyze HTP data by utilizing computational methods to process and extract meaningful insights from large biological datasets efficiently. Some key roles of bioinformatics in analysing HTP data include:

**Data Processing**: Bioinformatics tools are used to pre-process and clean raw data obtained from HTP experiments, ensuring data quality and reliability for downstream analysis (D'Argenio, 2018).

**Data Integration**: Bioinformatics techniques enable the integration of diverse datasets from different sources, allowing researchers to combine information for a comprehensive analysis (Braun, 2014).

**Data Mining**: Bioinformatics algorithms are employed to identify patterns, relationships, and tendencies within high-dimensional datasets, assisting in the discovery of valuable insights.

**Data Visualization**: Bioinformatics tools provide visualization methods to represent complex data in a more understandable format, facilitating the interpretation of results and communication of findings.

**Functional Analysis**: Bioinformatics is instrumental in conducting functional analyses of biological data, helping researchers understand the biological significance of the observed patterns and relationships (Braun, 2014).

Some common bioinformatics tools used in HTP data analysis for plant metabolic phenotyping include:

**CLC Genomics Workbench**: Software designed for exploring and visualizing next-generation sequencing information, offering functionalities like de novo assembly, read mapping, SNP, and DIP detection.

**Ingenuity Pathways Analysis**: A web-based bioinformatics application that permits investigators to upload information from HTP investigations like microarray and next-generation sequencing for functional analysis, integration, and further interpretation of gene expression, miRNA, SNP, proteomics, and metabolomics data.

**Lasergene**: Gives a range of devices for manipulating sequence data relevant to both protein and nucleic acid sequences, including functions like editing, annotating, virtual cloning, primer design, and SNP finding.

**Partek Genomics Suite**: Exploited to interpret HTP data from microarray investigations and next-generation sequencing experimentations, offering clustering and statistical algorithms for visualization and pattern discovery in large datasets.

**MATLAB (Bioinformatics Toolbox)**: A numerical computing environment that permits matrix manipulations, plotting processes and data, algorithm implementation, user interface creation, and interfacing with programs composed in various languages like Java, C++, Fortran, and Python. These bioinformatics implements are essential for processing, integrating, mining, visualizing, and functionally analysing the extensive amount of HTP data generated in plant metabolic phenotyping studies (*Analysis Tools & Databases | NIH Library*, n.d.).

Metabolomics integration for plants involves utilizing databases and tools to enhance research in plant metabolomics. Several resources are available to facilitate this integration:

**PlantMetabolomics.org**: This web portal and database offer access to plant metabolomics data, aiding in the exploration, visualization, and download of plant metabolites (Bais et al., 2010; Tian et al., 2024)

**PMhub 1.0**: PMhub is a complete plant metabolome database that provides detailed reports on plant metabolites, serving as a useful resource for scientists (Tian et al., 2024).

**Metabolome Databases**: Various databases like the Human Metabolome Project, Golm Database, and Madison Metabolomics Consortium Database offer comprehensive data on metabolites found in plants, aiding in research and analysis (Bais et al., 2010; Tian et al., 2024).

**Analytical Tools**: Advanced tools such as GC-MS, LC-MS, and nuclear magnetic resonance (NMR) are crucial for precise metabolic profiling in plants.

**Bioinformatics Tools**: Platforms like MeltDB, MetaboAnalyst, GNPS, and MZmine 2 provide essential support for data processing, statistical analysis, and metabolite identification in plant metabolomics (Patel et al., 2021).

## 17.5 CHALLENGES AND FUTURE DIRECTIONS

**Chemical Complexity and Heterogeneity**: The diverse range of metabolites present in plants leads to challenges in identifying and quantifying them accurately due to their chemical complexity and heterogeneity (Hall et al., 2022). The chemical diversity across metabolite pools in plants complicates the identification and quantification processes. This diversity requires sophisticated analytical methods capable of identifying and characterizing a comprehensive array of metabolites with varying chemical properties. Additionally, the heterogeneity of metabolites within plant samples further complicates the analysis, as different tissues or developmental stages may exhibit distinct metabolic profiles. To address these challenges, researchers in plant metabolomics must employ advanced analytical methods such as MS coupled with various separation techniques like LC or GC. These techniques enable the identification and quantification of a wide range of metabolites present in plants, helping to unravel the complex metabolic networks within plant systems.

**Extraction Efficiency**: Efficient extraction of metabolites from plant samples is crucial for obtaining reliable data, but variations in extraction methods can impact the quality and consistency of results. The choice of extraction method significantly influences the efficiency of metabolite recovery from plant samples. Different metabolites have varying solubility's and stabilities, requiring tailored extraction protocols to ensure comprehensive recovery. Variations in factors such as solvent composition, extraction time, temperature, and sample preparation techniques can lead to differences in metabolite yield and composition. Standardizing extraction protocols is crucial to minimize variability and ensure reproducibility in plant metabolomics studies. A comprehensive evaluation of extraction protocols is necessary to assess factors like extraction efficiency, repeatability, and the capability to extract a broad spectrum of metabolites. Researchers often compare different extraction methods to identify the most suitable approach for specific plant species or tissues.

**Analytical Dynamic Range**: The wide dynamic range of metabolite concentrations in plants requires analytical platforms to be sensitive enough to detect low-abundance metabolites while also being able to handle high concentrations without saturation. Low-abundance metabolites are often biologically significant and can provide valuable insights into plant metabolism. On the other hand, the analytical tools must also be able to accurately quantify highly abundant metabolites without reaching saturation, which could lead to inaccurate measurements and hinder the detection of less abundant compounds. Advanced analytical techniques such as MS and NMR spectroscopy are generally used in plant metabolomics due to their ability to cover a wide dynamic range of metabolite concentrations. MS offers high sensitivity and a wide dynamic range, allowing for the detection of both low and high-abundance metabolites. NMR, while less sensitive than MS, is valuable for its ability to provide structural information on metabolites (Hall et al., 2022).

**Measurement Throughput**: HTP phenotyping involves analysing a large number of plants quickly, which can pose challenges in terms of data processing, storage, and analysis, particularly when dealing with complex metabolic data sets (Poorter et al., 2023).

## 17.6 DEVELOPMENT OF NEW ANALYTICAL TOOLS AND SOFTWARE

These new analytical devices and software play an essential part in advancing plant metabolic phenotyping by enabling efficient data processing, automated analysis of plant traits, extraction of detailed measurements from images, and integration of multi-omics data for a complete understanding of plant metabolism and responses to environmental stimuli.

**IAP**: IAP is an open-source framework designed for HTP data processing. It endures further image sources, organizes phenotypic data efficiently, and allows users to create or modify analysis pipelines easily.

**Automated High-Throughput Plant Phenotyping Systems**: Diverse software applications have been developed for specific tasks in plant phenotyping, such as HYPOTrace, HTPheno, LAMINA, PhenoPhyte, Rosette Tracker, LeafAnalyser, RootNav, SmartGrain, and LemnaGrid. These tools extract a broad spectrum of measurements from plants, including height/length, width, area, volume, shape, growth rate, and colorimetric estimation.

**Open-Source Information System for High-Throughput Plant Phenotyping**: An open-source information system has been implemented for HTP data analysis. This system efficiently measures plant phenotypic characteristics from various sources and helps in organizing phenotypic data for analysis (Klukas et al., 2014).

**Advanced Phenotyping Platforms**: State-of-the-art phenomics centers worldwide utilize sensor platforms for HTP plant phenotyping. These platforms include RootReader3D, GROWSCREEN-Rhizo, Zeppelin NT aircraft, Phenocart, Phenovator, PHENOARCH, Field Scanalyzer, CropQuant, and MVS-Pheno. They offer capabilities for HTP phenotyping of traits related to stress tolerance/resistance in diverse plant species (Gill et al., 2022).

**Pathway Analysis and Reaction Flux Modelling Tools**: Model-based methods include integrative tools for pathway investigation and reaction flux modeling in plant metabolic phenotyping. These tools aid in understanding metabolic pathways and reactions within plants (Hall et al., 2022).

## 17.7 FUTURE PROSPECTS OF PLANT METABOLIC PHENOTYPING

The prospects of plant metabolic phenotyping are promising, with advancements in technology and methodologies enhancing our understanding of plant biology. Plant metabolomics, a promptly growing field, involves analysing small molecules in plant tissues to comprehend their active functions in physiology, development, and responses to stresses. Integrating metabolomics with other omics data gives a complete view of plant biology, aiding in understanding complex interactions within organisms. Metabolomics plays a necessary part in crop improvement by guiding breeding processes towards identifying promising traits and accelerating the development of elite crop varieties resistant to environmental challenges (Kumar et al., 2017; Manickam et al., 2023). HTP combined with metabolomics can facilitate the characterization of new biostimulants, contributing to sustainable agriculture and green economics. The use of non-invasive sensor-based phenotyping approaches in biostimulant investigation showcases the evolving interaction between these two domains and highlights the need for complex testing methods integrating various -omics approaches. Overall, the future of plant metabolic phenotyping holds great potential for enhancing crop quality, yield, stress tolerance, and sustainable agricultural practices (De Diego & Spíchal, 2022; Hall et al., 2022).

## 17.8 CONCLUSION

The fusion of HTP and high-resolution analytical platforms in plant research offers an exciting avenue for advancing phenotyping studies. These platforms, ranging from automated phenotyping systems to sophisticated imaging techniques and cutting-edge metabolomics technologies, provide researchers with unprecedented opportunities to delve into the intricacies of plant metabolism, physiology, and responses to environmental cues. By capitalizing on the strengths of both analytical systems, Researchers can get more in-depth details about the characteristics of plants, ultimately driving forward crop improvement initiatives and scientific breakthroughs in plant biology.

The speed, effectiveness, and data output of plant analysis have been completely transformed by recent advancements in HTP analytical systems, like MS and UHPLC. These advancements empower researchers to swiftly analyze large sample cohorts while maintaining high levels of selectivity and sensitivity, crucial for identifying and quantifying metabolites associated with key plant traits. Furthermore, the integration of bioinformatics tools and databases streamlines the processing, integration, mining, visualization, and functional analysis of vast datasets generated in plant metabolic phenotyping studies. Platforms like PM and MetaboAnalyst serve as invaluable resources, aiding researchers in unraveling the intricate metabolic networks within plant systems.

Despite notable progress, challenges such as the complexity of plant chemistry, extraction efficiency, dynamic range of analytical measurements, and throughput of data collection persist. Overcoming these hurdles necessitates continued innovation in analytical techniques, data processing algorithms, and experimental methodologies to ensure accurate and meaningful outcomes in plant phenotyping studies. In conclusion, the convergence of HTP and high-resolution analytical platforms, bolstered by advanced bioinformatics tools, holds immense potential for accelerating discoveries in plant biology and crop enhancement. By harnessing these technologies, researchers can unearth novel insights into plant metabolism, paving the way for the evolution of resilient and high-yielding crop types essential for global food safety amidst evolving environmental pressures.

## ACKNOWLEDGEMENT

All the authors are thankful to their respective Universities and Institutions for their valuable supports.

## DATA AVAILABILITY

The authors confirm that the data supporting the findings of this study are available within the book/ chapter.

## REFERENCES

Analysis Tools & Databases | NIH Library. (n.d.). https://www.nihlibrary.nih.gov/services/bioinformatics-support/analysis-tools

Bais, P., Moon, S. M., He, K., Leitao, R., Dreher, K., Walk, T., Sucaet, Y., Barkan, L., Wohlgemuth, G., Roth, M. R., Wurtele, E. S., Dixon, P., Fiehn, O., Lange, B. M., Shulaev, V., Sumner, L. W., Welti, R., Nikolau, B. J., Rhee, S. Y., & Dickerson, J. A. (2010). PlantMetabolomics.org: A web portal for plant metabolomics experiments. *Plant Physiology*, *152*(4), 1807–1816.

Bongomin, O., Lamo, J., Guina, J. M., Okello, C., Ocen, G. G., Obura, M., Alibu, S., Owino, C. A., Akwero, A., & Ojok, S. (2024). UAV image acquisition and processing for high-throughput phenotyping in agricultural research and breeding programs. *The Plant Phenome Journal*, *7*(1), e20096.

Braun, R. (2014). Systems analysis of high-throughput data. In S. J. Corey, M. Kimmel, & J. N. Leonard (Eds.), *A Systems Biology Approach to Blood* (Vol. 844, pp. 153–187). Springer New York. https://doi.org/10.1007/978-1-4939-2095-2_8

D'Argenio, V. (2018). The high-throughput analyses era: Are we ready for the data struggle? *High-Throughput*, *7*(1), 8.

De Diego, N., & Spíchal, L. (2022). Presence and future of plant phenotyping approaches in biostimulant research and development. *Journal of Experimental Botany*, *73*(15), 5199–5212.

Dong, Y., Sonawane, P., Cohen, H., Polturak, G., Feldberg, L., Avivi, S.H., Rogachev, I., & Aharoni, A. (2020) High mass resolution, spatial metabolite mapping enhances the current plant gene and pathway discovery toolbox. *New Phytologist*, *228,* 1986–2002.

Gill, T., Gill, S. K., Saini, D. K., Chopra, Y., De Koff, J. P., & Sandhu, K. S. (2022). A comprehensive review of high throughput phenotyping and machine learning for plant stress phenotyping. *Phenomics*, *2*(3), 156–183.

Hall, R. D., D'Auria, J. C., Silva Ferreira, A. C., Gibon, Y., Kruszka, D., Mishra, P., & Van De Zedde, R. (2022). High-throughput plant phenotyping: A role for metabolomics? *Trends in Plant Science*, *27*(6), 549–563.

Huck, C. (2015). Infrared spectroscopic technologies for the quality control of herbal medicines. In *Evidence-Based Validation of Herbal Medicine* (pp. 477–493). Elsevier. https://doi.org/10.1016/B978-0-12-800874-4.00022-2

Jangra, S., Chaudhary, V., Yadav, R. C., & Yadav, N. R. (2021). High-throughput phenotyping: A platform to accelerate crop improvement. *Phenomics*, *1*(2), 31–53.

Jud, W., Winkler, J. B., Niederbacher, B., Niederbacher, S., & Schnitzler, J.-P. (2018). Volatilomics: A non-invasive technique for screening plant phenotypic traits. *Plant Methods*, *14*(1), 109.

Kim, H. K., Choi, Y. H., & Verpoorte, R. (2010). NMR-based metabolomic analysis of plants. *Nature Protocols*, *5*(3), 536–549.

Klukas, C., Chen, D., & Pape, J.-M. (2014). Integrated analysis platform: An open-source information system for high-throughput plant phenotyping. *Plant Physiology*, *165*(2), 506–518.

Krishnan, P. (2004). Metabolite fingerprinting and profiling in plants using NMR. *Journal of Experimental Botany*, *56*(410), 255–265.

Kumar, R., Bohra, A., Pandey, A. K., Pandey, M. K., & Kumar, A. (2017). Metabolomics for plant improvement: Status and prospects. *Frontiers in Plant Science*, *8*, 1302.

Li, D., Quan, C., Song, Z., Li, X., Yu, G., Li, C., & Muhammad, A. (2021). High-throughput plant phenotyping platform (HT3P) as a novel tool for estimating agronomic traits from the lab to the field. *Frontiers in Bioengineering and Biotechnology*, *8*, 623705.

Li, L., Zhang, Q., & Huang, D. (2014). A review of imaging techniques for plant phenotyping. *Sensors*, *14*(11), 20078–20111.

Ludovisi, R., Tauro, F., Salvati, R., Khoury, S., Mugnozza Scarascia, G., & Harfouche, A. (2017). UAV-based thermal imaging for high-throughput field phenotyping of black poplar response to drought. *Frontiers in Plant Science*, *8*, 1681.

Ma, H., Huang, W., Dong, Y., Liu, L., & Guo, A. (2021). Using UAV-based hyperspectral imagery to detect winter wheat Fusarium head blight. *Remote Sensing*, *13*(15), 3024.

Manickam, S., Rajagopalan, V. R., Kambale, R., Rajasekaran, R., Kanagarajan, S., & Muthurajan, R. (2023). Plant metabolomics: Current initiatives and future prospects. *Current Issues in Molecular Biology*, *45*(11), 8894–8906.

Nakabayashi, R., Hashimoto, K., Mori, T., Toyooka, K., Sudo, H., & Saito, K. (2021). Spatial metabolomics using imaging mass spectrometry to identify the localization of asparaptine A in *Asparagus officinalis*. *Plant Biotechnology*, *38*(3), 311–315.

Patel, M., Pandey, S., Kumar, M., Haque, M., Pal, S., & Yadav, N. (2021). Plants metabolome study: Emerging tools and techniques. *Plants*, *10*(11), 2409.

Pieruschka, R., & Schurr, U. (2019). Plant phenotyping: Past, present, and future. *Plant Phenomics,* 1–6.

Poorter, H., Hummel, G. M., Nagel, K. A., Fiorani, F., Von Gillhaussen, P., Virnich, O., Schurr, U., Postma, J. A., Van De Zedde, R., & Wiese-Klinkenberg, A. (2023). Pitfalls and potential of high-throughput plant phenotyping platforms. *Frontiers in Plant Science*, *14*, 1233794.

Sohn, S.-I., Pandian, S., Oh, Y.-J., Zaukuu, J.-L. Z., Kang, H.-J., Ryu, T.-H., Cho, W.-S., Cho, Y.-S., Shin, E.-K., & Cho, B.-K. (2021). An overview of near infrared spectroscopy and its applications in the detection of genetically modified organisms. *International Journal of Molecular Sciences*, *22*(18), 9940.

Szymański, P., Markowicz, M., & Mikiciuk-Olasik, E. (2011). Adaptation of high-throughput screening in drug discovery-toxicological screening tests. *International Journal of Molecular Sciences*, *13*(1), 427–452.

Tian, Z., Hu, X., Xu, Y., Liu, M., Liu, H., Li, D., Hu, L., Wei, G., & Chen, W. (2024). PMhub 1.0: A comprehensive plant metabolome database. *Nucleic Acids Research*, *52*(D1), D1579–D1587.

Vervoort, N., Goossens, K., Baeten, M., & Chen, Q. (2021). Recent advances in analytical techniques for high throughput experimentation. *Analytical Science Advances*, *2*(3–4), 109–127.

Volpato, L., Pinto, F., González-Pérez, L., Thompson, I. G., Borém, A., Reynolds, M., Gérard, B., Molero, G., & Rodrigues, F. A. (2021). High throughput field phenotyping for plant height using UAV-based RGB imagery in wheat breeding lines: Feasibility and validation. *Frontiers in Plant Science*, *12*, 591587.

Wase, N., Abshire, N., & Obata, T. (2022). High-throughput profiling of metabolic phenotypes using high-resolution GC-MS. In A. Lorence & K. Medina Jimenez (Eds.), *High-Throughput Plant Phenotyping* (Vol. 2539, pp. 235–260). Springer US. https://doi.org/10.1007/978-1-0716-2537-8_19

Yamamoto, K, Takahashi K, Caputi L, Mizuno H, Rodriguez-Lopez CE, Iwasaki T, Ishizaki K, Fukaki H, Ohnishi M, Yamazaki M, et al. (2019) The complexity of intercellular localisation of alkaloids revealed by single-cell metabolomics. *New Phytologist*, *224*, 848–859

Zaman-Allah, M., Vergara, O., Araus, J. L., Tarekegne, A., Magorokosho, C., Zarco-Tejada, P. J., Hornero, A., Albà, A. H., Das, B., Craufurd, P., Olsen, M., Prasanna, B. M., & Cairns, J. (2015). Unmanned aerial platform-based multi-spectral imaging for field phenotyping of maize. *Plant Methods*, *11*(1), 35.

Zhang, Q. (Ed.). (2023). *Encyclopedia of Digital Agricultural Technologies*. Springer International Publishing. https://doi.org/10.1007/978-3-031-24861-0

# 18 High-Throughput Metabolomics for Agricultural Research

*Govindarajan Ramkumar, Sureshkumar Tharani, Siva Daniel Ajay Samuel, Chithra Devi Balasundarm Saraswathy, Sarah Jaison, Krishnan Kutty Sunitha Kumari, Muthukrishnan Arun, and Arumugam Vijaya Anand*

## LIST OF ABBREVIATIONS

**AI:** Artificial Intelligence
**GC:** Gas Chromatography
**LC:** Liquid Chromatography
**ML:** Machine Learning
**MS:** Mass Spectrometry
**NMR:** Nuclear Magnetic Resonance
**VOCs:** Volatile Organic Compounds

## 18.1 INTRODUCTION

### 18.1.1 Overview of Metabolomics

Metabolomics is a newly developing field within the broader area of systems biology, focusing on the comprehensive investigation of metabolites, and small molecules involved in metabolism within cells, tissues, or organisms (Jones, 2018). Metabolites denote the end products of cellular events, and their analysis provides a direct "snapshot" of the physiological state of an organism (Sengupta & Narad, 2018). Unlike genomics or proteomics, which offer indirect insights through genetic and protein blueprints, metabolomics captures the dynamic interactions within the metabolome, reflecting both intrinsic biological functions and responses to environmental stimuli (Katam et al., 2022). Metabolomics employs advanced analytical technologies, primarily mass spectrometry (MS) and nuclear magnetic resonance (NMR) spectroscopy, to identify and quantify metabolites (Marshall & Powers, 2017). These technologies are complemented by chromatographic techniques like gas chromatography (GC) and liquid chromatography (LC), which enhance the separation and identification of difficult metabolite mixtures (Allwood & Goodacre, 2010). The power of metabolomics lies in its ability to generate detailed metabolic profiles, facilitating the understanding of biochemical pathways and their alterations under various conditions (Johnson et al., 2016).

### 18.1.2 Importance in Agriculture

The application of metabolomics in agriculture has transformative potential, addressing critical challenges in crop improvement, food quality, and environmental sustainability (Kumar et al., 2017). Agriculture, the cornerstone of human civilization, faces mounting pressures from a growing

DOI: 10.1201/9781003505631-18

population, climate change, and declining natural resources (Mondal & Palit, 2022). Conventional agricultural practices, although effective, are often insufficient to meet these challenges without the support of advanced biotechnological approaches (Ruane & Sonnino, 2011). Metabolomics delivers a holistic view of plant metabolism, offering insights into plant development, and responses to abiotic and biotic stress (Piasecka et al., 2019). By deciphering the metabolic underpinnings of traits like yield, nutritional content, and stress toleration, metabolomics aids in the evolution of crops that are more productive, nutritious, and resilient (Kaur et al., 2021). Moreover, metabolomics can enhance our understanding of soil health and its impact on crop performance, paving the way for more sustainable agricultural practices (Parray & Shameem, 2019). In food quality and security, metabolomics plays a pivotal role in nutritional profiling, detection of food adulteration, and monitoring of pesticide residues (Li et al., 2021). It enables the characterization of food by-products at a molecular level, ensuring authenticity and compliance with safety standards (Oms-Oliu et al., 2013). This molecular insight supports the evolution of functional foods and contributes to public health by addressing issues related to diet and nutrition (Arun Kumar et al., 2018).

### 18.1.3 High-Throughput Technologies and Their Significance

The integration of high-throughput approaches into metabolomics has transformed agricultural research, enabling the analysis of large-scale metabolomic data with unprecedented speed and precision (Kumar et al., 2017). High-throughput metabolomics leverages automation, robotics, and advanced computational tools to handle the extensive datasets generated from metabolomic analyses (Coler et al., 2024). These technologies promote the immediate screening of numerous samples, making it feasible to conduct large-scale investigations and gain comprehensive insights into complex biological systems (Liu et al., 2019). Key components of high-throughput metabolomics include automated sample preparation, high-throughput MS and NMR platforms, and sophisticated data analysis pipelines (Cheng et al., 2018). Automation minimizes human error and variability, enhancing the reproducibility of results (González-Domínguez et al., 2024). High-throughput MS, with its capability to rapidly analyze and quantify numerous metabolites, coupled with the structural elucidation capabilities of NMR, provides a powerful toolkit for metabolomic investigations (Wolfender et al., 2018). Additionally, bioinformatics and machine learning (ML) algorithms an important in processing and analyzing the voluminous data, extracting meaningful patterns and correlations (Ren et al., 2015).

The significance of high-throughput approaches in agriculture is profound. They enable the scaling up of metabolomic studies to accommodate the diversity and complexity of agricultural systems (Parray & Shameem, 2019). This scalability supports the development of high-yielding and stress-resistant crops through large-scale phenotyping and metabolic profiling (Fu & Jiang, 2022). Furthermore, high-throughput approaches accelerate the discovery of metabolic markers associated with desirable traits, facilitating targeted breeding and genetic engineering efforts (Razzaq et al., 2022). High-throughput metabolomics offers a cutting-edge approach to addressing agricultural challenges (do Prado et al., 2018). Integrating comprehensive metabolomic analyses with advanced technological platforms, provides invaluable insights into plant biology, food quality, and environmental sustainability, driving innovation and progress in agricultural research (Alawiye & Babalola, 2021).

## 18.2 BASICS OF METABOLOMICS

Metabolomics is the complete study of metabolites, the small molecules that are the intermediates and end products of cellular metabolic processes (Muthubharathi et al., 2021). These are amino acids, lipids, sugars, nucleotides, and other small molecules, which play critical roles in cellular functions and provide a natural reflection of the physiological condition of an organism (Wishart, 2019). The scope of metabolomics extends across various biological systems, including plants, animals, and microorganisms, and it encompasses the identification, quantification, and understanding of metabolic profiles under different conditions (Goldansaz et al., 2017).

Metabolomics strives to provide a holistic view of the metabolic networks and pathways within a biological procedure, enabling researchers to understand how these networks are influenced by genetic factors, environmental changes, and interactions with other organisms (Oldiges et al., 2007). By incorporating metabolomics with other omics technologies like proteomics, transcriptomics, and genomics investigators can achieve a more extensive understanding of biological functions and regulatory mechanisms (Horgan & Kenny, 2011).

### 18.2.1 Metabolomic Techniques

To capture the complexity and diversity of metabolites in a biological sample, metabolomics relies on many advanced analytical methods (Zhang et al., 2012). The primary methods include MS, NMR, and Chromatography (Letertre et al., 2020).

#### 18.2.1.1 Mass Spectrometry (MS)

MS is a potent method for determining and quantifying metabolites established on their mass-to-charge ratio (Ren et al., 2018). It involves ionizing chemical compounds to develop charged molecules and measuring their mass-to-charge ratios (Medhe, 2018). MS can be coupled with chromatographic methods like Gas Chromatography (GC-MS) and Liquid Chromatography (LC-MS) to enhance the separation and analysis of complex mixtures (Williams & Burinsky, 2001). These techniques are also essential for elucidating complex metabolic networks and identifying biomarkers for disease, drug response, and environmental exposures (Johnson et al., 2016; Deng et al., 2019).

- **GC-MS**: GC-MS integrates the high-resolution separation capabilities of GS with the sensitive and specific detection provided by MS (Misra, 2021). In GC-MS, the sample is evaporated and transported through a chromatographic column by an inert carrier gas, typically helium (Sparkman et al., 2011). The column is covered with a stationary phase that interacts differentially with the components of the sample, separating volatile and semi-volatile compounds based on their boiling points and polarity (Ntlhokwe, 2016). As each compound elutes from the column at a different time, it enters the MS, where it is ionized, typically by electron impact ionization, producing charged fragments (Sparkman et al., 2011). These fragments are then separated and based on their mass-to-charge (m/z) ratios, generating a mass spectrum that provides both qualitative and quantitative information (Kitson et al., 1996). The mass spectrum is a distinctive fingerprint of the compound, allowing for its identification through comparison with spectral libraries and databases (Vinaixa et al., 2016). GC-MS is particularly effective for analyzing small, volatile organic compounds (VOCs), like fatty acids, and amino acids, making it indispensable in studies of plant metabolites, environmental pollutants, and breath analysis (Franchina et al., 2021). However, GC-MS is limited by its incapacity to examine non-volatile, thermally labile compounds without derivatization (Garcia & Barbas, 2011).
- **LC-MS**: LC-MS, on the other hand, overcomes these limitations by coupling LC with mass spectrometry. In LC-MS, the sample is liquefied in a liquid mobile phase, which is pumped through a chromatographic column packed with a solid stationary phase (Ventura et al., 1999). The separation in LC is established on the differential interchange of compounds with the stationary phase, exploiting differences in hydrophobicity, charge, and size (McCalley, 2023; Svrkota et al., 2023). LC-MS can handle a broader spectrum of compounds, including non-volatile, polar, and thermally labile metabolites, such as nucleotides, peptides, lipids, and complex carbohydrates (Ntlhokwe, 2016). After separation, the eluted compounds enter the mass spectrometer, where they are ionized, often by electrospray ionization or atmospheric pressure chemical ionization (Awad et al., 2015).

These soft ionization techniques generate intact molecular ions with minimal fragmentation, preserving the molecular information crucial for compound identification. The ionized metabolites are then analyzed by their m/z ratios, providing a mass spectrum that can be further interpreted using tandem mass spectrometry (MS/MS) for structural elucidation (Ma, 2022). LC-MS is renowned for its versatility, high sensitivity, and compatibility with high-throughput workflows, making it a cornerstone in metabolomics for biomarker discovery, pharmacokinetics, and systems biology studies (Jian, 2017). Both GC-MS and LC-MS have unique strengths and are often used complementarily in metabolomics to provide a comprehensive metabolic profile. While GC-MS excels in the analysis of small, volatile compounds with high resolution and reproducibility, LC-MS offers a broader metabolite coverage, capturing complex and non-volatile molecules with high sensitivity.

Advances in mass spectrometry, like high-resolution accurate mass analyzers and hybrid instruments combining quadrupole, time-of-flight, and orbitrap technologies, have further enhanced the capabilities of GC-MS and LC-MS, enabling more precise quantitation, structural characterization, and discovery of novel metabolites (Crutchfield & Clarke, 2017). GC-MS and LC-MS are indispensable tools in metabolomics, each with distinct advantages that together provide a robust platform for the comprehensive analysis of the metabolome, driving forward our understanding of biological processes and their implications for health and disease.

#### 18.2.1.2 Nuclear Magnetic Resonance (NMR)

NMR spectroscopy is a non-destructive analytical method that provides detailed details about the structure, dynamics, and environment of metabolites (Kumar, 2016). NMR measures the magnetic effects of atomic nuclei, particularly hydrogen and carbon, in the presence of a powerful magnetic field and radiofrequency pulses (Hatzakis, 2019). When positioned in a magnetic domain, certain atomic nuclei resonate at distinct frequencies depending on their chemical environment. By applying a series of radiofrequency pulses, NMR excites these nuclei and then detects the emitted signals as they return to their equilibrium states. The resulting NMR spectrum provides a unique fingerprint for each molecule, with peaks corresponding to different atomic environments (Krishnan et al., 2005). The position (chemical shift) and intensity of these peaks give insights into the molecular structure, including the identification of functional groups and the connectivity of atoms (Dias et al., 2016). Coupling patterns (spin-spin coupling) and relaxation times provide additional information on molecular dynamics and interactions (Cremer & Gräfenstein, 2007).

NMR is particularly valuable for its ability to analyze complex mixtures without prior separation, allowing for the direct study of biological fluids, tissue extracts, and other heterogeneous samples (Cremer & Gräfenstein, 2007). It can provide absolute quantification of metabolites by corresponding the integral of the NMR signals to those of known reference standards (Wishart, 2008b). This quantitative aspect, combined with the high-resolution structural information, makes NMR a powerful tool in metabolomics, drug discovery, and structural biology. Moreover, NMR's non-destructive nature allows for the repeated examination of the same sample under different situations, facilitating studies on metabolic changes over time or in response to environmental factors (Deborde et al., 2017). Advanced NMR methods, like multidimensional NMR and the application of hyperpolarization, further enhance the analytical capabilities, enabling the detailed study of larger and more complex biomolecules (Nagana Gowda & Raftery, 2019).

#### 18.2.1.3 Chromatography

Chromatography is a pivotal separation technique in metabolomics, essential for the detailed examination and description of a wide array of metabolites established on their distinct chemical properties, such as polarity, volatility, and molecular size (Zhang et al., 2012). By facilitating the isolation of individual metabolites from complex biological matrices, chromatography enables subsequent detection and quantification with high precision and accuracy (Zhou & Yin, 2016). Two primary forms of chromatography employed in metabolomics are GC and LC, each offering unique advantages and specific applications depending on the character of the metabolites under investigation.

- **Gas Chromatography**: GC is specifically designed for the separation of volatile and semi-volatile compounds that can be vaporized without decomposition (Stashenko & Martínez, 2014). In GC, the sample is introduced into the chromatographic system and vaporized in the injector port. An inert carrier gas, like helium or nitrogen, transports the vaporized sample via a capillary column coated with a liquid or polymer stationary phase (Falaki, 2019). The separation of compounds occurs as they interact differently with the stationary phase based on their volatility and polarity, resulting in distinct retention times for each compound. The separated compounds elute sequentially from the column and are often detected using MS, a technique that enhances the sensitivity and specificity of the analysis (Jandera, 2011). The combination of GC and MS permits the precise identification and quantification of small, volatile molecules, such as fatty acids, alcohols, esters, and VOCs (Makoś et al., 2019). GC-MS is highly valued in metabolomics for its ability to generate reproducible and high-resolution chromatograms, providing detailed insights into metabolic profiles and facilitating the discovery of biomarkers related to various physiological and pathological conditions (Beale et al., 2018).
- **Liquid Chromatography**: LC is an essential device in plant metabolomics, providing a robust method for the separation, identification, and quantification of a vast array of plant metabolites (Hill & Roessner, 2015). In plant metabolomics, LC is employed to analyze complex mixtures extracted from plant tissues, which are often rich in diverse compounds like alkaloids, flavonoids, terpenoids, and phenolic compounds (Salem et al., 2020). The sample preparation involves grinding the plant material and extracting the metabolites with suitable solvents (Mushtaq et al., 2014). The resulting extracts are then injected into the LC system, and it is separated based on their interactions with the stationary phase in the chromatographic column, exploiting differences in hydrophobicity, charge, and molecular size (François et al., 2009). LC is particularly effective for handling non-volatile, polar, and thermally labile compounds that are prevalent in plants (Nyiredy, 2004; Marston, 2007). Various detectors, such as UV, fluorescence, or refractive index detectors, are used to detect and quantify the separated compounds (Al-Sanea & Gamal, 2022). The adaptability of LC to use different stationary phases, such as reversed-phase and ion-exchange chromatography, enhances its capability to analyze a wide range of plant metabolites (Jin et al., 2016). This versatility and precision make LC indispensable for studying plant metabolic pathways, understanding plant physiology, and discovering biomarkers for plant breeding and stress response analysis (Katam et al., 2022).

### 18.2.2 Types of Metabolomics

Metabolomic studies in plants can be broadly classified into two main types: targeted and untargeted metabolomics, each with distinct goals and methodologies that are crucial for advancing our understanding of plant biology, improving crop productivity, and enhancing environmental sustainability (Allwood et al., 2021).

#### 18.2.2.1 Targeted Metabolomics

Targeted metabolomics concentrates on the quantitative analysis of a precise set of known metabolites. This approach is hypothesis-driven, aiming to estimate the concentrations of predefined metabolites that are relevant to a particular biological question or condition (Amer et al., 2023). For example, in plant metabolomics, targeted studies might focus on specific secondary metabolites, which play important roles in plant defense, signaling, and adaptation (Desmet et al., 2021). By using targeted LC methods, researchers can accurately quantify these metabolites under various conditions, such as stress responses or developmental stages. This approach is positively sensitive and specific, making it ideal for validating biomarkers, studying specific metabolic pathways, and monitoring the results of genetic modifications or environmental factors on plant metabolism (Chen et al., 2013). Targeted metabolomics provides an understanding of the functional roles of specific metabolites, aiding in the development of crops with enhanced nutritional profiles or enhanced resistance to pests and diseases (Razzaq et al., 2022).

#### 18.2.2.2 Untargeted Metabolomics

Untargeted metabolomics or global or discovery metabolomics, seeks to analyze as many metabolites as possible in a biological sample without prior knowledge of their identities (Gertsman & Barshop, 2018). This approach is exploratory and unbiased, enabling the finding of novel metabolites and the generation of comprehensive metabolic profiles (Bowman et al., 2006). In plant metabolomics, untargeted LC-MS is especially helpful for identifying biomarkers and comprehending the global metabolic modifications associated with different conditions, such as drought stress, pathogen attack, or nutrient deficiencies (Allwood et al., 2021). By comparing metabolic profiles across different samples, researchers can identify unique or significantly altered metabolites that can be related to specific physiological or environmental circumstances (Roessner et al., 2001). This can lead to the finding of new metabolic pathways and regulatory networks, providing a deeper knowledge of plant biology and the molecular mechanisms underlying plant responses to various stimuli (Arbona et al., 2013). Untargeted metabolomics is essential for advancing agricultural research, as it helps uncover the metabolic diversity and complexity of plants, which can be harnessed for crop progress and sustainable agriculture (Razzaq et al., 2022).

By combining targeted and untargeted metabolomics approaches, researchers can obtain a comprehensive view of plant metabolism (Allwood et al., 2021). Targeted metabolomics allows for precise quantification and validation of specific metabolites, while untargeted metabolomics delivers a broad overview of the metabolic landscape, facilitating the discovery of new metabolic pathways and potential biomarkers (Cajka & Fiehn, 2016). This integrative approach is crucial for understanding the complex biochemical networks that underpin plant biological systems. The foundational knowledge gained from metabolomics studies is important for developing innovative solutions to improve crop productivity, enhance food quality, and promote environmental sustainability.

## 18.3 HIGH-THROUGHPUT METABOLOMICS IN AGRICULTURE

High-throughput metabolomics is revolutionizing agricultural research by enabling the rapid and comprehensive analysis of metabolic profiles on a large scale (Mashabela et al., 2022). This approach leverages advanced techniques such as automation, robotics, multiplexing, and sophisticated data investigation to handle the extensive quantities of data generated, providing deeper insights into plant biology, crop improvement, and sustainable agricultural practices (Amer & Baidoo, 2021; Mochida et al., 2020).

### 18.3.1 High-Throughput Techniques in Plant Metabolomics

High-throughput techniques are integral to modern plant metabolomics, allowing for the efficient processing and analysis of numerous plant samples, which is crucial for comprehensive metabolic profiling in agricultural research (Manickam et al., 2023).

#### 18.3.1.1 Automation and Robotics

Automation and robotics play a crucial role in high-throughput plant metabolomics by streamlining sample preparation, processing, and analysis. Automated systems reduce human error, increase reproducibility, and enhance throughput, enabling researchers to handle large-scale plant studies with improved efficiency and accuracy (Abdi et al., 2024). Robotic platforms can perform tasks such as liquid handling, sample injection, and data acquisition, significantly speeding up the

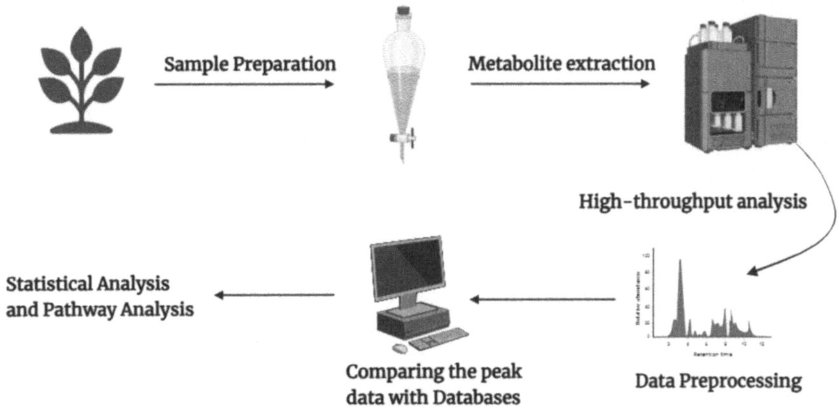


**FIGURE 18.1** Workflow of high-throughput metabolomics.

metabolomic workflow (Barrett, 2024). For instance, automated grinding and extraction systems can process large numbers of plant tissues simultaneously, ensuring uniform sample preparation and consistent results across studies (Gomez et al., 2010) (Figure 18.1).

#### 18.3.1.2 Multiplexing

Multiplexing involves the concurrent investigation of multiple metabolites or samples within a single experiment (Higashi & Ogawa, 2020). This technique enhances throughput and data richness by allowing the parallel processing of several biological samples, thus providing a comprehensive view of the plant metabolome (Foito & Stewart, 2018). Multiplexing is particularly useful in comparative studies, such as assessing metabolic differences across various plant varieties, developmental stages, or treatment conditions (Hong et al., 2016; Arbona et al., 2013). For example, multiplexing can facilitate the comparison of metabolic responses of different crop varieties to drought stress, providing insights into their adaptive mechanisms and aiding in the development of more resilient crops (Jain et al., 2019).

#### 18.3.1.3 High-Throughput Data Analysis

High-throughput data investigation utilizes advanced computational tools and bioinformatics techniques to manage and interpret the extensive quantities of data generated from plant metabolomic studies (Ren et al., 2015). ML algorithms, statistical methods, and network analysis are employed to identify patterns, correlations, and biomarkers within complex datasets (Liebal et al., 2020; Kisiel et al., 2023). These tools facilitate the incorporation of metabolomic data with other omics datasets, enhancing the interpretation of metabolic networks and pathways in plants (Töpfer et al., 2015). For instance, integrating metabolomic data with transcriptomic and proteomic data can reveal regulatory mechanisms controlling secondary metabolite biosynthesis in plants (Yang et al., 2014).

### 18.3.2 Data Management and Integration

Effective data governance and integration are essential for the prosperous application of high-throughput metabolomics in plant research (Ernst et al., 2014).

#### 18.3.2.1 Databases and Repositories

Databases and repositories serve as crucial resources for storing, organizing, and sharing plant metabolomic data (Go, 2010). Publicly accessible databases enable researchers to deposit their data, making it available for validation, comparison, and further analysis by the scientific community (Sud et al., 2016). Plant metabolome database and MetaboLights, which provide comprehensive repositories of plant metabolomic data (Salem et al., 2022; Ambrosino et al., 2020). These databases facilitate the recognition of metabolites and the comparison of metabolic profiles across different plant species and experimental conditions (Gaquerel et al., 2013).

#### 18.3.2.2 Data Standardization

Data standardization ensures consistency and comparability across different plant metabolomic studies (Rocca-Serra et al., 2016). Standardized methods for sample collection, preparation, analysis, and data reporting are important for reproducibility and dependability (Salem et al., 2020). Initiatives such as the metabolomics standards initiative promote best practices and guidelines to achieve uniformity in plant metabolomic research (Krajewski et al., 2015). Standardization helps in minimizing variability and enables more accurate comparisons of metabolomic data across various laboratories and studies (Fiehn et al., 2008).

### 18.3.3 Challenges and Solutions

Despite its potential, high-throughput plant metabolomics faces several challenges that are addressed to maximize its impact on agricultural investigation.

#### 18.3.3.1 Sample Preparation

Sample preparation is a necessary step in plant metabolomics that can significantly influence the quality and precision of the outcomes. Challenges include the need for precise and consistent extraction methods, the stability of metabolites, and the removal of interfering substances (Salem et al., 2020). Solutions involve optimizing extraction protocols, using stabilizing agents, and employing quality control measures to ensure reproducibility (Vuckovic, 2012). For example, the use of cold solvents and antioxidants can prevent the degradation of sensitive plant metabolites during extraction (Roessner & Dias, 2013).

#### 18.3.3.2 Data Overload

The extensive quantity of data developed by high-throughput plant metabolomics can lead to data overload, making it challenging to extract meaningful understandings (Wolfender et al., 2015). Effective data management strategies, such as robust data storage solutions, efficient data processing pipelines, and the use of advanced bioinformatics tools, are essential to handle and interpret the data efficiently (Cambiaghi et al., 2017). Developing standardized data formats and interoperable software platforms can also facilitate data sharing and integration across studies (Salek et al., 2015).

#### 18.3.3.3 Reproducibility

Reproducibility is a primary concern in plant metabolomics because of the variability introduced by different analytical platforms, sample preparation methods, and data analysis techniques (Ernst et al., 2014). Addressing this challenge requires the adoption of standardized protocols, rigorous quality control measures, and transparent reporting of methodologies and results. Collaborative efforts and inter-laboratory studies can also help in validating and harmonizing metabolomic methods (Anklam et al., 2022). High-throughput plant metabolomics offers a powerful approach to advancing agricultural research by enabling large-scale, comprehensive metabolic analyses (Narad et al., 2022). By leveraging automation, multiplexing, and advanced data analysis techniques, and addressing challenges related to sample preparation, data overload, and reproducibility, high-throughput plant metabolomics can drive significant innovations in crop improvement, food quality, and sustainable agricultural practices.

## 18.4 APPLICATIONS IN CROP IMPROVEMENT

High-throughput metabolomics is playing a transformative role in crop improvement by providing detailed insights into plant metabolism, identifying key metabolic traits, and aiding in the development of crops with enhanced qualities (Jamaloddin et al., 2021). The following sections highlight its applications in different characteristics of crop advancement.

### 18.4.1 Phenotyping and Trait Analysis

Metabolomics-based phenotyping involves the comprehensive analysis of metabolites to link specific metabolic profiles with phenotypic traits. This approach allows researchers to determine metabolic markers related to desirable traits like yield, growth rate, and resistance to environmental strains (Sharma et al., 2021). By comparing the metabolomes of different plant varieties, scientists can uncover the biochemical pathways underlying these traits, facilitating the selection and breeding of superior crop varieties (Alseekh et al., 2018).

### 18.4.2 Nutritional Enhancement

One of the significant applications of metabolomics in agriculture is the enhancement of nutritional content in crops (Dixon et al., 2006). By analyzing the metabolic reactions concerned with the synthesis of minerals, vitamins, and other essential nutrients, researchers can identify and manipulate key genes and enzymes to boost the nutritional value of crops (Singh et al., 2022). For example, metabolomics has been employed to increase the levels of essential amino acids, antioxidants, and other beneficial constituents in staple crops, addressing nutritional deficiencies and improving public health (Hamany Djande et al., 2020).

### 18.4.3 Stress Tolerance

Crops are constantly exposed to various stresses that can affect their growth and productivity. Metabolomics gives a useful understanding of the metabolic responses of plants to stress, aiding in the development of stress-tolerant crops (Parida et al., 2018).

#### 18.4.3.1 Abiotic Stress (Drought, Salinity)

Abiotic stresses are major challenges in agriculture. Metabolomics helps identify the metabolic shifts that occur in plants under these strain conditions, revealing key metabolites and paths interested in stress toleration (Khan et al., 2017). By understanding these mechanisms, researchers can develop crops that maintain productivity under unfavorable environmental circumstances. For instance, metabolites such as osmolytes and antioxidants are often upregulated in response to salinity and drought, and manipulating their biosynthesis can enhance stress tolerance (Qari & Tarbiyyah, 2021).

#### 18.4.3.2 Biotic Stress (Pests, Pathogens)

Biotic stresses from pests and pathogens can significantly impact crop yields. Metabolomics allows for the recognition of defense-related metabolites and the description of plant defense mechanisms (Castro-Moretti et al., 2020). This knowledge can be used to breed or engineer crops with enhanced resistance to pests and diseases. For example, the production of specific secondary metabolites can be increased to improve plant defense against biotic threats (Jha & Mohamed, 2022).

### 18.4.4 Metabolic Engineering

Metabolic engineering concerns the improvement of metabolic pathways to enhance the production of preferred compounds or to introduce new metabolic capabilities in plants (Pickens et al., 2011). Metabolomics provides the necessary information to identify target paths and monitor the outcomes of genetic modifications (Johnson et al., 2016).

#### 18.4.4.1 Secondary Metabolites

Secondary metabolites play important roles in plant defense and can have significant pharmaceutical and industrial applications (Roaa, 2020). Metabolomics enables the recognition and quantification of these metabolites, facilitating the engineering of plants to produce higher levels of valuable

**TABLE 18.1**
**Applications of High-Throughput Metabolomics in Agriculture**

| Domain | Specific Application | Impact | References |
|---|---|---|---|
| Crop enhancement | Identifying metabolites linked to yield and quality traits | Development of high-yield, nutritionally enhanced crops. | Fernie & Schauer (2009) |
| Disease resilience | Detecting biomarkers for disease resistance | Breeding disease-resistant varieties. | Kumar et al. (2024); Ontoy & Ham (2024) |
| Environmental stress tolerance | Analyzing plant responses to drought, salinity, etc. | Creating crops with improved tolerance to adverse circumstances. | Khan et al. (2017); Kumar et al. (2021) |
| Plant-microbe interactions | Understanding how microbes affect plant metabolism and nutrient uptake | Enhancing soil health, improving crop growth and sustainability. | Chen et al. (2021); Gupta et al. (2022) |

secondary metabolites (Yang et al., 2014). This approach can lead to the evolution of crops with enhanced resistance to pests and diseases, as well as the production of plant-based drugs and bioactive compounds (Jimenez-Garcia et al., 2013).

#### 18.4.4.2 Primary Metabolites

Primary metabolites are essential for plant growth and evolution (Bocso & Butnariu, 2022). Metabolomics helps elucidate the paths concerned in the synthesis and regulation of these compounds, providing targets for metabolic engineering (Sharma et al., 2021). By manipulating the levels of primary metabolites, researchers can improve crop yield, enhance nutritional content, and increase the efficiency of photosynthesis and nutrient utilization (Godoy et al., 2021). High-throughput metabolomics offers powerful tools for crop improvement by providing detailed insights into plant metabolism and identifying key metabolic traits (Khakimov et al., 2014). By leveraging these insights, researchers can develop crops with enhanced nutritional content, enhanced strain toleration, and increased production of beneficial metabolites, ultimately contributing to sustainable agricultural practices and global food security (Table 18.1).

## 18.5 APPLICATIONS IN FOOD QUALITY AND SAFETY

High-throughput metabolomics is instrumental in enhancing food quality and safety by providing comprehensive analyses of food composition, detecting adulteration, and monitoring contaminants (Li et al., 2021). The following sections discuss its applications in nutritional profiling, food authentication, adulteration detection, and monitoring pesticide residues.

### 18.5.1 Nutritional Profiling

Nutritional profiling involves the detailed analysis of the nutritional content of food products, including minerals, vitamins, fatty acids, amino acids, and other bioactive constituents (Astarita & Langridge, 2014). Metabolomics enables the accurate and comprehensive quantification of these nutrients, offering a detailed nutritional profile of food items (Emwas et al., 2021). This information is necessary for:

- **Improving Public Health**: By understanding the nutritional composition of different foods, health professionals can make more informed dietary recommendations to improve public health (German et al., 2003; Gibney et al., 2005).

- **Enhancing Food Products**: Food manufacturers can use metabolomic data to enhance the nutritional content of their products, fortify foods with essential nutrients and develop functional foods with health benefits (Cevallos-Cevallos et al., 2009).
- **Quality Control**: Nutritional profiling ensures that food products meet specific nutritional standards and regulations, maintaining quality and consumer trust (Wishart, 2008a).

### 18.5.2 Food Authentication and Adulteration Detection

Food authentication verifies the origin, quality, and authenticity of food products, ensuring that they meet specified standards and labeling claims (Danezis et al., 2016). Adulteration detection identifies the presence of unauthorized substances or lower-quality ingredients in food products (Reid et al., 2006). Metabolomics provides powerful tools for both purposes:

- **Authentication**: Metabolomic fingerprinting can identify unique metabolic profiles associated with specific food types, varieties, or geographic origins. This helps in verifying the authenticity of high-value products such as organic foods, speciality crops, and geographic indication products like wine, cheese, and olive oil (Cevallos-Cevallos et al., 2009; Danezis et al., 2016).
- **Adulteration Detection**: Metabolomics can detect unexpected metabolites indicative of adulteration, such as the addition of cheaper substitutes, contaminants, or undeclared ingredients. This ensures product integrity and protects consumers from fraud and potential health risks (Lees & Reimann, 2021).

### 18.5.3 Monitoring Pesticide Residues

The presence of pesticide remains in food is a major concern for food safety and consumer health (Hernández-Mesa & Moreno-González, 2022). Metabolomics offers advanced techniques for monitoring and quantifying pesticide residues in agricultural products:

- **Sensitive Detection**: High-throughput metabolomics can detect trace levels of pesticide residues, ensuring compliance with safety regulations and maximum residue limits (MRLs) (Romero-González & Frenich, 2017).
- **Comprehensive Analysis**: Metabolomic approaches can simultaneously monitor multiple pesticide residues, providing a comprehensive assessment of food safety (Castro-Puyana & Herrero, 2013).
- **Risk Assessment**: By analyzing the metabolic impact of pesticide residues on food by-products, metabolomics contributes to risk assessments and the consequence of secure agricultural practices (Aliferis & Chrysayi-Tokousbalides, 2011).

It significantly enhances food quality and safety by providing detailed nutritional profiling, ensuring food authenticity, detecting adulteration, and monitoring pesticide residues. These applications support the development of safer, healthier, and more reliable food products, ultimately benefiting consumers and the food industry.

## 18.6 APPLICATIONS IN SOIL AND ENVIRONMENTAL MANAGEMENT

High-throughput metabolomics has significant applications in soil and environmental control, offering detailed insights into soil health, fertility, and the impacts of environmental pollutants (Bhattacharjya et al., 2024). By analyzing the metabolic profiles of soil and environmental samples, investigators can develop approaches for endurable agriculture and effective environmental remediation.

### 18.6.1 Soil Health and Fertility

Soil health and fertility are critical for sustainable agriculture and crop productivity (Tahat et al., 2020). Metabolomics provides a comprehensive understanding of soil biochemical processes and microbial activity, which are essential for maintaining soil health (Withers et al., 2020).

- **Microbial Activity**: Metabolomics can profile the metabolites produced by soil microorganisms, offering an understanding of microbial diversity, activity, and their roles in nutrient cycling and soil fertility (Bhattacharjya et al., 2024).
- **Nutrient Availability**: By analyzing soil metabolites, researchers can assess the availability of vital nutrients like nitrogen, potassium, and phosphorus, helping to optimize fertilization practices and improve crop yields (Cheng et al., 2022).
- **Soil Organic Matter**: Metabolomic analysis of soil organic matter helps in understanding the decomposition processes and the formation of humus, which are vital for soil structure and fertility (Nikolaidis & Bidoglio, 2013).

### 18.6.2 Environmental Monitoring

Metabolomics is a potent device for environmental monitoring, enabling the estimation of pollution impacts and the effectiveness of bioremediation efforts (Chandran et al., 2020). Environmental contaminants, like heavy metals, industrial chemicals, and pesticides, can have detrimental effects on ecosystems. Metabolomics provides a thorough examination of the biochemical changes in soil and water caused by pollution (Liu et al., 2022). Metabolomics can identify and quantify pollutants and their metabolites in environmental samples, providing a comprehensive assessment of contamination levels. By examining the metabolic responses of plants, microorganisms, and other biota to pollutants, researchers can assess the ecological impacts and potential risks associated with environmental contamination (Matich et al., 2019). Metabolomic profiling helps in monitoring the overall health of ecosystems, detecting early signs of stress, and assessing the resilience and recovery of ecosystems after pollution events (Beale et al., 2022).

#### 18.6.2.1 Bioremediation

Bioremediation concerns the usage of biological agents, such as microorganisms and plants, to degrade or detoxify pollutants in the environment (Bhatnagar & Kumari, 2013). Metabolomics enhances the understanding and optimization of bioremediation processes:

- **Microbial Metabolism**: Metabolomics can profile the metabolic pathways utilized by microorganisms to degrade pollutants, identifying key enzymes and metabolites involved in bioremediation (Singh, 2006).
- **Phytoremediation**: By analyzing the metabolic responses of plants used in phytoremediation, researchers can identify the mechanisms by which plants uptake, translocate, and detoxify pollutants, optimizing the selection and use of plants for environmental cleanup (Yadav et al., 2023).
- **Efficiency Monitoring**: Metabolomic analysis can monitor the progress and efficiency of bioremediation efforts, ensuring the complete degradation of pollutants and assessing the restoration of soil and water quality.

High-throughput metabolomics offers valuable tools for soil and environmental management by providing detailed insights into soil health, fertility, and the impacts of environmental pollutants. By leveraging these insights, researchers and practitioners can develop sustainable agricultural practices, effective pollution monitoring, and bioremediation strategies, ultimately contributing to environmental conservation and the sustainability of agricultural ecosystems.

## 18.7 ADVANCES AND INNOVATIONS

The field of high-throughput metabolomics is rapidly evolving, driven by emerging technologies and innovative approaches that enhance its applications and impact in agriculture and environmental management.

### 18.7.1 Emerging Technologies

Emerging technologies are revolutionizing metabolomics by improving data acquisition, analysis, and integration, thereby expanding its potential applications.

#### 18.7.1.1 Machine Learning and AI in Metabolomics

ML and artificial intelligence (AI) are transforming metabolomics by offering advanced data analysis capabilities (Chi et al., 2024). These technologies enhance the interpretation of complex metabolomic datasets, facilitating the discovery of patterns, biomarkers, and metabolic pathways (Zhang et al., 2013).

- **Data Analysis and Pattern Recognition**: ML algorithms can manage large-scale metabolomic data, identifying intricate ways and correlations that might be missed by traditional analysis methods (Galal et al., 2022). AI can automate data preprocessing, peak detection, and compound identification, improving the accuracy and speed of metabolomic studies (Liebal et al., 2020).
- **Predictive Modeling**: ML models can predict metabolic responses to environmental changes, genetic modifications, or treatment conditions, aiding in the development of resilient crops and optimized agricultural methods (Naqvi et al., 2022).
- **Integration with Other Omics Data**: AI and ML can incorporate metabolomic data with genomic, transcriptomic, and proteomic datasets, furnishing a complete idea of biological systems and facilitating holistic understanding and discovery (Grapov et al., 2018).

#### 18.7.1.2 Integrative Omics

Integrative omics combines data from different omics fields to provide a multi-dimensional perspective of biological systems (Misra et al., 2019). This approach (Table 18.2) enhances the understanding of complex interactions and regulatory networks within organisms (Wani & Raza, 2019).

- **Holistic Insights**: By combining metabolomics with other omics data, researchers can uncover how genetic variations influence metabolic pathways and how these changes manifest at the molecular and phenotypic levels (Fang et al., 2019; Kumar et al., 2017).

**TABLE 18.2**
**Impact of High-Throughput Metabolomics on Agricultural Research**

| Aspect | Details | References |
|---|---|---|
| Speed of research | Accelerates finding of key metabolites and metabolic paths, reducing time from years to months | Johnson et al. (2016); Li et al. (2022); Wurtzel & Kutchan (2016) |
| Data generation | Produces vast amounts of data, enabling comprehensive analysis of plant biochemical composition | Hong et al. (2016); Wolfender et al. (2013) |
| Biomarker discovery | Facilitates the identification of biomarkers linked to specific traits, diseases, and environmental conditions | Manickam et al. (2023); Monteiro et al. (2013); Steinfath et al. (2010) |
| Crop improvement | Supports the development of crops with enhanced traits, resilience, and productivity | Razzaq et al. (2019); Yang et al. (2021) |
| Global food security | Contributes to breeding programs aimed at creating crops that can withstand climate change and ensure food security | Dubey et al. (2019); Razzaq et al. (2022) |

- **Systems Biology**: Integrative omics support systems biology approaches, enabling the modeling and simulation of biological processes, which can predict the outcomes of genetic modifications and environmental interventions (Anil Kumar et al., 2015).
- **Improved Crop Breeding**: This comprehensive approach can accelerate the identification of key genetic and metabolic traits, assisting in the evolution of crop varieties with enhanced yield, quality, and strain toleration (Mir et al., 2012).

### 18.7.2 Future Trends

Future trends in metabolomics point toward more precise and personalized approaches in agriculture, leveraging the latest technological advances to optimize crop production and management practices.

#### 18.7.2.1 Precision Agriculture

Precision agriculture involves the use of advanced technologies to monitor and manage agricultural practices at a highly detailed level, optimizing inputs and maximizing outputs.

- **Real-Time Monitoring**: High-throughput metabolomics, combined with sensors and IoT devices, enables real-time monitoring of plant health and soil conditions, allowing for timely interventions and precise application of water, fertilizers, and pesticides.
- **Site-Specific Management**: Metabolomic data can inform site-specific management practices, tailoring agricultural interventions to the unique conditions of different fields or even sections of a field, enhancing efficiency and sustainability.
- **Decision Support Systems**: AI-driven decision support systems can integrate metabolomic data with environmental and agronomic data, providing farmers with actionable insights and recommendations for optimizing crop management.

#### 18.7.2.2 Personalized Agriculture

Personalized agriculture tailors agricultural practices to the specific needs of individual plants or small groups of plants, taking precision agriculture to the next level.

- **Plant-Specific Treatments**: Metabolomics can identify the specific nutritional and health needs of individual plants, enabling the precise application of treatments to enhance growth and productivity.
- **Genotype-Tailored Practices**: By understanding the metabolic profiles of different plant genotypes, personalized agriculture can optimize cultivation practices to match the specific requirements and potentials of each genotype, maximizing yield and quality.
- **Customized Crop Development**: Personalized agriculture can support the development of customized crop varieties tailored to specific environmental conditions, consumer preferences, and nutritional needs, enhancing food security and quality.

Advances and innovations in metabolomics, driven by emerging technologies like ML, AI, and integrative omics, are expanding the capabilities and applications of this field. Future trends toward precision and personalized agriculture promise to revolutionize crop production, enhancing efficiency, sustainability, and resilience in the face of global challenges.

## 18.8 CONCLUSIONS

High-throughput metabolomics has appeared as a transformative device in agricultural investigation, showing a profound understanding of plant physiology, soil health, and environmental interactions. By enabling the comprehensive analysis of metabolites, this approach provides a detailed knowledge of the biochemical processes that underpin growth, development, and stress responses in plants. The integration of high-throughput approaches, like automation, robotics, and advanced data analysis, has significantly enhanced the efficiency and scope of metabolomic studies, allowing researchers to process large datasets and derive meaningful conclusions rapidly. In crop improvement, high-throughput metabolomics has enabled the recognition of key metabolic traits associated with desirable phenotypes, nutritional enhancement, and stress tolerance. This has led to the development of crops that are more nutritious, resilient, and productive. The application of metabolomics in food quality and security has improved nutritional profiling, authenticated food products, and detected adulteration and pesticide residues, ensuring the integrity and safety of the food supply chain.

Moreover, metabolomics has proven invaluable in soil and environmental management, providing insights into soil health, fertility, and the impacts of pollutants. It supports the evolution of endurable agricultural approaches and effective bioremediation strategies, contributing to environmental conservation. The advancements and innovations in metabolomics, particularly the integration of ML, AI, and other omics technologies, are poised to further revolutionize agricultural research. These technologies enhance data analysis, enable predictive modeling, and facilitate a holistic understanding of biological systems. Future trends toward precision and personalized agriculture promise to optimize crop production and management practices, tailoring interventions to specific plant needs and environmental conditions. In summary, high-throughput metabolomics is a potent and versatile device that drives significant advancements in agricultural investigation. Its applications span from improving crop yield and quality to ensuring food safety and managing environmental impacts. As technology continues to develop, metabolomics will undoubtedly play an increasingly vital function in addressing the global challenges of food security, sustainability, and environmental health.

## ACKNOWLEDGEMENT

All the authors are thankful to their respective Universities and Institutions for their valuable supports.

## DATA AVAILABILITY

The authors confirm that the data supporting the findings of this study are available within the book/chapter.

## REFERENCES

Abdi, G., Jain, M., Barwant, M., Tendulkar, R., Tendulkar, M., Tariq, M., & Amir, A. (2024). Unveiling the dynamic role of bioinformatics in automation for efficient and accurate data processing and interpretation. In V. Singh & A. Kumar (Eds.), *Advances in Bioinformatics* (pp. 279–319): Springer.

Alawiye, T. T., & Babalola, O. O. (2021). Metabolomics: Current application and prospects in crop production. *Biologia, 76*(1), 227–239.

Aliferis, K. A., & Chrysayi-Tokousbalides, M. (2011). Metabolomics in pesticide research and development: Review and future perspectives. *Metabolomics, 7*, 35–53.

Allwood, J. W., & Goodacre, R. (2010). An introduction to liquid chromatography–mass spectrometry instrumentation applied in plant metabolomic analyses. *Phytochemical Analysis: An International Journal of Plant Chemical and Biochemical Techniques, 21*(1), 33–47.

Allwood, J. W., Williams, A., Uthe, H., van Dam, N. M., Mur, L. A., Grant, M. R., & Pétriacq, P. (2021). Unravelling plant responses to stress—The importance of targeted and untargeted metabolomics. *Metabolites, 11*(8), 558.

Al-Sanea, M. M., & Gamal, M. (2022). Critical analytical review: Rare and recent applications of refractive index detector in HPLC chromatographic drug analysis. *Microchemical Journal, 178*, 107339.

Alseekh, S., Bermudez, L., De Haro, L. A., Fernie, A. R., & Carrari, F. (2018). Crop metabolomics: From diagnostics to assisted breeding. *Metabolomics, 14*, 1–13.

Ambrosino, L., Colantuono, C., Diretto, G., Fiore, A., & Chiusano, M. L. (2020). Bioinformatics resources for plant abiotic stress responses: State of the art and opportunities in the fast evolving-omics era. *Plants, 9*(5), 591.

Amer, B., & Baidoo, E. E. (2021). Omics-driven biotechnology for industrial applications. *Frontiers in Bioengineering and Biotechnology, 9*, 613307.

Amer, B., Deshpande, R. R., & Bird, S. S. (2023). Simultaneous quantitation and discovery (SQUAD) analysis: Combining the best of targeted and untargeted mass spectrometry-based metabolomics. *Metabolites, 13*(5), 648.

Anklam, E., Bahl, M. I., Ball, R., Beger, R. D., Cohen, J., Fitzpatrick, S., … Hirose, A. (2022). Emerging technologies and their impact on regulatory science. *Experimental Biology and Medicine, 247*(1), 1–75.

Arbona, V., Manzi, M., de Ollas, C., & Gómez-Cadenas, A. (2013). Metabolomics as a tool to investigate abiotic stress tolerance in plants. *International Journal of Molecular Sciences, 14*(3), 4885–4911.

Astarita, G., & Langridge, J. (2014). An emerging role for metabolomics in nutrition science. *Lifestyle Genomics, 6*(4–5), 181–200.

Awad, H., Khamis, M. M., & El-Aneed, A. (2015). Mass spectrometry, review of the basics: Ionization. *Applied Spectroscopy Reviews, 50*(2), 158–175.

Barrett, G. (2024). Development of high throughput metabolomics to aid the synthetic biology'design-build-test-learn'cycle. Thesis. The University of Edinburgh.

Beale, D. J., Jones, O. A., Bose, U., Broadbent, J. A., Walsh, T. K., van de Kamp, J., & Bissett, A. (2022). Omics-based ecosurveillance for the assessment of ecosystem function, health, and resilience. *Emerging Topics in Life Sciences, 6*(2), 185–199.

Beale, D. J., Pinu, F. R., Kouremenos, K. A., Poojary, M. M., Narayana, V. K., Boughton, B. A., … Dias, D. A. (2018). Review of recent developments in GC–MS approaches to metabolomics-based research. *Metabolomics, 14*, 1–31.

Bhatnagar, S., & Kumari, R. (2013). Bioremediation: A sustainable tool for environmental management–A review. *Annual Research & Review in Biology, 3*(4), 974–993.

Bhattacharjya, S., Ghosh, A., Sahu, A., Agnihotri, R., Pal, N., Sharma, P., … Singh, A. (2024). Utilizing soil metabolomics to investigate the untapped metabolic potential of soil microbial communities and their role in driving soil ecosystem processes: A review. *Applied Soil Ecology, 195*, 105238.

Bocso, N.-S., & Butnariu, M. (2022). The biological role of primary and secondary plants metabolites. *Journal of Nutrition and Food Processing, 5*(3), 1–7.

Bowman, M., Hilligoss, D., Rasmussen, S., & Thomas, R. (2006). Biodiesel: A renewable and biodegradable fuel. *Hydrocarbon Processing, 85*(2), 103.

Cajka, T., & Fiehn, O. (2016). Toward merging untargeted and targeted methods in mass spectrometry-based metabolomics and lipidomics. *Analytical Chemistry, 88*(1), 524–545.

Cambiaghi, A., Ferrario, M., & Masseroli, M. (2017). Analysis of metabolomic data: Tools, current strategies and future challenges for omics data integration. *Briefings in Bioinformatics, 18*(3), 498–510.

Castro-Moretti, F. R., Gentzel, I. N., Mackey, D., & Alonso, A. P. (2020). Metabolomics as an emerging tool for the study of plant–pathogen interactions. *Metabolites, 10*(2), 52.

Castro-Puyana, M., & Herrero, M. (2013). Metabolomics approaches based on mass spectrometry for food safety, quality and traceability. *TrAC Trends in Analytical Chemistry, 52*, 74–87.

Cevallos-Cevallos, J. M., Reyes-De-Corcuera, J. I., Etxeberria, E., Danyluk, M. D., & Rodrick, G. E. (2009). Metabolomic analysis in food science: A review. *Trends in Food Science & Technology, 20*(11–12), 557–566.

Chandran, H., Meena, M., & Sharma, K. (2020). Microbial biodiversity and bioremediation assessment through omics approaches. *Frontiers in Environmental Chemistry, 1*, 570326.

Chen, L., Schwier, M., Krumbach, J., Kopriva, S., & Jacoby, R. P. (2021). Metabolomics in plant-microbe interactions in the roots. *Advances in Botanical Research, 98*, 133–161.

Chen, W., Gong, L., Guo, Z., Wang, W., Zhang, H., Liu, X., … Luo, J. (2013). A novel integrated method for large-scale detection, identification, and quantification of widely targeted metabolites: Application in the study of rice metabolomics. *Molecular Plant, 6*(6), 1769–1780.

Cheng, H., Yuan, M., Tang, L., Shen, Y., Yu, Q., & Li, S. (2022). Integrated microbiology and metabolomics analysis reveal responses of soil microorganisms and metabolic functions to phosphorus fertilizer on semiarid farm. *Science of the Total Environment, 817*, 152878.

Cheng, J., Lan, W., Zheng, G., & Gao, X. (2018). Metabolomics: A high-throughput platform for metabolite profile exploration. *Computational Systems Biology: Methods and Protocols, 1754*, 265–292.

Chi, J., Shu, J., Li, M., Mudappathi, R., Jin, Y., Lewis, F., … Gu, H. (2024). Artificial intelligence in metabolomics: A current review. *TrAC Trends in Analytical Chemistry, 173*, 117852.

Coler, E. A., Chen, W., Melnik, A. V., Morton, J. T., & Aksenov, A. A. (2024). Metabolomics in the era of artificial intelligence. *Microbiota and Host, 2*(1), e230017.

Cremer, D., & Gräfenstein, J. (2007). Calculation and analysis of NMR spin–spin coupling constants. *Physical Chemistry Chemical Physics, 9*(22), 2791–2816.

Crutchfield, C., & Clarke, W. (2017). High resolution accurate mass (HRAM) mass spectrometry. In H. Nair & W. Clarke (Eds.), *Mass Spectrometry for the Clinical Laboratory* (pp. 247–259): Elsevier.

Danezis, G. P., Tsagkaris, A. S., Brusic, V., & Georgiou, C. A. (2016). Food authentication: State of the art and prospects. *Current Opinion in Food Science, 10*, 22–31.

Deborde, C., Moing, A., Roch, L., Jacob, D., Rolin, D., & Giraudeau, P. (2017). Plant metabolism as studied by NMR spectroscopy. *Progress in Nuclear Magnetic Resonance Spectroscopy, 102*, 61–97.

Deng, P., Li, X., Petriello, M. C., Wang, C., Morris, A. J., & Hennig, B. (2019). Application of metabolomics to characterize environmental pollutant toxicity and disease risks. *Reviews on Environmental Health, 34*(3), 251–259.

Desmet, S., Morreel, K., & Dauwe, R. (2021). Origin and function of structural diversity in the plant specialized metabolome. *Plants, 10*(11), 2393.

Dias, D. A., Jones, O. A., Beale, D. J., Boughton, B. A., Benheim, D., Kouremenos, K. A., … Wishart, D. S. (2016). Current and future perspectives on the structural identification of small molecules in biological systems. *Metabolites, 6*(4), 46.

Dixon, R. A., Gang, D. R., Charlton, A. J., Fiehn, O., Kuiper, H. A., Reynolds, T. L., … Ridley, W. P. (2006). Applications of metabolomics in agriculture. *Journal of Agricultural and Food Chemistry, 54*(24), 8984–8994.

do Prado, R. M., Porto, C., Nunes, E., de Aguiar, C. L., & Pilau, E. J. (2018). Metabolomics and agriculture: What can be done? *Msystems, 3*(2), doi:10.1128/msystems.00156-00117.

Dubey, A., Kumar, A., Abd_Allah, E. F., Hashem, A., & Khan, M. L. (2019). Growing more with less: Breeding and developing drought resilient soybean to improve food security. *Ecological Indicators, 105*, 425–437.

Emwas, A.-H. M., Al-Rifai, N., Szczepski, K., Alsuhaymi, S., Rayyan, S., Almahasheer, H., … Lachowicz, J. I. (2021). You are what you eat: Application of metabolomics approaches to advance nutrition research. *Foods, 10*(6), 1249.

Ernst, M., Silva, D. B., Silva, R. R., Vêncio, R. Z., & Lopes, N. P. (2014). Mass spectrometry in plant metabolomics strategies: From analytical platforms to data acquisition and processing. *Natural Product Reports, 31*(6), 784–806.

Falaki, F. (2019). Sample preparation techniques for gas chromatography. In P. Kusch (Ed.), *Gas Chromatography-Derivatization, Sample Preparation, Application*: IntechOpen.

Fang, C., Fernie, A. R., & Luo, J. (2019). Exploring the diversity of plant metabolism. *Trends in Plant Science, 24*(1), 83–98.

Fernie, A. R., & Schauer, N. (2009). Metabolomics-assisted breeding: A viable option for crop improvement? *Trends in Genetics, 25*(1), 39–48.

Fiehn, O., Wohlgemuth, G., Scholz, M., Kind, T., Lee, D. Y., Lu, Y., … Nikolau, B. (2008). Quality control for plant metabolomics: Reporting MSI-compliant studies. *The Plant Journal, 53*(4), 691–704.

Foito, A., & Stewart, D. (2018). Metabolomics: A high-throughput screen for biochemical and bioactivity diversity in plants and crops. *Current Pharmaceutical Design, 24*(19), 2043–2054.

Franchina, F. A., Zanella, D., Dubois, L. M., & Focant, J. F. (2021). The role of sample preparation in multidimensional gas chromatographic separations for non-targeted analysis with the focus on recent biomedical, food, and plant applications. *Journal of Separation Science, 44*(1), 188–210.

François, I., Sandra, K., & Sandra, P. (2009). Comprehensive liquid chromatography: Fundamental aspects and practical considerations—A review. *Analytica Chimica Acta, 641*(1–2), 14–31.

Fu, X., & Jiang, D. (2022). High-throughput phenotyping: The latest research tool for sustainable crop production under global climate change scenarios. In U. C. Jha & U. B. Singh (Eds.), *Sustainable Crop Productivity and Quality under Climate Change* (pp. 313–381): Elsevier.

Galal, A., Talal, M., & Moustafa, A. (2022). Applications of machine learning in metabolomics: Disease modeling and classification. *Frontiers in Genetics, 13*, 1017340.

Gaquerel, E., Kuhl, C., & Neumann, S. (2013). Computational annotation of plant metabolomics profiles via a novel network-assisted approach. *Metabolomics, 9*, 904–918.

Garcia, A., & Barbas, C. (2011). Gas chromatography-mass spectrometry (GC-MS)-based metabolomics. *Metabolic Profiling: Methods and Protocols, 708*, 191–204.

German, J. B., Roberts, M.-A., & Watkins, S. M. (2003). Personal metabolomics as a next generation nutritional assessment. *The Journal of Nutrition, 133*(12), 4260–4266.

Gertsman, I., & Barshop, B. A. (2018). Promises and pitfalls of untargeted metabolomics. *Journal of Inherited Metabolic Disease, 41*, 355–366.

Gibney, M. J., Walsh, M., Brennan, L., Roche, H. M., German, B., & Van Ommen, B. (2005). Metabolomics in human nutrition: Opportunities and challenges. *The American Journal of Clinical Nutrition, 82*(3), 497–503.

Go, E. P. (2010). Database resources in metabolomics: An overview. *Journal of Neuroimmune Pharmacology, 5*, 18–30.

Godoy, F., Olivos-Hernández, K., Stange, C., & Handford, M. (2021). Abiotic stress in crop species: Improving tolerance by applying plant metabolites. *Plants, 10*(2), 186.

Goldansaz, S. A., Guo, A. C., Sajed, T., Steele, M. A., Plastow, G. S., & Wishart, D. S. (2017). Livestock metabolomics and the livestock metabolome: A systematic review. *PLoS One, 12*(5), e0177675.

Gomez, L. D., Whitehead, C., Barakate, A., Halpin, C., & McQueen-Mason, S. J. (2010). Automated saccharification assay for determination of digestibility in plant materials. *Biotechnology for Biofuels, 3*, 1–12.

González-Domínguez, Á., Estanyol-Torres, N., Brunius, C., Landberg, R., & González-Domínguez, R. (2024). QC omics: Recommendations and guidelines for robust, easily implementable and reportable quality control of metabolomics data. *Analytical Chemistry, 96*(3), 1064–1072.

Grapov, D., Fahrmann, J., Wanichthanarak, K., & Khoomrung, S. (2018). Rise of deep learning for genomic, proteomic, and metabolomic data integration in precision medicine. *Omics: A Journal of Integrative Biology, 22*(10), 630–636.

Gupta, S., Schillaci, M., & Roessner, U. (2022). Metabolomics as an emerging tool to study plant–microbe interactions. *Emerging Topics in Life Sciences, 6*(2), 175–183.

Hamany Djande, C. Y., Pretorius, C., Tugizimana, F., Piater, L. A., & Dubery, I. A. (2020). Metabolomics: A tool for cultivar phenotyping and investigation of grain crops. *Agronomy, 10*(6), 831.

Hatzakis, E. (2019). Nuclear magnetic resonance (NMR) spectroscopy in food science: A comprehensive review. *Comprehensive Reviews in Food Science and Food Safety, 18*(1), 189–220.

Hernández-Mesa, M., & Moreno-González, D. (2022). Current role of mass spectrometry in the determination of pesticide residues in food. *Separations, 9*(6), 148.

Higashi, T., & Ogawa, S. (2020). Derivatization-based sample-multiplexing for enhancing throughput in liquid chromatography/tandem mass spectrometry quantification of metabolites: An overview. *Journal of Chromatography A, 1634*, 461679.

Hill, C. B., & Roessner, U. (2015). Advances in high-throughput untargeted LC-MS analysis for plant metabolomics. In De Vooght-Johnson, R. (Ed.), *Advanced LC-MS Applications for Metabolomics*, 2nd ed. (pp. 58–71). Future Science Ltd.

Hong, J., Yang, L., Zhang, D., & Shi, J. (2016). Plant metabolomics: An indispensable system biology tool for plant science. *International Journal of Molecular Sciences, 17*(6), 767.

Horgan, R. P., & Kenny, L. C. (2011). `Omic'technologies: Genomics, transcriptomics, proteomics and metabolomics. *Obstetrician & Gynaecologist, 13*(3), 189–195.

Jain, D., Ashraf, N., Khurana, J., & Shiva Kameshwari, M. (2019). The 'omics' approach for crop improvement against drought stress. In *Genetic Enhancement of Crops for Tolerance to Abiotic Stress: Mechanisms and Approaches* (Vol. I, pp. 183–204). Editors: Vijay Rani Rajpal, Deepmala Sehgal, Avinash Kumar, S.N. Raina. Publisher: Springer Cham.

Jamaloddin, M., Maliha, A., Gokulan, C., Gaur, N., & Patel, H. K. (2021). Metabolomics-assisted breeding for crop improvement: An emerging approach. *Omics Technologies for Sustainable Agriculture and Global Food Security, 1*, 241–279.

Jandera, P. (2011). Stationary and mobile phases in hydrophilic interaction chromatography: A review. *Analytica Chimica Acta, 692*(1–2), 1–25.

Jha, Y., & Mohamed, H. I. (2022). Plant secondary metabolites as a tool to investigate biotic stress tolerance in plants: A review. *Gesunde Pflanzen, 74*(4), 771–790.

Jian, W. (2017). Modern liquid chromatography and mass spectrometry for targeted biomarker quantitation. In N. Weng & W. Jian (Eds.), *Targeted Biomarker Quantitation by LC–MS* (pp. 45–63). John Wiley & Sons, Inc.

Jimenez-Garcia, S. N., Vazquez-Cruz, M. A., Guevara-Gonzalez, R. G., Torres-Pacheco, I., Cruz-Hernandez, A., & Feregrino-Perez, A. A. (2013). Current approaches for enhanced expression of secondary metabolites as bioactive compounds in plants for agronomic and human health purposes-A review. *Polish Journal of Food and Nutrition Sciences, 63*(2), 67–78.

Jin, H., Liu, Y., Guo, Z., Wang, J., Zhang, X., Wang, C., & Liang, X. (2016). Recent development in liquid chromatography stationary phases for separation of Traditional Chinese Medicine components. *Journal of Pharmaceutical and Biomedical Analysis, 130*, 336–346.

Johnson, C. H., Ivanisevic, J., & Siuzdak, G. (2016). Metabolomics: Beyond biomarkers and towards mechanisms. *Nature Reviews Molecular Cell Biology, 17*(7), 451–459.

Jones, O. A. (2018). Illuminating the dark metabolome to advance the molecular characterisation of biological systems. *Metabolomics, 14*, 1–11.

Katam, R., Lin, C., Grant, K., Katam, C. S., & Chen, S. (2022). Advances in plant metabolomics and its applications in stress and single-cell biology. *International Journal of Molecular Sciences, 23*(13), 6985.

Kaur, B., Sandhu, K. S., Kamal, R., Kaur, K., Singh, J., Röder, M. S., & Muqaddasi, Q. H. (2021). Omics for the improvement of abiotic, biotic, and agronomic traits in major cereal crops: Applications, challenges, and prospects. *Plants, 10*(10), 1989.

Khakimov, B., Bak, S., & Engelsen, S. B. (2014). High-throughput cereal metabolomics: Current analytical technologies, challenges and perspectives. *Journal of Cereal Science, 59*(3), 393–418.

Khan, N., Ali, S., Shahid, M. A., & Kharabian-Masouleh, A. (2017). Advances in detection of stress tolerance in plants through metabolomics approaches. *Plant Omics, 10*(3), 153–163.

Kisiel, A., Krzemińska, A., Cembrowska-Lech, D., & Miller, T. (2023). Data science and plant metabolomics. *Metabolites, 13*(3), 454.

Kitson, F. G., Larsen, B. S., & McEwen, C. N. (1996). *Gas Chromatography and Mass Spectrometry: A Practical Guide*: Academic Press.

Krajewski, P., Chen, D., Ćwiek, H., van Dijk, A. D., Fiorani, F., Kersey, P., … Nap, J. P. (2015). Towards recommendations for metadata and data handling in plant phenotyping. *Journal of Experimental Botany, 66*(18), 5417–5427.

Krishnan, P., Kruger, N., & Ratcliffe, R. (2005). Metabolite fingerprinting and profiling in plants using NMR. *Journal of Experimental Botany, 56*(410), 255–265.

Kumar, A., Mosa, K. A., Ji, L., Kage, U., Dhokane, D., Karre, S., … Pathania, N. (2018). Metabolomics-assisted biotechnological interventions for developing plant-based functional foods and nutraceuticals. *Critical Reviews in Food Science and Nutrition, 58*(11), 1791–1807.

Kumar, A., Pathak, R. K., Gupta, S. M., Gaur, V. S., & Pandey, D. (2015). Systems biology for smart crops and agricultural innovation: Filling the gaps between genotype and phenotype for complex traits linked with robust agricultural productivity and sustainability. *Omics: A Journal of Integrative Biology, 19*(10), 581–601.

Kumar, D. (2016). Nuclear magnetic resonance (NMR) spectroscopy for metabolic profiling of medicinal plants and their products. *Critical Reviews in Analytical Chemistry, 46*(5), 400–412.

Kumar, M., Kumar Patel, M., Kumar, N., Bajpai, A. B., & Siddique, K. H. (2021). Metabolomics and molecular approaches reveal drought stress tolerance in plants. *International Journal of Molecular Sciences, 22*(17), 9108.

Kumar, R., Bohra, A., Pandey, A. K., Pandey, M. K., & Kumar, A. (2017). Metabolomics for plant improvement: Status and prospects. *Frontiers in Plant Science, 8*, 1302.

Kumar, V., Prasad, A., Kashyap, P., Joshi, R., & Kumar, A. (2024). Integrated omics approaches for plant disease resistance. In S. M. Mohapatra & R. Das (Eds.), *Biotechnological Advances for Disease Tolerance in Plants* (pp. 211–237): Springer.

Lees, M., & Reimann, L. (2021). Analytical detection methods and strategies for food fraud. In M. Lipp & L. Reimann (Eds.), *Food Fraud* (pp. 45–67): Elsevier.

Letertre, M. P., Dervilly, G., & Giraudeau, P. (2020). Combined nuclear magnetic resonance spectroscopy and mass spectrometry approaches for metabolomics. *Analytical Chemistry, 93*(1), 500–518.

Li, J., Yan, G., Duan, X., Zhang, K., Zhang, X., Zhou, Y., … Hua, X. (2022). Research progress and trends in metabolomics of fruit trees. *Frontiers in Plant Science, 13*, 881856.

Li, S., Tian, Y., Jiang, P., Lin, Y., Liu, X., & Yang, H. (2021). Recent advances in the application of metabolomics for food safety control and food quality analyses. *Critical Reviews in Food Science and Nutrition, 61*(9), 1448–1469.

Liebal, U. W., Phan, A. N., Sudhakar, M., Raman, K., & Blank, L. M. (2020). Machine learning applications for mass spectrometry-based metabolomics. *Metabolites, 10*(6), 243.

Liu, L., Wu, Q., Miao, X., Fan, T., Meng, Z., Chen, X., & Zhu, W. (2022). Study on toxicity effects of environmental pollutants based on metabolomics: A review. *Chemosphere, 286*, 131815.

Liu, X., Zhou, L., Shi, X., & Xu, G. (2019). New advances in analytical methods for mass spectrometry-based large-scale metabolomics study. *TrAC Trends in Analytical Chemistry, 121*, 115665.

Ma, X. (2022). Recent advances in mass spectrometry-based structural elucidation techniques. *Molecules, 27*(19), 6466.

Makoś, P., Przyjazny, A., & Boczkaj, G. (2019). Methods of assaying volatile oxygenated organic compounds in effluent samples by gas chromatography—A review. *Journal of Chromatography A, 1592*, 143–160.

Manickam, S., Rajagopalan, V. R., Kambale, R., Rajasekaran, R., Kanagarajan, S., & Muthurajan, R. (2023). Plant metabolomics: Current initiatives and future prospects. *Current Issues in Molecular Biology, 45*(11), 8894–8906.

Marshall, D. D., & Powers, R. (2017). Beyond the paradigm: Combining mass spectrometry and nuclear magnetic resonance for metabolomics. *Progress in Nuclear Magnetic Resonance Spectroscopy, 100*, 1–16.

Marston, A. (2007). Role of advances in chromatographic techniques in phytochemistry. *Phytochemistry, 68*(22–24), 2786–2798.

Mashabela, M. D., Masamba, P., & Kappo, A. P. (2022). Metabolomics and chemoinformatics in agricultural biotechnology research: Complementary probes in unravelling new metabolites for crop improvement. *Biology, 11*(8), 1156.

Matich, E. K., Soria, N. G. C., Aga, D. S., & Atilla-Gokcumen, G. E. (2019). Applications of metabolomics in assessing ecological effects of emerging contaminants and pollutants on plants. *Journal of Hazardous Materials, 373*, 527–535.

McCalley, D. V. (2023). Understanding and managing peak shape for basic solutes in reversed-phase high performance liquid chromatography. *Chemical Communications, 59*(51), 7887–7899.

Medhe, S. (2018). Ionization techniques in mass spectrometry: A review. *Mass Spectrom Purif Tech, 4*(01), 1000126.

Mir, R. R., Zaman-Allah, M., Sreenivasulu, N., Trethowan, R., & Varshney, R. K. (2012). Integrated genomics, physiology and breeding approaches for improving drought tolerance in crops. *Theoretical and Applied Genetics, 125*, 625–645.

Misra, B. B. (2021). Advances in high resolution GC-MS technology: A focus on the application of GC-Orbitrap-MS in metabolomics and exposomics for FAIR practices. *Analytical Methods, 13*(20), 2265–2282.

Misra, B. B., Langefeld, C., Olivier, M., & Cox, L. A. (2019). Integrated omics: Tools, advances and future approaches. *Journal of Molecular Endocrinology, 62*(1), R21–R45.

Mochida, K., Nishii, R., & Hirayama, T. (2020). Decoding plant–environment interactions that influence crop agronomic traits. *Plant and Cell Physiology, 61*(8), 1408–1418.

Mondal, S., & Palit, D. (2022). Challenges in natural resource management for ecological sustainability. In M. K. Jhariya, R. S. Meena, A. Banerjee, & S. N. Meena (Eds.), *Natural Resources Conservation and Advances for Sustainability* (pp. 29–59): Elsevier.

Monteiro, M., Carvalho, M., Bastos, M., & Guedes de Pinho, P. (2013). Metabolomics analysis for biomarker discovery: Advances and challenges. *Current Medicinal Chemistry, 20*(2), 257–271.

Mushtaq, M. Y., Choi, Y. H., Verpoorte, R., & Wilson, E. G. (2014). Extraction for metabolomics: Access to the metabolome. *Phytochemical Analysis, 25*(4), 291–306.

Muthubharathi, B. C., Gowripriya, T., & Balamurugan, K. (2021). Metabolomics: Small molecules that matter more. *Molecular Omics, 17*(2), 210–229.

Nagana Gowda, G., & Raftery, D. (2019). Overview of NMR spectroscopy-based metabolomics: Opportunities and challenges. *NMR-Based Metabolomics: Methods and Protocols, 2037*, 3–14.

Naqvi, R. Z., Siddiqui, H. A., Mahmood, M. A., Najeebullah, S., Ehsan, A., Azhar, M., … Mukhtar, Z. (2022). Smart breeding approaches in post-genomics era for developing climate-resilient food crops. *Frontiers in Plant Science, 13*, 972164.

Narad, P., Gupta, R., & Sengupta, A. (2022). Plant metabolomics: A new era in the advancement of agricultural research. In P. Sharma (Ed.), *Bioinformatics in Agriculture* (pp. 139–160): Elsevier.

Nikolaidis, N. P., & Bidoglio, G. (2013). Soil organic matter dynamics and structure. *Sustainable Agriculture Reviews, 12*, 175–199.

Ntlhokwe, G. E. (2016). *Application of Comprehensive Two-Dimensional Gas Chromatography for the Characterisation of the Volatile Composition of Honeybush Tea*: Stellenbosch University.

Nyiredy, S. (2004). Separation strategies of plant constituents–current status. *Journal of Chromatography B, 812*(1–2), 35–51.

Oldiges, M., Lütz, S., Pflug, S., Schroer, K., Stein, N., & Wiendahl, C. (2007). Metabolomics: Current state and evolving methodologies and tools. *Applied Microbiology and Biotechnology, 76*, 495–511.

Oms-Oliu, G., Odriozola-Serrano, I., & Martín-Belloso, O. (2013). Metabolomics for assessing safety and quality of plant-derived food. *Food Research International, 54*(1), 1172–1183.

Ontoy, J. C., & Ham, J. H. (2024). Mapping and omics integration: Towards precise rice disease resistance breeding. *Plants, 13*(9), 1205.

Parida, A. K., Panda, A., & Rangani, J. (2018). Metabolomics-guided elucidation of abiotic stress tolerance mechanisms in plants. In N. Tuteja (Ed.), *Plant Metabolites and Regulation under Environmental Stress* (pp. 89–131): Elsevier.

Parray, J. A., & Shameem, N. (2019). *Sustainable Agriculture: Advances in Plant Metabolome and Microbiome*: Academic Press.

Piasecka, A., Kachlicki, P., & Stobiecki, M. (2019). Analytical methods for detection of plant metabolomes changes in response to biotic and abiotic stresses. *International Journal of Molecular Sciences, 20*(2), 379.

Pickens, L. B., Tang, Y., & Chooi, Y.-H. (2011). Metabolic engineering for the production of natural products. *Annual Review of Chemical and Biomolecular Engineering, 2*(1), 211–236.

Qari, S. H., & Tarbiyyah, I. (2021). The genetic regulation of secondary metabolic pathways in response to salinity and drought as abiotic stresses. *Applied Sciences, 11*(15), 6668.

Razzaq, A., Sadia, B., Raza, A., Khalid Hameed, M., & Saleem, F. (2019). Metabolomics: A way forward for crop improvement. *Metabolites, 9*(12), 303.

Razzaq, A., Wishart, D. S., Wani, S. H., Hameed, M. K., Mubin, M., & Saleem, F. (2022). Advances in metabolomics-driven diagnostic breeding and crop improvement. *Metabolites, 12*(6), 511.

Reid, L. M., O'donnell, C. P., & Downey, G. (2006). Recent technological advances for the determination of food authenticity. *Trends in Food Science & Technology, 17*(7), 344–353.

Ren, J.-L., Zhang, A.-H., Kong, L., & Wang, X.-J. (2018). Advances in mass spectrometry-based metabolomics for investigation of metabolites. *RSC Advances, 8*(40), 22335–22350.

Ren, S., Hinzman, A. A., Kang, E. L., Szczesniak, R. D., & Lu, L. J. (2015). Computational and statistical analysis of metabolomics data. *Metabolomics, 11*, 1492–1513.

Roaa, M. (2020). A review article: The importance of the major groups of plants secondary metabolism phenols, alkaloids, and terpenes. *International Journal for Research in Applied Sciences and Biotechnology (IJRASB), 7*(5), 354–358.

Rocca-Serra, P., Salek, R. M., Arita, M., Correa, E., Dayalan, S., Gonzalez-Beltran, A., … Haug, K. (2016). Data standards can boost metabolomics research, and if there is a will, there is a way. *Metabolomics, 12*, 1–13.

Roessner, U., & Dias, D. A. (2013). Plant tissue extraction for metabolomics. *Metabolomics Tools for Natural Product Discovery: Methods and Protocols, 1055*, 21–28.

Roessner, U., Luedemann, A., Brust, D., Fiehn, O., Linke, T., Willmitzer, L., & Fernie, A. R. (2001). Metabolic profiling allows comprehensive phenotyping of genetically or environmentally modified plant systems. *The Plant Cell, 13*(1), 11–29.

Romero-González, R., & Frenich, A. G. (2017). *Applications in High Resolution Mass Spectrometry: Food Safety and Pesticide Residue Analysis*: Elsevier.

Ruane, J., & Sonnino, A. (2011). Agricultural biotechnologies in developing countries and their possible contribution to food security. *Journal of Biotechnology, 156*(4), 356–363.

Salek, R. M., Neumann, S., Schober, D., Hummel, J., Billiau, K., Kopka, J., … Tenori, L. (2015). COordination of Standards in MetabOlomicS (COSMOS): Facilitating integrated metabolomics data access. *Metabolomics, 11*, 1587–1597.

Salem, M. A., Perez de Souza, L., Serag, A., Fernie, A. R., Farag, M. A., Ezzat, S. M., & Alseekh, S. (2020). Metabolomics in the context of plant natural products research: From sample preparation to metabolite analysis. *Metabolites, 10*(1), 37.

Salem, M. A., Wang, J. Y., & Al-Babili, S. (2022). Metabolomics of plant root exudates: From sample preparation to data analysis. *Frontiers in Plant Science, 13*, 1062982.

Sengupta, A., & Narad, P. (2018). Metabolomics. *Omics Approaches, Technologies and Applications: Integrative Approaches for Understanding OMICS Data*, 75–97.

Sharma, V., Gupta, P., Kagolla, P., Hangargi, B., Veershetty, A., Ramrao, D. P., … Kumar, A. (2021). Metabolomics intervention towards better understanding of plant traits. *Cells, 10*(2), 346.

Singh, D. P., Bisen, M. S., Shukla, R., Prabha, R., Maurya, S., Reddy, Y. S., … Chaturvedi, K. K. (2022). Metabolomics-driven mining of metabolite resources: Applications and prospects for improving vegetable crops. *International Journal of Molecular Sciences, 23*(20), 12062.

Singh, O. V. (2006). Proteomics and metabolomics: The molecular make-up of toxic aromatic pollutant bioremediation. *Proteomics, 6*(20), 5481–5492.

Sparkman, O. D., Penton, Z., & Kitson, F. G. (2011). *Gas Chromatography and Mass Spectrometry: A Practical Guide*: Academic Press.

Stashenko, E., & Martínez, J. R. (2014). Gas chromatography-mass spectrometry. *Advances in Gas Chromatography*, 1–38.

Steinfath, M., Strehmel, N., Peters, R., Schauer, N., Groth, D., Hummel, J., … Geigenberger, P. (2010). Discovering plant metabolic biomarkers for phenotype prediction using an untargeted approach. *Plant Biotechnology Journal, 8*(8), 900–911.

Sud, M., Fahy, E., Cotter, D., Azam, K., Vadivelu, I., Burant, C., … Nair, K. S. (2016). Metabolomics Workbench: An international repository for metabolomics data and metadata, metabolite standards, protocols, tutorials and training, and analysis tools. *Nucleic Acids Research, 44*(D1), D463–D470.

Svrkota, B., Krmar, J., Protić, A., & Otašević, B. (2023). The secret of reversed-phase/weak cation exchange retention mechanisms in mixed-mode liquid chromatography applied for small drug molecule analysis. *Journal of Chromatography A, 1690*, 463776.

Tahat, M. M., Alananbeh, K., Othman, Y., & Leskovar, D. I. (2020). Soil health and sustainable agriculture. *Sustainability, 12*(12), 4859.

Töpfer, N., Kleessen, S., & Nikoloski, Z. (2015). Integration of metabolomics data into metabolic networks. *Frontiers in Plant Science, 6*, 49.

Ventura, M. C., Farrell, W. P., Aurigemma, C. M., & Greig, M. J. (1999). Packed column supercritical fluid chromatography/mass spectrometry for high-throughput analysis. Part 2. *Analytical Chemistry, 71*(19), 4223–4231.

Vinaixa, M., Schymanski, E. L., Neumann, S., Navarro, M., Salek, R. M., & Yanes, O. (2016). Mass spectral databases for LC/MS-and GC/MS-based metabolomics: State of the field and future prospects. *TrAC Trends in Analytical Chemistry, 78*, 23–35.

Vuckovic, D. (2012). Current trends and challenges in sample preparation for global metabolomics using liquid chromatography–mass spectrometry. *Analytical and Bioanalytical Chemistry, 403*(6), 1523–1548.

Wani, N., & Raza, K. (2019). Integrative approaches to reconstruct regulatory networks from multi-omics data: A review of state-of-the-art methods. *Computational Biology and Chemistry, 83*, 107120.

Williams, J. D., & Burinsky, D. J. (2001). Mass spectrometric analysis of complex mixtures then and now: The impact of linking liquid chromatography and mass spectrometry. *International Journal of Mass Spectrometry, 212*(1–3), 111–133.

Wishart, D. S. (2008a). Metabolomics: Applications to food science and nutrition research. *Trends in Food Science & Technology, 19*(9), 482–493.

Wishart, D. S. (2008b). Quantitative metabolomics using NMR. *TrAC Trends in Analytical Chemistry, 27*(3), 228–237.

Wishart, D. S. (2019). Metabolomics for investigating physiological and pathophysiological processes. *Physiological Reviews, 99*(4), 1819–1875.

Withers, E., Hill, P. W., Chadwick, D. R., & Jones, D. L. (2020). Use of untargeted metabolomics for assessing soil quality and microbial function. *Soil Biology and Biochemistry, 143*, 107758.

Wolfender, J.-L., Marti, G., Thomas, A., & Bertrand, S. (2015). Current approaches and challenges for the metabolite profiling of complex natural extracts. *Journal of Chromatography A, 1382*, 136–164.

Wolfender, J.-L., Nuzillard, J.-M., Van Der Hooft, J. J., Renault, J.-H., & Bertrand, S. (2018). Accelerating metabolite identification in natural product research: Toward an ideal combination of liquid chromatography–high-resolution tandem mass spectrometry and NMR profiling, in silico databases, and chemometrics. *Analytical Chemistry, 91*(1), 704–742.

Wolfender, J.-L., Rudaz, S., Hae Choi, Y., & Kyong Kim, H. (2013). Plant metabolomics: From holistic data to relevant biomarkers. *Current Medicinal Chemistry, 20*(8), 1056–1090.

Wurtzel, E. T., & Kutchan, T. M. (2016). Plant metabolism, the diverse chemistry set of the future. *Science, 353*(6305), 1232–1236.

Yadav, R., Singh, G., Santal, A. R., & Singh, N. P. (2023). Omics approaches in effective selection and generation of potential plants for phytoremediation of heavy metal from contaminated resources. *Journal of Environmental Management, 336*, 117730.

Yang, D., Du, X., Yang, Z., Liang, Z., Guo, Z., & Liu, Y. (2014). Transcriptomics, proteomics, and metabolomics to reveal mechanisms underlying plant secondary metabolism. *Engineering in Life Sciences, 14*(5), 456–466.

Yang, Y., Saand, M. A., Huang, L., Abdelaal, W. B., Zhang, J., Wu, Y., … Wang, F. (2021). Applications of multi-omics technologies for crop improvement. *Frontiers in Plant Science, 12*, 563953.

Zhang, A., Sun, H., Wang, P., Han, Y., & Wang, X. (2012). Modern analytical techniques in metabolomics analysis. *Analyst, 137*(2), 293–300.

Zhang, A. H., Sun, H., Han, Y., Yan, G. L., Yuan, Y., Song, G. C., … Wang, X. J. (2013). Ultraperformance liquid chromatography–mass spectrometry based comprehensive metabolomics combined with pattern recognition and network analysis methods for characterization of metabolites and metabolic pathways from biological data sets. *Analytical Chemistry, 85*(15), 7606–7612.

Zhou, J., & Yin, Y. (2016). Strategies for large-scale targeted metabolomics quantification by liquid chromatography-mass spectrometry. *Analyst, 141*(23), 6362–6373.

# 19 Dissecting Plant-Pathogen Interactions by High-Throughput Metabolomics for Developing Biotic Stress-Resilient Crops

*Krishnan Kutty Sunitha Kumari, Sureshkumar Tharani, Sarah Jaison, Govindarajan Ramkumar, Siva Daniel Ajay Samuel, Chithra Devi Balasundarm Saraswathy, Muthukrishnan Arun, and Arumugam Vijaya Anand*

## LIST OF ABBREVIATIONS

**ABA**: Abscisic Acid
**BR**: Brassinosteroids
**$Ca^{2+}$**: Calcium
**CNGCs**: Cyclic Nucleotide-Gated Channels
**ET**: Ethylene
**ETI**: Effector-Triggered Immunity
**GC-MS**: Gas Chromatography-Mass Spectrometry
**GLRs**: Glutamate Receptor-Like Channels
**HPLC**: High-Performance Liquid Chromatography
**JA**: Jasmonic Acid
**LC-MS/MS**: Liquid Chromatography-Tandem Mass Spectrometry
**LC-MS**: Liquid Chromatography-Mass Spectrometry
**MALDI-MS**: Matrix-Assisted Laser Desorption/Ionization Mass Spectrometry
**NMR**: Nuclear Magnetic Resonance
**PR**: Pathogenesis-Related
**PTI**: Pattern-Triggered Immunity
**RNAi**: RNA Interference
**ROS**: Reactive Oxygen Species
**SA**: Salicylic Acid
**SAR**: Systemic Acquired Resistance
**UPLC-DAD**: Ultra-Performance Liquid Chromatography-Diode Array Detection
**VOCs**: Volatile Organic Compounds

## 19.1 INTRODUCTION

Plant-pathogen interactions are complex interactions involving various dynamics between hosts and multiple pathogens. These interactions may be competitive, cooperative, or coexisting, depending on the various strategies used by the pathogens in colonizing a plant host (Abdullah et al., 2017). The innate

DOI: 10.1201/9781003505631-19

immune mechanisms of plants, however, have evolved due to constant exposure to pathogenic stress. New sequencing technologies and 'omics' tools have since improved our knowledge of regulating plant defense and pathogen physical interaction (Imam et al., 2016). Systems biology therefore suggests an appropriate approach to integrate multi-omics data and model the outcome of plant-pathogen interaction (Peyraud et al., 2017). Such approaches across scales can reconstruct a mechanistic model and integrate host and pathogen models (Peyraud et al., 2017). In the future, research should address the integration of plant responses in time and space and how this could be translated into agricultural applications (Peyraud et al., 2017). Current studies into plant-pathogen interactions now encompass everything from molecular aspects to systems-level components (Singh & Singh, 2018).

Biotic stresses are making agriculture very difficult for global food security and crop productivity (Anmolpreet Kaur et al., 2024). They are caused by pathogens, Phyto parasitic organisms, or competition with plants and lead to substantial losses of yield and income (Gimenez et al., 2018; Ricroch et al., 2016). Some approaches taken by researchers to overcome such challenges include conventional breeding methods and advanced biotechnology (Hussain, 2015). However, there are wide gaps between farmers' needs and existing breeding investigations, several biotic stress issues still remain to be solved (Ricroch et al., 2016). On the scientific side, it has become important to study the plant mechanisms of defense against biotic stresses, regulation of gene expression, and phytohormone involvement in biotic strain responses (Gimenez et al., 2018). Molecular genetic approaches, including marker-assisted selection, genomics, and TILLING, have emerged as a powerful and more efficient approach to developing biotic stress-resistant crops compared to conventional breeding methods (Hussain, 2015). The integration of these strategies is crucial to harvester improvement and sustainable agriculture (Anmolpreet Kaur et al., 2024).

Metabolomics is emerging in the field of plant pathology as a powerful new tool with which to understand plant-pathogen interactions and defense mechanisms (Bhatt et al., 2022; Castro-Moretti et al., 2020). This allows for studying metabolic changes in plants infected by pathogens that reveal perturbations in signaling pathways and defensive responses (Castro-Moretti et al., 2020).

High-throughput metabolomics approaches allow for rapid identification and quantification of metabolites governing plant defense response (Kumari et al., 2024; Razzaq et al., 2019). This will enable investigations into the pathways and complex metabolic networks involved in stress tolerance, aiding in the identification of biomarkers and key metabolites linked to pest and pathogen resistance (Kumari et al., 2024). The integration of metabolomics with other omics technologies and sophisticated bioinformatic methods has greatly enhanced our understanding of genotype-metabolite-phenotype interactions, thus providing opportunities for metabolomics-assisted breeding of stress-resilient crop varieties (Razzaq et al., 2022). In wheat, metabolomics has been tremendously useful in solving distinctive metabolic pathways conferring biotic and abiotic stress toleration, helping to develop climate-smart varieties that are crucial in attaining global food security (Razzaq et al., 2021).

Secondary metabolites play a primary function in pathogen resistance and the recent improvements in mass spectrometry and analytical tools have facilitated their identification and validation (Muñoz-Hoyos & Stam, 2023). Metabolomic analysis reveals key changes in the signaling pathways and metabolic outputs that control plant-microbe interactions (Castro-Moretti et al., 2020). Applications in metabolomics can discern both plant resistance mechanisms and pathogen virulence strategies (Castro-Moretti et al., 2020; Chen et al., 2019). It contributes to understanding the interactions between fungal pathogens and plants with qualitative and quantitative data about pathogenesis and host defense (Chen et al., 2019). Recent studies have focused on primary metabolites, infection-related compounds, and hormones involved in plant immune responses (Williams et al., 2021). Metabolomics applications in plant pathology faces challenges, including distinguishing between host and pathogen metabolomes and metabolome annotation (Williams et al., 2021). The combination of metabolomics and further omics approaches, like genomics and transcriptomics, might speed up the study of chemical defenses against pathogens and improve the effect of markers for resistance breeding (Muñoz-Hoyos & Stam, 2023). The application of metabolo-proteomics offers possibilities to overcome the challenges of polygenic inheritance and identify candidate genes for improving plant resistance to biotic strain (Kushalappa & Gunnaiah, 2013). Despite its potential, metabolomics faces limitations, including ineffective 3D capturing of metabolites and a lack of cell-based sampling (Kumari et al., 2024).

## 19.2 BIOTIC STRESS

Plant productivity is greatly impacted by biotic strains induced by environmental factors like viruses, bacteria, fungi, insects, nematodes, arachnids, and weeds (Hashem et al., 2019). Plant-parasitic nematodes feed on plant parts, leading to soil-borne diseases, nutrient deficiencies, stunted growth, and wilting. While around 8,000 fungal species can cause plant diseases, only about 14 bacterial genera are known to affect plants (Chaha et al., 2022). Fungal pathogens are classified into three types established on their nutrient acquisition strategies: necrotrophic, biotrophic, and hemi biotrophic. Necrotrophs kill plant cells for nutrients, biotrophs feed on live cells, and hemi biotrophs start with living tissue before changing to a necrotrophic phase (McCombe et al., 2022). These pathogens cause leaf spots, cankers, and vascular wilts. Fungi can also infect plants through bacterial interactions, while viruses often kill their hosts and cause localized damage, stunting, deformities, and chlorosis (Olayinka & Aderibigbe, 2024). Insects and mites harm plants by feeding or laying eggs, and some serve as vectors for bacteria and viruses (Schumann & D'Arcy, 2006). Unlike vertebrates, plants lack an adaptive immune system but have evolved complex defense mechanisms to combat biotic stress (Singla et al., 2016).

## 19.3 PLANT–PATHOGEN INTERACTION AGAINST BIOTIC STRESS

Remarkably, only a limited number of pathogens and pests can successfully parasitize a specific plant species, despite the vast array of potential threats. This highlights the existence of highly complex plant defense mechanisms capable of recognizing biotic invaders and mitigating their harmful effects on growth, yield, and survival (Lamers, Van Der Meer, & Testerink, 2020). Pathogens like fungi, bacteria, and viruses use diverse strategies to infect and colonize plant tissues, while plants have acquired advanced defense systems to fend off these invasions. Plants have developed a range of defense systems (Table 19.1) to combat biotic stress, including described in the following sections:

### 19.3.1 Physical Barriers

Plant morphological and microstructural traits protect against herbivore invasions, serving as the first line of defense. Physical barriers like hairs, trichomes, thorns, spines, and cuticles covering aerial parts of plants deter insect herbivory (Kaur & Kariyat, 2020; Zhao et al., 2021). Cuticular waxes regulate plant-insect relations at multiple statuses (Zhao et al., 2021). Trichomes, including glandular types, enhance resistance by affecting insect feeding and oviposition (Ádám et al., 2018). Higher trichome density further strengthens defense, with both their structure and chemical composition contributing to protection. Trichomes are composed of compounds like phenolics, which are harmful to herbivores, while glandular trichomes store terpenes and volatile organic compounds (VOCs) that draw natural predators of insect herbivores (Monaghan & Zipfel, 2012; Spoel & Dong, 2012).

### 19.3.2 Chemical Defenses

Primary and secondary metabolites are two distinct types of organic constituents formed by plants through metabolic pathways. Primary metabolites, including carbohydrates, amino acids, proteins, lipids, and hormones, are necessary for key functions like photosynthesis, respiration, and growth (Zhao et al., 2021). In contrast, secondary metabolites, including flavonoids, alkaloids, lignans, quinones, peptides, phenolics, anthocyanins, and terpenoids, are synthesized through various pathways and recreate a necessary part in plant defense against pathogens, insects, and predators (Divekar et al., 2022). Around 200,000 plant secondary metabolites have been identified and studied (Satish et al., 2020).

**TABLE 19.1**
**Plant–Pathogen Interactions against Biotic Stress**

| S. No. | Biotic Stress | Plant–Pathogen Interactions | References |
|---|---|---|---|
| 1. | Fungal infection | Plants activate defense pathways such as hypersensitive response and systemic acquired resistance (SAR) | Noman et al. (2020); Patel et al. (2020); Thakur et al. (2019) |
| 2. | Bacterial infection | Pathogen's trigger PAMP-triggered immunity (PTI) and effector-triggered immunity (ETI); bacteria may secrete effector proteins | Hetmann & Kowalczyk (2019); Peng et al. (2018) |
| 3. | Viral infection | Plants employ RNA silencing mechanisms to degrade viral RNA and induce systemic resistance | Gaffar & Koch (2019) |
| 4. | Nematode infestation | Plants respond by producing secondary metabolites, thickening cell walls, and recruiting beneficial microbes | Meidani et al. (2019); Sikder & Vestergård (2020); Sikder et al. (2022) |
| 5. | Insect herbivory | Induction of jasmonic acid (JA) and ethylene (ET) signaling pathways, which activate defense-related genes and proteinase inhibitors | Figon et al. (2021); Hu et al. (2021); Qi et al. (2018) |
| 6. | Parasitic plants | Host plants activate physical barriers and produce chemical defenses to limit nutrient transfer to the parasite | Albert et al. (2021); Hegenauer et al. (2017) |
| 7. | Oomycete infection | Plants initiate both PTI and ETI, producing antimicrobial compounds and strengthening cell walls | Wang et al. (2019); Yuan et al. (2021) |
| 8. | Fungal root pathogens | Mycorrhizal associations may be disrupted; plants activate hormonal signaling such as JA and salicylic acid (SA) | Lemarié et al. (2015); Martínez-Medina et al. (2017) |
| 9. | Phytoplasma infection | Plants show symptoms like phyllody; immune response includes phloem protein degradation and systemic signaling | MacLean et al. (2011) |

### 19.3.3 Hormone-Mediated Signaling

The primary hormones produced by plants include auxins, gibberellins, cytokinins, ET, abscisic acid (ABA), SA, JA, brassinosteroids (BR), and strigolactones. Among these, SA, JA, and ET are particularly important in responding to biotic strain, as their levels rise during pathogen infections. The SA regulatory pathways are activated by biotrophic and hemibiotrophic pathogens, while JA and ET pathways respond to necrotrophic pathogens and chewing pests (De Vleesschauwer et al., 2013). SA triggers the SAR pathway, enhancing the expression of pathogenesis-related (PR) genes that increase tolerance to a wide array of pathogens (Ádám et al., 2018).

### 19.3.4 Molecular Recognition

Plants employ a two-tiered innate immune system to defend against pathogens. The first level, PTI, is activated when pattern recognition receptors (PRRs) detect pathogen-associated molecular patterns (Cheng et al., 2012; Hou et al., 2013). The second level, ETI, is initiated when intracellular nucleotide-binding leucine-rich repeat proteins recognize pathogen effectors (Hou et al., 2013; Yuan et al., 2021). While PTI and ETI involve different activation mechanisms and early signaling components, they converge on similar downstream responses, with ETI generally being more robust and prolonged (Tsuda & Katagiri, 2010; Yuan et al., 2021). Recent evidence suggests intricate interactions between PTI and ETI signaling cascades, with common components shared by both (Yuan et al., 2021). The plant immune system has evolved from PTI to ETI, as some pathogenic effectors can suppress PTI (Hou et al., 2013). Understanding the crosstalk between PTI and ETI will provide a more comprehensive view of plant defense (Yuan et al., 2021).

Calcium ($Ca^{2+}$) signaling is important in plant immune responses to biotic stresses. Rapid changes in cytosolic $Ca^{2+}$ concentrations occur upon pathogen perception, serving as early events in plant defense mechanisms (Aldon et al., 2018; Seybold et al., 2014). These $Ca^{2+}$ signals are decoded by various $Ca^{2+}$-binding proteins, like calmodulin, and $Ca^{2+}$-dependent protein kinases, which trigger downstream immune responses (Zhang et al., 2014). The generation of $Ca^{2+}$ signals involves ion channels, with glutamate receptor-like channels (GLRs) and cyclic nucleotide-gated channels (CNGCs) involved in immune reactions, although their precise roles require further investigation (Moeder et al., 2019). $Ca^{2+}$ signaling can have both positive and negative effects on plant defense responses and interacts with other cellular signaling procedures, like the ubiquitin-proteasome system, to fine-tune immune responses (Zhang et al., 2014). Understanding the complex $Ca^{2+}$ signaling network in plant immunity remains an ongoing challenge in the field.

CNGCs are important in plant immunity and stress responses. These tetrameric cation channels are triggered by cyclic nucleotides and can form heterotetrametric complexes with distinct properties (Jarratt-Barnham et al., 2021). Several Arabidopsis CNGCs, including AtCNGC2, 4, 11, and 12, have been implicated in defense responses through genetic studies (Moeder et al., 2011). CNGCs contribute to pathogen-induced cytosolic $Ca^{2+}$ elevations, which are essential for innate immunity signaling cascades (Ma et al., 2007). The $Ca^{2+}$ signal is then transduced through various $Ca^{2+}$ sensors and $Ca^{2+}$-dependent protein kinases, directing to the production of nitric oxide and stimulation of defense-related genes (Ma, 2011). While $Ca^{2+}$ signaling is upstream of many defense components, some downstream molecules can also regulate $Ca^{2+}$ homeostasis, highlighting the complexity of plant pathogen signaling networks (Ma, 2011).

Plants have developed defense mechanisms against herbivores and pathogens. Upon attack, plants rapidly activate mitogen-activated protein kinase reactions, which are important in signaling plant defense responses (Hettenhausen et al., 2013). These reactions possess the generation of reactive oxygen species (ROS), stimulation of defense-related genes, and biosynthesis of defensive metabolites (Bolwell & Wojtaszek, 1997; Hettenhausen et al., 2015). The plants can sense specific patterns of tissue damage and elicitors in insect oral secretions, triggering a complex signaling network (Wu & Baldwin, 2009). This network involves cell membrane depolarization, $Ca^{2+}$ influx, and modulation of phytohormone levels, particularly JA, ET, and SA (Hettenhausen et al., 2015; Wu & Baldwin, 2009). These defense responses are not restricted to the site of attack but can also be mounted in unattacked parts of the plant, demonstrating systemic defense activation (Wu & Baldwin, 2009).

PR proteins are important in plant defense against abiotic and biotic stresses. Classified into 19 families established on biochemical possessions, PR proteins are induced by pathogens and environmental stressors (Ali et al., 2018; Islam et al., 2023). They function in SAR and are utilized as molecular markers for defense signaling pathways (Ali et al., 2018; Seo et al., 2008). PR genes contain various cis-acting regulatory elements, including light-responsive motifs, suggesting their involvement in multiple stress responses and developmental processes (Amritpreet Kaur et al., 2017; Seo et al., 2008). Overexpression of PR genes has revealed promise in developing transgenic crops with improved resistance to multiple strains (Ali et al., 2018; Islam et al., 2023). For instance, PR-3 mediates ABA-dependent salt stress signals affecting seed germination in Arabidopsis (Seo et al., 2008). Understanding the regulation and versatile roles of PR proteins is crucial for enhancing crop tolerance to multiple strains (Ali et al., 2018).

Plants employ diverse defense mechanisms against herbivores and pathogens. Physical barriers like trichomes, cuticles, and waxes serve as constitutive defenses (Mostafa et al., 2022; Savatin et al., 2014). Chemical defenses include secondary metabolites and VOCs, which can repel or intoxicate insects (Mazid et al., 2011; Mostafa et al., 2022). Upon herbivory or wounding, plants activate induced defenses through complex signaling cascades involving ion fluxes, hormones like JA, and the formation of defense compounds (Mostafa et al., 2022; Savatin et al., 2014). These responses occur both locally and systemically (Savatin et al., 2014). Plants also release volatiles to attract predators of herbivores and offer extrafloral nectar or nesting sites to accommodate them (Fürstenberg-Hägg et al., 2013). Damage-associated molecular patterns are important in activating plant innate defense (Savatin et al., 2014). Despite these defenses, insects have evolved mechanisms to resist and sometimes even sequester plant defense compounds for their own protection (Fürstenberg-Hägg et al., 2013). To repel pathogens and insects, plants hold a passive first line of defense that consists of physical barriers including trichomes, wax, and cuticles (Iqbal et al., 2021).

Plants use secondary metabolites with antibacterial properties as an intrinsic defense against pathogenic microbes. In response to pathogen entry, plants trigger supplementary defense mechanisms like hypersensitivity response, cell wall strengthening, and production of proteins associated with pathogenesis such as nucleases, proteases, and chitinases. The majority of plants exhibit a variety of local and systemic stress reactions to herbivory, predatory behavior, and wounding; which includes the generation of ROS and the build-up of phenolics, tannins, and phytoalexins (Chahal et al., 2022).

## 19.4 METABOLOMICS

The exploration of biological systems has gained a new perspective as a result of its evolution (Vinayavekhin et al., 2010). An integrated systems biology method called metabolomics makes it possible to profile a biological system's metabolites both qualitatively and quantitatively, establishing a connection between phenotypes and genotypes (Fiehn et al., 2000). There are three different experimental steps in metabolomics analysis. Sample preparation comes first, followed by chemical analysis, data mining, and integration utilizing chemometrics. Every step is closely connected to the others, especially the first two, because different analytical techniques need different analyte properties, which dictate how the sample should be processed (Aliferis & Jabaji, 2012). It is imperative to harvest samples appropriately, quench them quickly, and extract them directly to prevent metabolite deterioration. In a metabolomics investigation, the separation of metabolites is a crucial stage in the preparation of a sample that should be carefully handled as it will determine which analytes are later detected depending on the solvent type. The final metabolite matrix is significantly influenced by the extraction solvent and technique selected. It is challenging to extract every metabolite from the sample using the single extraction technique because the variety of metabolites results in varying solubility (Montini et al., 2020). As a result, the experiment's goal and the range of metabolites to be studied must be taken into consideration when selecting the extraction technique. Primary metabolites including acids, carbohydrates, and organic acids can be effectively extracted using water, the universal solvent. Mixtures of immiscible organic solvents or single solvents like methanol can be employed for secondary metabolites (Pandey et al., 2022) (Table 19.2).

## 19.5 HIGH-THROUGHPUT METABOLOMICS

High-throughput metabolomics has become a vital tool for exploring plant-pathogen interactions, offering a detailed view of the metabolic changes during infection. By using advanced techniques including mass spectrometry and nuclear magnetic resonance (NMR) spectroscopy, investigators

**TABLE 19.2**
**Metabolites Produced by Plants and Solvents Used for Extraction**

| S. No. | Metabolites Produced by the Plants | Solvents Used for Extraction | References |
|---|---|---|---|
| 1 | Alkaloids | Methanol, Ethanol, Chloroform | Takla et al. (2018); Yubin et al. (2014) |
| 2 | Flavonoids | Methanol, Ethanol, Acetone | Anokwuru et al. (2011) |
| 3 | Terpenoids | Hexane, Petroleum ether, Ethyl acetate | Truong et al. (2021) |
| 4 | Phenolic compounds | Methanol, Acetone, Water | Reis et al. (2012) |
| 5 | Saponins | Methanol, Ethanol, Water | Bitencourt et al. (2014); Majinda (2012) |
| 6 | Tannins | Acetone, Methanol, Ethanol | Anokwuru et al. (2011); Downey & Hanlin (2010) |

can identify and measure a wide range of plant metabolites, including secondary metabolites crucial for plant defense (Piasecka et al., 2019; Ren et al., 2021; Zhang et al., 2019). The metabolome of a cell will change in reaction to alterations in its transcriptome, proteome, or genome (Heath, 2000). Metabolomics represents the most advanced level of post-genomic analysis as it considers both transcriptional and post-transcriptional regulation. Since metabolites have different chemical properties compared to transcripts and proteins, advanced tools like NMR and MS are necessary to detect them (Figure 19.1) (Fraser Jr, 2012; Montini et al., 2020).

### 19.5.1 Metabolite Profiling Using NMR Spectroscopy

NMR spectroscopy is a powerful technique for metabolite profiling in plants, offering a comprehensive study of complex extracts (Selegato et al., 2019). It provides insights into plant metabolism through in vitro and in vivo measurements, enabling the identification, quantification, and localization of metabolites (Ratcliffe & Shachar-Hill, 2001). NMR spectroscopy allows for metabolite fingerprinting and profiling, with the former using a multivariate study of unassigned 1H NMR spectra to compare the overall metabolic profile, and the latter identifying 20–40 metabolites in unfractionated extracts (Krishnan et al., 2005). Despite being slightly more sensitive than mass spectrometry, NMR offers advantages such as minimal sample preparation, non-destructive analysis, and the ability to identify a broad range of metabolites simultaneously (Kumar, 2016). These features make NMR spectroscopy a valuable tool for metabolomics, quality control of plant-derived products, and understanding the functional state of biological systems (Kumar, 2016; Selegato et al., 2019).

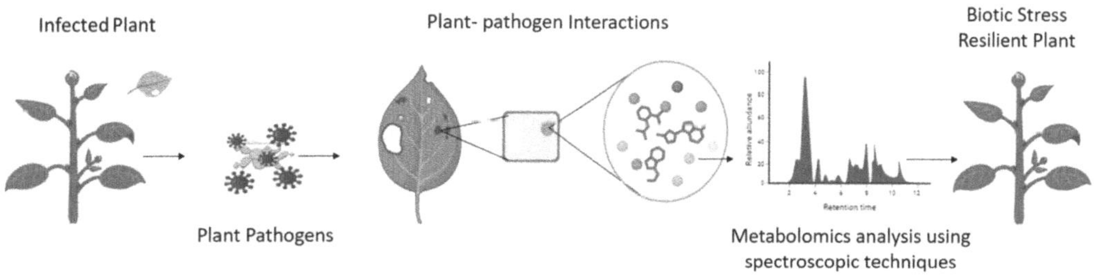


**FIGURE 19.1** Metabolomics analysis using spectrometry techniques.

LC-NMR enables NMR analysis without prior compound purification, recording spectra in continuous or stopped flow modes. In stopped-flow, fractions are diverted to a storage loop for quality spectrum collection. An advanced setup splits the HPLC output for simultaneous NMR and MS analysis, enhancing compound identification. This method has been employed to study xenobiotic metabolism in plants and identify natural products. Sensitivity is improved by integrating solid-phase extraction and using a cryogenic probe, which concentrates samples and reduces the need for expensive deuterated solvents during LC separation (Bailey et al., 2000; Exarchou et al., 2003). Despite limitations in sensitivity, $^{1}H$ NMR spectroscopy is effective for metabolite fingerprinting and profiling in plant samples. It complements high-throughput MS analyses, and combined NMR-MS techniques offer a promising approach to expand metabolome coverage. However, neither NMR nor MS has significantly advanced comprehensive system-wide metabolomics or the use of metabolic phenotypes to uncover gene function (Krishnan et al., 2005; Raamsdonk et al., 2001) (Table 19.3).

### 19.5.2 Metabolite Profiling Using GC-MS

In metabolomics, gas chromatography-mass spectrometry (GC-MS) is often considered the "gold standard" (Harrigan & Goodacre, 2003). This method is ideal for separating volatile metabolites or those that can be chemically changed to become volatile. GC-MS combines high-resolution separation with reliable identification and measurement of metabolites. Electron ionization spectrum libraries, which are publicly and commercially available, facilitate high-fidelity metabolite identification (Halket et al., 2005). GC-TOF-MS, which uses exact retention times, accurate masses, and MS fragmentation patterns, enhances compound identification and resolution of co-eluting peaks. In comparison, GC-QMS and GC-ITMS may require longer analysis times for improved separation. Innovations like 2D GC-MS enhance separation capacity and biomolecule detection through dual-column analysis (Rohloff, 2015).

Recent advancements in the GC-MS method have significantly improved transnational metabolomics, aiding in the understanding of biological processes, system concerns, diagnostics, and quality assessments. However, there are limitations. Strict control over processing, sample handling,

**TABLE 19.3**
**High-Throughput Metabolomics in Plant Species: Biotic Stress and Metabolite Identification**

| S. No. | Biotic Stress | Name of the Plants | Metabolite Profiled | Method Used for Profiling | References |
|---|---|---|---|---|---|
| 1 | Fungal infection | Tomato (*Solanum lycopersicum*) | Alkaloids, Phenolics | LC-MS | Galeano Garcia et al. (2018) |
| 2 | Bacterial infection | Rice (*Oryza sativa*) | Flavonoids, Terpenoids | HPLC (High-Performance Liquid Chromatography) | De Bruyne et al. (2016); Lai et al. (2024); Wu et al. (2011) |
| 3 | Viral infection | Clove (*Syzygium aromaticum* L) | Terpenoids, Glycosides | GC-MS | El-Saber Batiha et al. (2020) |
| 4 | Insect herbivory | Arabidopsis (*Arabidopsis thaliana*) | Jasmonates, Glucosinolates | NMR | Hendrawati et al. (2006) |
| 5 | Oomycete infection | Potato (*Solanum tuberosum*) | Glycoalkaloids, Saponins | LC-MS/MS | Baur et al. (2022) |
| 6 | Fungal root pathogens | Wheat (*Triticum aestivum*) | Phenolic acids, Lignins | UPLC-DAD (Ultra-Performance Liquid Chromatography-Diode Array Detection) | Sharma et al. (2018) |

and analysis is essential to minimize data variation and ensure accurate quantitation. General extraction and derivatization protocols can underrepresent specific metabolites, affecting recovery rates. Additionally, complex mixtures of derivatized metabolites can tarnish separation and identification due to co-eluting peaks and similar MS spectra. Despite these challenges, automation and derivatization techniques have broadened the range of noticeable compounds, enabling high-throughput profiling of both abundant and minor metabolites, including plant secondary compounds (Rohloff, 2015) (Table 19.3).

### 19.5.3 Metabolite Profiling Using LC-MS

Liquid chromatography associated with mass spectrometry (LC-MS) is a pivotal technique in plant metabolomics, particularly for profiling secondary metabolites involved in plant-pathogen interactions. Unlike GC-MS, this technique is better suited for analyzing polar, non-volatile metabolites such as phytoalexins and phytoanticipins without the need for derivatization. Its high selectivity and sensitivity make it ideal for detecting low-abundance compounds essential for metabolic reprogramming. This makes LC-MS indispensable for monitoring metabolic changes in plant-pathogen interactions and other stress responses (Marr et al., 2021).

LC-MS enables the comprehensive characterization of metabolite accumulation patterns without relying on original reference compounds or isolating particular metabolites. Structural assignment is typically based on reference database searches using MS and MS/MS data. However, characterizing unexplored metabolites is difficult since they cannot be linked to a relational database. The fragmentation design, parent ions, and daughter ions in Liquid Chromatography-Tandem Mass Spectrometry (LC-MS/MS) are key parameters that enable the identification of both targeted and untargeted metabolites based on m/z values and reference databases. For validation, m/z values are matched with data from mass data banks under defined tolerances (Smith et al., 2006). Data accession relies on passions to score the chance between query and respect data, with tolerance adjustments for untargeted metabolites, considering m/z values and chromatographic retention times, aligned using specialized software (Pluskal et al., 2009). Advanced web-based platforms streamline data processing, peak alignment, annotation, and statistical analysis (Sawada & Hirai, 2013).

LC-Q-TOF-MS high-speed scanning can detect over one million untargeted MS/MS data points, but only a fraction is confirmed with respective data, widening the gap between detected metabolites and respective evidence due to advances in MS sensitivity (Tohge & Fernie, 2010). More than 90% of noticeable metabolites remain unspecified, highlighting the need for better identification methods for phytoconstituents and their metabolic patterns (Sawada & Hirai, 2013). Numerous LC-MS studies have revealed the diversity of plant metabolites, like biomolecules, organic acids, and phytohormones. Thus, MS-based technologies are vital for qualitative and quantitative metabolite validation, with metabolite-targeted biological assessments playing a significant role in screening for biological interactions (Fernie, 2003) (Table 19.3).

### 19.5.4 Metabolite Profiling Using MSI

Matrix-assisted laser desorption/ionization mass spectrometry (MALDI-MS) is becoming popular in plant science because of its high sensitivity, rapid scan speeds (up to 10 kHz laser excitation), and minimal contamination during ionization (Kompauer et al., 2017; Silva et al., 2016). Enhancements in MALDI matrices and mass spectrometers have greatly improved its capability to analyze plant compounds (He et al., 2019; Wang et al., 2015). MALDI-MS imaging (MALDI-MSI) allows for the simultaneous spatial and quantitative analysis of both exogenous and endogenous compounds in thin plant tissue slices (Liu et al., 2021; Liu et al., 2019). However, its effectiveness is limited by plant structures like the cuticle, epidermal wax, and cell wall (Qin et al., 2018). Methods like cryo-sectioning and porous material-aided imprinting have been developed to address these challenges, but they still face issues like incomplete metabolite release and potential metabolite loss during imprinting (Bjarnholt et al., 2014; Montini et al., 2020). Along with these advancements, further improvements in MALDI-MSI methods are needed for complete, high-throughput investigation of plant metabolites (Table 19.3).

## 19.6 DEVELOPING BIOTIC STRESS-RESILIENT CROPS

In order to satisfy the food demands of a growing global population, tolerable agriculture aims to decrease crop yield loss and quality decline without understanding biotic conditions. Traditional methods, such as breeding resistant cultivars and using agrochemicals, have limitations. Breeding relies on germplasm resources and is sufficient for specific biotrophic pathogens, though resistance can be easily bypassed. Developing broad-spectrum resistance against non-specific pathogens remains challenging due to limited germplasm and genetic diversity. While agrochemicals control pests, they increase costs and cause pollution. Biological control remains largely experimental, with few successful implementations (Yu et al., 2022).

Transgenic technology addresses hybridization barriers by utilizing a wide range of genes from distant species, enhancing crop biotic resistance precisely while minimally impacting beneficial soil microbes and the environment. Often seen as a second Green Revolution, this technology aims to improve crop traits, significantly contributing to food security. The process involves cloning wanted genes, making expression vectors, transforming recipient cultivars, and identifying transgenic lines. It also allows for modifying or knocking out suboptimal genes to enhance phenotypes. After rigorous safety assessments, transgenic cultivars with improved biotic resistance are approved for cultivation and market release (Farre et al., 2010). Transgenic improvement employs three primary strategies: heterologous expression of exogenous genes from remote species, overexpression of endogenous genes from the identical or sexually consistent species, and silencing undesirable endogenous genes or those associated with pathogens and pests using RNA interference (RNAi) or CRISPR/Cas technology (Yu et al., 2022) (Table 19.4).

Heterologous transgenes are typically driven by constitutive promoters, with codon optimization reducing usage bias. Research on transgenic insect-resistant cotton and rice shows a nonlinear correlation between Cry protein accumulation and transcription classes of exogenous Cry genes, suggesting potential for broad-spectrum resistance to disease-causing fungi with deeper insights into hypersensitive responses (Heath, 2000). Biochemical responses are reversible and operate in complex networks, so overexpressing an enzyme gene may not increase overall reaction flux. ISAAA data shows that 93 of 118 authorized transgenic events used bacterial insect-resistant genes, while 25 involved synthetic sequences for silencing viral housekeeping genes, raising questions about the commercial viability of gene overexpression (Fraser Jr, 2012).

RNAi uses antisense or double-stranded RNAs to protect crops from viruses, pests, and nematodes. Its advantages include specific targeting and no foreign proteins. While 25 transgenic virus-resistant events are authorized, careful dsRNA design is essential to avoid off-target effects, supported by advances in genomics and bioinformatics (Pandey et al., 2022). CRISPR/Cas approach, a specific and cost-effective genome-editing method, enhances biotic resistance in crops without integrating foreign DNA, boosting public acceptance. Recent advancements have improved efficiency and reduced off-target effects. CRISPR/Cas13 allows for precise targeting of plant RNA viruses, aiding in the development of transgenic potato lines with wide-spectrum resistance (Noureen et al., 2022).

**TABLE 19.4**
**Biotic Stress and Biotechnological Interventions**

| S. No. | Biotic Stress | Biotechnology Interventions | References |
|---|---|---|---|
| 1 | Fungal infection | Development of transgenic crops expressing antifungal proteins and pathogen-related (PR) genes | Jain & Khurana (2018); Kaur et al. (2022) |
| 2 | Bacterial infection | CRISPR/Cas-mediated genome editing for resistance genes and bacterial recognition receptors | Rizwan et al. (2021); Wu et al. (2022) |
| 3 | Viral infection | RNAi to target viral RNA and block replication | Akbar et al. (2022); Tabassum et al. (2012) |
| 4 | Nematode infestation | Use of RNAi to silence nematode effector genes and introduction of nematode-resistant genes | (Dinh et al. (2014); Dinh et al. (2015) |
| 5 | Insect herbivory | Bacillus thuringiensis toxin-expressing crops to target insect pests | Paddock et al. (2021); Van Rie & Jansens (2019) |
| 6 | Parasitic plants | Genetic engineering of host plants to enhance resistance against parasitic plant attacks | Albert et al. (2021); Kaiser et al. (2015) |
| 7 | Oomycete infection | Overexpression of R genes (resistance genes) and genes involved in signal transduction | Capriotti et al. (2020); Fawke et al. (2015) |
| 8 | Fungal root pathogens | Engineering mycorrhizae-plant associations to enhance resistance to root pathogens | Kadam et al. (2020); Prasad (2021) |
| 9 | Phytoplasma infection | Use of transgenic plants expressing phloem-specific genes to limit pathogen spread | Bertamini et al. (2018); Li et al. (2013) |

## 19.7 CONCLUSION

Metabolomics is a critical mechanism for analyzing biotic strains in plants, providing insights into the complex interactions between plants and pathogens. Achieving a comprehensive understanding of the metabolome, requires advanced chemometrics, analytical techniques, and computational methods with high sensitiveness and reproducibility. Unlike further -omics fields, metabolomics faces unique challenges, particularly due to the metabolome's dynamic nature, chemical diversity, and varying metabolite concentrations. These challenges become more pronounced when studying plants, which host a vast microbiome.

While considerable improvement has been made in understanding the plant microbiome, there is still a need to explore the processes governing microbial community formation and function in plants. Earlier interactions between plants and microorganisms are crucial, as they determine whether a microorganism is beneficial or pathogenic. Environmental factors can further alter these interactions, making it essential to explore the mechanisms at play during the early stages of contact. Prospective studies should integrate systematic sampling of inoculated plants with advanced approaches like MALDI-MSI, which allows researchers to pinpoint metabolite distribution within plant tissues.

Metabolomics serves as a crucial connection between transcriptomics and phenomics, enabling researchers to determine genes reliable for phenotypic differences. For plants inoculated with PGP microorganisms, metabolites linked to enhanced phenotypes could be traced back to specific genes, which could be further studied or genetically edited to enhance crop resilience. Finally, while metabolomic studies often begin in controlled environments, validating hypotheses in more natural conditions is essential to ensure the applicability of findings to real-world agriculture.

## ACKNOWLEDGEMENT

All the authors are thankful to their respective Universities and Institutions for their valuable supports.

## DATA AVAILABILITY

The authors confirm that the data supporting the findings of this study are available within the book/ chapter.

## REFERENCES

Abdullah, A. S., Moffat, C. S., Lopez-Ruiz, F. J., Gibberd, M. R., Hamblin, J., & Zerihun, A. (2017). Host–multi-pathogen warfare: Pathogen interactions in co-infected plants. *Frontiers in Plant Science, 8*, 1806.

Ádám, A. L., Nagy, Z. Á., Kátay, G., Mergenthaler, E., & Viczián, O. (2018). Signals of systemic immunity in plants: Progress and open questions. *International Journal of Molecular Sciences, 19*(4), 1146.

Akbar, S., Wei, Y., & Zhang, M.-Q. (2022). RNA interference: Promising approach to combat plant viruses. *International Journal of Molecular Sciences, 23*(10), 5312.

Albert, M., Axtell, M. J., & Timko, M. P. (2021). Mechanisms of resistance and virulence in parasitic plant–host interactions. *Plant Physiology, 185*(4), 1282–1291.

Aldon, D., Mbengue, M., Mazars, C., & Galaud, J.-P. (2018). Calcium signalling in plant biotic interactions. *International Journal of Molecular Sciences, 19*(3), 665.

Ali, S., Ganai, B. A., Kamili, A. N., Bhat, A. A., Mir, Z. A., Bhat, J. A., … Yadav, P. (2018). Pathogenesis-related proteins and peptides as promising tools for engineering plants with multiple stress tolerance. *Microbiological Research, 212*, 29–37.

Aliferis, K. A., & Jabaji, S. (2012). FT-ICR/MS and GC-EI/MS metabolomics networking unravels global potato sprout's responses to Rhizoctonia solani infection. *PLoS One, 7*, e42576.

Anokwuru, C., Anyasor, G., Ajibaye, O., Fakoya, O., & Okebugwu, P. (2011). Effect of extraction solvents on phenolic, flavonoid and antioxidant activities of three Nigerian medicinal plants. *Nature and Science, 9*(7), 53–61.

Bailey, N. J., Stanley, P. D., Hadfield, S. T., Lindon, J. C., & Nicholson, J. K. (2000). Mass spectrometrically detected directly coupled high performance liquid chromatography/nuclear magnetic resonance spectroscopy/mass spectrometry for the identification of xenobiotic metabolites in maize plants. *Rapid Communications in Mass Spectrometry, 14*(8), 679–684.

Baur, S., Bellé, N., Frank, O., Wurzer, S., Pieczonka, S. A., Fromme, T., … Hückelhoven, R. (2022). Steroidal saponins—new sources to develop potato (Solanum tuberosum L.) Genotypes Resistant against Certain Phytophthora infestans Strains. *Journal of Agricultural and Food Chemistry, 70*(24), 7447–7459.

Bertamini, M., Muthuchelian, K. & Grando, M. (2002). Effects of Phytoplasma Infection on Growth and Photosynthesis in Leaves of Field Grown apple (Malus Pumila Mill. cv. Golden Delicious). *Photosynthetica* 40, 157–160 (2002). https://doi.org/10.1023/A:1020156021629.

Bhatt, B., Shikha, S., Mathpal, S., Sharma, G., & Pandey, A. (2022). Metabolomics and its role in plant pathology. *Phytoprotection, 102*(1), 6–14.

Bitencourt, R. G., Queiroga, C. L., Junior, Í. M., & Cabral, F. A. (2014). Fractionated extraction of saponins from Brazilian ginseng by sequential process using supercritical $CO_2$, ethanol and water. *The Journal of Supercritical Fluids, 92*, 272–281.

Bjarnholt, N., Li, B., D'Alvise, J., & Janfelt, C. (2014). Mass spectrometry imaging of plant metabolites–principles and possibilities. *Natural Product Reports, 31*(6), 818–837.

Bolwell, G. P., & Wojtaszek, P. (1997). Mechanisms for the generation of reactive oxygen species in plant defence – A broad perspective. *Physiological and Molecular Plant Pathology, 51*(6), 347–366.

Capriotti, L., Baraldi, E., Mezzetti, B., Limera, C., & Sabbadini, S. (2020). Biotechnological approaches: Gene overexpression, gene silencing, and genome editing to control fungal and oomycete diseases in grapevine. *International Journal of Molecular Sciences, 21*(16), 5701.

Castro-Moretti, F. R., Gentzel, I. N., Mackey, D., & Alonso, A. P. (2020). Metabolomics as an emerging tool for the study of plant–pathogen interactions. *Metabolites, 10*(2), 52.

Chahal, G. K., Ghai, M., & Kaur, J. (2022). Stress mechanisms and adaptations in plants. *Journal of Pharmacognosy and Phytochemistry, 11*(2), 109–114.

Chen, F., Ma, R., & Chen, X.-L. (2019). Advances of metabolomics in fungal pathogen–plant interactions. *Metabolites, 9*(8), 169.

Cheng, X., Tian, C.-J., Li, A.-N., & Qiu, J.-L. (2012). Advances on molecular mechanisms of plant-pathogen interactions. *Yi Chuan = Hereditas, 34*(2), 134–144.

De Bruyne, L., Van Poucke, C., Di Mavungu, D. J., Zainudin, N. A. I. M., Vanhaecke, L., De Vleesschauwer, D., … Höfte, M. (2016). Comparative chemical screening and genetic analysis reveal tentoxin as a new virulence factor in C ochliobolus miyabeanus, the causal agent of brown spot disease on rice. *Molecular Plant Ppathology, 17*(6), 805–817.

De Vleesschauwer, D., Gheysen, G., & Höfte, M. (2013). Hormone defense networking in rice: Tales from a different world. *Trends in Plant Science, 18*(10), 555–565.

Dinh, P. T., Brown, C. R., & Elling, A. A. (2014). RNA interference of effector gene Mc16D10L confers resistance against Meloidogyne chitwoodi in Arabidopsis and potato. *Phytopathology, 104*(10), 1098–1106.

Dinh, P. T., Zhang, L., Mojtahedi, H., Brown, C. R., & Elling, A. A. (2015). Broad Meloidogyne resistance in potato based on RNA interference of effector gene 16D10. *Journal of Nematology, 47*(1), 71.

Divekar, P. A., Narayana, S., Divekar, B. A., Kumar, R., Gadratagi, B. G., Ray, A., … Singh, A. K. (2022). Plant secondary metabolites as defense tools against herbivores for sustainable crop protection. *International Journal of Molecular Sciences, 23*(5), 2690.

Downey, M. O., & Hanlin, R. (2010). Comparison of ethanol and acetone mixtures for extraction of condensed tannin from grape skin. *South African Journal of Enology and Viticulture, 31*(2), 154–159.

El-Saber Batiha, G., Alkazmi, L. M., Wasef, L. G., Beshbishy, A. M., Nadwa, E. H., & Rashwan, E. K. (2020). Syzygium aromaticum L.(Myrtaceae): Traditional uses, bioactive chemical constituents, pharmacological and toxicological activities. *Biomolecules, 10*(2), 202.

Exarchou, V., Godejohann, M., van Beek, T. A., Gerothanassis, I. P., & Vervoort, J. (2003). LC-UV-solid-phase extraction-NMR-MS combined with a cryogenic flow probe and its application to the identification of compounds present in Greek oregano. *Analytical Chemistry, 75*(22), 6288–6294.

Farre, G., Ramessar, K., Twyman, R. M., Capell, T., & Christou, P. (2010). The humanitarian impact of plant biotechnology: Recent breakthroughs vs bottlenecks for adoption. *Current Opinion in Plant Biology, 13*(2), 219–225.

Fawke, S., Doumane, M., & Schornack, S. (2015). Oomycete interactions with plants: Infection strategies and resistance principles. *Microbiology and Molecular Biology Reviews, 79*(3), 263–280.

Fernie, A. R. (2003). Metabolome characterisation in plant system analysis. *Functional Plant Biology, 30*(1), 111–120.

Fiehn, O., Kopka, J., Dörmann, P., Altmann, T., Trethewey, R. N., & Willmitzer, L. (2000). Metabolite profiling for plant functional genomics. *Nature Biotechnology, 18*(11), 1157–1161.

Figon, F., Baldwin, I. T., & Gaquerel, E. (2021). Ethylene is a local modulator of jasmonate-dependent phenolamide accumulation during Manduca sexta herbivory in Nicotiana attenuata. *Plant, Cell & Environment, 44*(3), 964–981.

Fraser Jr, M. J. (2012). Insect transgenesis: Current applications and future prospects. *Annual Review of Entomology, 57*(1), 267–289.

Fürstenberg-Hägg, J., Zagrobelny, M., & Bak, S. (2013). Plant defense against insect herbivores. *International Journal of Molecular Sciences, 14*(5), 10242–10297.

Gaffar, F. Y., & Koch, A. (2019). Catch me if you can! RNA silencing-based improvement of antiviral plant immunity. *Viruses, 11*(7), 673.

Gai, Y.-P., Yuan, S.-S., Liu, Z.-Y., Zhao, H.-N., Liu, Q., Qin, R.-L., … Ji, X.-L. (2018). Integrated phloem sap mRNA and protein expression analysis reveals phytoplasma-infection responses in mulberry. *Molecular & Cellular Proteomics, 17*(9), 1702–1719.

Galeano Garcia, P., Neves dos Santos, F., Zanotta, S., Eberlin, M. N., & Carazzone, C. (2018). Metabolomics of Solanum lycopersicum infected with Phytophthora infestans leads to early detection of late blight in asymptomatic plants. *Molecules, 23*(12), 3330.

Gimenez, E., Salinas, M., & Manzano-Agugliaro, F. (2018). Worldwide research on plant defense against biotic stresses as improvement for sustainable agriculture. *Sustainability, 10*(2), 391.

Halket, J. M., Waterman, D., Przyborowska, A. M., Patel, R. K., Fraser, P. D., & Bramley, P. M. (2005). Chemical derivatization and mass spectral libraries in metabolic profiling by GC/MS and LC/MS/MS. *Journal of Experimental Botany, 56*(410), 219–243.

Harrigan, G. G., & Goodacre, R. (2003). *Metabolic Profiling: Its Role in Biomarker Discovery and Gene Function Analysis: Its Role in Biomarker Discovery and Gene Function Analysis*: Springer Science & Business Media.

Hashem, A., Tabassum, B., & Abd_Allah, E. F. (2019). Bacillus subtilis: A plant-growth promoting rhizobacterium that also impacts biotic stress. *Saudi Journal of Biological Sciences, 26*(6), 1291–1297.

He, H., Qin, L., Zhang, Y., Han, M., Li, J., Liu, Y., ... Zeng, M. (2019). 3, 4–Dimethoxycinnamic acid as a novel matrix for enhanced in situ detection and imaging of low-molecular-weight compounds in biological tissues by MALDI-MSI. *Analytical Vhemistry, 91*(4), 2634–2643.

Heath, M. C. (2000). Nonhost resistance and nonspecific plant defenses. *Current opinion in plant biology, 3*(4), 315–319.

Hegenauer, V., Körner, M., & Albert, M. (2017). Plants under stress by parasitic plants. *Current Opinion in Plant Biology, 38*, 34–41.

Hendrawati, O., Yao, Q., Kim, H. K., Linthorst, H. J., Erkelens, C., Lefeber, A. W., ... Verpoorte, R. (2006). Metabolic differentiation of Arabidopsis treated with methyl jasmonate using nuclear magnetic resonance spectroscopy. *Plant Science, 170*(6), 1118–1124.

Hetmann, A., & Kowalczyk, S. (2019). Suppression of PAMP-triggered immunity (PTI) by effector proteins synthesized by phytopathogens and delivered into cells of infected plant. *Postepy Biochemii, 65*(1), 58–71.

Hettenhausen, C., Schuman, M. C., & Wu, J. (2015). MAPK signaling: A key element in plant defense response to insects. *Insect science, 22*(2), 157–164.

Hou, S., Zhang, C., Yang, Y., & Wu, D. (2013). Recent advances in plant immunity: Recognition, signaling, response, and evolution. *Biologia Plantarum, 57*, 11–25.

Hu, C., Wei, C., Ma, Q., Dong, H., Shi, K., Zhou, Y., ... Yu, J. (2021). Ethylene response factors 15 and 16 trigger jasmonate biosynthesis in tomato during herbivore resistance. *Plant Physiology, 185*(3), 1182–1197.

Hussain, B. (2015). Modernization in plant breeding approaches for improving biotic stress resistance in crop plants. *Turkish Journal of Agriculture and Forestry, 39*(4), 515–530.

Imam, J., Singh, P. K., & Shukla, P. (2016). Plant microbe interactions in post genomic era: Perspectives and applications. *Frontiers in Microbiology, 7*, 1488.

Iqbal, Z., Iqbal, M. S., Hashem, A., Abd_Allah, E. F., & Ansari, M. I. (2021). Plant defense responses to biotic stress and its interplay with fluctuating dark/light conditions. *Frontiers in Plant Science, 12*, 631810.

Islam, M. M., El-Sappah, A. H., Ali, H. M., Zandi, P., Huang, Q., Soaud, S. A., ... Liang, Y. (2023). Pathogenesis-related proteins (PRs) countering environmental stress in plants: A review. *South African Journal of Botany, 160*, 414–427.

Jain, D., & Khurana, J. P. (2018). Role of pathogenesis-related (PR) proteins in plant defense mechanism. *Molecular Aspects of Plant-Pathogen Interaction*, 265–281.

Jarratt-Barnham, E., Wang, L., Ning, Y., & Davies, J. M. (2021). The complex story of plant cyclic nucleotide-gated channels. *International Journal of Molecular Sciences, 22*(2), 874.

Kadam, S. B., Pable, A. A., & Barvkar, V. T. (2020). Mycorrhiza induced resistance (MIR): A defence developed through synergistic engagement of phytohormones, metabolites and rhizosphere. *Functional Plant Biology, 47*(10), 880–890.

Kaiser, B., Vogg, G., Fürst, U. B., & Albert, M. (2015). Parasitic plants of the genus Cuscuta and their interaction with susceptible and resistant host plants. *Frontiers in Plant Science, 6*, 45.

Kaur, A., Kaur, S., Kaur, A., Sarao, N. K., & Sharma, D. (2022). *Pathogenesis-Related Proteins and their Transgenic Expression for Developing Disease-Resistant Crops: Strategies Progress and Challenges*: IntechOpen.

Kaur, A., Pati, P. K., Pati, A. M., & Nagpal, A. K. (2017). In-silico analysis of cis-acting regulatory elements of pathogenesis-related proteins of Arabidopsis thaliana and Oryza sativa. *PLoS One, 12*(9), e0184523.

Kaur, A., Singh, G., Menon, S., Kumari, K., Ngairangbam, H., & Choudhary., S. (2024). Elucidating the molecular mechanisms of plant stress responses: Deciphering the impacts of abiotic and biotic stressors and developing innovative mitigation strategies for sustainable agriculture. *International Journal of Research in Agronomy, 7*(6). doi:10.33545/2618060X.2024.v7.i6h.922

Kaur, J., & Kariyat, R. (2020). Role of trichomes in plant stress biology. *Evolutionary Ecology of Plant-Herbivore Interaction*, 15–35.

Kompauer, M., Heiles, S., & Spengler, B. (2017). Autofocusing MALDI mass spectrometry imaging of tissue sections and 3D chemical topography of nonflat surfaces. *Nature Methods, 14*(12), 1156–1158.

Krishnan, P., Kruger, N., & Ratcliffe, R. (2005). Metabolite fingerprinting and profiling in plants using NMR. *Journal of Experimental Botany, 56*(410), 255–265.

Kumar, D. (2016). Nuclear magnetic resonance (NMR) spectroscopy for metabolic profiling of medicinal plants and their products. *Critical Reviews in Analytical Chemistry, 46*(5), 400–412.

Kumari, M., Yagnik, K. N., Gupta, V., Singh, I. K., Gupta, R., Verma, P. K., & Singh, A. (2024). Metabolomics-driven investigation of plant defense response against pest and pathogen attack. *Physiologia Plantarum, 176*(2), e14270.

Kushalappa, A. C., & Gunnaiah, R. (2013). Metabolo-proteomics to discover plant biotic stress resistance genes. *Trends in Plant Science, 18*(9), 522–531.

Lai, J., Peng, W., Song, S., Jiang, J., & Liu, B. (2024). Transcriptome analysis reveals the inhibitory mechanism of 3, 4-Dimethoxyphenol from Streptomyces albidoflavus strain ML27 against Xanthomonas oryzae pv. oryzae. *Pesticide Biochemistry and Physiology, 202*, 105913.

Lamers, J., Van Der Meer, T., & Testerink, C. (2020). How plants sense and respond to stressful environments. *Plant Physiology, 182*(4), 1624–1635.

Lemarié, S., Robert-Seilaniantz, A., Lariagon, C., Lemoine, J., Marnet, N., Jubault, M., … Gravot, A. (2015). Both the jasmonic acid and the salicylic acid pathways contribute to resistance to the biotrophic clubroot agent Plasmodiophora brassicae in Arabidopsis. *Plant and Cell Physiology, 56*(11), 2158–2168.

Li, Y., Chen, W., Ma, D., & Wu, Y. (2013). cDNA-AFLP analysis revealed genes potentially implicated in Catharanthus roseus flowers during wheat blue dwarf phytoplasma infection. *Physiological and Molecular Plant Pathology, 84*, 1–9.

Liu, H., Han, M., Li, J., Qin, L., Chen, L., Hao, Q., … Han, H. (2021). A caffeic acid matrix improves in situ detection and imaging of proteins with high molecular weight close to 200,000 Da in tissues by matrix-assisted laser desorption/ionization mass spectrometry imaging. *Analytical Chemistry, 93*(35), 11920–11928.

Liu, Y., Chen, L., Qin, L., Han, M., Li, J., Luo, F., … Wang, X. (2019). Enhanced in situ detection and imaging of lipids in biological tissues by using 2, 3-dicyanohydroquinone as a novel matrix for positive-ion MALDI-MS imaging. *Chemical Communications, 55*(83), 12559–12562.

Ma, W. (2011). Roles of $Ca^{2+}$ and cyclic nucleotide gated channel in plant innate immunity. *Plant Science, 181*(4), 342–346.

Ma, W., Yoshioka, K., & Berkowitz, G. A. (2007). Cyclic nucleotide gated channels and $Ca^{2+}$-mediated signal transduction during plant innate immune response to pathogens. *Plant Signaling & Behavior, 2*(6), 548–550.

MacLean, A. M., Sugio, A., Makarova, O. V., Findlay, K. C., Grieve, V. M., Tóth, R., … Hogenhout, S. A. (2011). Phytoplasma effector SAP54 induces indeterminate leaf-like flower development in Arabidopsis plants. *Plant Physiology, 157*(2), 831–841.

Majinda, R. R. (2012). Extraction and isolation of saponins. *Natural Products Isolation*, 415–426.

Marr, S., Hageman, J. A., Wehrens, R., van Dam, N. M., Bruelheide, H., & Neumann, S. (2021). LC-MS based plant metabolic profiles of thirteen grassland species grown in diverse neighbourhoods. *Scientific Data, 8*(1), 52.

Martínez-Medina, A., Appels, F. V., & van Wees, S. C. (2017). Impact of salicylic acid-and jasmonic acid-regulated defences on root colonization by Trichoderma harzianum T-78. *Plant Signaling & Behavior, 12*(8), e1345404.

Mazid, M., Khan, T., & Mohammad, F. (2011). Role of secondary metabolites in defense mechanisms of plants. *Biology & Medicine, 3*(3), 232–249.

McCombe, C. L., Greenwood, J. R., Solomon, P. S., & Williams, S. J. (2022). Molecular plant immunity against biotrophic, hemibiotrophic, and necrotrophic fungi. *Essays in Biochemistry, 66*(5), 581–593.

Meidani, C., Ntalli, N. G., Giannoutsou, E., & Adamakis, I.-D. S. (2019). Cell wall modifications in giant cells induced by the plant parasitic nematode Meloidogyne incognita in wild-type (Col-0) and the fra2 Arabidopsis thaliana katanin mutant. *International Journal of Molecular Sciences, 20*(21), 5465.

Meng, X., & Zhang, S. (2013). MAPK cascades in plant disease resistance signaling. *Annual Review of Phytopathology, 51*(1), 245–266.

Moeder, W., Phan, V., & Yoshioka, K. (2019). $Ca^{2+}$ to the rescue–$Ca^{2+}$ channels and signaling in plant immunity. *Plant Science, 279*, 19–26.

Moeder, W., Urquhart, W., Ung, H., & Yoshioka, K. (2011). The role of cyclic nucleotide-gated ion channels in plant immunity. *Molecular Plant, 4*(3), 442–452.

Monaghan, J., & Zipfel, C. (2012). Plant pattern recognition receptor complexes at the plasma membrane. *Current Opinion in Plant Biology, 15*(4), 349–357.

Montini, L., Crocoll, C., Gleadow, R. M., Motawia, M. S., Janfelt, C., & Bjarnholt, N. (2020). Matrix-assisted laser desorption/ionization-mass spectrometry imaging of metabolites during sorghum germination. *Plant Physiology, 183*(3), 925–942.

Mostafa, S., Wang, Y., Zeng, W., & Jin, B. (2022). Plant responses to herbivory, wounding, and infection. *International Journal of Molecular Sciences, 23*(13), 7031.

Muñoz-Hoyos, L., & Stam, R. (2023). Metabolomics in plant pathogen defense: From single molecules to large-scale analysis. *Phytopathology®, 113*(5), 760–770.

Noman, A., Aqeel, M., Qari, S. H., Al Surhanee, A. A., Yasin, G., Alamri, S., … Al-Saadi, A. M. (2020). Plant hypersensitive response vs pathogen ingression: Death of few gives life to others. *Microbial Pathogenesis, 145*, 104224.

Noureen, A., Zuhaib Khan, M., Amin, I., Zainab, T., Ahmad, N., Haider, S., & Mansoor, S. (2022). Broad-spectrum resistance against multiple PVY-strains by CRSIPR/Cas13 system in Solanum tuberosum crop. *GM Crops & Food, 13*(1), 97–111.

Olayinka, A. O., & Aderibigbe, O. R. (2024). Unveiling phenotypic diversity among tomato (Solanum lycopersicum L.) accessions: A comprehensive analysis of agronomic and physico-chemical traits. *Technology in Horticulture, 4*(1), 8.

Paddock, K. J., Pereira, A. E., Finke, D. L., Ericsson, A. C., Hibbard, B. E., & Shelby, K. S. (2021). Host resistance to Bacillus thuringiensis is linked to altered bacterial community within a specialist insect herbivore. *Molecular Ecology, 30*(21), 5438–5453.

Pandey, P., Mysore, K. S., & Senthil-Kumar, M. (2022). Recent advances in plant gene silencing methods. *Plant Gene Silencing: Methods and Protocols*, 1–22.

Patel, Z. M., Mahapatra, R., & Jampala, S. S. M. (2020). Role of fungal elicitors in plant defense mechanism. In V. Sharma, R. Salwan, & L. K. T. Al-Ani (Eds.), *Molecular Aspects of Plant Beneficial Microbes in Agriculture* (pp. 143–158): Elsevier.

Peng, Y., van Wersch, R., & Zhang, Y. (2018). Convergent and divergent signaling in PAMP-triggered immunity and effector-triggered immunity. *Molecular Plant-Microbe Interactions, 31*(4), 403–409.

Peyraud, R., Dubiella, U., Barbacci, A., Genin, S., Raffaele, S., & Roby, D. (2017). Advances on plant–pathogen interactions from molecular toward systems biology perspectives. *The Plant Journal, 90*(4), 720–737.

Piasecka, A., Kachlicki, P., & Stobiecki, M. (2019). Analytical methods for detection of plant metabolomes changes in response to biotic and abiotic stresses. *International Journal of Molecular Sciences, 20*(2), 379.

Pluskal, T., Nakamura, T., Villar-Briones, A., & Yanagida, M. (2009). Metabolic profiling of the fission yeast S. pombe: Quantification of compounds under different temperatures and genetic perturbation. *Molecular Biosystems, 6*(1), 182–198.

Prasad, K. (2021). Arbuscular mycorrhizal fungi and plant collaborations influences ecology and environmental changes for global sustainable development. *Journal of Ecology and Natural Resources, 5*(1), 1–16.

Qi, J., ul Malook, S., Shen, G., Gao, L., Zhang, C., Li, J., … Wu, J. (2018). Current understanding of maize and rice defense against insect herbivores. *Plant Diversity, 40*(4), 189–195.

Qin, L., Zhang, Y., Liu, Y., He, H., Han, M., Li, Y., … Wang, X. (2018). Recent advances in matrix-assisted laser desorption/ionisation mass spectrometry imaging (MALDI-MSI) for in situ analysis of endogenous molecules in plants. *Phytochemical Analysis, 29*(4), 351–364.

Raamsdonk, L. M., Teusink, B., Broadhurst, D., Zhang, N., Hayes, A., Walsh, M. C., … Rowland, J. J. (2001). A functional genomics strategy that uses metabolome data to reveal the phenotype of silent mutations. *Nature Biotechnology, 19*(1), 45–50.

Ratcliffe, R. G., & Shachar-Hill, Y. (2001). Probing plant metabolism with NMR. *Annual Review of Plant Biology, 52*(1), 499–526.

Razzaq, A., Guul, W., Khan, M. S., & Saleem, F. (2021). Metabolomics: A powerful tool to study the complexity of wheat metabolome. *Protein and Peptide Letters, 28*(8), 878–895.

Razzaq, A., Sadia, B., Raza, A., Khalid Hameed, M., & Saleem, F. (2019). Metabolomics: A way forward for crop improvement. *Metabolites, 9*(12), 303.

Razzaq, A., Wishart, D. S., Wani, S. H., Hameed, M. K., Mubin, M., & Saleem, F. (2022). Advances in metabolomics-driven diagnostic breeding and crop improvement. *Metabolites, 12*(6), 511.

Reis, S. F., Rai, D. K., & Abu-Ghannam, N. (2012). Water at room temperature as a solvent for the extraction of apple pomace phenolic compounds. *Food Chemistry, 135*(3), 1991–1998.

Ren, L., Zhang, T., Wu, H., Ge, Y., Zhao, X., Shen, X., … Ma, D. (2021). Exploring the metabolic changes in sweet potato during postharvest storage using a widely targeted metabolomics approach. *Journal of Food Processing and Preservation, 45*(2), e15118.

Ricroch, A., Harwood, W., Svobodová, Z., Sági, L., Hundleby, P., Badea, E. M., … Marfa Riera, V. (2016). Challenges facing European agriculture and possible biotechnological solutions. *Critical Reviews in Biotechnology, 36*(5), 875–883.

Rizwan, M., Arshad, M., Kashif, M., Durrani, A. Z., Abbas, A., Ahmad, T., … Khan, K. (2021). CRIPSR case system: Biological role in bacterial virulence, genome editing and in antimicrobial resistance. *Punjab University Journal of Zoology, 36*(1), 111–118.

Rohloff, J. (2015). Analysis of phenolic and cyclic compounds in plants using derivatization techniques in combination with GC-MS-based metabolite profiling. *Molecules, 20*(2), 3431–3462.

Satish, L., Shamili, S., Yolcu, S., Lavanya, G., Alavilli, H., & Swamy, M. K. (2020). Biosynthesis of secondary metabolites in plants as influenced by different factors. *Plant-derived Bioactives: Production, Properties and Therapeutic Applications*, 61–100.

Savatin, D. V., Gramegna, G., Modesti, V., & Cervone, F. (2014). Wounding in the plant tissue: The defense of a dangerous passage. *Frontiers in Plant Science, 5*, 470.

Sawada, Y., & Hirai, M. Y. (2013). Integrated LC-MS/MS system for plant metabolomics. *Computational and Structural Biotechnology Journal, 4*(5), e201301011.

Schumann, G. L., & D'Arcy, C. J. (2006). *Essential Plant Pathology*. American Phytopathological Society.

Selegato, D. M., Pilon, A. C., & Carnevale Neto, F. (2019). Plant metabolomics using NMR spectroscopy. *NMR-Based Metabolomics: Methods and Protocols*, 345–362.

Seo, P. J., Lee, A.-K., Xiang, F., & Park, C.-M. (2008). Molecular and functional profiling of Arabidopsis pathogenesis-related genes: Insights into their roles in salt response of seed germination. *Plant and Cell Physiology, 49*(3), 334–344.

Seybold, H., Trempel, F., Ranf, S., Scheel, D., Romeis, T., & Lee, J. (2014). $Ca^{2+}$ signalling in plant immune response: From pattern recognition receptors to $Ca^{2+}$ decoding mechanisms. *New Phytologist, 204*(4), 782–790.

Sharma, S., Sahu, R., Navathe, S., Mishra, V. K., Chand, R., Singh, P. K., … Pandey, S. P. (2018). Natural variation in elicitation of defense-signaling associates to field resistance against the spot blotch disease in bread wheat (Triticum aestivum L.). *Frontiers in Plant Science, 9*, 636.

Sikder, M. M., & Vestergård, M. (2020). Impacts of root metabolites on soil nematodes. *Frontiers in Plant Science, 10*, 1792.

Sikder, M. M., Vestergård, M., Kyndt, T., Topalović, O., Kudjordjie, E. N., & Nicolaisen, M. (2022). Genetic disruption of Arabidopsis secondary metabolite synthesis leads to microbiome-mediated modulation of nematode invasion. *The ISME Journal, 16*(9), 2230–2241.

Silva, R., Lopes, N. P., & Silva, D. B. (2016). Application of MALDI mass spectrometry in natural products analysis. *Planta Medica, 82*(08), 671–689.

Singh, A., & Singh, I. K. (2018). *Molecular Aspects of Plant-Pathogen Interaction* (Vol. 10): Springer.

Singla, J., Krattinger, S., Wrigley, C., Faubion, J., Corke, H., & Seetharaman, K. (2016). Biotic stress resistance genes in wheat. In *Encyclopedia of Food Grains* (pp. 388–392). Elsevier. Editors: Colin Wrigley, Harold Corke, Koushik Seetharaman, Jon Faubion.

Smith, C. A., Want, E. J., O'Maille, G., Abagyan, R., & Siuzdak, G. (2006). LC/MS Preprocessing and Analysis with xcms. *Analytical Chemistry, 78*(779.10), 1021.

Spoel, S. H., & Dong, X. (2012). How do plants achieve immunity? Defence without specialized immune cells. *Nature Reviews Immunology, 12*(2), 89–100.

Tabassum, B., Nasir, I. A., Aslam, U., & Husnain, T. (2012). How RNA interference combat viruses in plants. In G. Meroni & F. Petrera (Eds.), *Functional Genomics* (pp. 113–130): InTech.

Takla, S. S., Shawky, E., Hammoda, H. M., & Darwish, F. A. (2018). Green techniques in comparison to conventional ones in the extraction of Amaryllidaceae alkaloids: Best solvents selection and parameters optimization. *Journal of Chromatography A, 1567*, 99–110.

Thakur, A., Verma, S., Reddy, V. P., & Sharma, D. (2019). Hypersensitive responses in plants. *Agricultural Reviews, 40*(2), 113–120.

Tohge, T., & Fernie, A. R. (2010). Combining genetic diversity, informatics and metabolomics to facilitate annotation of plant gene function. *Nature Protocols, 5*(6), 1210–1227.

Truong, D. H., Ta, N. T. A., Pham, T. V., Huynh, T. D., Do, Q. T. G., Dinh, N. C. G., … Bui, A. V. (2021). Effects of solvent–solvent fractionation on the total terpenoid content and in vitro anti-inflammatory activity of Serevenia buxifolia bark extract. *Food Science & Nutrition, 9*(3), 1720–1735.

Tsuda, K., & Katagiri, F. (2010). Comparing signaling mechanisms engaged in pattern-triggered and effector-triggered immunity. *Current Opinion in Plant Biology, 13*(4), 459–465.

Van Rie, J., & Jansens, S. (2019). Midgut-Transgenic Crops Expressing Bacillus thuringiensis Cry Proteins. *Modern Crop Protection Compounds, 3*, 1103–1136.

Vinayavekhin, N., Homan, E. A., & Saghatelian, A. (2010). Exploring disease through metabolomics. *ACS Chemical Biology, 5*(1), 91–103.

Wang, X., Han, J., Yang, J., Pan, J., & Borchers, C. H. (2015). Matrix coating assisted by an electric field (MCAEF) for enhanced tissue imaging by MALDI-MS. *Chemical Science, 6*(1), 729–738.

Wang, Y., Tyler, B. M., & Wang, Y. (2019). Defense and counterdefense during plant-pathogenic oomycete infection. *Annual Review of Microbiology, 73*(1), 667–696.

Williams, A., Gamir, J., Gravot, A., & Pétriacq, P. (2021). Untangling plant immune responses through metabolomics. In E. J. Beckles (Ed.), *Advances in Botanical Research* (Vol. 98, pp. 73–105): Elsevier.

Wu, J., & Baldwin, I. T. (2009). Herbivory-induced signalling in plants: Perception and action. *Plant, Cell & Environment, 32*(9), 1161–1174.

Wu, Q., Cui, L., Liu, Y., Li, R., Dai, M., Xia, Z., & Wu, M. (2022). CRISPR-Cas systems target endogenous genes to impact bacterial physiology and alter mammalian immune responses. *Molecular Biomedicine, 3*(1), 22.

Wu, Y., Hillwig, M. L., Wang, Q., & Peters, R. J. (2011). Parsing a multifunctional biosynthetic gene cluster from rice: Biochemical characterization of CYP71Z6 & 7. *FEBS Letters, 585*(21), 3446–3451.

Yu, H., Wang, Y., Fu, F., & Li, W. (2022). Transgenic improvement for biotic resistance of crops. *InternationalJjournal of Molecular Sciences, 23*(22), 14370.

Yuan, M., Ngou, B. P. M., Ding, P., & Xin, X.-F. (2021). PTI-ETI crosstalk: An integrative view of plant immunity. *Current Opinion in Plant Biology, 62*, 102030.

Yubin, J., Miao, Y., Bing, W., & Yao, Z. (2014). The extraction, separation and purification of alkaloids in the natural medicine. *Journal of Chemical and Pharmaceutical Research, 6*(1), 338–345.

Zhang, L., Du, L., & Poovaiah, B. (2014). Calcium signaling and biotic defense responses in plants. *Plant Signaling & Behavior, 9*(11), e973818.

Zhang, X., Qu, X., Liu, C., Chen, G., Wu, C., & He, X. (2019). Optimization of extraction technology and oxidation resistance analysis of alkaloids in jujube. *Mol. Plant Breed, 17*(3), 972–977.

Zhao, H., Ni, S., Cai, S., & Zhang, G. (2021). Comprehensive dissection of primary metabolites in response to diverse abiotic stress in barley at seedling stage. *Plant Physiology and Biochemistry, 161*, 54–64.

# 20 Applied Metabolomics for Plant Disease Management

*Ganesan Megha, Domesa Monisha, Kempanna Sushmitha, Manoharan Rajesh, Natarajan Pushpa, Chenniappan Anchana Devi, Kandasamy Vijayalakshmi, and Arumugam Vijaya Anand*

## LIST OF ABBREVIATIONS

| | |
|---|---|
| **ANCOVA**: | Analysis of Covariance |
| **ANOVA**: | Analysis of Variance |
| **CE**: | Capillary Electrophoresis |
| **CE-MS**: | Capillary Electrophoresis-Mass Spectrometry |
| **CGAs**: | Chlorogenic Acids |
| **D3G**: | Deoxynivalenol-3 Glucoside |
| **DON**: | Deoxynivalenol |
| **GC-MS**: | Gas Chromatography-Mass Spectrometry |
| **HLB**: | Huanglongbing |
| **HPLC**: | High-Performance Liquid Chromatography |
| **IR**: | Infrared |
| **LC-MS**: | Liquid Chromatography-Mass Spectrometry |
| **MALDI**: | Imaging Matrix-Assisted Laser Desorption Ionization |
| **NIR**: | Near-Infrared |
| **OPLS**: | Orthogonal PLS |
| **PCA**: | Principal Component Analysis |
| **PLS-DA**: | Partial Least Squares-Discriminant Analysis |
| **ROS**: | Reactive Oxygen Species |
| **TOF-MS**: | Time-of-flight mass spectrometry |
| **UPLC/ESI-HR-Q-TOF-MS**: | Ultra Performance Liquid Chromatography//Electron Ionization-High-Resolution-Quadrupole-TOF-MS |

## 20.1 INTRODUCTION

Sustainable agriculture has gained significant importance and top priority within the agricultural sector. Plant diseases persist as serious barriers to agricultural productivity, leading to decreased crop production and financial losses worldwide. Therefore, monitoring and detecting diseases in plants are essential for promoting sustainable agriculture (Dordas, 2008; Sankaran et al., 2010). Traditional plant pathology methods, which analyze individual metabolites and phenotypes, like symptoms development and molecular responses such as callose deposition, reactive oxygen species (ROS) production, and antimicrobial secretion, had limitations in fully understanding plant diseases. So in recent times, plant pathologists have turned to metabolomics, a more comprehensive approach to develop deeper insights into plant diseases by studying collective metabolites (Castro-Moretti et al., 2020).

DOI: 10.1201/9781003505631-20

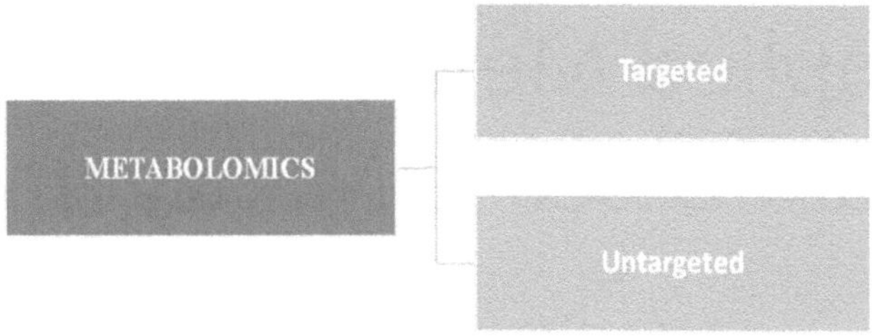


**FIGURE 20.1** Classification of metabolomic approaches.

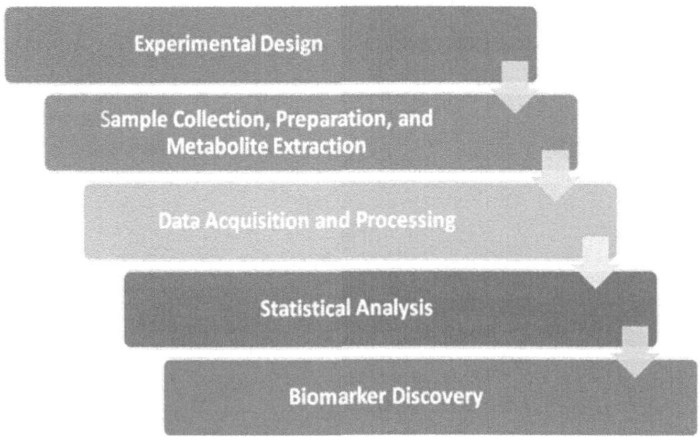


**FIGURE 20.2** Steps involved in conducting a metabolomic study.

Metabolomics is a detailed and unbiased analysis of complex mixtures of metabolites, conducted quickly and efficiently, to ideally identify and quantify every single metabolite (Gomez-Casati et al., 2013). It is a potent tool in plant biology, broadening our understanding of how metabolic and molecular mechanisms control growth, development, and responses to stress, thereby enhancing crop quality and yield. Analyzing the metabolomic data helps identify metabolite groups undergoing concentration shifts during disease progression (Terentev& Dolzhenko, 2023).

As represented in Figure 20.1, experimental designs often utilize two main strategies in metabolomics: targeted and untargeted. Targeted metabolomics focuses on measuring specific metabolites crucial in stress responses, while untargeted metabolomics analyzes overall metabolic profiles, allowing for investigating defense mechanisms and biomarkers. These methods have provided valuable insights by uncovering unique metabolic patterns and assisting plants in adapting to stress, thereby developing strategies for managing plant diseases more effectively (Allwood et al., 2021).

To conduct a successful targeted or untargeted metabolomic investigation, researchers follow several phases as illustrated in Figure 20.2. Initially, they design an experiment based on factors like the type of biological sample and the chosen metabolomic approach. Questions regarding sample type and the specific approach, targeted or untargeted, are essential considerations. Following this, sample collection, preparation, and metabolite extraction play a key role. This phase may entail employing various techniques depending on the analytical method used, such as liquid-liquid extraction for MS-based approaches. Next, data accession and processing were conducted utilizing analytical platforms like mass spectrometry (MS), nuclear magnetic resonance (NMR), or Fourier-transform infrared spectroscopy. Statistical analysis is then conducted, utilizing both univariate and multivariate tools to identify significant differences. The final step is biomarker discovery which involves identifying metabolites with altered expression levels using standardized bioinformatics resources (Aderemi et al., 2021).

Metabolomics traces its origins back to the 1970s when it first emerged with a focus on medical analyses using gas chromatography-mass spectrometry (GC-MS). Since then, it has undergone significant development, expanding its applications to include the profiling of metabolites in humans for diagnostics and drug evolution. In the 1990s, metabolomics found its way into plant investigation, initially through investigations into herbicide mechanisms. During the late 1990s, the integration of functional genomics with metabolomics, driven by advancements in chromatography, MS, and imaging technologies was noticed (Crandall et al., 2020). With the view of this introduction, this chapter strives to concern the potential implications of metabolomics in plant disease management.

## 20.2 METABOLIC PROFILING TECHNIQUES FOR PLANT DISEASE DIAGNOSIS

Metabolic profiling approaches are important in diagnosing plant diseases by investigating metabolites within biological systems, typically in response to a specific therapy or condition. In most cases, this analytical approach emphasizes the analysis of small molecules or metabolites available in an extract, providing valuable insights into the metabolome at a specific moment (Weston et al., 2015).

The fundamental equipment required for assessing plant metabolomes typically involves direct spectroscopy, chromatography, or advanced chromatography methods like GC or High-performance liquid chromatography (HPLC) coupled with spectroscopy methods such as infrared (IR), near-infrared (NIR), or MS (Gomez-Casati et al., 2013). Using advanced separation methods like GC, HPLC, Ultra HPLC (UHPLC), or CE combined with MS significantly improved metabolome coverage compared to other direct analysis techniques such as MS or NMR (Bedair&Sumner, 2008).

GC-MS and liquid chromatography-mass spectrometry (LC-MS) are the primary strategies employed for metabolite profiling due to their cost-effectiveness, extensive historical use, and ability to provide comprehensive data per analysis (Roessner & Beckles, 2012). GC–MS, an early hybrid technique, offers established methodologies, machine setup, data mining protocols, and accessible EI ionization spectral libraries (Hill& Roessner, 2013), while LC-MS is preferred for separating and detecting semipolar secondary metabolites in plants due to its ability to handle large and unique compound groups. Both techniques have unique strengths; with GC-MS providing extensive qualitative data from volatilized samples and LC-MS offering concise quantitative data from minimally processed samples (Castro-Moretti et al., 2020).

Advanced technique such as direct injection-MS involves introducing a gas sample directly into a mass analyzer without any pre-treatment or separation steps (Romano et al., 2023). This non-separative method provides a metabolic fingerprint of a biological specimen. Field asymmetric ion mobility spectrometry, a next-generation technology paired with MS, selects specific ions to minimize ion suppression and matrix effects, allowing the separation of isobaric compounds alongside particular MS (Fernie & Schauer, 2009).

A high-capacity method known as Fourier transform ion cyclotron MS was developed by Aharoni et al. (2002) for analyzing metabolites in biological samples, such as plant mutant populations, allowing for rapid screening. This technique separates metabolites with ultra-high mass resolution and identifies them by precise mass measurement of their elemental composition. Relative quantities are assessed by comparing absolute intensities using internal calibration.

Soga et al. (2003) later described a method for the complete and quantitative assessment of charged metabolites using capillary electrophoresis-mass spectrometry (CE-MS). In this approach, metabolites are first separated by their charge and size using capillary electrophoresis (CE), observed by selective detection with MS, which monitors ions over a broad spectrum of $m/z$ values.

Furthermore, Vibrational NIR-Fourier transform (NIR-FT) Raman microspectroscopy allows for the examination of the microstructure and chemical composition of plant tissues, facilitating the identification of crucial secondary metabolites. This technique has been applied to study fennel fruits, chamomile inflorescence, and Curcuma roots, enabling the identification of important secondary metabolites in these plant samples (Baranska et al., 2004).

NMR, a widely used tool in metabolomics, illustrates how studying metabolites can greatly advance our understanding of how plants resist pests (Leiss et al., 2011). NMR is most effective for specific uses like real-time imaging of flux in living organisms, determining compound structures, and metabolite fingerprinting. However, there are emerging methods with superior resolving power that may surpass NMR in certain applications. An instance of this is the utilization of imaging matrix-assisted laser desorption ionization MS (I-MALDI) combined with time-of-flight mass spectrometry (TOF-MS). Such a method was employed to identify and quantify various metabolites, even at micromolar levels, within distinct subcellular components of plant cells (Roessner& Beckles, 2009).

## 20.3 METABOLOMICS IN DISEASE DIAGNOSIS AND MONITORING

Metabolomics is utilized in plant biology to identify biomarkers related to specific physiological necessities or diseases, offering potential implementation in plant breeding and accuracy in agriculture. This metabolic biomarker screening tool aids in identifying markers for complex traits, including disease resistance (Manickam et al., 2023), as detailed in Table 20.1. For instance, Sauter et al. (1991) developed a novel method for analyzing plant metabolites, enabling the quantification of a diverse spectrum of metabolites in a single chromatogram through extraction, silylation, and capillary GC. By treating barley seedlings with different bioregulators and herbicides, they were able to identify distinct response patterns characteristic of each treatment. This not only facilitated compound classification but also revealed new physiological responses, providing valuable insights into understanding and managing plant diseases by uncovering previously unknown modes of action. In a complementary effort, Pontes et al. (2016) focused on developing methods for the early detection of Huanglongbing (HLB), a disease that poses significant economic challenges to the citrus industry. By employing, NMR spectroscopy and chemometrics, they identified metabolic biomarkers in *Citrus sinensis* for early detection of HLB. Utilizing 2D NMR data, they pinpointed approximately 20 key biomarkers, highlighting promising prospects for effective disease management.

## 20.4 METABOLOMICS FOR UNDERSTANDING PLANT-PATHOGEN INTERACTIONS

Plants defend against most microbial attacks by strengthening cell walls, producing antimicrobial agents, and generating ROS. However, pathogens can also bypass these defenses and extract nutrients from the plant. Alterations in the plant's energy and chemical usage during infections are critical in determining whether it becomes diseased or not. Metabolomic studies can compare the chemical compositions of healthy, just infected, and diseased or resistant plants, revealing these changes and improving our understanding of plant defense mechanisms. Although this potential exists, metabolomics is underutilized compared to genetic or chemical studies. Thus, it is crucial to use metabolomics to understand how plants can fight off diseases or be vulnerable to them (Castro-Moretti et al., 2020).

The study of how plants' chemical networks respond to threats like pathogens is an emerging field within metabolomics. This field holds promise for discovering new understandings of plant functions, their interactions with pathogens, and ways to protect and enhance crops. Despite the widespread use of metabolomics in various fields, its application in studying plant-pathogen interactions is still in its early stages (Manickam et al., 2023). Aliferis andJabaji (2012) summarized a complete method for investigating these interactions, focusing on potato plants and their relationship with *Rhizoctonia solani* by utilizing advanced techniques such as Fourier transform-ion

**TABLE 20.1**
**Metabolomics Biomarkers Showing Plant Disease Resistance in Crop Plants**

| Plant Name | Sample Tissue | Disease | Micro-organism | Metabolites/Biomarker | Diagnostic Technique | Data Analysis | Effect of Metabolite in Plant | References |
|---|---|---|---|---|---|---|---|---|
| Strawberry | Leaves | Strawberry anthrac-nose | *Colletotrichum theobromicola* (Fungus) | 20 metabolites namely phosphorate, glyceric acid, malic acid, erythrose, tartaric acid, D-xylose, D-ribose, shikimic acid, citric acid, galactose, methyl α-D galactopyranoside, glucitol, gallic acid, phytol, D-mannopyranose, β-D-glucopyranose, sucrose, rhodioloside, galacturonic acid, melibiose | GC-MS | Partial least squares -discriminant analysis (PLS-DA), Analysis of Variance (ANOVA) and OPLS | Before anthracnose symptoms, certain metabolites including gallic acid, citric acid, malic acid, and erythrose initially increased and then decreased, hinting at early plant defense. The presence of this fungus in strawberry plants decreased shikimic acid, galactose, and D-xylose, and increased sucrose and inorganic phosphate content | Dai et al. (2019) |
| Watermelon (*Citrullus vulgaris*) | Leaves | Fusarium wilt | *Fusarium oxysporum f.* sp. *Niveum* | Jasmonic acid–isoleucine, indole-3-acetic acid, abscisic acid, arginine, aspartic acid, melatonin, citrulline, valine, histidine, lysine and phthalic acid | Ultra performance liquid chromatography (UPLC)/Electron ionization (ESI)-high-resolution (HR)-Quadrupole (Q)-TOF-MS, HPLC | Principal component analysis (PCA), MetaboAnalyst 3.0 | Increased levels of the metabolites in plants affected by Fusarium wilt compared to healthy plants | Kasote et al. (2020) |
| Olive tree (*Olea europaea*) | Leaves | Olive quick decline syndrome | *Xylella fastidiosa* subsp. *pauca* ST53 (Bacteria) | Oleic acid | HPLC-ESI-Q-TOF-MS | SIMCA (version 16) for PCA, PLS-DA and OPLS-DA, XCMS | A significant six-fold rise in oleic acid, believed to act as a signal for bacterial communication, indicates *Xylella fastidiosa* infection and hints at potential disease advancement | Di Masi et al. (2022) |

(*Continued*)

**TABLE 20.1 (*Continued*)**
**Metabolomics Biomarkers Showing Plant Disease Resistance in Crop Plants**

| Plant Name | Sample Tissue | Disease | Micro-organism | Metabolites/Biomarker | Diagnostic Technique | Data Analysis | Effect of Metabolite in Plant | References |
|---|---|---|---|---|---|---|---|---|
| Wheat (*Triticum aestivum L.*) | Seedling | Stem rust disease | *Puccinia graminis f.* sp. *Tritici* | Flavonoids, carboxylic acids, and alkaloids, octadecatrienoyl-glycerol | LC-qTOF-MS | PCA, OPLS-DA | These metabolites are highly expressed, whereas octadecatrienoyl-glycerol, a lipid that supports the cell wall's physical barrier, was only detected at 14 days post-infection, indicating it may be suppressed as the infection progresses | Maserumule et al. (2023) |
| Rice | Rice grains | Rice sheath blight disease | *Rhizoctonia solani* | Pipecolic acid, amino acids like glutamate, histidine, glycine, γ-aminobutyric acid, phenylalanine, serine, tryptophan, and tyrosine | CE/TOF-MS | Mass profiler professional | After *R. solani* inoculation, various metabolites displayed changes: chlorogenic acid exhibited a positive reaction in 32R, pipecolic acid exhibited the highest fold change in 29S, and eight other amino acids also increased in 29S | Suharti et al. (2016) |
| Sweet oranges (*Citrus sinensis*) | Fruit | HLB | *Candidatus-Liberibacter asiaticus* | Sugars, organic acids, amino acids, limonin glucoside, adenosine, and limonin | NMR | SIMCA-P for PCA, OPLS-DA | Symptomatic fruit showed a 3-fold reduction in sucrose, a 3-fold addition in phenylalanine, a nearly 2-fold gain in citrate, decreased fructose, glucose, amino acids, choline, malate, and adenosine, and increased asparagine, histidine, ascorbate, formate, synephrine, limonin, and limonin glucoside | Slisz et al. (2012) |

(*Continued*)

**TABLE 20.1 (*Continued*)**
**Metabolomics Biomarkers Showing Plant Disease Resistance in Crop Plants**

| Plant Name | Sample Tissue | Disease | Micro-organism | Metabolites/Biomarker | Diagnostic Technique | Data Analysis | Effect of Metabolite in Plant | References |
|---|---|---|---|---|---|---|---|---|
| Maize (*Zea mays* L.) | Kernel | Ear rot | *Fusarium verticillioides* | Sugars, polyols, amino acids, acids, and polyamines | GC-MS | Mixed-model ANOVA | After fungal inoculation, the resistant maize inbred showed minimal metabolite changes, suggesting it relies on constitutive defense mechanisms. Conversely, the susceptible inbred exhibited significant changes, including decreased levels of amino acids, sugars, and organic acids, indicating an induced but inadequate defense response | Campos-Bermudez et al. (2013) |
| Tomato (*Solanum lycopersicum*) | Leaves | Tomato mosaic disease | Tomato Mosaic Virus | Thirty-two metabolites, like sugars, amino acids, organic acids, flavonoids, phenylpropanoids, and other miscellaneous compounds | NMR | Multivariate data analysis methods, such as PLS and (PLS-DA) | A time-dependent reduction in amino acids (excluding asparagine), organic acids, phenylpropanoids, and rutin; metabolite levels varied with leaf development, showing higher concentrations of certain compounds in lower leaves and elevated sugars in upper leaves; metabolite content changed with asymptomatic ToMV infection, with flavonoids increasing in virus-inoculated leaves and phenylpropanoids in non-inoculated leaves | López-Gresa et al. (2012) |

(*Continued*)

**TABLE 20.1** (*Continued*)
**Metabolomics Biomarkers Showing Plant Disease Resistance in Crop Plants**

| Plant Name | Sample Tissue | Disease | Micro-organism | Metabolites/Biomarker | Diagnostic Technique | Data Analysis | Effect of Metabolite in Plant | References |
|---|---|---|---|---|---|---|---|---|
| | | Late blight | *Phytophthora infestans* | α-tomatine, pheophytin α and 1,2-di-O-palmitoyl 3-O-(6-sulfoquinovopyranosyl) glycerol | LC-MS, MALDI-MS | MetaboAnalyst 3.0 for PCA, OPLS-DA, PLS-DA | α-tomatine serves as an infection biomarker, while saponins indicate early infection, and isocoumarin detects both early and late asymptomatic infection stages | Galeano Garcia et al. (2018) |
| | | Exocortis disease | Citrus exocartis viroid | Gentisic acid | LC-MS, NMR | PCA, PLS-DA using SIMCA-P software | Glycosylated gentisic acid was identified as the most significant induced metabolite | López-Gresa et al. (2010) |
| | | Tissue necrosis | *Pseudomonas syringae* (Bacteria) | Amino and organic acids, as well as phenylpropanoids | LC-MS, NMR | PCA, PLS-DA with SIMCA-P software | Elevated levels of amino and organic acids, rutin, as well as phenylpropanoids | López-Gresa et al. (2010) |
| | | Tomato chlorosis virus infection | Tomato curly stunt virus | Key biomarkers: hydroxycinnamic acid esters to quinic acid (chlorogenic acids, CGAs), glucose, and galactaric acid, alongside negligible components such as flavonoids, benzenoids, and steroidal glycoalkaloids | UHPLC-ESI-MS | SIMCA-P+ software version 13.0 for PCA | In whitefly mock-inoculated plants, there was a notable decrease in CGA levels approximated to control plants, while virus-inoculated plants exhibited a slight additional decline in CGA | Rossouw et al. (2019) |

(*Continued*)

**TABLE 20.1 (*Continued*)**
**Metabolomics Biomarkers Showing Plant Disease Resistance in Crop Plants**

| Plant Name | Sample Tissue | Disease | Micro-organism | Metabolites/Biomarker | Diagnostic Technique | Data Analysis | Effect of Metabolite in Plant | References |
|---|---|---|---|---|---|---|---|---|
| Barley (*Hordeum vulgare*) | Spikes (grain) | Fusarium head blight | *Fusarium culmorum* | Jasmonic acid, dihydro-7-hydroxymyoporone, 4-methoxycinnamic acid, kaempferol 3-O-glucoside 7-O-rhamnoside, deoxynivalenol (DON), and deoxynivalenol-3 glucoside (D3G) | UHPLC-Q-TOF-MS in ESI (–) mode | OPLS-DA, PCA | An increase in plant stress-related metabolites like kaempferol 3-O-glucoside7-O-rhamnoside,dihydro-7-hydroxymyoporone, jasmonic acid, and 4-methoxycinnamic acid in infected samples associated entirely with higher levels of DON and D3G, which are crucial resistance-indicator metabolites in barley | Cajka et al. (2014) |
| Sweet orange (*Citrus sinensis* L.) | Leaves | HLB | *Candidatus-Liberibacter asiaticus* | Hydroxycinnamate compound, isosinensetin, desmethyl polymethoxyflavones | HPLC-MS | PCA, Analysis of Covariance (ANCOVA) | Around 27 weeks post-inoculation, metabolite profiles of leaves from *Candidatus Liberibacter asiaticus* infected trees exhibited noticeable changes, including a significant rise in HCAs by over 10-fold, while others including isosinensetin and desmethyl PMF decreased by more than 3-fold | Hijaz et al. (2013) |

(*Continued*)

**TABLE 20.1 (*Continued*)**
**Metabolomics Biomarkers Showing Plant Disease Resistance in Crop Plants**

| Plant Name | Sample Tissue | Disease | Micro-organism | Metabolites/Biomarker | Diagnostic Technique | Data Analysis | Effect of Metabolite in Plant | References |
|---|---|---|---|---|---|---|---|---|
| Potato (*Solanum tuberosum*) | Leaves | Late blight disease | *Phytophthora infestans* | Amino acids, fatty acids, organic acids, sugars | GC-MS | Statistical analysis systemfor ANOVA | After pathogen infection, the resistant cultivar had higher levels of diverse metabolites, especially amino acids, fatty acids, organic acids, and sugars | Abu-Nada (2006) |
| Potato (*Solanum tuberosum*) | Roots and root exudates | Root infection | *Spongospora subterranea f.* sp. *subterranea* (Sss) | Fatty acids, organic acids, amino acids, sugars, alkaloids, and phenolics | UPLC-Q-TOF/MS | SIMCA-P software version 12.0 for PCA, OPLS-DA, ClustVis for HCA | The non-protein amino acid, gamma-aminobutyric acid was much higher in inoculated tolerant potato roots than in susceptible ones, making it a potential marker for tolerance | Lekota et al. (2020) |
| Narrow Leaf lupin (*Lupinusan-gustifolius*) | Leaves | Anthra-cnose | *Colletotrichum lupini* fungus | 20-hydroxygenistein, wighteone, and luteone | GC-MS and LC-MS | Microsoft Excel and Marker Lynx (Waters) | Three isoflavone aglycones, including 20-hydroxygenistein, wighteone, and luteone, were significantly elevated in tissues across all time points compared to controls, while genistein increased notably only at the experiment's end. Notably, luteone levels in infected plant leaves after seven days were over 50 times more elevated than controls | Wojakowska et al. (2013) |

(*Continued*)

**TABLE 20.1 (*Continued*)**
**Metabolomics Biomarkers Showing Plant Disease Resistance in Crop Plants**

| Plant Name | Sample Tissue | Disease | Micro-organism | Metabolites/Biomarker | Diagnostic Technique | Data Analysis | Effect of Metabolite in Plant | References |
|---|---|---|---|---|---|---|---|---|
| Rice (*Oryza sativa*) | Leaves | Bacterial leaf blight (BLB) | *Xanthomonas oryzae pv. oryzae* | Ornithine, citrulline, tyrosine, phenylalanine, lysine, and arginine | LC-MS, GC-MS | Student t-test, one-way ANOVA, PCA, PLS, HCA, GeneSpring-MS 1.2 software, StatisticaDataMiner and Statistica 7.0 | Ornithine, citrulline, tyrosine, phenylalanine, lysine, and arginine exhibited elevated levels in the resistant group compared to mock-treated controls, with no such increase observed in other susceptible comparisons | Sana et al. (2010) |
| Grapevine (*Vitis vinifera* L.) Berry | Fruit | Noble rot disease | *Botrytis cinerea* (Fungus) | N-acetylglucosamine, oxylipin 13-KODE, stilbene dimers, stilbene trimers, stilbene tetramers, N-(3-methylbutyl) acetamide, as well as sherry lactone 1, benzaldehyde, p-cresol, 1-octen-3-ol, trans-8-dihydroxylinalool, ethyl isoamyl succinate, ethyl vanillate, diethyl succinate, ho-diendiol, 4-terpineol, ethyl phenylacetate, and γ-nonalactone. | LC-MS and GC-MS | PLS-DA, OPLS-DA/O2PLS-D | Fungal metabolites like N-acetylglucosamine and plant defense compounds such as oxylipins and stilbene oligomers accumulated, along with D-glucose-6-phosphate, pantothenic acid, and two unidentified lipids. Higher levels of VOCs, including N-(3-methylbutyl)-acetamide, sherry lactones, benzaldehyde, and 4-terpineol, were noted in botrytized fruits | Negri et al. (2017) |

(*Continued*)

**TABLE 20.1 (*Continued*)**
**Metabolomics Biomarkers Showing Plant Disease Resistance in Crop Plants**

| Plant Name | Sample Tissue | Disease | Micro-organism | Metabolites/Biomarker | Diagnostic Technique | Data Analysis | Effect of Metabolite in Plant | References |
|---|---|---|---|---|---|---|---|---|
| | Shoots and cordon | Dying-arm disease | *Eutypalatab* | Only six chemicals were produced in noticeable amounts namely, Eutypinol and its chromene derivative, eulatachromene, eulatinol, eutypine and its cyclization product, benzofuran together with the quinol, siccayne and its monomethyl analogue. | GC-MS, HPLC | - | *Eutypalata* infection is characterized by the presence of eutypinol and eulatachromene, with eutypinol being the main metabolite indicating infection. Detection involves culturing infected cordon sections, where eutypinol is readily found in the solution, revealing its role in disrupting vascular function | Mahoney et al. (2005) |
| Apple (*Malus domestica*) | Fruit | Blue mold disease | *Penicillium expansum* (*P. expansum*) | Sugars, amino acids, organic acids, and peptides | UPLC-Q-TOF/MS | PLS-DA, OPLS-DA were achieved by SIMCA-P V13.0, PCA | During *P. expansum* infection, most sugars decreased while amino acids, organic acids, and peptides increased, with phenolics and fatty acids showing varied changes. Early infection disturbed pathways like secondary metabolite biosynthesis and metabolism of amino acid, identifying 140 core metabolites, 27 of which could be biomarkers for the infection | Shen et al. (2021) |

cyclotron resonance/mass spectrometry (FT-ICR/MS) with GC-MS to analyze a broad spectrum of plant metabolites, including amino acids, fatty acids, and carbohydrates. These findings strive to comprehend how plants' metabolic networks adapt to pathogen invasions and link these changes to gene activity.

There has been massive progress in unravelling the plant-microbe interactions. For example, flavonoids in root exudates activate bacterial nod genes, initiating nodule growth (Mhlongo et al., 2018). Organic acids, the primary liberated metabolites, are absorbed by bacterial enhancement in the substrate (Zhalnina et al., 2018). Additionally, metabolomic analysis of various maize genotypes interacting with nitrogen-fixing plant growth-promoting bacteria species reveals changes in plant metabolites due to bacterial nitrogen fixation (Brusamarello-Santos et al., 2017). Integrated investigation of transcript and metabolite profiles uncovers quantitative trait loci linked to volatile compound production induced by spider mite infestation in cucumber (Mercke et al., 2004). Untargeted metabolomics characterized unique lipids, including ceramides and by-products of eicosapentaenoic and arachidonic acid in *Plasmopara viticola*, enabling early detection of downy mildew and improving disease management. Its potential extended to developing high-sensitivity LC-MS-based techniques applicable to various plant pathogens, facilitating accurate examination of host-pathogen interactions and resistance to fungicides (Negrel et al., 2018).

Additionally, LC-MS-based untargeted metabolomics studies show how wheat plants respond to *P. striiformis* infection. It reveals changes in metabolism over time, highlighting alterations in phenolic compounds and other metabolites in resistant and sensitive wheat varieties (Mashabela et al., 2023). Metabolomics, as evidenced by these studies, provides valuable insights into the intricate biochemical mechanisms underlying plant-pathogen interactions. By revealing metabolic signatures, metabolomics can empower precision disease prevention and control methods, enhancing crop health and productivity.

## 20.5 METABOLOMICS GUIDED BREEDING FOR PLANT DISEASE MANAGEMENT

By 2050, with an expected population of 9 billion, there will be a huge challenge to meet the future food demand. Climate change, urbanization, and land degradation will exacerbate the pressure on the food supply. To tackle these issues, there's an urgent need to speed up the development and distribution of improved crops to farmers. This is vital because historical evidence shows that crop breeding has been essential in reducing poverty and increasing global food production (Lenaerts et al., 2019). So, Plant breeders have utilized modern analytical techniques such as metabolomics to create unique crops that can adapt to environmental differences, resist pests and diseases, and retain their nutritional value. Despite variations in their metabolome forced by different factors including time, season, and temperature, these varieties can still offer health benefits (Alawiye & Babalola, 2021).

Environmental stressors like salinity and heavy metal contamination can harm plants and make them more vulnerable to diseases. Rouphael et al. (2016) revealed the utility of metabolomics in elucidating plant responses to environmental stresses. Their research highlighted the pivotal roles of osmotic stress and redox balance in lettuce root metabolism under conditions of NaCl and zinc-induced stress, providing valuable insights for breeders aiming to enhance crop resilience to salinity and heavy metal contamination. Steinfath et al. (2010) utilized metabolomics to guide plant breeding toward disease management by employing an effective, broad screening technique to identify metabolic biomarkers for predicting crop quality and vulnerability to black spot bruising in potato (*Solanum tuberosum*) tubers. Their approach facilitates the choice of superior progeny from crossbreeding populations and highlights the potential for using multiple metabolic markers, including fructose, glucose, and tyrosines simultaneously. This method enhances knowledge of biochemical responses underlying complex traits, facilitating advancements in crop breeding for disease resistance.

Metabolomic profiling indicated significant disparities in gene expression and hormonal changes in the leaves and roots of maize in response to anthracnose. Understanding these organ-specific defense systems helped plant breeders to produce maize plants with anthracnose resistance (Balmer et al., 2013). Metabolomics-guided breeding for plant disease management employs pathway analysis, as demonstrated by Beckles et al. (2012), to identify optimal treatment steps for fruits and vegetables. Metabolomics-based investigations of volatiles have been utilized for postharvest disease detection in stored onions (Prithiviraj et al., 2004), mangoes (Moalemiyan et al., 2006), and apples (Li et al., 2007). Integrating metabolomics with other omics approaches holds significant promise for advancing the efficiency and precision of future breeding projects (Razzaq et al., 2022).

## 20.6 INTEGRATION OF OMICS APPROACHES IN PLANT DISEASE RESEARCH

Breeding crops is a multifaceted process that involves the selection of desirable traits like higher yield, and disease resistance, traditionally limited by genetic diversity. However, the emergence of high-throughput omics technologies as depicted in Figure 20.3 highlights crop breeding by supplying extensive molecular data on plant growth and responses to surrounding challenges (Chao et al., 2024).

Advancements in single-cell genomics and epigenomics offer powerful tools for understanding cellular differences and molecular mechanisms in plants (Luo et al., 2020). Analyzing the transcriptomics of plants enables altered gene expression, which helps in developing improved varieties by revealing downstream gene activities and regulatory networks. (Agarwal et al., 2014). Though still advancing, proteomics presents promising tools for plant breeders, complementing genomics and other biotechnological methods. Integrating proteomics with genomics accelerates gene discovery

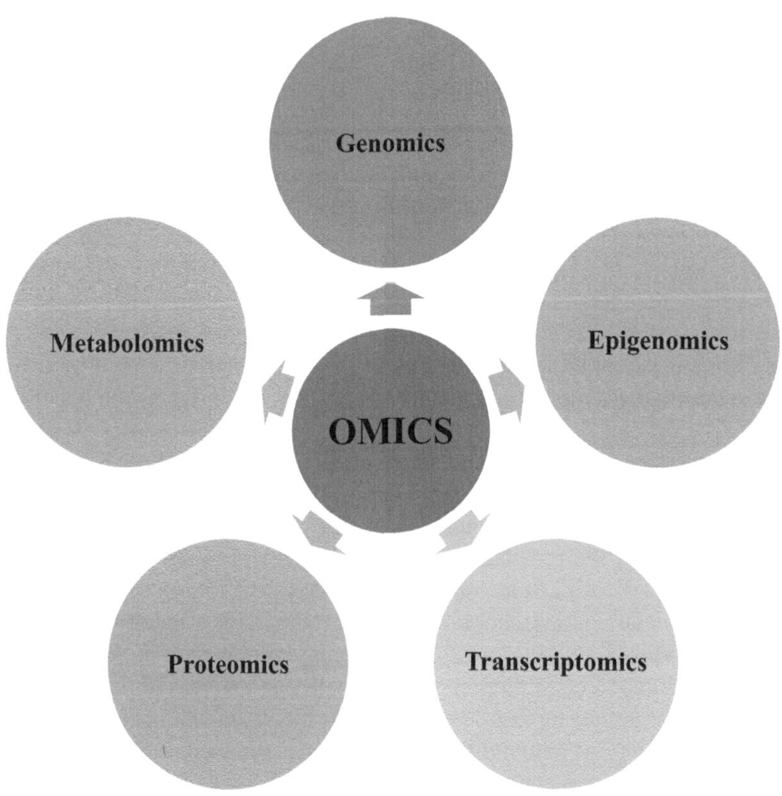


**FIGURE 20.3** Types of omics applied in plant research.

and enhances understanding of stress tolerance mechanisms for sustainable agriculture (Okoli et al., 2021). Metabolomics provides a valuable understanding of plant chemical composition, enhancing genomics-assisted selection for crop advancement through integrated high-throughput technologies (Fernie & Schauer, 2009). Integrating metabolic and genomic research in workflows helps us to understand efficiently the differences between organisms, which gene expression studies alone cannot fully achieve (Sana et al., 2010).

Integrative studies aim to predict gene function and characterize systemic interactions within biological systems. Omics studies gather vast data on cellular components and interactions, requiring computational tools to simplify it. Metabolic network construction facilitates the integration of omics data, enabling the exploration of complex host-microbial interactions (Mhlongo et al., 2018). However, integrating proteomics, transcriptomics, and metabolomics data for understanding host-pathogen interactions faces challenges due to limited direct gene-metabolite links. Platforms like MetaboAnalyst, accessible online or via R, are expanding omics data integration capabilities, aiding plant-microbe interaction research. Advancements in merging multi-omics data are crucial, requiring collaborative efforts to merge existing genomic and proteomic data with newly obtained metabolomic data. This collaboration will enable the creation of comprehensive databases and platforms, enhancing our knowledge of host-pathogen interactions and aiding in the evolution of innovative approaches to strengthen plant resistance against diseases (Castro-Moretti et al., 2020).

## 20.7 BIOINFORMATICS AND DATA ANALYSIS IN METABOLOMICS

Bioinformatics and data analysis play a key position in the field of metabolomics, especially in the context of plant disease management. The recognition of plant metabolites is a complex and challenging assignment because of the vast array of primary and secondary metabolites produced by plants, their extensive range, low abundance, and constantly changing profiles (Hegeman, 2010). Additionally, the complex matrix of plant specimens, harsh sample preparation conditions, and limitations of analytical instruments further hinder the structural characterization of these metabolites (Casas & Matamoros, 2021). The spatiotemporal variation in metabolite dispersal across various plant tissues and over time also complicates the understanding of their physiological and ecological functions (Peters et al., 2018).

Despite these challenges, metabolomics holds great importance in plant disease management. Plant metabolite profiling offers high sensitivity, multiplexing capability, and high-throughput analysis, enabling the detection of subtle changes in metabolite levels that can indicate earlier phases of disease development (Yang et al., 2023. This approach provides a complete view of the plant's metabolic rate and allows for the specific identification of plant pathogens, aiding in timely disease control measures. Moreover, the study of pathogen metabolites through metabolite profiling offers specificity to the pathogen of interest, facilitating reliable detection and diagnosis of plant pathogens (Castro-Moretti et al., 2020). By analyzing changes in metabolite levels, investigators can gain an understanding of plant defense mechanisms, specific antimicrobial compounds, monitor disease progression, evaluate treatment efficacy, and study plant-microbiome interactions, all of which are essential for effective plant disease management. Metabolomics, as a new tool in plant pathology, provides useful details on metabolic transformations during plant-pathogen interactions, offering a more profound understanding of plant resistance/susceptibility and the virulence mechanisms employed by pathogens (Verma et al., 2020).

Different data analysis platforms to facilitate metabolite identification are utilized for plant disease management (Table 20.2). Mass Spectral Repositories, such as MassBank, METLIN, and NIST, house comprehensive mass spectral libraries that enable the precise annotation of metabolites implicated in plant defense reactions and pathogen virulence mechanisms. These repositories serve as invaluable resources for accurate metabolite identification, which is pivotal for uncovering the intricate metabolic signatures associated with plant-pathogen interactions and developing targeted interventions (Horai et al., 2010; Smith et al., 2005). Plant Metabolome Databases, including GMD@CSB.DB, Medicinal Plant Metabolomics Resource, and PlantMetabolomics.org, provide

**TABLE 20.2**
**Different Data Analysis Platforms Utilized for Plant Disease Management**

| S. No | Category | Name | Description | Reference |
|---|---|---|---|---|
| 1. | Plant metabolome databases | PlantMetabolomics.org | Web-based Arabidopsis metabolomics data repository | Bais et al. (2012) |
| 2. | | Chloroplast 2010 Project | Metabolite profiles of >10,000 Arabidopsis SALK lines | Lu et al. (2011) |
| 3. | | Medicinal Plant Metabolomics Resource | Metabolome database for medicinal plants | Wurtele et al. (2012) |
| 4. | | ARMeC | ESI-MS metabolomics database, focused on potato | Overy et al. (2008) |
| 5. | | GMD@CSB.DB | Comprehensive plant metabolome database; GC-MS profiles and retention indices (GC-specific) | Kopka et al. (2005) |
| 6. | | MoTo DB | Tomato metabolic database based on LC-MS | Moco et al. (2006) |
| 7. | Mass spectral repositories | GMDB | Glycan mass spectral database (MALDI-specific) | Kameyama et al. (2005) |
| 8. | | MS/MS Spectral Tag Viewer | LC-MS MS2 spectral collection | Matsuda et al. (2005) |
| 9. | | AtMetExpress Development | Phytoconstituents atlas of Arabidopsis (LC-specific) | Matsuda et al. (2010) |
| 10. | | MaConDa | Database of MS contaminants | Weber et al. (2012) |
| 11. | | FiehnLib | GC-MS mass spectra and retention index library | Kind et al. (2009) |
| 12. | | MetaboLights | Repository for metabolomics data and metadata | Steinbeck et al. (2012) |
| 13. | | MetaboSearch | Software for metabolite identification | Zhou et al. (2012) |
| 14. | | ReSpect | MSn spectra from literature (LC-specific) | Sawada et al. (2012) |
| 15. | | VocBinBase | Volatile compound library for GC-MS | Skogerson et al. (2011) |
| 16. | | METLIN | LC-MS and MS2 mass spectral database | Smith et al. (2005) |
| 17. | | NIST | Mass spectral database for LC-MS and MS2 | NA |
| 18. | | PlantMetabolomics.org | Web-based Arabidopsis metabolomics data repository | Bais et al. (2012) |
| 19. | | MetabolomeExpress | GC-MS metabolomics data repository | Carroll et al. (2010) |
| 20. | Metabolomics data analysis platforms | MeltDB | Web system for data analysis and management | Neuweger et al. (2008) |
| 21. | | MBMole | Web-based enrichment analysis tool | Chagoyen & Pazos (2011) |
| 22. | | metaP-Server | Web-based metabolomics information investigation platform | Kastenmüller et al. (2011) |
| 23. | | MetaboAnalyst | Web application for metabolomics data analysis | Xia et al. (2012) |
| 24. | | MetaMapp | Network graph generation from metabolomics data | Barupal et al. (2012) |

access to curated metabolite profiles and metabolomic data specific to various plant species, facilitating the interpretation of metabolic perturbations caused by pathogens and the identification of potential biomarkers for disease susceptibility or resistance (Kopka et al., 2005; Wurtele et al., 2012; Bais et al., 2012).

Furthermore, Metabolomics Data Analysis Platforms, such as MetaboAnalyst, MetaMapp, and metaP-Server, offer powerful computational tools and workflows for processing, analyzing, and visualizing complex metabolomics datasets. These platforms enable the detection of statistically significant metabolic modifications and the illustration of metabolic pathways dysregulated during plant-pathogen interactions (Xia et al., 2012; Barupal et al., 2012; Kastenmüller et al., 2011). By incorporating metabolomics with other -omics approaches, plant pathology investigation can be further advanced, enhancing our ability to protect plants from harmful pathogens and ensuring a sustainable and secure food supply.

## 20.8 STATISTICAL METHODS FOR METABOLOMICS DATA ANALYSIS

Statistical analysis plays a vital role in extracting valuable insights from metabolomics data. Data preprocessing and normalization techniques are crucial steps before statistical analysis. Preprocessing steps, including normalization techniques like total signal normalization, vector length normalization, and peak intensity normalization, are essential for obtaining meaningful information (Guet et al., 2009; Misra, 2020). These normalization approaches can be executed employing various statistical packages and programming languages (e.g., R, Python) or dedicated metabolomics platforms like MetaboAnalyst (Chonget al., 2018), XCMS Online (Smith et al., 2006), and Workflow4Metabolomics (Guitton et al., 2017).

After proper data preprocessing and normalization, univariate statistical tools like Student's t-tests and ANOVA are essential methods for analyzing metabolite datasets (Sanchez et al., 2008). t-tests assess the significance of discrepancies in mean metabolite levels between two groups, while ANOVA extends this comparison to multiple groups or conditions. These methods consider assumptions like normality, variance homogeneity, and degrees of freedom. ANOVA can handle multiple experimental factors simultaneously, account for interactions, and employ variations like repeated measures or ANCOVA to control for confounding variables. Despite the power of multivariate approaches, univariate analysis remains crucial for uncovering significant metabolites that may be masked in multivariate models (Percivalet al., 2020).

Multivariate analysis techniques are essential for metabolomics data, as they can uncover patterns and correlations that may be missed by univariate approaches. PCA is a widely used unsupervised strategy for data exploration, outlier detection, and visualizing metabolic signatures. It reduces high-dimensional data into principal components capturing the maximum variance. PLS-DA, a directed technique, focuses on maximizing the separation between predefined groups or classes (Anwardeen et al., 2023). It identifies the most discriminant metabolites and highlights whether they are up- or downregulated. Both PCA and PLS-DA enable the detection of metabolic patterns and the identification of potential biomarkers, complementing the insights gained from univariate methods. By combining robust data preprocessing, normalization techniques, and appropriate statistical tools (univariate and multivariate), researchers can effectively analyze complex metabolomics datasets, uncover metabolic patterns, identify potential biomarkers, and gain insights into biological mechanisms underlying plant diseases. This holistic approach, integrating data handling and statistical analyses, is necessary for advancing our experience of plant health and developing effective disease management strategies.

## 20.9 CONCLUSION

Applied metabolomics serves as a shield, protecting our agriculture and empowering plants to combat diseases, ensuring their growth into resilient crops. By studying how plants respond to threats, we can able to uncover methods to fortify them without using harmful chemicals. With the fusion of metabolomics and cutting-edge technologies, the future of agriculture shines brighter than ever. It's not merely about cultivation; it's about cultivating a future where crops thrive, resources are conserved, and global food security is fortified.

## ACKNOWLEDGEMENT

All the authors are thankful to their respective Universities and Institutions for their valuable supports.

## DATA AVAILABILITY

The authors confirm that the data supporting the findings of this study are available within the book/chapter.

## REFERENCES

Abu-Nada, Y. (2006). Metabolic profiling of potato cultivars varying in horizontal resistance to late blight, Phytophthora infestans. Thesis. McGill University.

Aderemi, A. V., Ayeleso, A. O., Oyedapo, O. O., & Mukwevho, E. (2021). Metabolomics: A scoping review of its role as a tool for disease biomarker discovery in selected non-communicable diseases. *Metabolites, 11*(7), 418.

Agarwal, P., Parida, S. K., Mahto, A., Das, S., Mathew, I. E., Malik, N., & Tyagi, A. K. (2014). Expanding frontiers in plant transcriptomics in aid of functional genomics and molecular breeding. *Biotechnology Journal, 9*(12), 1480–1492.

Aharoni, A., Ric de Vos, C. H., Verhoeven, H. A., Maliepaard, C. A., Kruppa, G., Bino, R., & Goodenowe, D. B. (2002). Nontargeted metabolome analysis by use of Fourier transform ion cyclotron mass spectrometry. *Omics: A Journal of Integrative Biology, 6*(3), 217–234.

Alawiye, T. T., & Babalola, O. O. (2021). Metabolomics: Current application and prospects in crop production. *Biologia, 76*(1), 227–239.

Aliferis, K. A., & Jabaji, S. (2012). Deciphering plant–pathogen interactions applying metabolomics: Principles and applications. *Canadian Journal of Plant Pathology, 34*(1), 29–33.

Allwood, J. W., Williams, A., Uthe, H., van Dam, N. M., Mur, L. A., Grant, M. R., & Pétriacq, P. (2021). Unravelling plant responses to stress—The importance of targeted and untargeted metabolomics. *Metabolites, 11*(8), 558.

Anwardeen, N. R., Diboun, I., Mokrab, Y., Althani, A. A., & Elrayess, M. A. (2023). Statistical methods and resources for biomarker discovery using metabolomics. *BMC Bioinformatics, 24*(1), 250.

Bais, P., Moon-Quanbeck, S. M., Nikolau, B. J., & Dickerson, J. A. (2012). Plantmetabolomics.org: Mass spectrometry-based Arabidopsis metabolomics—Database and tools update. *Nucleic Acids Research, 40*(D1), D1216–D1220.

Balmer, D., de Papajewski, D. V., Planchamp, C., Glauser, G., & Mauch-Mani, B. (2013). Induced resistance in maize is based on organ-specific defence responses. *The Plant Journal, 74*(2), 213–225.

Baranska, M., Schulz, H., Rösch, P., Strehle, M. A., & Popp, J. (2004). Identification of secondary metabolites in medicinal and spice plants by NIR-FT-Raman microspectroscopic mapping. *Analyst, 129*(10), 926–930.

Bartel, J., Krumsiek, J., & Theis, F. J. (2013). Statistical methods for the analysis of high-throughput metabolomics data. *Computational and Structural Biotechnology Journal, 4*(5), e201301009.

Barupal, D. K., Haldiya, P. K., Wohlgemuth, G., Kind, T., Kothari, S. L., Pinkerton, K. E., & Fiehn, O. (2012). MetaMapp: Mapping and visualizing metabolomic data by integrating information from biochemical pathways and chemical and mass spectral similarity. *BMC Bioinformatics, 13*, 1–15.

Bedair, M., & Sumner, L. W. (2008). Current and emerging mass-spectrometry technologies for metabolomics. *TrAC Trends in Analytical Chemistry, 27*(3), 238–250.

Brusamarello-Santos, L. C., Gilard, F., Brulé, L., Quilleré, I., Gourion, B., Ratet, P., …& Hirel, B. (2017). Metabolic profiling of two maize (Zea mays L.) inbred lines inoculated with the nitrogen fixing plant-interacting bacteria Herbaspirillumseropedicae and Azospirillumbrasilense. *PLoS One, 12*(3), e0174576.

Cajka, T., Vaclavikova, M., Dzuman, Z., Vaclavik, L., Ovesna, J., & Hajslova, J. (2014). Rapid LC-MS-based metabolomics method to study the Fusarium infection of barley. *Journal of Separation Science, 37*(8), 912–919.

Campos-Bermudez, V. A., Fauguel, C. M., Tronconi, M. A., Casati, P., Presello, D. A., & Andreo, C. S. (2013). Transcriptional and metabolic changes associated to the infection by Fusarium verticillioides in maize inbreds with contrasting ear rot resistance. *PLoS One, 8*(4), e61580.

Carroll, A. J., Badger, M. R., & Harvey Millar, A. (2010). The MetabolomeExpress Project: Enabling web-based processing, analysis and transparent dissemination of GC-MS metabolomics datasets. *BMC Bioinformatics, 11*, 1–13.

Casas, M. E., & Matamoros, V. (2021). Analytical challenges and solutions for performing metabolomic analysis of root exudates. *Trends in Environmental Analytical Chemistry, 31*, e00130.

Castro-Moretti, F. R., Gentzel, I. N., Mackey, D., & Alonso, A. P. (2020). Metabolomics as an emerging tool for the study of plant–pathogen interactions. *Metabolites, 10*(2), 52.

Chagoyen, M., & Pazos, F. (2011). MBRole: Enrichment analysis of metabolomic data. *Bioinformatics, 27*(5), 730–731.

Chao, H., Zhang, S., Hu, Y., Ni, Q., Xin, S., Zhao, L., …& Chen, M. (2024). Integrating omics databases for enhanced crop breeding. *Journal of Integrative Bioinformatics, 20*(4), 20230012.

Chong, J., Soufan, O., Li, C., Caraus, I., Li, S., Bourque, G., …& Xia, J. (2018). MetaboAnalyst 4.0: Towards more transparent and integrative metabolomics analysis. *Nucleic acids Research, 46*(W1), W486–W494.

Crandall, S. G., Gold, K. M., Jiménez-Gasco, M. D. M., Filgueiras, C. C., & Willett, D. S. (2020). A multi-omics approach to solving problems in plant disease ecology. *PLoS One, 15*(9), e0237975.

Dai, T., Chang, X., Hu, Z., Liang, L., Sun, M., Liu, P., & Liu, X. (2019). Untargeted metabolomics based on GC-MS and chemometrics: A new tool for the early diagnosis of strawberry anthracnose caused by Colletotrichum theobromicola. *Plant Disease, 103*(10), 2541–2547.

De Livera, A. M., Olshansky, G., Simpson, J. A., & Creek, D. J. (2018). NormalizeMets: Assessing, selecting and implementing statistical methods for normalizing metabolomics data. *Metabolomics, 14*, 1–5.

Di Masi, S., De Benedetto, G. E., Malitesta, C., Saponari, M., Citti, C., Cannazza, G., & Ciccarella, G. (2022). HPLC-MS/MS method applied to an untargeted metabolomics approach for the diagnosis of "olive quick decline syndrome". *Analytical and Bioanalytical Chemistry, 414*, 465–473.

Dordas, C. (2008). Role of nutrients in controlling plant diseases in sustainable agriculture. A review. *Agronomy for Sustainable Development, 28*, 33–46.

Ernest, B., Gooding, J. R., Campagna, S. R., Saxton, A. M., & Voy, B. H. (2012). MetabR: An R script for linear model analysis of quantitative metabolomic data. *BMC Research Notes, 5*, 1–10.

Fernie, A. R., & Schauer, N. (2009). Metabolomics-assisted breeding: A viable option for crop improvement?. *Trends in Genetics, 25*(1), 39–48.

Galeano Garcia, P., Neves dos Santos, F., Zanotta, S., Eberlin, M. N., & Carazzone, C. (2018). Metabolomics of Solanum lycopersicum infected with Phytophthora infestansleads to early detection of late blight in asymptomatic plants. *Molecules, 23*(12), 3330.

Garrett, K. A., Madden, L. V., Hughes, G., & Pfender, W. F. (2004). New applications of statistical tools in plant pathology. *Phytopathology, 94*(9), 999–1003.

Gomez-Casati, D. F., Zanor, M. I., & Busi, M. V. (2013). Metabolomics in plants and humans: Applications in the prevention and diagnosis of diseases. *BioMed Research International, 2013*, 792527.

Gu, H., Pan, Z., Xi, B., Hainline, B. E., Shanaiah, N., Asiago, V., …& Raftery, D. (2009). 1H NMR metabolomics study of age profiling in children. *NMR in Biomedicine:An International Journal Devoted to the Development and Application of Magnetic Resonance In Vivo, 22*(8), 826–833.

Guitton, Y., Tremblay-Franco, M., Le Corguillé, G., Martin, J. F., Pétéra, M., Roger-Mele, P., …& Thévenot, E. A. (2017). Create, run, share, publish, and reference your LC–MS, FIA–MS, GC–MS, and NMR data analysis workflows with the Workflow4Metabolomics 3.0 Galaxy online infrastructure for metabolomics. *The International Journal of Biochemistry & Cell Biology, 93*, 89–101.

Hegeman, A. D. (2010). Plant metabolomics-meeting the analytical challenges of comprehensive metabolite analysis. *Briefings in Functional Genomics, 9*(2), 139–148.

Hijaz, F. M., Manthey, J. A., Folimonova, S. Y., Davis, C. L., Jones, S. E., & Reyes-De-Corcuera, J. I. (2013). An HPLC-MS characterization of the changes in sweet orange leaf metabolite profile following infection by the bacterial pathogen CandidatusLiberibacterasiaticus. *PLoS One, 8*(11), e79485.

Hill, C. B., & Roessner, U. (2013). Metabolic profiling of plants by GC–MS. In W. Weckwerth & G. Kahl (Eds.), *The handbook of plant metabolomics* (pp. 1–23). Weinheim, Germany: Wiley-VCH Verlag GmbH & Co. KGaA.

Horai, H., Arita, M., Kanaya, S., Nihei, Y., Ikeda, T., Suwa, K., …& Nishioka, T. (2010). MassBank: A public repository for sharing mass spectral data for life sciences. *Journal of Mass Spectrometry, 45*(7), 703–714.

Kameyama, A., Kikuchi, N., Nakaya, S., Ito, H., Sato, T., Shikanai, T., …& Narimatsu, H. (2005). A strategy for identification of oligosaccharide structures using observational multistage mass spectral library. *Analytical Chemistry, 77*(15), 4719–4725.

Kasote, D. M., Jayaprakasha, G. K., Singh, J., Ong, K., Crosby, K. M., & Patil, B. S. (2020). Metabolomics-based biomarkers of Fusarium wilt disease in watermelon plants. *Journal of Plant Diseases and Protection, 127*, 591–596.

Kastenmüller, G., Römisch-Margl, W., Wägele, B., Altmaier, E., & Suhre, K. (2011). metaP-server: A web-based metabolomics data analysis tool. *BioMed Research International,* 2011, 839862.

Kind, T., Wohlgemuth, G., Lee, D. Y., Lu, Y., Palazoglu, M., Shahbaz, S., & Fiehn, O. (2009). FiehnLib: Mass spectral and retention index libraries for metabolomics based on quadrupole and time-of-flight gas chromatography/mass spectrometry. *Analytical Chemistry, 81*(24), 10038–10048.

Kopka, J., Schauer, N., Krueger, S., Birkemeyer, C., Usadel, B., Bergmüller, E., …& Steinhauser, D. (2005). GMD@ CSB. DB: The Golm metabolome database. *Bioinformatics*, *21*(8), 1635–1638.

Leiss, K. A., Choi, Y. H., Verpoorte, R., & Klinkhamer, P. G. (2011). An overview of NMR-based metabolomics to identify secondary plant compounds involved in host plant resistance. *Phytochemistry Reviews*, *10*, 205–216.

Lekota, M., Modisane, K. J., Apostolides, Z., &van der Waals, J. E. (2020). Metabolomic fingerprinting of potato cultivars differing in susceptibility to Spongosporasubterranea f. sp. subterranea root infection. *International Journal of Molecular Sciences*, *21*(11), 3788.

Lenaerts, B., Collard, B. C., & Demont, M. (2019). Improving global food security through accelerated plant breeding. *Plant Science*, *287*, 110207.

Li, B., Tang, J., Yang, Q., Li, S., Cui, X., Li, Y., …& Zhu, F. (2017). NOREVA: Normalization and evaluation of MS-based metabolomics data. *Nucleic Acids Research, 45*(W1), W162–W170.

Li, C., Heinemann, P. H., & Irudayaraj, J. (2007). Detection of apple deterioration using an electronic nose and zNoseTM. *Transactions of the ASABE*, *50*(4), 1417–1425.

López-Gresa, M. P., Lisón, P., Kim, H. K., Choi, Y. H., Verpoorte, R., Rodrigo, I., …& Bellés, J. M. (2012). Metabolic fingerprinting of tomato mosaic virus infected Solanum lycopersicum. *Journal of Plant Physiology, 169*(16), 1586–1596.

López-Gresa, M. P., Maltese, F., Bellés, J. M., Conejero, V., Kim, H. K., Choi, Y. H., & Verpoorte, R. (2010). Metabolic response of tomato leaves upon different plant–pathogen interactions. *Phytochemical Analysis:An International Journal of Plant Chemical and Biochemical Techniques, 21*(1), 89–94.

Lu, Y., Savage, L. J., Larson, M. D., Wilkerson, C. G., & Last, R. L. (2011). Chloroplast 2010: A database for large-scale phenotypic screening of Arabidopsis mutants. *Plant Physiology, 155*(4), 1589–1600.

Luo, C., Fernie, A. R., & Yan, J. (2020). Single-cell genomics and epigenomics: Technologies and applications in plants. *Trends in Plant Science*, *25*(10), 1030–1040.

Mahoney, N., Molyneux, R. J., Smith, L. R., Schoch, T. K., Rolshausen, P. E., & Gubler, W. D. (2005). Dying-arm disease in grapevines: Diagnosis of infection with Eutypalata by metabolite analysis. *Journal of Agricultural and Food Chemistry, 53*(21), 8148–8155.

Manickam, S., Rajagopalan, V. R., Kambale, R., Rajasekaran, R., Kanagarajan, S., & Muthurajan, R. (2023). Plant Metabolomics: Current Initiatives and Future Prospects. *Current Issues in Molecular Biology, 45*(11), 8894–8906.

Maserumule, M., Rauwane, M., Madala, N. E., Ncube, E., & Figlan, S. (2023). Defence-related metabolic changes in wheat (Triticum aestivum L.) seedlings in response to infection by Puccinia graminis f. sp. tritici.*Frontiers in Plant Science, 14*, 1166813.

Mashabela, M. D., Tugizimana, F., Steenkamp, P. A., Piater, L. A., Dubery, I. A., & Mhlongo, M. I. (2023). Metabolite profiling of susceptible and resistant wheat (Triticum aestivum) cultivars responding to Puccinia striiformis f. sp. tritici infection. *BMC Plant Biology*, *23*(1), 293.

Matsuda, F., Hirai, M. Y., Sasaki, E., Akiyama, K., Yonekura-Sakakibara, K., Provart, N. J., …& Saito, K. (2010). AtMetExpress development: A phytochemical atlas of Arabidopsis development. *Plant Physiology, 152*(2), 566–578.

Matsuda, F., Yonekura-Sakakibara, K., Niida, R., Kuromori, T., Shinozaki, K., & Saito, K. (2009). MS/MS spectral tag-based annotation of non-targeted profile of plant secondary metabolites. *The Plant Journal, 57*(3), 555–577.

Mercke, P., Kappers, I. F., Verstappen, F. W., Vorst, O., Dicke, M., & Bouwmeester, H. J. (2004). Combined transcript and metabolite analysis reveals genes involved in spider mite induced volatile formation in cucumber plants. *Plant Physiology*, *135*(4), 2012–2024.

Mhlongo, M. I., Piater, L. A., Madala, N. E., Labuschagne, N., & Dubery, I. A. (2018). The chemistry of plant–microbe interactions in the rhizosphere and the potential for metabolomics to reveal signaling related to defense priming and induced systemic resistance. *Frontiers in Plant Science*, *9*, 314703.

Misra, B. B. (2020). Data normalization strategies in metabolomics: Current challenges, approaches, and tools. *European Journal of Mass Spectrometry, 26*(3), 165–174.

Moalemiyan, M., Vkram, A., Kushalappa, A. C., & Yaylayan, V. (2006). Volatile metabolite profiling to detect and discriminate stemend rot and anthracnose diseases of mango fruits. *Plant Pathology, 55*, 792–802

Moco, S., Bino, R. J., Vorst, O., Verhoeven, H. A., De Groot, J., Van Beek, T. A., …& De Vos, C. R. (2006). A liquid chromatography-mass spectrometry-based metabolome database for tomato. *Plant Physiology, 141*(4), 1205–1218.

Negrel, L., Halter, D., Wiedemann-Merdinoglu, S., Rustenholz, C., Merdinoglu, D., Hugueney, P., & Baltenweck, R. (2018). Identification of lipid markers of Plasmoparaviticola infection in grapevine using a non-targeted metabolomic approach. *Frontiers in Plant Science*, *9*, 328879.

Negri, S., Lovato, A., Boscaini, F., Salvetti, E., Torriani, S., Commisso, M., …& Guzzo, F. (2017). The induction of noble rot (Botrytis cinerea) infection during postharvest withering changes the metabolome of grapevine berries (Vitis vinifera L., cv. Garganega).*Frontiers in Plant Science, 8*, 1002.

Neuweger, H., Albaum, S. P., Dondrup, M., Persicke, M., Watt, T., Niehaus, K., …& Goesmann, A. (2008). MeltDB: A software platform for the analysis and integration of metabolomics experiment data. *Bioinformatics, 24*(23), 2726–2732.

Okoli, E. E., Nworji, M. J., & Okoronkwo, C. M. (2021). Potentials of proteomics in plant improvement. *International Journal of Scientific Advances*, *2*(1), 80–84.

Overy, D. P., Enot, D. P., Tailliart, K., Jenkins, H., Parker, D., Beckmann, M., & Draper, J. (2008). Explanatory signal interpretation and metabolite identification strategies for nominal mass FIE-MS metabolite fingerprints. *Nature Protocols, 3*(3), 471–485.

Percival, B., Gibson, M., Leenders, J., Wilson, P. B., & Grootveld, M. (2020). *Univariate and Multivariate Statistical Approaches to the Analysis and Interpretation of NMR-Based Metabolomics Datasets of Increasing Complexity*. Royal Society of Chemistry.

Peters, K., Worrich, A., Weinhold, A., Alka, O., Balcke, G., Birkemeyer, C., …& Dam, N. M. V. (2018). Current challenges in plant eco-metabolomics. *International Journal of Molecular Sciences, 19*(5), 1385.

Pontes, J. G. M., Ohashi, W. Y., Brasil, A. J., Filgueiras, P. R., Espíndola, A. P. D., Silva, J. S., …& Tasic, L. (2016). Metabolomics by NMR spectroscopy in plant disease diagnostic: Huanglongbing as a case study. *ChemistrySelect*, *1*(6), 1176–1178.

Prithiviraj, B., Vikram, A., Kushalappa, A. C., & Yaylayan, V. (2004). Volatile metabolite profiling for the discrimination of onion bulbs infected by Erwinia carotovora ssp. carotovora, Fusarium oxysporum and Botrytis allii. *European Journal of Plant Pathology*, *110*, 371–377.

Razzaq, A., Wishart, D. S., Wani, S. H., Hameed, M. K., Mubin, M., & Saleem, F. (2022). Advances in metabolomics-driven diagnostic breeding and crop improvement. *Metabolites*, *12*(6), 511.

Roessner, U., & Beckles, D. M. (2009). *Metabolite Measurements* (pp. 39–69). Springer.

Roessner, U., & Beckles, D. M. (2012). Metabolomics for salinity research. Plant Salt Tolerance*:* Methods and Protocols, 203–215.

Romano, A., Capozzi, V., Khomenko, I., & Biasioli, F. (2023). Advances in the application of direct injection mass spectrometry techniques to the analysis of grape, wine and other alcoholic beverages. *Molecules, 28*(22), 7642.

Rossouw, L. T., Madala, N. E., Tugizimana, F., Steenkamp, P. A., Esterhuizen, L. L., & Dubery, I. A. (2019). Deciphering the resistance mechanism of tomato plants against whitefly-mediated tomato curly stunt virus infection through ultra-high-performance liquid chromatography coupled to mass spectrometry (UHPLC-MS)-based metabolomics approaches. *Metabolites, 9*(4), 60.

Rouphael, Y., Colla, G., Bernardo, L., Kane, D., Trevisan, M., & Lucini, L. (2016). Zinc excess triggered polyamines accumulation in lettuce root metabolome, as compared to osmotic stress under high salinity. *Frontiers in Plant Science*, *7*, 187271.

Sana, T. R., Fischer, S., Wohlgemuth, G., Katrekar, A., Jung, K. H., Ronald, P. C., & Fiehn, O. (2010). Metabolomic and transcriptomic analysis of the rice response to the bacterial blight pathogen Xanthomonas oryzaepv. oryzae. *Metabolomics, 6,*451–465.

Sanchez, D. H., Siahpoosh, M. R., Roessner, U., Udvardi, M., & Kopka, J. (2008). Plant metabolomics reveals conserved and divergent metabolic responses to salinity. *Physiologia Plantarum, 132*(2), 209–219.

Sankaran, S., Mishra, A., Ehsani, R., & Davis, C. (2010). A review of advanced techniques for detecting plant diseases. *Computers and Electronics in Agriculture*, *72*(1), 1–13.

Sauter, H., Lauer, M., & Fritsch, H. (1991). *Metabolic Profiling of Plants: ANew Diagnostic Technique*. Semantic Scholar.

Sawada, Y., Nakabayashi, R., Yamada, Y., Suzuki, M., Sato, M., Sakata, A., …& Saito, K. (2012). RIKEN tandem mass spectral database (ReSpect) for phytochemicals: A plant-specific MS/MS-based data resource and database. *Phytochemistry, 82*, 38–45.

Shen, Y., Liu, M., Nie, J., Ma, N., Xu, G., Zhang, J., …& Li, Z. (2021). Metabolite changes of apple Penicillium expansum infection based on a UPLC-Q-TOF metabonomics approach. *Postharvest Biology and Technology, 181*, 111646.

Skogerson, K., Wohlgemuth, G., Barupal, D. K., & Fiehn, O. (2011). The volatile compound BinBase mass spectral database. *BMC Bioinformatics, 12*, 1–15.

Slisz, A. M., Breksa III, A. P., Mishchuk, D. O., McCollum, G., & Slupsky, C. M. (2012). Metabolomic analysis of citrus infection by 'Candidatus Liberibacter'reveals insight into pathogenicity. *Journal of Proteome Research, 11*(8), 4223–4230.

Smith, C. A., O'Maille, G., Want, E. J., Qin, C., Trauger, S. A., Brandon, T. R., …& Siuzdak, G. (2005). METLIN: A metabolite mass spectral database. *Therapeutic Drug Monitoring, 27*(6), 747–751.

Smith, C. A., Want, E. J., O'Maille, G., Abagyan, R., & Siuzdak, G. (2006). XCMS: Processing mass spectrometry data for metabolite profiling using nonlinear peak alignment, matching, and identification. *Analytical Chemistry, 78*(3), 779–787.100

Soga, T., Ohashi, Y., Ueno, Y., Naraoka, H., Tomita, M., & Nishioka, T. (2003). Quantitative metabolome analysis using capillary electrophoresis mass spectrometry. *Journal of Proteome Research, 2*(5), 488–494.

Steinbeck, C., Conesa, P., Haug, K., Mahendraker, T., Williams, M., Maguire, E., …& Griffin, J. L. (2012). MetaboLights: Towards a new COSMOS of metabolomics data management. *Metabolomics, 8*, 757–760.

Steinfath, M., Strehmel, N., Peters, R., Schauer, N., Groth, D., Hummel, J., …& Van Dongen, J. T. (2010). Discovering plant metabolic biomarkers for phenotype prediction using an untargeted approach. *Plant Biotechnology Journal, 8*(8), 900–911.

Suharti, W. S., Nose, A., & Zheng, S. H. (2016). Metabolite profiling of sheath blight disease resistance in rice: In the case of positive ion mode analysis by CE/TOF-MS. *Plant Production Science, 19*(2), 279–290.

Terentev, A., & Dolzhenko, V. (2023). Can Metabolomic approaches become a tool for improving early plant disease detection and diagnosis with modern remote sensing methods? A review. *Sensors, 23*(12), 5366.

Thonusin, C., IglayReger, H. B., Soni, T., Rothberg, A. E., Burant, C. F., & Evans, C. R. (2017). Evaluation of intensity drift correction strategies using MetaboDrift, a normalization tool for multi-batch metabolomics data. *Journal of Chromatography A, 1523*, 265–274.

Tsugawa, H., Cajka, T., Kind, T., Ma, Y., Higgins, B., Ikeda, K., …& Arita, M. (2015). MS-DIAL: Data-independent MS/MS deconvolution for comprehensive metabolome analysis. *Nature Methods, 12*(6), 523–526.

Verma, C., Jandaik, S., Gupta, B. K., Kashyap, N., Suryaprakash, V. S., Kashyap, S., & Kerketta, A. (2020). Microbial metabolites in plant disease management: Review on biological approach. *International Journal of Chemical Studies, 8*, 2570–2581.

Weber, R. J., Li, E., Bruty, J., He, S., & Viant, M. R. (2012). MaConDa: A publicly accessible mass spectrometry contaminants database. *Bioinformatics, 28*(21), 2856–2857.

Weston, L., Skoneczny, D., Weston, P., & Weidenhamer, J. (2015). Metabolic profiling: An overview-new approaches for the detection and functional analysis of biologically active secondary plant products. *Journal of Allelochemical Interactions, 2*(1), 15–27.

Willforss, J., Chawade, A., & Levander, F. (2018). NormalyzerDE: Online tool for improved normalization of omics expression data and high-sensitivity differential expression analysis. *Journal of Proteome Research, 18*(2), 732–740.

Wojakowska, A., Muth, D., Narożna, D., Mądrzak, C., Stobiecki, M., & Kachlicki, P. (2013). Changes of phenolic secondary metabolite profiles in the reaction of narrow leaf lupin (Lupinusangustifolius) plants to infections with Colletotrichum lupini fungus or treatment with its toxin. *Metabolomics, 9*, 575–589.

Wurtele, E. S., Chappell, J., Daniel Jones, A., Celiz, M. D., Ransom, N., Hur, M., …& Nikolau, B. J. (2012). Medicinal plants: A public resource for metabolomics and hypothesis development. *Metabolites, 2*(4), 1031–1059.

Xi, B., Gu, H., Baniasadi, H., & Raftery, D. (2014). Statistical analysis and modeling of mass spectrometry-based metabolomics data. *Methods in Molecular Biology, 1198*, 333–353.

Xia, J., Mandal, R., Sinelnikov, I. V., Broadhurst, D., & Wishart, D. S. (2012). MetaboAnalyst 2.0-a comprehensive server for metabolomic data analysis. *Nucleic Acids Research, 40*(W1), W127–W133.

Yang, P., Zhao, L., Gao, Y. G., & Xia, Y. (2023). Detection, diagnosis, and preventive management of the bacterial plant pathogen pseudomonas syringae. *Plants, 12*(9), 1765.

Zhalnina, K., Louie, K. B., Hao, Z., Mansoori, N., Da Rocha, U. N., Shi, S., …& Brodie, E. L. (2018). Dynamic root exudate chemistry and microbial substrate preferences drive patterns in rhizosphere microbial community assembly. *Nature Microbiology, 3*(4), 470–480.

Zhou, B., Wang, J., &Ressom, H. W. (2012). MetaboSearch: Tool for mass-based metabolite identification using multiple databases. *PLoSOne, 7*(6), e40096.

# Index

Note: **Bold** page numbers refer to tables and *italic* page numbers refer to figures.